Astrophilosophy, Exotheology, and Cosmic Religion

CONTEMPORARY WHITEHEAD STUDIES

Edited by Roland Faber, Claremont School of Theology,
and Brian G. Henning, Gonzaga University

Contemporary Whitehead Studies, cosponsored by the Whitehead Research Project, is an interdisciplinary book series that publishes manuscripts from scholars with contemporary and innovative approaches to Whitehead studies by giving special focus to projects that: explore the connections between Whitehead and contemporary Continental philosophy, especially sources like Heidegger, or contemporary streams like poststructuralism; reconnect Whitehead to pragmatism, analytical philosophy and philosophy of language; explore creative East/West dialogues facilitated by Whitehead's work; explore the interconnections of the mathematician with the philosopher and the contemporary importance of these parts of Whitehead's work for the dialogue between sciences and humanities; reconnect Whitehead to the wider field of philosophy, the humanities, the sciences and academic research with Whitehead's pluralistic impulses in the context of a pluralistic world; address Whitehead's philosophy in the midst of contemporary problems facing humanity, such as climate change, war and peace, race, and the future development of civilization.

Recent Titles in this Series

Astrophilosophy, Exotheology, and Cosmic Religion: Extraterrestrial Life in a Process Universe, edited by Andrew M. Davis and Roland Faber

Virginia Woolf as a Process-Oriented Thinker: Parallels Between Woolf's Fiction and Process Philosophy, by Veronika Krajíčková

Unearthing the Unknown Whitehead, by Joseph Petek

Whitehead and the Pittsburgh School: Preempting the Problem of Intentionality, by Lisa Landoe Hedrick

On Philosophy, Intelligibility, and the Ordinary: Going the Bloody Hard Way, by Randy Ramal

Untying the Gordian Knot: Process, Reality and Context, by Timothy E. Eastman

Mind, Value, and the Cosmos: On the Relational Nature of Ultimacy, by Andrew M. Davis

Whitehead's Radically Temporalist Metaphysics: Recovering the Seriousness of Time, by George Allan

Propositions in the Making: Experiments in a Whiteheadian Laboratory, edited by Roland Faber, Michael Halewood, and Andrew M. Davis

Whitehead and Continental Philosophy in the Twenty-First Century: Dislocations, edited by Jeremy D. Fackenthal

Astrophilosophy, Exotheology, and Cosmic Religion

Extraterrestrial Life in a Process Universe

Edited by Andrew M. Davis
and Roland Faber

LEXINGTON BOOKS
Lanham • Boulder • New York • London

Published by Lexington Books
An imprint of The Rowman & Littlefield Publishing Group, Inc.
4501 Forbes Boulevard, Suite 200, Lanham, Maryland 20706
www.rowman.com

86-90 Paul Street, London EC2A 4NE

British Library Cataloguing in Publication Information Available

Library of Congress Cataloging-in-Publication Data

Names: Astrobiology, Exo-Philosophy, and Cosmic Religion (Conference) (2022 : Salem, Or.), author. | Davis, Andrew M., 1987- editor. | Faber, Roland, 1960- editor.
Title: Astrophilosophy, exotheology, and cosmic religion : extraterrestrial life in a process universe / edited by Andrew M. Davis and Roland Faber.
Description: Lanham : Lexington Books, [2024] | Series: Contemporary Whitehead studies | Includes bibliographical references and index.
Identifiers: LCCN 2023051304 (print) | LCCN 2023051305 (ebook) | ISBN 9781666944365 (cloth) | ISBN 9781666944372 (ebook)
Subjects: LCSH: Exobiology--Religious aspects--Congresses. | Exobiology--Philosophy--Congresses. | Life on other planets--Religious aspects--Congresses. | Life on other planets--Philosophy--Congresses.
Classification: LCC BL254 .A87 2022 (print) | LCC BL254 (ebook) | DDC 113--dc23/eng/20231207
LC record available at https://lccn.loc.gov/2023051304
LC ebook record available at https://lccn.loc.gov/2023051305

Contents

Figures and Tables

CHAPTER 2

CHAPTER 4

CHAPTER 8

Foreword

Timothy E. Eastman

Growing up on a farm in western Minnesota during the 1950s, I frequently marveled at the night sky, especially at the majestic Milky Way overhead during summer nights. I wondered if there were other worlds out there. Are we alone in the cosmos? What does it all mean?

For several decades, I left those questions behind and advanced in this world (family, career, publications, etc.). Along with essentially all fellow passengers on this "blue boat home,"[1] practically without awareness, I became increasingly blinded both visually and spiritually. The visual blinding began as I moved to large urban areas for employment opportunities but which immersed me in light pollution. The vivid Milky Way experience of youth became just a distant memory while I pursued increasingly abstract computerized data and representations in my chosen field of space plasma physics. I had simply become an armchair astronomer, however professional in external clothing. Long past was any immediate, visceral experience of distant astronomical realms.

Over these decades, two modes of pollution had increasingly blinded me and most of my fellow "blue boat" passengers, light pollution and "thing" pollution.

Light Pollution: It's now reported that "more than 80 percent of the world's population lives under light-polluted skies. It is even worse in the United States and Europe with 99 percent of their denizens experiencing skyglow at night."[2] Over 80 percent of the US population lives in urban areas and many people rarely, if ever, experience the Milky Way as I did in my youth. This is tragic.

Thing Pollution: The dominant forms of external pollution are pollutions of air, water, and land. These pollutants all impact human resilience and sustainability through disease and environmental degradation. Some sources of pollution are beyond human control (e.g., volcanoes), but many sources are driven by human patterns of living and consumption: fossil-fuel usage, waste streams, plastics, soil contamination, thermal pollution, water pollution, and radioactive contamination. These patterns of consumption are driven in turn, at least in part, by a materialist worldview in which the pursuit of wealth, property, and power displaces attention from deep human values and meaning. Scaled up to civilizational levels, such materialism and its nihilistic modes of competition have led to multiple levels of "thing" pollution, both external and internal, external wars and social strife (psychological and cultural maladies), and so forth.[3]

WORLDVIEWS, IMAGINATION, AND POSSIBILITY

To restore our health (personal, social, civilizational) we need to recognize, confront, and address the pollution. We need to become more self-aware and rethink our worldviews. When confronted with the wonders of the Milky Way so many years ago, I remember my profound sense of unknowing, a radical humility—a humility that contrasts starkly with the unthinking arrogance of materialism, competitiveness, domination over others, and excessive control over environments. If light pollution could go away and everyone were to truly (and deeply, viscerally) experience the glories of the night sky, deep humility could become newly commonplace, questions of meaning would no longer be disparaged, and the excesses of our materialistic culture would be fundamentally reconsidered. Sadly, even if we can make substantial progress in reducing light pollution, perfectly clear views of the Milky Way are a thing of the past for us earth-bound travelers.

However, with imagination and the fullness of context and history, we have the potential for transformation, to build on the wealth of new observations and understandings that point beyond the materialist reductionisms of the past. The new and amazing space observations, from solar and space plasmas, to planets, stars, and galaxies, and associated theory, modeling, and testing, demonstrate that several assumptions of the standard worldview are demonstrably false. Space plasma research has discovered that the notion of empty space is a myth—space environments are full of interactive particles, fields, electromagnetism, and plasmas everywhere without limit. Planetary science research has found unexpected complexity and interactivity of all planetary systems, and has revealed planetary systems to be commonplace—almost every star in the night sky is likely to have planets around it![4] And finally, major space observatories, the Hubble Space Telescope and now the James Webb Space Telescope, reveal fully formed galaxies at all observable distances (and times), indicating the need for new humility even for our models of the cosmos at large.

NEW NATURAL PHILOSOPHY FOR THE TWENTY-FIRST CENTURY

Until recently, science has often been seen as fundamentally reductionist, a search for ultimate causes and/or parts. As Nobel Prize–winner Robert Laughlin points out, however, "science has now moved from an Age of Reductionism to an Age of Emergence, a time when the search for ultimate causes of things shifts from the behavior of parts to the behavior of the collective."[5] Science is focused on the systematic coordination of observations of the natural world. Such observations or data are outputs of physical processes, which involve complex overlapping systems at multiple scales. They neglect the emerging understanding that, the real is both the actual and the possible,[6] which (in contemporary logic) can be expressed in both Boolean and non-Boolean description.[7] Scientific "laws" are mathematically expressed constraints on possibility involving, in quantum physics, transitions from possibility, to probability, to actuality.

Mechanisms are systems so highly constrained that they provide high reliability for specific functions. The reductionist modeling of all systems as simply mechanistic effectively makes the metaphysical claim that the real is nothing but the actual and that only Boolean logic applies—a false claim. Mechanical models can be very useful and successful; however, materialistic mechanism as a philosophy turns good science upside down by claiming that high abstractions (particular highly constrained systems), instead of being derivative, are themselves the basic reality. Whitehead famously called this "the fallacy of misplaced concreteness."[8] Similarly, scientism effectively claims that Boolean descriptions and "causal closure of the physical" apply without limit. Again, this is a problematic, metaphysical claim. The full range of meaningful propositions about reality, most inclusively, goes beyond such nominalistic actualism.

For many scientific applications, it is sufficient to focus simply on system input-output; however, more generally all systems, at multiple levels, involve input-output-*context*. The full treatment of such contextuality and history goes beyond scientific description alone, through complex systems theory,[9] to include semiotics[10] and philosophy. Contemporary research in the foundations of physics, especially quantum physics, philosophy of chemistry, and the philosophy of biology, illustrate the importance of both science and philosophy in such work. To truly address the fullness of reality we need *both* science *and* philosophy. Whitehead's philosophy of organism and this volume's various discussions of the biological universe are examples of efforts toward a new natural philosophy for the twenty-first century.[11]

ARE WE ALONE?

We now understand that planetary systems are common and that habitable planets, however small the associated probability, are of sufficient number just within our own galaxy to enable numerous opportunities for the emergence of life. As David Grinspoon states, "It took 4.5 billion years for Earth to go from dead rock to space walk, from molten ball to shopping mall, from sea to me, from goo to you."[12] He then follows this with an extensive discussion on the history and complexity of the extraterrestrial intelligence (ETI) question. Such questions remain unresolved.

In the final chapter of this volume, Roland Faber confronts the Fermi paradox about the inferred high probability that ETI exists and the lack of evidence to date for such. My view is that ETI almost certainly exists, especially given new research results showing the commonality of planetary systems as Derek Malone-France argues in the first essay of this collection. Any ETI exemplar must have successfully resolved two fundamental challenges: (1) how to live sustainably long term within available resources of their host planet's habitable zone and (2) how to avoid self-destruction through competition and war. Our particular species, *Homo sapiens* (literally "wise men"), is having great difficulty in managing both of these challenges essential for long-term survival. We appear to be in the early childhood of civilization and are hardly wise. If we can evolve to become responsible adults, then perhaps we can become sustainable long term. Assuming, for example, that the median timescale for a successful civilization is 50 million years (50 My),[13] then our 0.01 My civilization is not yet out of infancy! A truly mature civilization needs to both avoid self-destruction and be fully sustainable. These needs suggest features like an emphasis on cooperation versus competition, ethics versus narrow self-interest, and self-reflective wisdom versus materialistic greed. In my view, to be sustainable long term, a fully mature civilization would almost certainly have these features. In speculating that ETIs will be hostile, as many do, are we projecting our fears, based on who we are, onto potential companions in the cosmos?[14] Civilizations that fail to achieve genuine maturity will likely self-destruct. Is that our future?

In my view, the Fermi paradox is not really a paradox because, by several measures, we have barely initiated our investigations. To date, observations via the search for extraterrestrial intelligence (SETI), have focused on low-data-rate radio communications. In contrast, any truly advanced civilization would likely use very high-data-rate, highly collimated (low-leakage) communication systems (e.g., optical laser communication) such that any signal coming our way would be either very weak or nonexistent. Further, any such contact will likely be with a civilization that is more advanced both ethically and technologically, the latter suggested by the very short time since the beginnings of our own electronic communications capabilities.

WHAT DOES IT ALL MEAN?

Books like this about the origins of life, human, and cosmic evolution, and speculations about other worlds very often either avoid the question of meaning, or indulge in speculative claims about untestable multiverses. Such speculations are useful for movies like *Everything Everywhere All at Once*,[15] but involve unstated metaphysical claims such as actualism, which is the doctrine that the real is only the actual, or reducible to only Boolean logical description. However, in keeping with the book's title of *Astrophilosophy, Exothelogy, and Cosmic Religion: Extraterrestrial Life in a Process Universe*, contributors to this edited volume are generally explicit and well-reasoned about their philosophical arguments, many of which are informed by (and relevant to) process philosophies from Bergson to Whitehead, Teilhard, and beyond. Further, questions about the possibility of extraterrestrial life are artfully used to, shall we say, "stress test" various approaches to metaphysical analysis, theological reflection, and religious imagination. In his chapter on cosmic pluralism, for example, Jeffrey Long argues that, in handling these cosmic questions, "religion . . . will be transformed by this process, hopefully in a pluralistic direction."[16] At least these questions can now be openly debated. Long points out that "Giordano Bruno, after all, was burned at the stake in 1600, at least in part for his belief that the universe contains countless worlds, and that these worlds could be inhabited."[17] If you're worried about your worldview or religion being transformed, then you better collect your wood and matches. Otherwise, read on and enjoy this adventure of ideas.[18]

Timothy E. Eastman, PhD
Consultant in Space Physics and Plasma Science (retired)
April 9, 2023
Silver Spring, MD

NOTES

1. "Blue Boat Home" by singer/songwriter Peter Mayer, http://www.thegreatstory.org/songs/blueboat.html.

2. International Dark-Sky Association, "80% of World Population Lives Under Skyglow," https://www.darksky.org/80-of-world-population-lives-under-skyglow-new-study-finds/.

3. For an analysis of cultural and social maladies, see Singer and Kimbles, *The Cultural Complex*.

4. See Malone-France, this volume; also Grinspoon, *Earth in Human Hands*.

5. Laughlin, *A Different Universe*, 208.

6. Eastman, *Untying the Gordian Knot*, Epperson and Zafiris, *Foundations of Relational Realism*.

7. Primas, "Non-Boolean Descriptions for Mind-Matter Problems."

8. Max Tegmark has stated that "our physical world not only is *described* by mathematics, but that it *is* mathematics, making us self-aware parts of a giant mathematical objects." Tegmark, *Our Mathematical Universe*, 6. Such Platonism clearly illustrates Whitehead's fallacy.

9. Thomas, "Robert Rosen in the Age of Systems Biology."

10. Hoffmeyer, *Biosemiotics*.

11. There are various examples of recent works that are developing such a new natural philosophy. For example: McGilchrist, *The Matter with Things*; Lent, *The Web of Meaning*; Davis, *Mind, Value, and Cosmos*; Faber, *The Divine Manifold*, and *The Cosmic Spirit*; Lupisella, *Cosmological Theories of Value*; Eastman, *Untying the Gordian Knot*; Marshall, *A Complex Integral Realist Perspective*; Kastner, *Understanding Our Unseen Reality*; Kauffman, *Humanity in a Creative Universe*; Koutroufinis, *Life and Process*; Epperson and Zafiris, *Foundations of Relational Realism*; Brier, *CyberSemiotics*; Henning and Scarfe, *Beyond Mechanism*; Cahoone, *The Orders of Nature*; Zalamea, *Peirce's Logic of Continuity*; Niemoczynski, *Charles Sanders Peirce and a Religious Metaphysics of Nature*.

12. Grinspoon, *Earth in Human Hands*, 283.

13. Here I adopt 50 My as a proxy for potential cataclysmic events of the type illustrated by the asteroid impact ending the age of dinosaurs 65 My ago.

14. Faber, this volume; also Steven Spielberg envisions an ETI that is curious, not fear-based or hostile, in his 1997 movie *Close Encounters of the Third Kind*.

15. Kwan and Scheinert, *Everything Everywhere All at Once*.

16. Long, this volume.

17. Ibid.

18. See Whitehead's great masterpiece, *Adventures of Ideas*.

BIBLIOGRAPHY

Brier, S. *CyberSemiotics: Why Information is Not Enough*. Toronto: University of Toronto Press, 2013.

Cahoone, L. *The Orders of Nature*. Albany: SUNY Press, 2013.

Davis, A. M. *Mind, Value, and Cosmos: On the Relational Nature of Ultimacy*. Lanham, MD: Lexington, 2020.

Eastman, T. E. *Untying the Gordian Knot: Process, Reality, and Context*. Lanham, MD: Lexington, 2020.

Epperson, M., and E. Zafiris. *Foundations of Relational Realism*. Lanham, MD: Lexington, 2013.

Faber, R. *The Cosmic Spirit: Awakenings at the Heart of All Religions, the Earth, and the Multiverse*. Eugene, OR: Cascade, 2021.

———. *The Divine Manifold*. Lanham, MD: Lexington, 2014.

Grinspoon, D. *Earth in Human Hands: Shaping Our Planet's Future*. New York: Grand Central Publishing, 2016.

Henning, B., and A. Scarfe, eds. *Beyond Mechanism: Putting Life Back into Biology*. Lanham, MD: Lexington, 2013.

Hoffmeyer, J. *Biosemiotics: An Examination into the Signs of Life and the Life of Signs*. Chicago: University of Scranton Press, 2008.

International Dark-Sky Association. "80% of World Population Lives Under Skyglow, New Study Finds." *International Dark-Sky Association* (2016). Accessed March 16, 2023. https://www.darksky.org/80-of-world-population-lives-under-skyglow-new-study-finds/

Kastner, R. *Understanding Our Unseen Reality: Solving Quantum Riddles*. London: Imperial College Press, 2015.

Kauffman, S. *Humanity in a Creative Universe*. Oxford: Oxford University Press, 2016.

Koutroufinis, S., ed. *Life and Process: Towards a New Biophilosophy*. Berlin: de Gruyter, 2014.

Kwan, D., and D. Scheinert, directors. *Everything Everywhere All at Once*. A24 Films, 2022. 2hrs., 19mins.

Laughlin, R. *A Different Universe: Reinventing Physics from the Bottom Down*. New York: Basic Books, 2005.

Lent, J. *The Web of Meaning: Integrating Science and Traditional Wisdom to Find Our Place in the Universe*. Gabriola Island, BC: New Society Publishers, 2022.

Lupisella, M. *Cosmological Theories of Value: Science, Philosophy, and Meaning in Cosmic Evolution*. Cham: Springer, 2020.

Malone-France, D. "Prospects for a Universal Philosophy of Organism." current volume.

Marshall, P. *A Complex Integral Realist Perspective: Towards a New Axial Vision*. New York: Routledge, 2016.

Mayer, P. "Blue Boat Home." Track 4 on *Earth Town Square*, 2002. Accessed March 16, 2023. http://www.thegreatstory.org/songs/blueboat.html

McGilchrist, I. *The Matter with Things*, Vol 1 & 2. London: Perspectiva Press, 2021.

Niemoczynski, L. *Charles Sanders Peirce and a Religious Metaphysics of Nature*. Lanham, MD: Lexington, 2011.

Primas, H. "Non-Boolean Descriptions for Mind-Matter Problems." *Mind & Matter* 5, no. 1 (2007): 7–44.

Singer, T., and S. Kimbles, eds. *The Cultural Complex: Contemporary Jungian Perspectives on Psyche and Society*. New York: Routledge, 2004.

Tegmark, M. *Our Mathematical Universe*. New York: Knopf, 2014.

Thomas, S. R. "Robert Rosen in the Age of Systems Biology." *Chemistry & Biodiversity* 4, no. 10 (2007): 2407–14. https://doi.org/10.1002/cbdv.200790196

Whitehead, A. N. *Adventures of Ideas*. New York: Macmillan, 1933.

Zalamea, F. *Peirce's Logic of Continuity*. Boston: Docent Press, 2012.

Acknowledgments

Roland Faber

It is due to the enthusiasm and energy of my student, assistant, colleague, and friend Andrew M. Davis that this unique collection has come about, diving deep into the tradition of cosmological reflection by process and related thinkers, and reaching far into the future of our understanding of the cosmos in which we live and, at some point, may find ourselves as part of a cosmic community of life. It is time to prepare ourselves for a biotic universe that—in so far as we will not destroy ourselves as humanity and the planet we call home—may surprise us, as it not only feeds the breeze under the wings of our imagination, but, besides unbound ingenuity or scientific boldness, also releases a sensitivity to a wider cosmology that can catapult us beyond the simplifications of anthropic exclusivism and merely Earth-centered propositions regarding our future in the cosmos.

There are many worthy voices to be considered in such an endeavor; but it is especially the preparation of this sea change in thought and action by prophets of such a vast cosmology, primary among them, Bergson, Whitehead, and Teilhard, that will enable us to advance a more empathic worldview which values life in its myriads of abilities and possibilities. And it is their expansive, unrestricted, and bold thought that has opened my own senses, thoughts, and imaginations to the unthinkable and unexpected strangeness of a potentially infinite variety of life in this universe and, perhaps, of yet unimagined, decentered multiverse-contexts for our own existence as humanity and the precariousness and value of our small Earth.

To imagine a biotic universe is not only a mind-bending exercise, then, but a vital enterprise by which, as Whitehead says, humanity "by means of its elaborate system of symbolic transference can achieve miracles of sensitiveness to a distant environment, and to a problematic future,"[1] and by which we can "provide a rational understanding of the rise of civilization, and of the tenderness of mere life itself, in a world which superficially is founded upon the clashings of senseless compulsion."[2]

It is in this sense that our cosmological adventure becomes an investment in the future of humanity through a vision of peace that is animated by multiplicities of life and mind and their diverse embodiments in the universe. And such a vision may provoke a new sensitivity to the ancient promise of deeper insight into an unceasing creativity through which we may feel that we are companions in the movements of an "Eros" that is always driving life beyond itself and its limitations in a grand "Adventure in the Universe as One"[3] for which our imaginations are always too small, but to which our existence in its fragility is still worth stiving for.

On the occasion of this book, and the contribution it may make in this regard, I find it appropriate to dedicate it to a humanity—or its successors—that will be inspired to live into a far future that will gratefully and with compassion "look back to the queer, contracted three-dimensional universe from which the nobler, wider existence has emerged."[4]

NOTES

1. Whitehead, *Symbolism*, 87.
2. Whitehead, *Adventures of Ideas*, 170.
3. Ibid., 295.
4. Whitehead, *Modes of Thought*, 57.

BIBLIOGRAPHY

Whitehead, A. N. *Adventures of Ideas*. New York: The Free Press, 1967.

———. *Modes of Thought*. New York: The Free Press, 1966.

———. *Symbolism: Its Meaning and Effect*. New York: Fordham University Press, 1985.

Introduction

Process Philosophy and Extraterrestrial Life: Past, Present, and Future

Andrew M. Davis

The possibility of a truly biological universe has not ceased to haunt the human imagination. Over the past three decades, the discovery of thousands of exoplanets has spurred novel research programs integrating science, philosophy, and theology in exciting new ways. A variety of stimulating proposals have drawn together convergent insights in physics, cosmology, and astrobiology, metaphysics and the philosophy of mind, the philosophy of religion, philosophical theology, and religiosity in a living cosmos. Ever-more-relevant questions as to the status and implications of wide-ranging life in the universe continue to be raised. These questions in turn spark deeper questions about the necessary philosophical assumptions or presuppositions of a biocentric universe, and wider constructive considerations as to how theology, religion, and society might change in light of the impact of discovery.

Little sustained attention, however, has been directed toward the resources inherent in the process philosophy of Alfred North Whitehead (1861–1947), and the associated process philosophies of Henri Bergson (1859–1941), Teilhard de Chardin (1881–1955), and others when considering the interdisciplinary layers of astrobiological research. What are the implications of universal creative evolution for life and intelligence beyond earth? How might the metaphysical primacy of organism, temporality, novelty, value, and mind enrich current discussions across disciplines? What is the potential relevance of process philosophy and theology to current debates surrounding the ability or inability of religions and theologies to incorporate the reality of extraterrestrial intelligence?

This volume is the first collective scholarly exploration of the relevance of process philosophy to various scientific, philosophical, and theological concerns found in current discussions of astrobiology and extraterrestrial life.

Featuring experts in their fields, contributors are informed by process conceptualities, but not confined to them. As the principal name associated with modern process philosophy, Whitehead naturally figures prominently, but the insights of Bergson, Teilhard, and others are also central to the discussion. Current issues and debates are also explored without direct relevance (but certainly not irrelevance) to these figures and the modern process tradition more generally.

It is important to state that this volume makes no claim to be exhaustive in nature. There are a variety of other process philosophers the examinations of which might have also been included, and whose insights may be of great value to the discussion. In addition to Whitehead, Bergson, and Teilhard, these might include George W. F. Hegel (1770–1831), Friedrich W. J. Schelling (1775–1854), Charles Sanders Peirce (1839–1914), William James (1842–1910), John Dewey (1859–1952), Samuel Alexander (1859–1938), Charles Hartshorne (1897–2000) and Gilles Deleuze (1925–1995)—to name only a few. It must be remembered that the process tradition remains wider than Whitehead and retains a depth and diversity signaling that further research can (and should be) conducted with respect to the subject at hand.

Still, this volume does make an important contribution to current literature which has, until now, neglected to achieve a collective scholarly exploration of the multivalent resources of process philosophy and theology to the topic. In filling this scholarly lacuna, it is my hope that this volume constructively advances current discussions with the various possibilities afforded by process thought, as well as canalizes certain philosophical and theological trajectories of further research in the future.

My goal in this introduction is fourfold in nature. First, I will briefly review the backstory which led to the emergence of this volume, including an innovative seminar at Claremont School of Theology and conference at Willamette University, as well as some of the key voices involved. Second, I will selectively review some of the admittedly surface-level references made to process philosophy and theology in recent literature endeavoring theological responses to astrobiology and extraterrestrial life. Third, I will then look to a variety of important, but neglected, statements made by past and present process philosophers and theologians on the topic of extraterrestrial life and its theological implications. In highlighting some core themes undergirding these statements, the need for this volume and further volumes on the relevance of process philosophy and theology to current discussions will become apparent. Fourth and finally, I will give an overview of this book and the unfolding of its chapters through its three respective parts: Astrophilosophy, Exotheology, and Cosmic Religion.

EXTRATERRESTRIALS IN CLAREMONT AND SALEM

This volume has a backstory without which it would not have come into being. In the spring of 2017, Roland Faber offered a seminar at Claremont School of Theology titled "Religions and Exo-Life." The first of its kind, the course aimed to "explore the history of the integration or exclusion, embrace or limitation" of extraterrestrial life by different religious traditions, including "the constraints it puts on religious worldviews and doctrines, and the insights the horizon of exo-life can offer when . . . applied to religious self-understanding."[1]

The course drew in a variety of graduate students who were working at the intersections of philosophy, science, and religion. I was a research assistant for Faber at the time and aided the organization and execution of the seminar. Throughout the course, students were given ample opportunity to acquaint themselves with the wide array of scientific, philosophical, and theological literature relevant to the topic. Discussions ranged widely from the historical debate surrounding the plurality of worlds and intelligent extraterrestrial life, to the metaphysics of exo-life and the nature of both biological and cosmological evolution, to the impact of discovery on religious and theological worldviews.

The course literature included several readings from the corpus of Steven J. Dick (b. 1949), former NASA chief historian and Blumberg Chair in Astrobiology at the Library of Congress. For decades, Dick has been one of the leading experts on the long-standing historical debate regarding the plurality of worlds and extraterrestrial life.[2] In reading Dick, it was clear that he was not only concerned with *the past*, but also with *the future* and the kind of impact the discovery of extraterrestrial life might have on society and culture at large, especially on religion and theology.[3] How might religions respond to the discovery of extraterrestrial life, and with what resources? In what ways might theology need to change in order to incorporate the potential reality of billions of inhabited planets? Dialogue and debate on this question has persisted throughout history and continues today.[4]

It was Faber's seminar that first exposed me to Dick's proposal of "naturalistic cosmotheology" and its six associated principles. In its strong rejection of supernaturalism and anthropocentrism, Dick's cosmotheology seeks to be fully consistent with all that the sciences (cosmology in particular) have come to know about our deeply evolutionary universe. This, for Dick, must include the likely fact that ours is a "biological universe" where life, rather than being restricted to our planet alone, is cosmically common in nature.[5] What then of religion and theology? In his first ever chapter on cosmotheology, Dick said the following:

> I suggest the time is ripe for us to take cosmotheology seriously, to consider how religions and their accompanying theologies should change in light of what we now know about the universe and what we are likely to know in the future: we are not the only intelligent creatures in the universe, most likely not the most superior. . . . The question is how to proceed. . . . No Thomas Aquinas for cosmotheology has yet appeared to reconcile current doctrine with new world views.[6]

I discerned a sense of opportunity in these early statements by Dick. I saw that Whitehead himself offered a cosmotheological metaphysics that appeared at least prima facie consistent with Dick's rejection of supernaturalism and anthropocentrism, as well as with his desire to incorporate our ever-expanding scientific knowledge and the potential of pervasive extraterrestrial life. Could Whitehead be the "Thomas Aquinas of cosmotheology" for the twenty-first century? Perhaps, yet Whitehead was not a professional theologian. Aquinas, it must be remembered, drew heavily from the metaphysical well of Aristotle. Thus, I thought it might be better to say that Whitehead is the "Aristotle of cosmotheology" because he offers a metaphysics that has proved helpful for theological use in our cosmic evolutionary context, not least by process theologians. Indeed, process theologians have often drawn analogies between their use of Whitehead's metaphysics and the ways in which Augustine and Aquinas used the metaphysics of Aristotle and Plato.[7]

I was intrigued to see that Dick later came to recognize the potential importance of Whitehead for his cosmotheology. "In its emphasis on evolutionary becoming," he sates, "cosmotheology resonates well with Whitehead's process theology."[8] I found this promising, but further questions remained: Just how deep was this resonance between Whitehead's philosophical theology and Dick's cosmotheology? What are the implications of Whitehead's philosophy of organism for astrobiology and extraterrestrial life? Did he ever say anything on the topic? What about other process philosophers like Bergson, arguably a competing candidate for the position of "Aristotle of cosmotheology," or Teilhard de Chardin, the great paleontologist-priest, who is perhaps most deserving of the title "Aquinas of cosmotheology?" I was abducted by these questions just as Faber's exo-life seminar was drawing to a close. It was clear that a deeper investigation of the relevance of process philosophy to astrobiology and extraterrestrial life was required.

I graduated in 2020 and accepted a position as the Program Director for Center for Process Studies, a faculty research center of Claremont School of Theology founded by John B. Cobb Jr. and David Ray Griffin in 1973. Following Griffin, my principal role as program director is to lead research and organize conferences relating Whitehead's philosophy—and process philosophy and theology more generally—to all manner of contemporary disciplines and discussions, especially those at the intersection of philosophy, science, and religion.[9]

With support from colleagues Wm. Andrew Schwartz and Philip Clayton (and a modest Templeton grant), Faber and I began planning for a 2022 conference that would mend existing lacunae between process philosophy and theology and current discussions of astrobiology and the impact of discovering life beyond earth. Once the conference vision and rationale were in place, Dick was the first to receive an invitation. He happily agreed to participate. He also suggested other invitees such as Mark Lupisella, an astrophilosopher from NASA, whose work on relationality and connectedness as foundations for cosmic creativity and value is heavily influenced by Whitehead.[10] Lupisella received the second invitation and accepted.

Co-organized by the Center for Process Studies and the Whitehead Research Project, the "Astrobiology, Exo-Philosophy, and Cosmic Religion" conference took place at Willamette University in Salem, Oregon on May 5–7, 2022. Subtitled "Toward a Constructive Process Cosmotheology," the conference aimed to be wider than Whitehead and include papers on the contributions of other key process philosophers and theologians like Henri Bergson and Teilhard De Chardin. The conference was also wider than the contributions of Dick, including the work of a variety of invited scientists, philosophers, theologians, and religious scholars relevant to a truly interdisciplinary conversation. Alongside Faber, Dick, Lupisella, and myself were also Bruce Damer, Matthew Segall, Derek Malone-France, Wahida Khandker, Chelsea Haramia, Ted Peters, Constance Bertka, Jeffery Long, Brianne Donaldson, Ilia Delio, Noreen Herzfeld, Brian Henning, and Kelly Smith. Theodore Walker Jr. was also invited, but later declined due to COVID-19-related cautions.

The two-and-a-half-day discussion was vast in reach and rich in both content and collegiality. Conference papers targeted a variety of discussions and discoveries in astrobiology, cosmology, and evolution, as well as considered a constellation of philosophical topics from shared extraterrestrial knowledge and values, to the possibility or limitations afforded by A.I. technology, the Fermi paradox, the Drake equation, and the increasing need to nurture the cosmic dimensions of our Eastern and Western theological and religious traditions when facing the likelihood of a biological universe.

It is the revised and expanded papers first presented at the "Astrobiology, Exophilosophy, and Cosmic Religion" conference which now appear in the current volume.[11] A debt of gratitude is due to all participants who made both the Salem conference and this resulting volume a success. A special thanks in particular to Steven J. Dick, whose work functioned as one of the chief inspirations for my organization of the event. Thanks of course are also due to Roland Faber for his catalytic exo-life seminar which ignited the entire conversation and led the way toward both the conference and this book.

PROCESS THEOLOGY IN RECENT LITERATURE

I mentioned above the scholarly neglect that process philosophy and theology has sustained in current discussions of astrobiology and extraterrestrial life. This is unfortunate. As we will see below, key process philosophers and theologians have made a variety of important statements on the topic that integrate core scientific, philosophical, and theological themes at the forefront of current conversations. I do not mean to suggest that current literature has been wholly ignorant of process thought. To the contrary, various insights of process thought (Whitehead's and beyond) continue to inform wide-ranging interdisciplinary discussions from physics, to bio-philosophy, psychology, neuroscience, and cosmology.[12] These and other texts, however, rarely offer any sustained attention to extraterrestrial life.

In fact, much of the discussion of extraterrestrial life and intelligence has come from the domains of theology and religion, and in the form of a variety of texts endeavoring theological responses to the topic. While process philosophy and theology have not figured prominently in these texts, a select survey of recent literature does reveal that references to "process theology" have not been fully absent, nor has mention of Whitehead, Teilhard (far less so Bergson), and some significant process theologians such as Norman Pittenger (1905–1997) and Lewis S. Ford (1933–2018) who offered some of the earliest treatment of the topic. The work of some current Teilhardian process theologians also finds mention, although often without reference to process theology. Teilhard certainly figures more than Whitehead in current literature, no doubt because of his explicit statements on the topic and his overt theological considerations (after all, he was a Jesuit priest).

Whitehead in fact makes no sustained statements on extraterrestrial life in his own corpus (although much can be inferred and extended from his cosmological wonder) and, as mentioned already, he was not a professional theologian. The journalist Lucien Price (1883–1964), however, does record some fascinating and imaginative statements Whitehead apparently makes in conversations with him, though Whitehead scholars continue to debate the merit and trustworthiness of Price's *Dialogues of Alfred North Whitehead*.[13] Despite ongoing debate, for the purposes of this volume, I find nothing in such statements that is utterly inconsistent with Whitehead's philosophy, and in fact much that is worthwhile and (I hope) trustworthy given other hints Whitehead offers in his actual works.

It is worth noting that the contributions of Charles Hartshorne, the principal originator of "process theology" after Whitehead, are virtually absent from current theological literature on astrobiology and extraterrestrial life (with one important exception noted below). We will see in the next section, however, that Hartshorne in fact made various important offhand statements throughout his corpus that have yet to be adequately acknowledged or studied. I admit fully that it remains a limitation of the current volume that it neglects Hartshorne's important work as a process philosopher and theologian. Let it be said for now that there remains a book to be written on Hartshorne and extraterrestrial life.

Let us look then to some of this recent literature in quest of relevant references to process theology and process theologians. While in no way exhaustive, I have chosen a selection of relevant works that have appeared over some two decades and up to our current time. What is revealed aims to be historically informative and set the stage for considering some important, although neglected, statements process philosophers and theologians have made in the following section.

In an early chapter titled "Exo-Theology: Speculations on Extraterrestrial Life" (1995), Ted Peters reviews key historical and contemporary perspectives on many worlds, God, and extraterrestrial life. Among a spectrum of contemporary theologians, Peters includes Lewis Ford, "a spokesperson for the school of process theology"[14] who, in a very early article titled "Theological Reflections on Extraterrestrial Life" (1968), wrote what appears to be the first sustained consideration of the topic from the perspective of process philosophy and theology. "Disciples of the philosophy of Alfred North Whitehead," Peters clarifies, "process theologians usually find themselves on the liberal end the Protestant spectrum."[15] This much Ford himself admits in sketching what he calls a "liberal theology suitable for the cosmos."[16] Peters notes that Ford "embraces the concept of evolution, applies it to every location in the universe, and then asserts that God is always and everywhere drawing the evolutionary process toward greater complexity and higher value," and thus toward abundant life throughout the universe.[17]

In considering "the issue of the universality of Earth's Christ event," Peters quotes Ford saying "salvation is not just limited to men but applies to all intelligent beings wherever they may dwell."[18] Peters also quotes Ford's later book *The Lure of God: A Biblical Background for Process Theism* (1978) which includes a variety of repeated and expanded statements in the context of process Christology. While we will look deeper to Ford's statements from his early article and *The Lure of God* in the next section, his inclusion by Peters is one of the first that can be found in the literature, and through Peters, it has found its way into later publications on the topic. The fact that Ford himself was likely the first process philosopher and theologian to publish an article on the topic is also significant in the context of this volume. We will see that he anticipates many of the interdisciplinary contours of current discussions.

In *Vast Universe: Extraterrestrials and Christian Revelation* (2012), Thomas F. O'Meara, being "led by discoveries in the universe," notes that "The history of philosophy and theology gives examples—Albertus Magnus, Friedrich Schelling, Pierre Teilhard de Chardin—of how science stimulates Christian reflection to find new paradigms and insights."[19] While O'Meara's mention of Schelling is not insignificant,[20] he offers a more sustained treatment of Teilhard. "An advocate of evolution, [Teilhard] sought to understand and express religious patterns on a cosmic scale," he states. "The Jesuit's principles—evolution, variety, development of complexity with a climax in intelligence—argue for extraterrestrials," and "cosmic redemption" in ways that were not anthropocentric, but theocentric and, indeed, Christocentric.[21] There is no mention, however, of Teilhard's important influence on Catholic process theology.[22]

In *Science, Religion, and the Search for Extraterrestrial Intelligence* (2013), David Wilkinson speaks of Teilhard as one of "the most imaginative thinkers in the Catholic engagement of science and religion."[23] What is more, Wilkinson cites Whitehead's book *Science and the Modern World* (1925) with respect to own his claim that "a God who freely creates the Universe and welcomes the enquiring mind gives a strong basis for the empirical method."[24] In passing, he also mentions Bergson and William James with respect to their belief "in a force which represented the continuously creative nature of reality."[25]

Of particular note is Wilkinson's mention of the important process theologian Norman Pittenger—although without connecting him to process theology.[26] Pittenger was an outspoken Cambridge process theologian and Anglican priest who spent his career advocating Christian appropriation of process thought in the vein of Whitehead, Teilhard, and Hartshorne. As we will see in the next section, Pittenger held to the reality of other worlds and extraterrestrial life on both philosophical and theological grounds. Although he did not publish a full-length article, he did include an early piece in *The Christian Century* (1956) the significance of which is noted below. He would also include some important statements in his early (and later) books. As such we will see that he rivals Lewis Ford in being the earliest commentator in the context of process theology.

More so than most texts, David A. Weintraub's, *Religions and Extraterrestrial Life: How Will We Deal With It?* (2014), knows of at least some of the contributions process theologians have made to the conversation. Teilhard figures prominently in his chapter on "Roman Catholicism" (ch. 8), as does the work of Ilia Delio (b. 1955) in reference to her article "Christ and Extraterrestrial Life" (2008) and John F. Haught (b. 1942) in reference to his earlier article, "Theology After Contact: Religion and Extraterrestrial Intelligent Life" (2001).[27] No mention, however, is made to Delio and Haught as long-standing advocates of Teilhardian process theology (as also informed by Whitehead). In reference to Haught's chapter for example, Weintraub neglects to include his very important closing remarks:

> I would suggest . . . that the cosmic vision of Teilhard de Chardin as well as process theology (based on concepts of the philosopher Alfred North Whitehead) are both already inherently open to being developed into a 'theology after contact.' Not the least of the reasons for their adaptability is that they have already enthusiastically embraced the Darwinian portrait of life as well as the notion that the entire universe is still in the process of being created.
>
> Contemporary 'process theology' with its vision of cosmic purpose is also expansive enough to accommodate the discovery of ETI. For the 'process philosopher' Alfred North Whitehead and his theological followers, the purpose of the cosmos consists of its aim toward the intensification of beauty. Since—at least for Whitehead—beauty is an intrinsic value, any process that leads toward its establishment could be called 'teleological,' at least in a loose sense. 'Beauty,' in Whitehead's thought, means the 'harmony of contrasts' or the 'ordering of novelty,' many diverse instances of which have appeared in the evolution of the cosmos and in the emergence of life, mind, and culture in our terrestrial setting.[28]

In his chapter on "Mainline Protestant Christianity" (ch. 11), Weintraub does in fact speak of "The school of process theology, which in the early twentieth century emerged and built on the theological ideas of Alfred North Whitehead." "From the perspective of process theology," he states, "Christianity is a terrestrial religion meant for humans alone, though all of God's creatures require some form of salvation."[29] He briefly considers important statements made by Lewis Ford on Christology, soteriology, and incarnation in a cosmic context in *The Lure of God* and also (via the work of Ted Peters) references Ford's early article "Theological Reflections on Extraterrestrial Life" in a footnote.[30]

Weintraub also makes important mention of "the physicist and process theologian" Ian Barbour (1923–2013) whose work in science and religion was heavily influenced by Whitehead, Teilhard, and Hartshorne. For Barbour, there is no reason why life—and even superior life—should not emerge in other stellar regions of the cosmos. What is called into question is naïve anthropocentrism and exclusive terrestrial claims as to the finality of Christ's redemption—both of which can be overcome with the resources of process theology.[31]

In *Cosmology in Theological Perspective: Understanding Our Place in the Universe* (2018), Olli-Pekka Vainio makes slim reference to the work of Ilia Delio, Ian Barbour, and Keith Ward, although without engagement or reference to their influence by process theology. Throughout his career, Keith Ward (b. 1938) has made consistent use of Whitehead for his work in science and religion and his cosmological reconsiderations of many traditional Christian doctrines, often in light of the possibility of other worlds and extraterrestrial life.[32] In drawing from, among others, Whitehead's triadic conceptuality of God as primordial, consequent, and superjective for example, Ward develops a cosmic trinitarianism that "leads to the postulate that God will act in alien worlds and among alien forms of life, if there are any, with a similar threefold form, manifesting in other modes of finite personal reality, acting in their innermost lives, and leading them to union with the divine."[33] Moreover, this threefold conceptuality (and others like it) is not confined to Christianity, but can also be found in various religious and philosophical expressions, whether Bradley's Absolute Idealism, Plotinus' Neo-Platonism, or some versions of Vedantic thought.[34]

In *Astrotheology: Science and Religion Meet Extraterrestrial Life* (2018), edited by Ted Peters, multiple contributors weigh in on various contours of the current discussion of theology and extraterrestrial life. While no process theologian contributes a chapter, passing reference is made to Whitehead, Teilhard, Ilia Delio, and John Haught, although with little sustained engagement.[35] Peters, however, who attended the Salem conference and appears in this volume, has recently invited my contribution of a chapter on "Process Astrotheology" in an effort to expand the work begun in his *Astrotheology* volume and further recognize the important contributions process philosophy and theology can make to the discussion.[36] I've also accepted a generous invitation from Peters and his college Maynard Moore to join the 2024 IRAS (Institute on Religion in the age of Science) summer conference on Star Island where I will speak further to the contributions of Whitehead and process theologians.

In his recent book *Exotheology: Theological Explorations of Intelligent Extraterrestrial Life* (2021), Joel Parkyn notes the "considerable attention" given by "modern Protestant theologians," to "the efforts of exotheology in its engagement with the interdisciplinary field of astrobiology." In a footnote, he includes "Alfred Whitehead" and "Lewis Ford" alongside other theologians who have made these contributions.[37] He speaks of Ford as a "process theologian" and notes both Ford's *Lure of God* as well as his early article (again via Peters) on "Theological Reflections on Extraterrestrial Life." He speaks of Ian Barbour too as "another process theologian" who has made important contributions.[38] Norman Pittenger is also included although not in reference to process theology[39] and Teilhard, Ilia Delio, and (slimly) John

Haught also figure in the Catholic context, albeit in ways disconnected from their process conceptualities.[40]

In *Extraterrestrial Intelligence and the Catholic Faith: Are We Alone in the Universe with God and the Angels?* (2022), Paul Thigpen treats Teilhard briefly as well as the work of Ilia Delio.[41] In *Astrobiology and Christian Doctrine: Exploring the Implications of Life in the Universe* (2023), Andrew Davison also makes scant footnote reference of Teilhard and Delio.[42] For both texts, however, no connection is made to process theology

What can we take from this brief review of recent literature and the status of process philosophy and theology in the current discussion? We have seen that process philosophy and theology have not been absent, and that some important mentions do exist. However, deep engagement of process theology as a tradition of cosmic theological reflection is strongly lacking. This remains a lacuna that this book (and I hope future books) aims to fill, and it is a lacuna that is hardly justified given a constellation of past and present statements offered by process philosophers and theologians. Reviewing a selection of these statements below will orient us to multiple themes emerging in the current volume.

PROCESS PERSPECTIVES ON OTHER WORLDS AND EXTRATERRESTRIAL LIFE

A brief appraisal of statements made by process philosophers and theologians reveals varieties of openness, wonder, and affirmation of the reality of other worlds and extraterrestrial life. As based in the trans-planetary creativity of the cosmos, there is no ultimate philosophical, scientific, or theological reason that life—and intelligent life—should be restricted to our planet alone. To the contrary, these tangled disciplines, although contingent and terrestrial, point to various possibilities of life and intelligence beyond earth. For all the thinkers below, it is clear that the cosmic life and activity of God transcends, yet includes, any particular planetary habitat or species.

Henry Nelson Wieman (1884–1975) was for a time a contemporary of Whitehead and is credited as one of the "founders of process theology" in its creative use of Whitehead's philosophical theology.[43] He admits to having studied Bergson intensively and was greatly influenced by him. He points to Bergson, Dewy, Alexander and Whitehead as "all having something in common" which "they all approach from different directions and in different ways," he states. "Creativity is as good a name for it as any, and they all use that word with different shades of meaning. I am deeply indebted to them all." In studying Whitehead's early works, Wieman notes that, "These books seemed to open a door which I had vaguely felt was there to be opened but which Whitehead first swung wide. . . . I consider Whitehead one of my greatest teachers."[44]

Indeed, Whitehead's influence figures prominently in Wieman's notable book *The Source of Human Good* (1946) which details the workings of the "creative event" and the "creative good" as the means by which entropy is transcended and various shapes and shades of value are attained in both evolution and human experience. For Wieman, "this increase of good that has occurred upon the planet is not an accident. It is the working of a known and definite reality we have described as creative good. This creative good works against obstacles and overcomes them with a power qualitatively and immeasurably different from man's."[45]

What we call "inanimate matter" is really the "statistical average of innumerable atomic activities that cancel out in their total impact upon us," Wieman states. Nevertheless, the "particular atoms and molecules seem to be highly dynamic, even organized."[46] Wieman notes that "the organization and creative process at the subatomic level does not seem to be able to rise to higher levels except in those rare instances in which life has emerged." He points to the first triumph of the "creative event" as well as signals his openness to such triumphs beyond earth:

> The living cell is the first triumph of the creative event over the inertia due to this canceling-out of the diverse activities in the atom and molecule. Throughout most of the universe, creativity seems to have been halted at the level of matter's inertia. Since our knowledge of remote space is very limited, it would be rash to state that only on this planet has creative fulfilment risen above the molecule.[47]

Despite his openness to other creative fulfilments beyond our planet, and in potentially "far higher" forms than the human, Wieman discourages such futile speculations and directs our attention to what has in fact occurred on *this* planet.

> The form of the creative event working at the higher levels of value is created by its working at the lower levels. Whether or not it ever did begin at some lowest level, mounting upward ever since; whether or not there are, even now, levels far higher than the human; whether there are a rise and fall of levels throughout the cosmic whole of things or an upward march forever—these and many other such matters we do not know and consider it futile to speculate about. All we know is what has happened upon this planet, and even that we know but slightly.[48]

Wieman describes the creative event or good at work in nature as "supra-human" and "absolute good." It is "a good that is not relative to time or place or person or race or class or need or hope or desire or belief." Rather, the creative good "works outside of human life, but in the sense that it creates the good of the world in a way that man cannot do."[49]

Ultimately, Wieman identifies the "creative event" with "God," emphasizing that "God, according to this interpretation, is immediately accessible to human living and human feeling in all the fullness of his concrete reality." The divine creative process, he states, "makes qualities more appreciable by creative sensitive organisms; by making them progressively more sensitive; by generating signs with meanings and thus producing mind; by bringing forth communication of meanings and thus establishing society; and by sustaining the continuity of history and thus accumulating meanings through a sequence of generations."[50] Despite his attention to this planet, there is no reason to restrict these claims of the "creative good" to our planet alone. Indeed, the creative good is at work in and among the "cosmic whole of things."

Wieman had an important influence upon his students, among them Bernard Meland (1899–1993), Bernard Loomer (1912–1985), and Daniel Day Williams (1910–1973). All would make important contributions as process theologians. For the purposes of this volume, it is noteworthy that Bernard Loomer would later be quoted in a *Time* magazine article titled "Theology: Challenge in the Heavens" (1969) as saying that space exploration "may reinforce the idea that man may not be the most important thing in creation. Say that out there we find persons superior to us, as we consider ourselves superior to dogs?"[51] Loomer's statement, no doubt uttered in wake of the moon landing, is an early indication of the openness of process theology to the topic. We will see this further confirmed below in Lewis Ford's earlier article "Theological Reflections on Extraterrestrial Life" (1968).

I mentioned above that Charles Hartshorne (1897–2000) became the principal figure associated with the development of process philosophy and theology after Whitehead. Although Hartshorne was influenced heavily by Whitehead (being his assistant for a time), it is important to also state that he developed aspects of his own process philosophy and theology independent of Whitehead.[52] Hartshorne's work remains vast and it is not well known how truly affirmative he was of other worlds and extraterrestrial life. To use Haught's language above, it would not be false to say that Hartshorne's various insights are also "inherently open to being developed into a 'theology after contact.'" This has yet to happen, however. While Hartshorne offers no extended discussion of the topic (like Ford below), he does make a variety of important statements throughout his corpus.[53]

For example, in his early book *Beyond Humanism* (1937), Hartshorne states that panpsychism and theism (at least his form) "implies the eternal existence of finite minds of some kind in the universe," adding that "this could be verified . . . by the discovery of other inhabited planets."[54] In *A Natural Theology for Our Time* (1967) Hartshorne speaks of "other inhabited planets, the nonexistence of which we have no right to assume" as also being included and loved by God.[55] In *Creativity in American Philosophy* (1984), he speaks of "the present richly variegated universe, with its probability of many inhabited planets," and in *Wisdom as Moderation* (1987), he inquires about a "presumptive species . . . on far-off planets," including parenthetically: "I believe they exist."[56] He would state this in stronger fashion in his closing remarks to a famous debate between Gary Habermas and Anthony Flew on the resurrection of Jesus. His statements make it clear that his theology transcends both anthropocentrism and terracentrism. "I strongly incline to believe in rational inhabitants of some other planets" he states. "My view is not humanistic and earth-centered, but cosmic and God-centered; theomorphic, not anthropomorphic."[57]

These are only a few relevant examples from Hartshorne's corpus. As Don Viney confirms, Hartshorne uses "life" or "intelligent life" on "other planets" in passing throughout his work. According to Viney, "The almost casual references in Hartshorne's writings . . . to the specifics of life on 'this' planet, show pretty clearly that he thought of life on earth, including intelligent life, as one of many examples in the universe."[58]

We saw above that there has been some citation (though no substantial engagement with) Lewis Ford (1933–2018) and his pioneering article "Theological Reflections on Extraterrestrial Life" (1968). As mentioned, this appears to be the first sustained consideration of the topic by a well-respected process philosopher and theologian.[59] In many ways, it is a remarkable article in that Ford, presumably with the moon landing in view, offers a sophisticated consideration of a variety of themes and trajectories that would later blossom in subsequent decades of discussion, including today.

Ford begins with a discussion of the nature of theological method and the inability of existential theology (specifically that of Rudolf Bultmann) to be of service to the theological question of extraterrestrial life. Rather, this question necessitates the relevance of cosmological statements to theology. He then speaks in detail to various concerns including the importance of the biblical shift from a God of a particular people (Israel) to the God of all creation, the scientific history concerning the regularity of planetary development and evolution and, thus, the likelihood of at least "one million life-worlds in our galaxy," and the nature of evolution, particularly the insufficiency of natural selection and chance variation alone to explain the rising complexity and novelty of the evolutionary advance. He also considers the nature of God's "cosmological function" in evolution as persuasion toward greater complexification, and the nature of life and the "decisive difference between living and lifeless matter." In a series of imaginative paragraphs, Ford also comments on extraterrestrial culture, technology, ethics, morality, economics, politics, interstellar communication, "extraterrestrial comparative religion," and even extraterrestrial religious experience and expression. It is noteworthy that these are all now issues that are being raised in current discussions.[60]

In such a cosmic context Ford defines God "as that dynamic source of value which lures the evolutionary process to an ever-richer complexity productive of increasing freedom and intensity of experience." He insists that God is "necessarily operative in the development of every life and in every culture, whether terrestrial or extraterrestrial," although this may or may not be known by extraterrestrials. After sketching a "liberal theology suitable for the cosmos," Ford then looks to particular theological issues raised for Christian faith. These include scriptural hermeneutics, multiple incarnations, soteriology, resurrection and the body of Christ, ecclesiology, and the possibility of a kind of ultimate Teilhardian convergence of interstellar "world-civilizations." This bevy of considerations leads Ford to his final cosmotheological comment: "The possibilities an everlasting God can actualize in due time through countless cycles of expansion and contraction [in the universe] are simply inexhaustible."

Ford's detailed considerations are repeated and expanded in the context of "Recent Process Christology" in his later book *The Lure of God*. As first expressed in his early article, however, Ford's contributions remain thoroughly original in the context of process metaphysics and theology and eminently relevant to current conversations in the field. It is for this reason that Ford's article is republished (with minor edits) in the epilogue of the current volume.

Another vital figure mentioned above is Norman Pittenger (1905–1997) who rivals Ford in first raising the question of other worlds and extraterrestrial life in the context of process philosophy and theology. Although his considerations are less sustained (and expansive) than those of Ford, his short article titled "Christianity and the Man on Mars" (1956) is included in *The Christian Century*, which Brenda Denzler describes as "the first religious periodical to deal with the question of extraterrestrial life."[61] If true, this statement is of utmost significance, for it locates process theology at the origins of overt religious considerations of extraterrestrial life in the mid-20th century. In fact, this appears to be confirmed by both the 1969 annotated bibliography on "UFOs and Related Subjects" prepared by the Library of Congress Science and Technology Division, and NASA's 1978 "Bibliography on the Search for Extraterrestrial Intelligence," both of which cite Pittenger's article as among the earliest of theological considerations of the topic.[62]

Pittenger begins by commenting on the role that science fiction literature has had on our imagination of other inhabited planets. "But that it would create a new problem for Christian apologists did not occur to many of us," he states. "Yet the fact is that it has done precisely this." He notes that he encountered the problem of extraterrestrials on several college campuses and among theological students. He phrases the issue in the following way:

> How does the Christian gospel, concerned with the salvation of men in this world, have any universal significance when we know that there may well be intelligent life on other planets? Can Jesus Christ have more than an earthly significance? Is it not preposterous to make such extravagant claims for the Christian revelation when we are now sure of a vast universe in which other planets, which have never heard of that revelation, may well have conscious inhabitants?[63]

"For the real question," Pittenger continues, "is not simply the enormous reaches of time and space with which astronomy and astrophysics have acquainted us. It is rather the problem of the relationship between the Divine Reality, God, and the inhabitants of other parts of the created order. Do such 'people' know God? In what sense is salvation offered to them? How is Jesus Christ central to our God-man relationship if he is not known to these 'people'?"

In reflecting upon these questions, Pittenger is quick to note that he is not concerned to discuss—nor, he admits, is he competent to do so—the question of "whether or not there are in fact conscious intelligent inhabitants, 'people,' on other planets or elsewhere in the universe." Surveying the scientific literature at the time leaves the issue still "unsettled." While he notes that the possibility (and probability) of life beyond earth "is increasingly held these days," he stresses that his interest, "as befits a theologian, is with the problem that is raised by the possibility or probability," and "not with the solution of the scientific problem itself."[64]

The theological problem as Pittenger encounters it has been "seriously misconceived" because it has been based in a "Jesu-centrism" that restricts divine activity solely to our planet, instead of the cosmic "Christocentrism" affirmed by the "fathers of Christian theology." He demonstrates this through appeal to statements in the Nicene creed and the opening verses of the gospel of John, both of which allow one to say that "for Christian faith Jesus *defines* but does not *confine* God in his relationship to the created world." Thus, Pittenger suggests the right "orthodox" approach to the "possibility of conscious intelligent life on other planets":

> We may therefore say that if there should be conscious intelligent life on any other planet, God has not left himself without witness there, since by his Word [Logos] he both creates and sustains, and also wills to make himself known in some fashion wherever he does create and sustain. We can say this because we know, from our Christian faith and our own experience of God in Christ,

> that it is the divine nature ever to give itself generously and insistently to its creatures.[65]

Pittenger concludes his article with the claim that we will not know in what ways God might reveal Godself on other planets, whether through multiple incarnations in other "conscious intelligent life," or by other means. As concerns the "decisive nature of Jesus Christ," however, he quotes Hamilton King: "God may have other Words for other worlds, But for *this* world the Word of God is Christ."[66] Like Ford, Pittenger would also repeat and expand these early considerations in his later book *The Word Incarnate* (1959).[67]

These are not the only places in Pittenger's corpus where he engages the question of extraterrestrials. For example, in his underappreciated introduction to process theology titled *Picturing God* (1982), he includes "A Note on God and Other Worlds."[68] In this note, Pittenger includes repeated and new material. One of his principal claims is that among various problems with "the older and conventional models of God" is their thoroughly "anthropocentric quality."[69] It is this quality that is expressed when religious believers take offense at the prospects of other inhabited worlds. In affirming Whitehead's "new model" of God that avoids this "quality," Pittenger indirectly relates Whitehead to his discussion of other inhabited worlds. Although he does not name Whitehead in this note, his affirmation of God as "cosmic Lover," moving the sun and other stars, comes directly from Whitehead (and Hartshorne).[70]

In *Process Theology: An Introductory Exposition* (1976), John B. Cobb Jr. (b. 1925) and David Ray Griffin (1939–2022) emphasize that "Process theology intends to think through the meaning for our existence and actions of the space-time scales that scientific cosmology suggests."[71] While they do not include any overt treatment of extraterrestrial life, they do indicate their openness to this possibility within Whitehead's philosophy. They insist for example that "there was real value and enjoyment in the eons of time before high forms of life appeared anywhere in the universe" and "that the level and importance of enjoyment increased greatly when, on this planet (and wherever else a similar development may have occurred) animals and finally humans emerged."[72]

Griffin would make similar comments in his later text *God and Religion in the Postmodern World* (1989), speaking in the context of human persons with their "personal qualities." Griffin stresses that human persons are part of the created order, and the "crown of creation" in the sense that they are the "chief exemplification, on our planet anyway, of the qualities that all creatures embody." This view according to Griffin (and Cobb) presupposes a hierarchy of values in nature and "provides the basis for rejecting anthropocentrism, according to which other creatures are emptied of all importance except their usefulness to human beings, while at the same time holding that human beings are intrinsically more important than other earthly creatures."[73]

This claim of value and importance, moreover, presupposes deeper standards of value; indeed, primordial values that (for both Griffin and Cobb) belong to the divine nature and, as such, are necessarily applicable across the cosmos. As Griffin later insists in *Panentheism and Scientific Naturalism* (2014), "on other planets with the conditions for life to emerge and to evolve for many billions of years, we should expect there to be some with creatures that, no matter how different in physical constitution and appearance, would share some of our capacities, such as those for mathematics, music and morality, or more generally, truth, beauty and goodness."[74]

In *The Liberation of Life* (1981), John Cobb and the geneticist-theologian Charles Birch (1918–2009) endeavor a sustained discussion of life from the cell to community to the cosmos at large. Deeply indebted to both Whitehead and Wieman, they speak of "Life as a cosmic power" and "cosmic principle" that "works for higher order in the midst of entropy." "When conditions allowed, it brought forth living forms upon this planet and multiplied them until it transformed the whole surface of the earth," they state. "It brought forth human intelligence and . . . transformed the earth once more. Probably it has worked similar miracles elsewhere in the universe, perhaps at many places." While life is only known on our planet, they insist that "it can be assumed that where conditions allow in other planets circling other stars in this or other galaxies, Life has worked its magic. Having observed Life here one can assume that it is a cosmic principle."[75] "Somewhere in the universe there may be wholly different conditions that also make entirely different forms of life possible," they state, "perhaps even in empty space, or in dimensions different from our own. Of such possibilities nothing is known."[76]

Ultimately, Cobb and Birch speak of "Life as God." They do so in reference to Whitehead. "The Whiteheadian idea of God is appropriately called Life not only because the immanence of God in the world is the life-giving principle, but also because the life-giving principle is itself alive." This follows because "A lifeless principle could not ground or explain the urge to aliveness that permeates the universe." Indeed, God as "Life" does not "aim specifically at the creation of human beings . . . on our planet," but rather achieves "rich value in dolphins as well as human beings. We cannot guess the forms it may have achieved on other worlds," they state.[77]

Recently, Cobb has endeavored yet another discussion of "Cosmic Life" that, while clearly drawing on these early collaborative insights with Birch, nevertheless apply Whitehead's insights anew. "Living" is a "matter of degree" in nature, such that the fundamental units of nature can be said to "embody pre-life," Cobb states. In speaking of the possibilities of novelty associated with the future, and the problem of past history, both terrestrial and cosmic, Cobb affirms the reality of "super-life." "The cosmos as a whole includes fundamental features that make life possible and actual," he states, "It may not be alive in the same way that we are, but it grounds life by its eternal inclusion of novel potentiality . . . it is characterized by super-life." Cobb emphasizes this by capitalizing "Life." "The cosmos, or a crucial dimension of the cosmos is Life Itself." The cosmos thus favors life and common sense for Cobb ultimately suggests that "there was a cosmic decision in favor of life."[78]

In a robust chapter titled "Extraterrestrial Science" (1985), Nicholas Rescher (b. 1928) considers whether science in another planetary setting might overcome various limitations associated with our merely human science. "Our science as we humans cultivate it here on earth is limited and imperfect," he states, "and is bound to remain so. It thus becomes tempting to wonder whether another, astronomically remote civilization might be scientifically more advanced than ourselves. Is it not plausible to suppose that an alien civilization might overcome the limitations of our science and manage to surpass us in the furtherance of this enterprise?"[79] Rescher's discussion ranges widely from considerations of "the potential diversity of science" in the cosmos, to the "one world, one science" argument, first principles and the nature of evolution, and whether or not "the mission of intelligence in the cosmos is uniform or diversified?"[80] Rescher remains consistently aware of the "anthropocentric supposition" that is behind such discourse, and concludes by pointing to the unique and peculiar nature of human science in a more-than-human cosmos.

> The upshot of these deliberations is thus relatively straightforward. The prospect that some astronomically remote civilization is "scientifically more advanced" than ourselves—that somebody else is doing "our sort of science" better than we ourselves—requires in the first instance that they be doing our sort of science at all. And this deeply anthropomorphic supposition is extremely unlikely. To endorse the idea of a scientifically superior extraterrestrial civilization is to step beyond the realm of scientifically plausible fact. The quest for superhuman science should bring home to us the uniqueness of that artifact which is our human science. It should be stressed, however, that this consideration that "our sort of natural science" may well be unique is not so much a celebration of our intelligence as a recognition of our peculiarity.[81]

Rescher would again treat the question of shared science and epistemology among extraterrestrials in *Pluralism: Against the Demand for Consensus* (1993), where he revisits and expands his earlier considerations.[82]

Rescher's rich corpus through the years has presupposed not only our current cosmological setting, but also a process conceptuality of this setting. His aversion to any "anthropocentric supposition" is consistently shown by the fact that he speaks of "beings," "intelligent beings," and "intelligent creatures," without ever restricting being, intelligence, or creaturehood to *human beings* on this planet alone. In *Process Metaphysics: An Introduction to Process Philosophy* (1996), for example, he reviews how God is conceived to interact with the cosmos according to process theological conception: "God's processual involvement with the world is of many sorts, preeminently productive, cognitive, and affective," he states. "God is linked to the world of created beings through a reciprocity of affective appreciation, and responds to the world's eventuation by way of approbation or disapproval." What is more, he highlights that this context "would have to include the doings of autonomous agents that 'go their own way,' including most importantly, some of the free actions of intelligent creatures."[83]

In *Metaphysical Perspectives* (2017), Rescher further discusses rationality, intelligence, and personhood, all of which are much wider than one species on one planet. "To be human is to belong to a particular animal species through the operations of biological and physiological evolution. As such, humans exist only here on earth," he states. "However, to be a person is something quite different and far broader, something determined by one's capacities and possibilities for action. Persons are beings that make their way in the world by means of thought, acting on the basis of their beliefs and choices. In principle, alien creatures very different from humans could be persons."[84]

That these "persons" exist on other worlds appears to be part and parcel of Rescher's own proposal of axiological optimism in the universe; yet, questions still remain: "If the development of intelligent beings is the telos of cosmic evolution, why are there not more of us on innumerable planets?" he asks. "And why are those presently here not a great deal more intelligent?"[85] These open questions notwithstanding, a universe that is axiologically governed is one that is eminently friendly not only to intelligence as a value-achievement on this and other planets, but also to conceptions of divine intelligence operating ubiquitously throughout the cosmos.[86]

In *The Big Bang and God: An Astro-Theology* (2015), the astronomer Chandra Wickramasinghe (b. 1939) and process theologian Theodore Walker Jr. (b. 1953) endeavor a detailed investigation of various interdisciplinary convergences relevant to current discussions in the field. Wickramasinghe was for a time a close collaborator with Fred Hoyle (1915–2021) and Walker remains a long-standing advocate of "neoclassical metaphysics" in the tradition of both Whitehead and Hartshorne. The insights of both Hoyle and Hartshorne in particular function as the conceptual backbone of their investigations. "This study of scientific literature by and about Sir Fred Hoyle and his collaborators shows how *astronomy, biology, astrobiology, astrophysics, and cosmology converge with natural theology*," they state. "And, as advanced by constructive postmodern scholars instructed by mathematician-philosopher

Alfred North Whitehead and logician-metaphysician-philosopher Charles Hartshorne, *natural theology yields biology and psychology with cosmological scope*."[87]

Although Walker and Wickramasinghe offer no extended discussion of extraterrestrial intelligence, they do argue for various constructive integrations of scientific, philosophical, and theological domains that are antecedently relevant to the discussion. For example, Hoyle and Wickramasinghe argued that "biology cannot remain exclusively an Earth science. The study of life is extraterrestrial (cometary), astronomical (stellar, interstellar), galactic, and cosmological."[88] Walker and Wickramasinghe connect this biocentrism to Hartshorne, who "agrees with physicists who dare to say 'all nature is in some sense life-like.' No 'absolutely new principle of life' emerges from nonlife 'at some point in cosmic evolution.'" Rather, "All life emerges from previous life and ultimately from the everlasting divine life."[89] As required by mereological analysis, they further argue that this divine life must be deemed the "one all-inclusive whole of reality" and thus, the ultimate super-intelligent context without which there would be no life or universe at all. This "panentheism" they equated with the "process theology" of Whitehead and the "neoclassical theology" of Hartshorne.[90]

The Big Bang and God has the merit of being one of the only book-length, interdisciplinary explorations that is overtly indebted to process philosophy and theology. The fact that the book shows deeper embeddedness to Hartshorne than to Whitehead is also a distinction. It is thus an exception to my claim above that Hartshorne's process philosophy and theology are *virtually* absent from recent literature.

Finally, the recent work of Roland Faber (b. 1960) has shown consistent engagement with various philosophical, theological, and religious issues related to astrobiology, other worlds, and extraterrestrial life from within (but also beyond) Whitehead's philosophy. In fact, next to Lewis Ford's early considerations, Faber is the most recent process theologian to consider the topic in both depth and detail. Faber's contributions also remain unique in that while most of the preceding considerations of extraterrestrial life by process theologians have come from the context of Christian theology, Faber speaks from within the Bahá'í Faith and Bábí religions which retain expansive cosmological and theological visions necessitating a manifold of worlds and intelligent extraterrestrial life. In Faber's recent work, we thus see a unique integration of Whitehead's philosophical cosmology with that of the Bahá'í and Bábí traditions.

Several of Faber's statements with respect to Whitehead and his cosmology are worthy of note with respect to the current volume. In *The Ocean of God: On the Transreligious Future of Religion* (2019), Faber weaves together a variety of ancient and modern sources in discussing "the earth and other worlds" as "a story of cosmic magnitude" essential to "the future of religions."[91] Faber insists that "In Whitehead's organismic philosophy, both the ecological and the interstellar horizon for determining the place of humanity in the cosmos and on earth intersect intensely." Whitehead, he stresses, "does not assume any beginning or end" to the "revolutions and oscillations" of the universe. Instead, "we can either expect a series of cosmic epochs, in themselves relatively consistently interrelated, but different from other cosmic epochs of unimaginable patterns" of organization, "or a multiverse of such relatively independent and diverging cosmoi."[92]

In contrast to the dead and lifeless universe of reductive materialism, Whitehead followed the revolutionary insights of the new physics, laying what Faber describes as "a new metaphysical basis . . . beyond the senseless spaces of scientific materialism" that embraces the paradoxical entanglement of life, mind, and matter as "expressions of the very fabric of the universe itself."[93] Indeed, "the fact of life and mind on Earth," not only presupposes that they are possible, but that they are in fact "actualizations of universal potentials of cosmic . . . history," Faber states. What is more, Faber insists that there is no rule restraining such possibilities "to be necessarily only realized once, only contingent reasons of serendipitous coalescence of causes" such that, whenever such causes or constellations come together in the universe, "we can expect such potentials to be realized—again." For Faber, this is "implied in Whitehead's cosmology of multiple worlds."[94]

As conceived in Whitehead (and the Bahá'í universe), life and mind are ultimately "divine processes" revealing that divine immanence is the "motive force of the whole cosmic movement toward spiritualization, that is," Faber clarifies "the emergence not only of life and mind, but their transcendence into a process of always-new unifications of [an] ultimately religious nature" and an "anticipatory future of ever-new and ever-vaster fulfillments of this . . . cosmic drive in life-forms" and "their actualization of divine attributes, such as compassion and love, justice and peace, freedom and creativity in the physical, personal, and social organizations of the universe(s)." In light of such vast cosmological landscapes and the indefatigable workings of "divine wisdom," Faber insists that "humanity can hardly be the only species expressing its mind in spirituality and religion." Rather, "If God is the God of the universe (and being of apophatic nature, but infinitely dispersing of grace, ever-creative,) we must assume that other cosmic sites would also have been set in motion to developing sentience and life, and Reality consciousness."[95]

In *The Cosmic Spirit: Awakenings at the Heart of All Religions, the Earth, and the Multiverse* (2021), Faber draws together the interspiritual and the interstellar in a series of philosophical and theological meditations provoked by "staring at the stars" and imagining "the life of the other days."[96] Here Faber stresses that "the recognition of our *spiritual* nature means that the evolution of the universe" is aimed toward the "overcoming of opposition and mutual destruction" and the embrace of "relationality and cooperation, the 'better life' instead of naked survival, and the realization of the *common religious values* of compassion and love, justice and mutuality, instead of mere self-preservations and beyond any special interests." It is this point for Faber that "harbors one of the most exciting side-effects of the cosmological question of alien *spiritual life*: namely, that, if it is meant to develop *purposefully* in the cosmos *once*, it can happen *more than once*—especially, if it happens to some such being, like humanity," whose "own spiritual development and purpose recognize that such evolution cannot be meant to have already come to perfection" in themselves.[97]

In Faber's most recent book *The Mind of Whitehead: Adventure of Ideas* (2023), he insists that "Life" for Whitehead is one of the "universals of the universe."[98] While Faber admits that "Whitehead has not left us with many clues as to his vision on these matters," he states that he has instead given us "profound parameters for a reasonable projection." Whitehead certainly knew of the discovery of other planets and of the inherently speculative nature of "scientific methodology in interpreting such discoveries," Faber states. Thus, to the question as to whether any alien environment allows for life beyond Earth, Faber answers: "Probably, because Life is a Universal of the Universe; but it will be environmentally and socially diversified." To the question as to whether there may be "*constraints* on such a diversification of life," Faber again answers: "Yes—especially if we take into account Whitehead's insistence that the teleology of the universe is aesthetic in nature, that is, is meant to generate life through the creation of multiplicities of integrations of intensities and harmonies; the emergence of mind, consciousness, and spirit; communication, intelligence, and symbolism; art, culture, and civilization; the universals of life, environment, evolution, and cosmic magnitude, and their particular interactions and constraints . . . " All of this for Faber contributes to an "informed" and "affirmative answer."[99]

While Faber remains skeptical of the relevant statements recorded by Price in his *Dialogues*, he nevertheless speaks to the "openness of [Whitehead's] own cosmology to such a multiplicity of diverse appearance of forms of life that add to a variety of *other* streams of manifestations . . . in the wider cosmic organizations of individual or social emergences of sentience, intelligence, and civilizations." As reported in Price's *Dialogues*, Whitehead was open to "diverse magnitudes at which such other life-forms may appear: of whole universes within infinite dimensions even in minute quantum events; or of infinite worlds full of life; or of life-forms of the magnitude of nebulae."[100]

What can be said of these various statements uttered by notable past and present process philosophers and theologians? As signaled at the outset of this survey, these statements have demonstrated varieties of openness, wonder, and affirmation with respect to other worlds and extraterrestrial life. It is this fact that demands more deliberate inclusion of the resources and/or convictions of process philosophy and theology in current conversations on the topic. Indeed, the principal purpose of this volume is to launch this inclusive endeavor, and in doing so, ensure that both process philosophy and theology are henceforth part of the discussion.

Before turning to the structure and flow of the current volume, it is instructive to conclude the foregoing survey with a series of fifteen thematic affirmations that emerge in the context of process philosophy and theology:

- Matter (and its antecedents) are dynamic, organic, and experiential in nature.
- Mind and mental activity are inherently part of nature.
- The differences between life and nonlife are matters of degree, organization, and complexification, rather than kind.
- Cosmological and biological evolution are oriented toward achievements of value as expressed in conscious life and intelligence.
- What has occurred on this planet can (and likely will) occur on other planets where conditions are ripe.
- Value and hierarchies of value are a reality in the universe.
- Superior extraterrestrial intelligence is a live potentiality.
- Inexhaustible possibility and structures of possibility are being explored in the universe.
- Cosmocentrism overthrows naïve anthropocentrism and terracentrism.
- Terrestrial human disciplines (whether science, philosophy, theology, or religion) are contingent, but not irrelevant.
- Metaphysics, cosmology, and theology are relevant to each other.
- God is organically (as opposed to mechanically/supernaturally) related to the cosmic process and its achievements.
- Cosmic divine influence is immanent, persuasive, non-anthropocentric and, non-terracentric in nature.
- Theological doctrines (e.g., incarnation, revelation, salvation) are open to be expanded and extended to other worlds and extraterrestrial life.
- God's telic embrace of the universe is aesthetic, beatific, and all-inclusive.

These fifteen affirmations are not exhaustive in nature, nor are they equally affirmed by all the figures encountered above. However, as this volume will demonstrate, they do delineate broad arenas of conviction, debate, and ongoing research within (and beyond) process philosophy and theology. They in fact remain integrally wedded to the contributions process philosophers and theologians can make, both now and into the future.

THE JOURNEY AHEAD: ASTROPHILOSOPHY, EXOTHEOLOGY, AND COSMIC RELIGION

This volume is structured around three distinct, but interrelated parts. Part 1 concerns various dimensions of "Astrophilosophy" as informed by the resources of process philosophy and other relevant domains of philosophical and scientific research. Its five chapters differently contribute to current conversations and debates in astrobiology and cosmology, from the origin and function of life, to the place of cosmic creativity and value in evolution, and the prospects of shared axiological and biological knowledge with extraterrestrial others.

In chapter 1, "Prospects for a Universal Philosophy of Organism: Some Initial Reflections," Derek Malone-France considers several recent discoveries and developments in biology that appear congenial from various perspectives and concerns in process philosophy and theology. He looks to different domains of inquiry involved in NASA's definition of astrobiology and maps the historical context dating back to ancient Greek debates surrounding atomism, pluralism, and the definition of "life." In dialogue with different contours of ancient and current debates, Malone-France points to both challenges and opportunities for process thought. He argues that the resources of Whitehead's metaphysics and ontology, particularly the category of "organism," is not only a better candidate than "life" for universalizability, but also carries additional advantages, not least the convergence of both "physical" and "biological" phenomena. This, in turn, may allow for the ultimate clarification of both common and disparate elements characterizing "living" and "nonliving" phenomena. Malone-France also points to productive theological implications that emerge when God too is framed in terms of the Whiteheadian concept of organism. Malone-France's own proposal of "Theorganism" wherein God is the "final organism of organisms" carries important advantages, and indeed, holds promise for further development in the future.

In chapter 2, "The Cosmological Context of the Origin of Life: Process Philosophy and the Hot Spring Hypothesis," Matt Segall and Bruce Damer endeavor a sustained discussion of astrobiology and its cosmological/metaphysical underpinnings with shared convictions that a Kuhnian paradigm shift is occurring across multiple disciplines and methodologies relevant to origins of life research. In particular, they examine the "Hot Spring Hypothesis" as a leading candidate for the origins of life, with attention to both the empirical and metaphysical domains which are often kept separate. The empirical, as elucidated by Damer, explicates the emerging, experimentally grounded Hot Spring Hypotheses with the "progenitor" of living cells situated among wet-dry cycling in fresh water hot springs. The metaphysical, as elucidated by Segall, leverages the conceptual innovations of Whitehead's organic realism to overcome Kantian epistemic quandaries by better defining the cosmological conditions of life's origin. In bringing the empirical and the metaphysical together, Segall and Damer thus achieve an integrative collaboration that makes an important contribution to origins of life research.

In chapter 3, "The Organic Universe and Otherworldly Lives: Bergson and Sagan," Wahida Khandker articulates integrative convergences between Carl Sagan's insights in *Varieties of Scientific Experience* and Henri Bergson's *Creative Evolution*, shedding particular light on the ways in which complementarity can be achieved between their shared interests in cosmic and biological evolution. From these perspectives, Khandker considers features of life that have evolved in the deepest parts of our oceans, and that have only recently become accessible with the development of technologies able to navigate ecosystems at otherwise completely inhospitable depths. Despite the challenges of traversing vast distances in space, the scientific and philosophical insights of Sagan and Bergson, coupled with new data and research on "extremophiles," make for a truly exciting time of both discovery and speculation.

In chapter 4, "The Connection-Action Principle: A Basis for Process Philosophy, Cosmic Creativity, and Value?" Mark Lupisella explores a relational-action metaphysics in the form of what he calls the "connection-action principle." This principle suggests, in its simplest form, that the universe's (or multiverse's) property of connectedness is instantiated as relations and actions. In dialogue with a host of voices, and in deep resonance with Whitehead's own emphasis on creativity, process, and togetherness, Lupisella reveals the ways in which the connection-action principle may provide a basis for process philosophy and how it may relate to the universe's apparent creative tendencies. Focusing in particular on the emergence of value, life, intelligence, and meaning in the cosmos, Lupisella draws out important ethical and moral considerations based on the connection-action principle. He suggests not only that we might recognize and even revere a highly relational, dynamic, and diverse universe, but also that we can encourage these characteristics in our contemplations and actions as cocreators. Ultimately, our actions may help give rise to a "morally creative cosmos" where we (and potentially extraterrestrial others) take action on behalf of each other.

In chapter 5, "Astrobiological Searches for Shared Knowledge," Chelsea Haramia presents different arguments for the claim that humans and extraterrestrials will share common axiological and biological domains. She begins with a "partners in crime" argument that explores the partnership among biology, mathematics, and ethics, and then outlines important parallels between the scientific search for extraterrestrial life and pursuits of mathematical and ethical knowledge. Showing that these parallels justify an appeal for consistency in our reasoning, Haramia identifies and explains shared epistemic challenges in each of these domains, arguing that value is as real and detectable as life and mathematical truths. But how real are these things? To answer this crucial question, Haramia first addresses demarcation concerns regarding the difference between life and nonlife. She notes that the puzzles produced by these concerns mirror some of the puzzles faced by those who posit robust intrinsic value in the universe or the objective truth of mathematical claims. Haramia thus proposes the following avenue of response to these puzzles: We may assume that life and value are genuinely present in other areas of the universe in the same way that we may assume mathematics is genuinely present in other areas of the universe. That is, numbers, life, and value are "non-observables" the effects of which we may nonetheless recognize. For Haramia, reorienting extraterrestrial searches with this in mind reveals overlooked commonalities in these domains, and indicates that we may reasonably assume that at least some extraterrestrials could recognize these commonalities as well.

Part 2 situates four chapters that differently contribute to both developments and debates in "Exotheology" or, more specifically, "astrotheology" and "cosmotheology" as informed by key dimensions of process philosophy and theology. From interdisciplinary convergences between astrobiology, cosmology, and theology, to astrobiology as a religious science welcoming of astrotheology, and the development of naturalistic cosmotheologies, these chapters touch upon a variety of scientific, philosophical, and theological themes relevant to current debates.

In chapter 6, "Evolution Connected to Theory of Value by Process-Relational Astrotheology and Cosmotheology," Theodore Walker Jr. reviews a variety of interdisciplinary convergences emerging from relevant literature in the field, including the need for an ultimate context provided by astro- or cosmotheology wherein God is panentheistically conceived as the "one all-inclusive whole of reality." Following John B. Cobb Jr.'s call for a "richer account of evolution" in biology, Walker journeys through the ways in which evolution is also shown to embrace behavioral, cultural, stellar, and cosmic domains. Noting key contributions in the literature, Walker shows how connecting cosmic evolution to value and ethics is another enrichment that can be aided by process-relational thought and the neoclassical metaphysics and theology of both Whitehead and Hartshorne. In drawing out the related implications of creativity and co-creation in evolution, Walker also draws from relevant Eastern perspectives. He stresses in particular the problematic nature of "variable referencing" among various disciplines, and appeals to the need for "univocal references" to further advance the discussion. Proper mereological analysis concerning the relations between parts and wholes is helpful in articulating a variety of meanings associated with different conceptions of "God," "cosmos," and "universe." In quest of univocal referencing, Walker thus insists that the word "God" should refer to the transcendent whole of reality, while the word "cosmos" or "universe" should refer to the nontranscendent sum of all parts of reality.

In chapter 7, "Astrobiology, Astrotheology, and Cosmic Consciousness," Ted Peters considers the mission of astrobiology as "the almost but not quite religious science." Despite its emphasis on objective data, he draws out the "spiritual valence" of astrobiology in dialogue and debate with key voices, stressing that this discourse should be of strong interest to astrotheologians and the further development of both astrotheology and astroculture. In reviewing the tasks of astrotheology, principal among them, to bridge the divide between the strictly objectivist epistemology of astrobiology and the subjectivity of human experience with its built-in spiritual sensibilities, Peters draws from the physics of David Bohm and the metaphysics of Whitehead. Whiteheadian and Bohmian cosmological schemes include both objective science and subjective human experience in their respective worldviews, and these are of great value to the astrotheologian. By retrieving subjectivity in both Whiteheadian and Bohmian holism, astobiology, astrotheology, and astroculture might begin to mutually inform each other in truly meaningful ways. As societies continue to grow in cosmic consciousness and we anticipate our discovery and/or interaction with extraterrestrials, Peters also looks to the cosmic vision of Teilhard de Chardin as holding both promise and caution for the future.

In chapter 8, "Astrobiology, Cosmotheology, and the Biological Universe: Implications for Religion and Theology," Steven J. Dick considers the ways in which recent revelations in astronomy and astrobiology strongly indicate the need for a transformation of established theologies and suggest possibilities for new cosmically oriented theologies such as "naturalistic cosmotheology." In particular, the biological universe, the idea that intelligent life in the universe is common, necessitates a reconciliation of this new universe with dogmas of the Abrahamic religions in the same way that Thomas Aquinas tried to reconcile natural philosophy and Christianity in thirteenth-century Europe. Other religions and their associated theologies will be less affected, Dick argues, but they will still need to incorporate the cosmic perspective. Recent discoveries in astronomy and astrobiology resonate with the dynamism of process theology in the sense that all theologies must take into account cosmic evolution and the possibilities of a biological universe in which life may be part of the very fabric of the universe. Indeed, Dick shows the ways in which process theology also resonates with discoveries suggesting the denial of supernaturalism, the criticism of revelation, and the need to rethink the nature of God and the sacred in the tradition of religious naturalism. In contrast to traditional theologies, Dick argues that human destiny must be couched in cosmic terms. The creation of new cosmic theologies, whether characterized as cosmotheology or astrotheology, is a new and increasingly robust discipline that embraces the possibility of a more universal theology common to all intelligence in the cosmos. For Dick, these efforts are a necessary part of restructuring our worldviews in light of the new universe.

In chapter 9, "From Negation to Exemplification: A Deeper Whiteheadian Cosmotheology," I argue that Dick's admirable project of "naturalistic cosmotheology" can be considerably deepened when inversely approached through Whitehead's organic metaphysics. Endeavoring this requires showing the ways in which the core cosmological negations at the heart of Dick's six cosmotheological principles can instead be inverted into metaphysical exemplifications. For Whitehead, human experience, while certainly cosmologically peripheral, is nevertheless metaphysically central. Human existence and experience exemplify fundamental metaphysical principles of an organic, experiential, and aesthetic character that are essential to a creative evolutionary universe at all scales. Rather than being the great supernatural "exception" to metaphysical conditions, "invoked to save their collapse," Whitehead's cosmotheology affirms God as their "chief exemplification." Whitehead's deeper naturalism coupled with his theological realism thus allow for imaginative metaphysical continuity in our reflection all the way "down" and all the way "up" the biological universe: from terrestrial and extraterrestrial life to the cosmic life of God. What is more, Whitehead's cosmotheology reintegrates and resolves outstanding metaphysical problems surrounding necessity and contingency, possibility and actuality, objective value, truth, and meaning, that arguably remain philosophically unsubstantiated in Dick's anti-realist cosmotheology. Whitehead therefore offers a deeper cosmotheology with six alternative principles that include Dick's naturalistic intuitions while also transcending his conceptual and explanatory limitations.[101]

Part 3 contains six chapters differently oriented toward "cosmic religion" and the cosmic resources of religions, religious philosophies, and key religious figures in dialogue with process philosophers and theologians.

In chapter 10, "Religious Belief and the Discovery of Extraterrestrial Life: What's Worldview Got to Do with It?," Constance M. Bertka considers how our first extraterrestrial discovery might impact worldviews, particularly theological perspectives and religious belief, noting that academics within theological traditions (including process theology) answer the question quite differently than academics outside these traditions. Bertka reviews the nature of astrobiology as an interdisciplinary science asking three general questions as to our cosmic origins, whether we are alone in the universe, and where we are ultimately headed. She stresses that these questions are not new and that they continue to appeal to scholars, theologians, and nonscholars alike. While we cannot currently answer the question "Are we alone?" with a definitive "No," Bertka shows why the reality is that extraterrestrial life might be an increasingly reasonable expectation given recent discoveries. The question of worldview impact falls clearly within the context of how the relationship between science and religion is viewed. In reviewing different perspectives on the relationship, Bertka shows that process theologians have already been at work for decades proposing revisions of the classical view of God. To the question as to whether the discovery of extraterrestrial life might increase the popularity of process theology, Bertka insists the answer will hinge in part on normalizing science and religion discussions that break down siloed worldviews. Not only theologians but also the scientific community writ large needs to be concerned with a broader conversation that reaches beyond the ivory tower of the academe.

In chapter 11, "'Worlds as Numerous as the Grains of Sand of the Ganges': Cosmic Pluralism and Swami Vivekananda's Religion of the Future," Jeffery D. Long argues that responding to the question of whether there is life in the universe from the perspective of philosophies of process prompts another question, namely, "What is life?" From within a process conceptuality, it is not obvious that there is anything which is not, in some sense, alive. Long argues that the emergence of philosophies of process, as well as a growing sense among thinkers in the West that consciousness (or at least experience) is in some sense foundational to existence, puts the question of life in the universe on a different level. It becomes not only an empirical question, but a fundamental question of ontology. Yet such a situation is already envisioned in many Indian philosophical traditions that both take consciousness to be foundational to existence and take for granted that there are many planes of existence inhabited by intelligent beings. Over a century ago, Swami Vivekananda envisioned this convergence when he spoke of religions to come in the future. In discussing the interrelated questions of the nature of life, experience, and consciousness as seen in the context of the search for extraterrestrial life, Long develops the kind of religious sensibility that emerges from reflections of this kind. In contrast to premature claims of some scholars, he integrates Vedanta and process philosophy to argue that religion will not be destroyed or superseded as humanity engages with the possibility, and eventually the reality, of extraterrestrial life, but that it will be radically transformed by this process.

In chapter 12, "Multiplicity without Tyranny: The Nonviolent Aim in Jainism and Whitehead's Process Metaphysics," Brianne Donaldson uses the metaphysical framework of Jainism, with its ethical experiments in nonviolence, as a comparative case study to provide an account of Whitehead's ethical aim as a telos of nonviolence. She shows that ethical-aesthetic terms within process philosophy, such as "beauty" and "intensity" often aim to preserve the role of contrast, conflict, and inevitable loss as essential aspects of an ecological planetary multiplicity. In these views, exclusionary valuations, and even destruction, are accepted as a necessity, even if a tragic one, of a worldview that ascribes some level of self-determined creativity to all actual entities. If "Life is robbery," as Whitehead states, then suggesting nonviolence as a valuable aim could smuggle in a loophole of exceptionalism wherein the morality of humans is somehow different from that of "nature." Moreover, any kind of telos could suggest a predestined future, and nonviolence, if interpreted as passivity or sacrificial paralysis, could result in a stultified, rather than dynamic, universe. While these concerns, and the ecological sensibilities that motivate them, offer an important metaphysical and realist intervention in anthropocentric accounts of the universe, Donaldson notes how proponents often overlook a normative aspect of Whitehead's religious and metaphysical vision of a possible future characterized by less, or perhaps no, loss. In reviewing many of Whitehead's technical concepts Donaldson thus reveals how they present a vision of social and ecological nonviolence for an unfolding future. In light of both Jain and process philosophy, harm reduction and social ecology need not be at metaphysical odds. Rather, nonviolence is part of the structure of becoming, not only for certain exceptional beings, but ultimately for all existent entities.

In chapter 13, "Extending the Noosphere into Intergalactic Life: Teilhard de Chardin and the Third Axial Age," Ilia Delio examines Teilhard's vision of the evolution of consciousness toward Omega by using the axial age as a paradigm of development. In 1949, Karl Jaspers coined the phrae "the axial age" to described a global breakthrough in human consciousness, a term which has served as a heuristic marker of consciousness and human development, including the depth dimension of religion. Ewert Cousins followed and described a second axial age ushered in by the new science and mass communication. This new age is marked by ecology, community, divine immanence, and global consciousness. Teilhard de Chardin argued that evolution is the rise of consciousness and described a process of theistic evolution grounded in second axial consciousness. He proposed that life is unfolding from simple, biological material life, to complex interconnected life, empowered by a divine presence, following Henri Bergson's élan vital. Evolution continues in and through human development and is accelerating with technology toward the maximization of conscious life marked by the symbol of Omega. In unfolding Teilhard's ideas on consciousness and matter, Delio argues that the inextricable link between consciousness and matter, the drive to complex consciousness, and the development of the noosphere and computerized planetary consciousness, render the search for conscious intergalactic life the next logical step in the evolution of the Christic toward Omega. In this respect, she stresses that Teilhard's notion of the Noosphere marks the beginning of the third axial age, which will continue with interstellar exploration and the expansion of human consciousness into other life-forms, a development consistent with the fecund symbol of Omega.

In chapter 14, "Astrobiology without Biology: Will AI be Our Emissary or Our Bottleneck?" Noreen Herzfeld reviews various dimensions involved in current discussions of artificial intelligence (AI) as it relates to the search for extraterrestrial intelligence, including its pseudo-religious eschatology. Given the vast distances of space and the impossibilities of space travel for biological beings, it is likely that our first contact with another intelligence will be through their and/or our technology. Indeed, AI offers the opportunity for a presence in space that is both dynamic and functional. Thus, AI might be the best emissary to other worlds. Herzfeld points to the persistence of the Fermi paradox and analyzes the Drake equation in detail. If intelligent life is plentiful, why have we not encountered their technology? In dialogue with key voices, Herzfeld argues that the answer may lie in a paradox of evolutionary biology. The very nature of biological evolution may mean that technological advancement always outstrips a society's ability to control that technology. While AI might be our best avenue to space exploration, it might also be the technological bottleneck that precludes that exploration. Nevertheless, millenarian orientations with respect to superintelligent AI continue to captivate researchers, philosophers, and the general public as they project a variety of eschatological hopes onto the future. Herzfeld notes that apocalyptic scenarios usually involve the creation of an AI that outstrips human intelligence and control, and that could turn on its creators or be used as a weapon by those creators. However, she insists that we hardly need conscious or superintelligent AIs to produce programs that can be utterly damaging to the fabric of human culture and society, an obvious fact in today's world of ransomware attacks, autonomous weapon development, social media dispersal of falsehoods, and machine learning programs with inherent biases. Such mundane uses of computers, coupled with our natural tendencies toward competition and domination, may keep us, and potentially all other technologically developed societies, from anything more than menial flirtations with space.

In chapter 15, "A Darker Forest? The Fermi Paradox and Extraterrestrial Spiritual Beings," Roland Faber discusses various layers, assumptions, and implications of the "Dark Forrest" hypothesis as one the more fascinating solutions to the Fermi paradox. The Dark Forest hypothesis states that the universe is not only biotic (ETL), but full of intelligent life (ETI). However, since the probability that the encounter of ETIs will lead to mutual destruction rather than cooperation is "astronomical," everyone hides as though behind a tree in a dark forest, with any contact passively or actively leading to the eradication of the communicator. Is the Earth thus doomed? Faber argues that this conjecture is based on an assumption that evolution is locked, in Darwinian terms, in the survival of the fittest, and that even cooperation, as on Earth, is, if not the exception, merely a means for self-survival and promotion. However, in a detailed analysis, Faber shows how the philosophical resources found in Teilhard, Whitehead, and 'Abdu'l-Baha, specifically their religious intuitions and desires concerning the function of cosmic religiosity in the universe, can actually counter this dismal assumption. The question for Faber then becomes: How does the Fermi paradox impact not only their acceptance that life is ubiquitous (ETL) and suggestive of the appearance of intelligent life-forms (ETI) in our universe, but also support the claim that evolution tends to foster the emergence of spiritual life-forms (ETS) that might overcome this biotic, evolutionary condition of self-assertion, competition, and expansion? In layout of several propositions, and the various possibilities at play, Faber answers the Dark Forest hypothesis and the Fermi paradox from the perspective of the assumption of extraterrestrial spiritual life (ETS) and the Great Spiritual Barrier for star-faring civilizations. For Faber, humanity is the test subject whose open future may or may not further the emergence of the cosmic spirit of life.

As indicated earlier, for the first time since 1968, the epilogue to this volume republishes Lewis Ford's pioneer article "Theological Reflections on Extraterrestrial Life" as the first sustained consideration of the topic from within the resources of process philosophy and theology.

Let it be finally stated again that the contributions to this volume are hardly exhaustive in nature. They seek not to end conversation, but to expand and deepen it in ways that have yet to be accomplished in current literature. Such conversations, occurring on a small planet circling around a small sun, will continue indefinitely. I trust they are also occurring (in some fashion) on other worlds and by the light of other suns. My modest hope with respect to this volume, and the further research it necessitates, is captured succinctly by Whitehead: "In the end—though there is no end—what is being achieved, is width of view, issuing in greater opportunities."[102]

NOTES

1. Faber's longer course description as found in the course syllabus is as follows: "Since the discovery of the massive presence of exoplanets in our galaxy, the old question whether there may be life on other worlds than our earth has sparked new interest not only in scientific fields such as exo-biology, but also regarding the views of diverse religions on this question. While in many cultures this assumption was part of their expansive worldview and while philosophies in the East and West have speculated on life and even human-like or strange forms of intelligence in the universe for

ages, the impact this possibility might have on religious identities has become a more pressing issue today: How to think about creation, the human predicament, salvation, and eschatological visions in light of many worlds full of life and maybe even intelligent life? This seminar will explore the history of the integration or exclusion, embrace or limitation of such a vision throughout different religious traditions, the constraints it puts on religious worldviews and doctrines, and the insights the horizon of exo-life can offer when it is applied to religious self-understanding."

2. See for example, Dick, *Plurality of Worlds* and *Life on Other Worlds*. Next to Dick, the work of Michael J. Crowe has also led the discussion. See Crowe, *The Extraterrestrial Life Debate 1750–1900* and *Extraterrestrial Life Debate, Antiquity to 1915*. For a more recent contribution to the discussion, see Connes, *History of the Plurality of Worlds*.

3. Refer to Dick's two edited volumes *Many Worlds* and *The Impact of Discovering Life Beyond Earth*. More recently, see *Astrobiology, Discovery, and Societal Impact*.

4. See for example, Davison, *Astrobiology and Christian Doctrine*; Thigpen, *Extraterrestrial Intelligence and the Catholic Faith*; Pearce and Adair, *Aliens and Religion*; Parkyn, *Exotheology*; Pryor, *Living with Tiny Aliens*; Peters, *Astrotheology*; Malone-France, "Hell Is Other Planets"; Weintraub, *Religions and Extraterrestrial Life*; Bertka, *Exploring the Origin, Extent, and Future of Life*; O'Meara, *Vast Universe*; Wilkinson, *Science, Religion, and the Search for Extraterrestrial Intelligence*; Consolmagno and Mueller, *Would You Baptize an Extraterrestrial?*.

5. See Dick, *The Biological Universe*.

6. Dick, "Cosmotheology," 206.

7. Consider Norman Pittenger's comments in this regard: "In a way not dissimilar to that of Augustine with his use of Neo-Platonism, or of Thomas Aquinas with his similar use of the newly recovered Aristotelianism of his day, the exponents of 'process theology' have found in Whitehead's vision of the world material which in their judgment provides a context for Christian faith and a conceptuality with which Christian theologians can work. But it must be adapted to that purpose." Pittenger, *Alfred North Whitehead*, 45. Robert B. Mellert is even more adamant than Pittenger regarding the adequacy of these analogies: "This is precisely what Augustine did with the philosophy of Plato and what Thomas Aquinas did with the philosophy of Aristotle. Each sought to integrate his Christian faith with the best available understanding of reality as a whole. . . . What process theologians are attempting to do is essentially the same as what Augustine and Thomas did: to express their Christian faith in the conceptual language of a philosophy that makes sense to their age." Mellert, *What is Process Theology?*, 17.

8. Dick, "Toward a Constructive Naturalistic Cosmotheology," 200.

9. As Griffin put in 1989: "My major task for the center has been to plan conferences to relate Whiteheadian philosophy to various areas of thought. Most the conferences have involved either Eastern religious philosophies or the natural sciences, especially physics and biology, which enabled me to become acquainted with some of the most creative thinkers in these areas." Griffin and Smith, *Primordial Truth and Postmodern Theology*, 5.

10. See Lupisella, *Cosmological Theories of Value*.

11. Although Theodor Walker was unable to attend the conference, his contribution does appear in the current volume. While Brian Henning and Kelly Smith attended and presented at the conference, their contributions, as part of a wider research agenda, do not appear in this volume.

12. A variety of texts could be listed in this regard. See for example, Eastman, *Physics and Whitehead*; *Untying the Gordian Knot*; Epperson, *Quantum Mechanics and the Philosophy of A.N. Whitehead*; *Foundations of a Relational Realism*; Van Dijk, "Process Physics"; Khandker, *Process Metaphysics and Mutative Life*; Dick and Lupisella, *Cosmos and Culture*; Reiss and Ruse, *The New Biology*; Henning and Scarfe, *Beyond Mechanism*; Bartosh et al., *Towards a Philosophy of Cosmic Life*; Cobb, *Back to Darwin*; Koutroufinis, *Life and Process*; Segall, *Physics of the World-Soul*; Malone-France, *Deep Empiricism*; Dupré and Nicholson, *Everything Flows*; Dupré, *The Metaphysics of Biology*; Mix, "Philosophy and Data in Astrobiology"; Chela-Flores, "Fitness of the Cosmos" ;Thacker, "After Life"; Webber and Weeks, *Process Approaches to Consciousness in Psychology, Neuroscience, and Philosophy of Mind*; Weber, *Handbook of Whiteheadian Process Thought*; McGilchrist, *The Matter With Things*; Grandpierre, "Extending Whiteheadian Organic Cosmology"; "The Fundamental Biological Activity of the Universe"; Lupisella, *Cosmological Theories of Value*; Davis et al. *Process Cosmology*; Davis, *Mind, Value and Cosmos*.

13. For some of these statements as recorded by Price, refer to my chapter in the current volume (ch. 9). See also my recent book *Metaphysics of Exo-Life*. For a brief review of the case for and against Price's *Dialogues*, see Henning, "Revisiting Lucien Price's Dialogues."

14. Peters, "Exo-Theology," 196.

15. Ibid.

16. Ford, Theological Reflections on Extraterrestrial Life," 10.

17. Peters, "Exo-Theology," 197.

18. Ibid., 196.

19. O'Meara, *Vast Universe*, 16.

20. Ibid., 16, 26.

21. Ibid., 88, 89n. 25.

22. For a wider discussion refer to Pugliese and Becker, *Process Thought and Roman Catholicism*.

23. Wilkinson, *Science, Religion, and the Search for Extraterrestrial Intelligence*, 155.

24. Ibid., 24.

25. Ibid., 71.

26. Ibid., 151, 156–57.

27. See Delio, "Christ and Extraterrestrial Life"; Haught, "Theology After Contact."

28. Haught, "Theology After Contact," 306–7.

29. Weintraub, *Religions and Extraterrestrial Life*, 129.

30. Weintraub, *Religions and Extraterrestrial Life*, 130n3.

31. For an informative overview of Teilhard's process vision, see Barbour, "Teilhard's Process Metaphysics."

32. See for example Ward, *Religion and Revelation*; *Christ and the Cosmos*; and *The Christian Idea of God.*

33. Ward, *Christ and the Cosmos*, xiii, 259.

34. Ibid., 260–61.

35. See Peters, *Astrotheology*, 239, 290, 306, 324, 326–27, 15, 203.

36. See Peters, *Astroanthropology: Science, Ethics, and Religion*, forthcoming.

37. Parkyn, *Exotheology*, 3n11.

38. Ibid., 117.

39. Ibid., 99.

40. Ibid., 87–90; 102–6; 2.

41. Thigpen, *Extraterrestrial Intelligence the Catholic Faith*, 194–96; 335–36.

42. Davison, *Astrobiology and Christian Doctrine*, 205n29, 314n39, 300, 325.

43. Faber, *God as Poet*, 26.

44. Wieman, "Confessions of a Religious Seeker." https://www.harvardsquarelibrary.org/theology-philosophy/appendix-henry-nelson-wieman-the-confessions-of-a-religious-seeker/.

45. Wieman, *The Source of Human Good*, 90.

46. Ibid., 110.

47. Ibid.

48. Ibid., 192.

49. Ibid., 76, 79.

50. Ibid., 307.

51. *Time*, "Theology: Challenge in the Heavens," https://content.time.com/time/subscriber/article/0,33009,900595–1,00.html.

52. Refer to Viney and Shield's, *The Mind of Charles Hartshorne*.

53. I thank Don Viney for his valuable aid in pointing out many of Hartshorne's relevant statements on the topic, only a selection of which appear here.

54. Hartshorne, *Beyond Humanism*, 256.

55. Hartshorne, *A Natural Theology for Our Time*, 16.

56. Hartshorne, *Creativity in American Philosophy*, 270; *Wisdom as Moderation*, 133.

57. Quoted in Miethe, *Did Jesus Rise from the Dead?*, 142.

58. Viney, Personal email correspondence, March 23, 2023.

59. From 1971 to 1996, Ford was the editor of the *Process Studies* journal. For a special focus tribute to Ford's work, see *Process Studies* vol, 27, no. 1–2 (1998).

60. See for example, Smith and Mariscal, *Social and Conceptual Issues in Astrobiology*; Ashkenazi, *What We Know About Extraterrestrial Intelligence*; Cockell, *The Meaning of Liberty Beyond Earth*; *Human Governance Beyond Earth*; *Dissent, Revolution and Liberty Beyond Earth*; Chon et al., *Astrobiology*; Crawford, *Expanding Worldviews*; Fuller et al., *Issues in Science and Theology*; Faber, *The Ocean of God*; *The Cosmic Spirit*; Pearce and Adair, *Aliens and Religion*; Peters, *Astrotheology*; Parkyn, *Exotheology*; Playford, *Exophilosophy*; Davis, "Whiteheadian Cosmotheology."

61. Denzler, *The Lure of the Edge*, 182n52.
62. See Catoe, "UFOs and Related Subjects," 244. https://apps.dtic.mil/sti/pdfs/AD0688332.pdf and Mallove et al., "A Bibliography on the Search for Extraterrestrial Intelligence," 100. https://ntrs.nasa.gov/api/citations/19780013076/downloads/19780013076.pdf
63. Pittenger, "Christianity and the Man on Mars," 747.
64. Ibid.
65. Ibid., 748.
66. Ibid.
67. Pittenger, *The Word Incarnate*, 248–51.
68. Pittenger, *Picturing God*, 86–89.
69. Ibid., 86.
70. Ibid., ch. 4.
71. Cobb and Griffin, *Process Theology*, 146.
72. Ibid.
73. Griffin, *God and Religion*, 23.
74. Griffin, *Panentheism and Scientific Naturalism*, 88.
75. Cobb and Birch, *Liberation of Life*, 189.
76. Ibid., 192.
77. Ibid, 195, 197.
78. Cobb, "Cosmic Life," forthcoming.
79. Rescher, "Extraterrestrial Science," 361.
80. Ibid., 386.
81. Ibid., 388–89.
82. Rescher, *Pluralism*, 67–76ff.
83. Rescher, *Process Metaphysics*, 159.
84. Rescher, *Metaphysical Perspectives*, 172.
85. Ibid., 94.
86. See for example, Rescher, *Axiogensis*.
87. Walker and Wickramasinghe, *The Big Bang and God*, xi. Emphasis original.
88. Ibid., 18.
89. Ibid., 52.
90. Ibid., 115, 41.
91. See Faber, *The Ocean of God*, chs. 13–14.
92. Ibid., 153–54.
93. Ibid., 155.
94. Ibid, 160n52.
95. Ibid.,161–62.
96. See Faber, *The Cosmic Spirit*, chs. 8–9.
97. Ibid., 242–43, emphasis original.
98. See Faber, *The Mind of Whitehead*, chs. 23, 26.
99. Ibid., 594–95.
100. Ibid., 597. Again, refer to my chapter (ch. 9) in the current volume for some of these statements as recorded by Price.

101. For my expanded engagement of Dick's "naturalistic cosmotheology" from within Whitehead's metaphysics see my recent book *Metaphysics of Exo-Life*.

102. Whitehead, *Adventures of Ideas*, 159.

BIBLIOGRAPHY

Ashkenazi, M. *What We Know About Extraterrestrial Intelligence: Foundations of Xenology.* Cham: Springer, 2017.

Barbour, I. G., "Teilhard's Process Metaphysics." *The Journal of Religion* 49, no. 2 (1969): 136–59.

Bartosh, D. et al. *Towards a Philosophy of Cosmic Life: New Discussions and Interdisciplinary Views*. Cham: Springer, 2023.

Bertka, C. M. *Exploring the Origin, Extent, and Future of Life: Philosophical, Ethical and Theological Perspectives*. Cambridge: Cambridge University Press, 2009.

Birch, C., and J. B. Cobb Jr. *Liberation of Life: From the Cell to the Community.* Cambridge: Cambridge University Press, 1981.

Catoe, L. E. "UFOs and Related Subjects: An Annotated Bibliography." Library of Congress Science and Technology Division (1969). Accessed April 30, 2023. https://apps.dtic.mil/sti/pdfs/AD0688332.pdf.

Chela-Flores, J. "Fitness of the Cosmos for the Origin and Evolution of Life: From Biochemical Fine-Tuning to the Anthropic Principle." In *Fitness of the Cosmos for Life: Biochemistry and Finetuning*, edited by J. D. Barrow et al., 151–65. Cambridge: Cambridge University Press, 2007.

Cobb, J. B. Jr. "Cosmic Life." In *Towards a Philosophy of Cosmic Life: New Discussions and Interdisciplinary Views*, edited by D. Bartosh, D. Cham: Springer, 2023.

———. *Back to Darwin: A Richer Account of Evolution*. Grand Rapids: Eerdmans, 2008.

Cobb, J. B. Jr., and D. R. Griffin. *Process Theology: An Introductory Exposition*. Louisville: Westminster Press, 1976.

Cockell, C. S., ed. *Dissent, Revolution and Liberty Beyond Earth*. Cham: Springer, 2016.

———. *Human Governance Beyond Earth: Implications for Freedom*. Cham: Springer, 2015.

———. *The Meaning of Liberty Beyond Earth*. Cham: Springers, 2014.

Connes, P., and J. Lequeux. *History of the Plurality of Worlds: The Myths of Extraterrestrials Through the Ages*. Cham: Springer, 2020.

Consolmagno, G., and P. Mueller. *Would You Baptize an Extraterrestrial?: . . . and Other Questions from the Astronomers' In-box at the Vatican Observatory*. New York: Image, 2018.

Crawford, I. *Expanding Worldviews: Astrobiology, Big History and Cosmic Perspectives*. Cham. Springer, 2021.

Crowe, M. J. *The Extraterrestrial Life Debate: Antiquity to 1915.* Notre Dame: University of Notre Dame Press, 2008.

———. *The Extraterrestrial Life Debate, 1750–1900: The Idea of a Plurality of Worlds from Kant to Lowell*. Mineola, NY: Dover Publications, 1999.

Davis, A. M. *Metaphysics of Exo-Life: Toward a Constructive Whiteheadian Cosmotheology*. Grasmere: SacraSage Press, 2023.

———. *Mind, Value, and Cosmos: On the Relational Nature of Ultimacy*. Lanham, MD: Lexington, 2020.

———. "Whiteheadian Cosmotheology: Platonic Entities, Divine Realities and Shared Extraterrestrial Values." In *Process Cosmology: New Integrations in Science and Philosophy*, edited by A. M. Davis et al., 423–52. Cham: Palgrave, 2022.

Davis, A. M., et al. *Process Cosmology: New Integrations in Science and Philosophy*. Cham: Palgrave, 2022.

Davison, A. *Astrobiology and Christian Doctrine: Exploring the Implications of Life in the Universe*. Cambridge: Cambridge University Press, 2023.

Delio, I. "Christ and Extraterrestrial Life." *Theology and Science* 5, no. 3 (2007): 249–65. https://doi.org/10.1080/14746700701622008.

Denzler, B. *The Lure of the Edge: Scientific Passions, Religious Beliefs, and the Pursuit of UFOs*. Berkely: University of California Press, 2003.

Dick, S.J. *Astrobiology, Discovery, and Societal Impact*. Cambridge: Cambridge University Press, 2020.

———. "Cosmotheology: Theological Implications of the New Universe." In *Many Worlds: The New Universe, Extraterrestrial Life, and ils Theological Implications*, edited by S. J. Dick. Philadelphia: Templeton Press, 2000, 191–210.

———. "Toward a Constructive Naturalistic Cosmotheology." In *Astrotheology: Science and Theology Meet Extraterrestrial Life*, edited by T. Peters, 228–44. Eugene: Wipf and Stock: , 2018.

———. *Plurality of Worlds: The Extraterrestrial Life Debate from Democritus to Kant*. Cambridge: Cambridge University Press, 1982.

———. *The Biological Universe: The Twentieth Century Extraterrestrial Life Debate and the Limits of Science*. Cambridge: Cambridge University, 1996.

———. *The Impact of Discovering Life Beyond Earth*. Cambridge: Cambridge University Press, 2015.

Dick, S. J., ed. *Many Worlds: The New Universe, Extraterrestrial Life & the Theological Implications*. West Conshohocken, PA: Templeton Foundation Press, 2000.

Dick, S. J., and M. Lupisella, eds. *Cosmos and Culture: Cultural Evolution in a Cosmic Context*. NASA, 2011.

Dupré, J. *The Metaphysics of Biology*. Cambridge: Cambridge University Press, 2021.

Eastman, T. *Untying the Gordian Knot: Process, Reality, and Context*. Lanham, MD: Lexington, 2020.

Eastman, T. E., and H. Keeton. *Physics and Whitehead: Quantum, Process, and Experience*. Albany: SUNY Press, 2009.

Epperson, M. *Quantum Mechanics and the Philosophy of A. N. Whitehead*. New York: Fordham University Press, 2004.

Epperson, M., and E. Zafiris. *Foundations of Relational Realism: A Topological Approach to Quantum Mechanics and the Philosophy of Nature*. Lanham, MD: Lexington, 2013.

Faber, R. *God as Poet of the World: Exploring Process Theologies*. Louisville: Westminster John Knox Press, 2008.

———. *The Cosmic Spirit: Awakenings at the Heart of All Religions, the Earth, and the Multiverse*. Eugene: Cascade, 2021.

———. *The Ocean of God: On the Transreligious Future of Religions*. London: Anthem Press, 2019.

———. *The Mind of Whitehead: Adventure in Ideas.* Eugene: Pickwick, 2023.

Ford, L. "Theological Reflections on Extraterrestrial Life." *The Raymond Review* 2, no. 2 (1968): 1–14.

Fuller, M. et al., *Issues in Science and Theology: Nature—and Beyond*. Cham: Springer, 2020.

Grandpierre, A. "Extending Whiteheadian Organic Cosmology to a Comprehensive Science of Nature." In *Process Cosmology: New Integrations in Science and Philosophy*, edited by A. M. Davis et al., 59–91. Cham: Palgrave, 2022.

———. "The Fundamental Biological Activity of the Universe." In *Eco-Phenomenology: Life, Human Life, Post-Human Life in the Harmony of the Cosmos*, edited by W. Smith, et al., 115–40. Analecta Husserliana, vol. 121. Cham: Springer, 2018.

Griffin D. R. *Panentheism and Scientific Naturalism: Rethinking Evil, Morality, Religious Experience, Religious Pluralism, and the Academic Study of Religion*. Claremont: Process Century Press, 2014.

———. *God and Religion in the Postmodern World: Essays in Postmodern Theology* Albany: SUNY Press, 1988.

Griffin, D. R., and H. Smith, *Primordial Truth and Postmodern Theology*. Albany: SUNY Press, 1989.

Hartshorne, C. *A Natural Theology for Our Time*. Chicago: Open Court, 1967.

———. *Beyond Humanism: Essays in the Philosophy of Nature*. Eugene: Pickwick, 2017. Originally published by Willett, Clark and Company, 1937.

———. *Creativity in American Philosophy*. Albany: SUNY Press, 1984.

———. *Wisdom as Moderation: A Philosophy of the Middle Way*. Albany: SUNY Press, 1987.

Haught, J. "Theology After Contact: Religion and Extraterrestrial Intelligent Life." *Annals of the New York Academy of Sciences*, 950 (2001): 296–308.

Henning, B. G. "Revisiting Lucien Price's Dialogues." whiteheadresearch.org. Accessed April 30, 2023. http://whiteheadresearch.org/2020/01/03/revisiting-lucien-prices-dialogues/

Henning, B., and A. Scarfe, *Beyond Mechanism: Putting Life Back Into Biology*. Lanham, MD: Lexington, 2013.

Khandker, W. *Process Metaphysics and Mutative Life: Sketches of Lived Time*. Cham: Palgrave, 2020.

Koutroufinis, S. A. *Life and Process: Towards a New Biophilosophy*. Berlin: de Gruyter, 2014.

Lupisella, M. *Cosmological Theories of Value: Science, Philosophy, and Meaning in Cosmic Evolution*. Cham: Springer, 2020.

Mallove, E. F. et al. "A Bibliography on the Search for Extraterrestrial Intelligence." NASA (1978). Accessed April 30, 2023. https://ntrs.nasa.gov/api/citations/19780013076/downloads/19780013076.pdf.

Malone-France, D. "Hell is Other Planets: Extraterrestrial Life in the Western Theological Imagination." In *Social and Conceptual Issues in Astrobiology*, edited by K. C. Smith and C. Mariscal, 21–53. Oxford: Oxford University Press, 2020.

Malone-France, D. *Deep Empiricism: Kant, Whitehead, and the Necessity of Philosophical Theism*. Lanham, MD: Lexington, 2007.

McGilchrist, I. *The Matter With Things: Our Brains, Our Delusions, and the Unmaking of the World*, Vol 1 & 2. London: Perspectiva Press, 2021.

Mellert, R. B. *What is Process Theology?* Mahwah: Paulist Press, 1975.

Miethe, T. L. ed. *Did Jesus Rise from the Dead?: The Resurrection Debate,* edited by Gary Habermas and Antony Flew. San Francisco: Harper & Row, 1987.

Mix, L. J. "Philosophy and Data in Astrobiology." *International Journal of Astrobiology* 17, no. 1 (2018): 189–200. https://doi.org/10.1017/S1473550417000192.

Nicholson, D. J., and J. Dupre. *Everything Flows: Towards a Processual Philosophy of Biology*. Oxford: Oxford University Press, 2018.

O'Meara, T. F. *Vast Universe: Extraterrestrials and Christian Revelation*. Collegeville, MN: Liturgical Press, 2012.

Parkyn, J. *Exotheology: Theological Explorations of Intelligent Extraterrestrial Life*. Eugene: Pickwick, 2021.

Pearce, P. M. S., and A. Adair. *Aliens and Religion: Where Two Worlds Collide: Assessing the Impact of Discovering Extraterrestrial Life on Religion and Theology*. Onus Books, 2022.

Peters, T. "Exo-Theology: Speculations on Extraterrestrial Life." In *The Gods Have Landed: New Religions from Other Worlds*, edited by J. R. Lewis, 187–206. Albany: SUNY Press, 1995.

———. *Astroanthropology: Science, Ethics, and Religion*, forthcoming.

———. *Astrotheology: Science and Theology Meet Extraterrestrial Life*. Eugene: Cascade, 2018.

Pittenger, N. "Christianity and the Man on Mars." *The Christian Century* (June 20, 1956): 747–48. Available in Christian Century archives: https://archive.org/details/sim_christian-century_1956-06-20_73_25/page/746/mode/2up?view=theater.

———. *Alfred North Whitehead*. Makers of contemporary theology. Richmond: John Knox Press, 1969.

———. *Picturing God*. London: SCM Press 1982.

———. *The Word Incarnate: A Study of the Doctrine of the Person of Christ*. New York: Harper & Brothers, 1959.

Playford, R. *Exophilosophy: The Philosophical Implications of Alien Life*. London: Routledge, forthcoming.

Price, L. *Dialogues of Alfred North Whitehead*. Jaffrey: David R. Godine, 2001.

Pryor, *Living with Tiny Aliens: The Image of God for the Anthropocene*. New York: Fordham University Press, 2020.

Pugliese, M. A. and J. Becker. *Process Thought and Roman Catholicism: Challenges and Promises*. Lanham, MD: Lexington, 2022.

Reiss, M. J., and M. Ruse. *The New Biology: A Battle between Mechanism and Organicism*. Cambridge: Harvard University Press, 2023.
Rescher, N. *Axiogensis: An Essay in Metaphysical Optimalism*. Lanham, MD: Lexington, 2010.
———. "Extraterrestrial Science." In *Change and Progress in Modern Science*, edited by J. C. Pitt, 361–92. Dordrecht: D. Reidel Publishing, 1985.
———. *Metaphysical Perspectives*. Notre Dame: University of Notre Dame Press, 2017.
———. *Pluralism: Against the Demand for Consensus*. Oxford: Clarendon Press, 1993.
———. *Process Metaphysics: An Introduction to Process Philosophy*. Albany: SUNY Press, 1996.
Segall, M. D. *Physics of the World-Soul: Alfred North Whitehead's Adventure in Cosmology*. SacraSage Press, 2021.
Smith, K. C., and C. Mariscal. *Social and Conceptual Issues in Astrobiology*. Oxford: Oxford University Press, 2020.
Thacker, E. "After Life: De Anima and Unhuman Politics." *Radical Philosophy* 155 (2009). Accessed April 30, 2023. https://www.radicalphilosophy.com/article/after-life.
Thigpen, P. *Extraterrestrial Intelligence and the Catholic Faith: Are We Alone in the Universe with God and the Angels?* Gastonia: Tan Books, 2022.
TIME Magazine. "Theology: Challenge in the Heavens." *TIME* (January 24, 1969) .
Torres, O. A. C, et al. *Astrobiology: Science, Ethics, and Public Policy*. Beverly: Scrivener, 2021.
Van Dijk, "Process Physics: Toward an Organismic, Neo-Whiteheadian Physics." In *Process Cosmology: New Integrations in Science and Philosophy*, edited by A. M. Davis et al., 27–57. Cham: Palgrave, 2022.
Viney, D. W., and G. W. Shields. *The Mind of Charles Hartshorne: A Critical Examination*. Anoka: Process Century Press, 2020.
Walker, T. Jr., and C. Wickramasinghe, *The Big Bang and God: An Astro-Theology*. New York: Palgrave, 2015.
Ward, K. *Christ and the Cosmos: A Reformulation of Trinitarian Doctrine*. Cambridge: Cambridge University Press, 2015.
———. *Religion and Revelation: A Theology of Revelation in the World's Religions*. Oxford: Clarendon Press, 1994.
———. *The Christian Idea of God: A Philosophical Foundation for Faith*. Cambridge: Cambridge University Press, 2017.
Webber, M., and A. Weeks, *Process Approaches to Consciousness in Psychology, Neuroscience, and Philosophy of Mind*. Albany: SUNY Press, 2009.
Weber, M., and W. Desmond, eds. *Handbook of Whiteheadian Process Thought*. Frankfurt: De Gruyter, 2014.
Weintraub, D. A. *Religions and Extraterrestrial Life: How Will We Deal with It?* Cham: Springer, 2014.
Whitehead, A. N. *Adventures of Ideas*. New York: The Free Press, 1967.

Wieman, H. N. "Confessions of a Religious Seeker." Accessed April 30, 2023. https://www.harvardsquarelibrary.org/theology-philosophy/appendix-henry-nelson-wieman-the-confessions-of-a-religious-seeker/.

———. *The Source of Human Good.* Chicago: University of Chicago Press, 1946.

Wilkinson, D. *Science, Religion, and the Search for Extraterrestrial Intelligence.* Oxford: Oxford University Press, 2013.

PART I

Astrophilosophy

Chapter 1

Prospects for a Universal Philosophy of Organism

Some Initial Reflections

Derek Malone-France

There are a number of recent discoveries and developments in biology that seem evidently congenial from process philosophical and theological perspectives. Take for example, the discovery of "quorum sensing" functions which enable *anticipatory, non-random-adaptive* environmental response phenomena in microbial societies, which may be interpreted as verifying essentially *teleological* (non-Darwinian, *directed*-adaptative) and *cooperative-altruistic* dynamics in the environmental responses of these individual and collective organisms.[1] Increasingly—and often with real cogency—process thinkers are engaging with this new biology in order to make the case that process metaphysics can better explain these observational realities than can the dominant alternative philosophical perspectives. (See, for example, the growing "biophilosophy" movement that is an outgrowth from contemporary Continental process thought.) This is an appropriate and productive response to this ongoing science.

Yet, if we take seriously Whitehead's conception of the proper critical function and mode of philosophical inquiry, we must also discipline ourselves, to maintain a certain degree of theoretical distance from these developments. And we must acknowledge that there are some other recent scientific discoveries and developments that may be seen as potentially cutting against common principles or presuppositions of process thought. Take for example, elegant mechanistic modeling of the developmental pathways of the molecular regimes associated with the evolution of particular planetary chemical systems.[2]

Moreover, if the ultimate aim is to demonstrate the ongoing relevance of process thought as providing a *universalizable metaphysical perspective*, then we must also recognize the fundamental ways in which our biological and physical, as well as our cultural, philosophical, and theological concepts and categories necessarily, to some degree, reflect the *contingent particularities of the evolution of the particular planet* (*and solar system*) on (and in) which we have emerged and by which we have been conditioned.[3] Just as philosophy and theology have always needed to be mindful of the problem of *anthropocentrism*, we now understand, better and more specifically than ever before, the extent to which our thinking is also prone to the more generic problem of "*terracentrism.*"

Indeed, if we live in a "process universe," then the nature of "process" must be more general than any form thereof that would depend—even at the level of quantum particle formation and relation—on conditions specific to our planet, our solar system, our galaxy, our neighborhood of galaxies. . . . So, we must proceed with caution as we seek to extend our conceptual frameworks to encompass a cosmic totality that we have still, as a species, only begun to explore.

One concrete implication of this recognition of our *terracentric* perspective is that we may need to *look beyond "life"* as a category around which we construct our general claims and inquiries about the phenomena of complexity, reproductivity, sociality, sentience, and intelligence. Over time, the concept of "life" has become, if anything, theoretically more, not less, problematic, especially in light of recent developments in a number of diverse (but increasingly intersecting) scientific areas ranging from prebiotic chemical evolution and evolutionary virology, to planetary chemistry and astrobiology.[4]

My sense is that the more general category of "organism," as constructed in Whitehead's thought, is a better candidate than "life" for universalizability. In addition, the concept of "organism"—again, as Whitehead constructs it—has the advantage of reflecting the ultimate convergence (nondualism) of both "non-biological" and "biological" phenomena, under the rubric of a more generic ontological category. This, in turn, *might* allow for the ultimate clarification of both the *common* and the *disparate* elements characterizing what we commonly refer to here on Earth as "living" and "nonliving" phenomena. It also suggests a speculative cosmic application with what I take to be productive theological implications.

In this chapter, my aim is simply to offer some initial constructive thoughts regarding the prospects for such a universalization of basic Whiteheadian/neo-Whiteheadian ideas regarding the metaphysical character of reality, with a particular focus on the concept of "organism." My discussion will assume some basic familiarity with process philosophy and/or theology, but no more than that. And I will briefly introduce some current scientific context, developments, and ideas that may not be so familiar for all those who are coming to the chapter from philosophy or theology.

CONTEXT

The background for much of what I will say here is my own increasing involvement over the last several years with the rapidly developing scientific and inter- and transdisciplinary field of *astrobiology* (sometimes referred to as *exobiology*). I was drawn into this field through an unexpected engagement with a program cocreated and sponsored by the Library of Congress and NASA—the Baruch S. Blumberg Program in Astrobiology, named in honor of the founder of NASA's Astrobiology Program. And this engagement has significantly affected the trajectory of my thinking as a process philosopher.[5]

NASA defines "astrobiology" as encompassing all the following expansive domains of inquiry:

1. the exploration of the origins and evolution of life on Earth (as inextricable from the search for, identification, and successful study of extraterrestrial life);
2. the search for (and possible study of) extraterrestrial life;
3. the consideration of the future conditions and evolution of life, both on and beyond Earth

Now, although astrobiology is a relatively new and still developing field, in terms of its scientific demarcation and justification, its conceptual roots stretch deeply into human intellectual history throughout the world.

Among those thinkers traditionally associated with the development of "Western" philosophical and religious perspectives, the first recorded affirmation of the hypothesis of extraterrestrial life *representing something like what we mean by that term today* was made by the first generation of Greek pre-Socratic Atomists, Leucippus and Democritus, and expanded upon later by Epicurus. And the core concepts of these original atomistic thinkers continue to have both profound and pervasive relevance in contemporary thought, across a range of issues that are fundamental to our evolving understandings of the nature of reality; the origins and evolution of "life" and "living systems" and the cosmic context of human existence, experience, and thought.

New and emerging *scientific discoveries* and insights, combined with the gradual ongoing revitalization of *metaphysical speculation* (even in Anglo-American philosophy), and more critical attention to the proper bounds of the *distinction between those two*, have once again placed atomistic ideas at the very center of many of the most important inquiries and debates of our time.

In the physical sciences, of course, the belated triumph of the classic atomistic hypothesis, per se, helped forge the modern understanding of the universe and its constituent elements, and enabled the breathtaking and world-reshaping technological advances that have derived therefrom. Yet, the continuing advance of our understanding of "physical" reality, especially at the quantum level, now contravenes the ultimate accuracy of the Atomists' ontological conception of an "atom," per se.

Similarly—and for reasons that are no doubt ultimately related at an ontological level—in the biological sciences, the *reductivist physicalism* and *ontological mechanism* that have historically been associated with *atomistic materialism*—which undeniably drove generations of fundamental discoveries into the chemical structures of life—now increasingly appears insufficient to the task of explaining the actual complexity of even microbial organic societies and the dynamics by which they relate to and with one another and their larger environments.

At the same time, the ancient Atomist notion of an actually infinite physical universe—that is, a universe with *infinite actual content*—not only resonates powerfully within the context of contemporary debates over cosmic topography and multiversal modeling, but also often now manifests itself as a sort of (notably nonempirical) first-order theoretical *assertion* about reality that is made by some of the world's most culturally prominent theoretical scientists (and self-proclaimed "defenders" of science), in connection with their own (not entirely scientific) battles against the perceived enemy of religion, per se, and/or theism more generally.

Moreover, even for those, like myself, who deny the sensical-ness of the notion of an *actually infinite* physical universe—in the sense of an *infinite set of actual things*—the modern re-contextualization of human existence within a universe the expanse of which (both spatially and temporally) we might reasonably describe as *infinite for all practical (human) purposes* makes the existential and moral ruminations of the early Atomists highly relevant to discussions of the meaning, value, and future of human life and society; the possibilities to be considered regarding trans- or post-human forms of existence; the ever-ongoing process of cultural and collective identity construction among human beings; and the definition of and search for "life" elsewhere in the universe.

In their temporarily losing argument with Plato and Aristotle, the school of the Atomists maintained that the fundamental "stuff" out of which the "*kosmos*" is constructed is an infinite proliferation of absolutely "simple" ("indivisible"), inert, and imperishable material particles—"atoms"—that *move without intent* and *collide without any permanent attachment* within an infinite "void" space. The physical world—that which the first generally recognized process philosopher, Heraclitus, had already termed the "*koinos kosmos*" ("shared world," "common world")—that we observe with our senses is, according to the Atomists, the product of the eternal, random *jostling* of an infinite number of atoms across the infinite void. Moreover, contra Plato and Aristotle, the pre-Socratic Atomists had asserted that there is not merely a single "world," bounded by the heavenly spheres and centered on this little terrestrial orb of ours, but, rather, an infinite multiplicity of "worlds" (and, hence, the distinction arose between the respective notions of "kosmos" and "world"), both like and unlike ours, spread out through infinite time, across an infinite "void" space that is *filled to brimming* with an equally infinite plenum of "atoms"—and with the *macroscopic forms* to which these atoms' *purely fortuitous* physical *entanglements* give rise, including the largest macroscopic forms, which are precisely the "worlds" within the Atomists' now infinite *meta-kosmos*. Hence, according to the Atomists, the universe contains *worlds upon worlds upon worlds*.[6]

Today, astronomers and astrophysicists have attained what is believed to be a fairly accurate estimate of the age of the observable universe, placing it at around 13.8 billion years. Our solar system is a relative newcomer, at not quite 5 billion years old, meaning that our sun is a second- (or possibly even third-) generation star. And this is a crucial fact, because without the prior implosions of first-generation stars in the earlier universe, there would be no heavy metallic elements—in which case the origin and evolutionary development of life *as we know it* would not have been possible. *In other words, we had to come into being where and when we did, in the universe, to be anything like what we are.*

Now looking forward, rather than backward, if one imagines an advanced intelligence emerging in the universe 100 billion years in the future, the universe will have, by then, expanded so far that it will have become impossible—not just practically, but *in principle*—for that intelligence to become aware of *any other galaxies than its own*. This is an implication of the observational horizon-limit imposed by the speed of light. Thus, at least given our current best scientific understandings of physical reality, such an intelligence would be *guaranteed to have a false conception of the universe*, by virtue of nothing other than the fact of *where and when in the universe it emerged.*

In other words, as intelligent beings, our location, both temporally and spatially, within the universe, can place definitive limits around our capacity accurately to assess our ultimate situation, and in ways that may well be permanently invisible to us.

And in terms of this assessment of our situation, it is worth noting that just *thirty years ago* we did not know for sure if there was a single planetary body outside of our solar system. *Let that one sink in. . . .* The very existence of any planets outside our solar system was nothing more than a *debatable conjecture* until the first confirmed discovery was made in *1993*.

Since then, astronomers have not only cataloged *thousands* of "exoplanets," they have determined that our solar system alone has between *100 and 400 billion stars*, and that *most stars host multiple planets*. That means there are likely *at least several hundred billion planets* in our galaxy alone. And I'll note here that, from this point forward, the numbers I will use to describe the content of the observable universe and our galaxy represent the most conservative mainstream estimates—beginning with a default to the current low-end estimate of 100 billion stars in the Milky Way. There is some reason to believe the actual numbers could be not just higher, but much higher, regarding some of these variables. While there is almost no chance, at this point, that they are lower, as they are founded on measurements of observable galactic mass.

Moreover, recent analyses of mission data from the Kepler satellite—which has a primary mission of exoplanet discovery and study—suggest that "terrestrial" planets—that is, planets similar to ours in their general composition and size—are more common than other categories of planets—such as gas giants. Based on these analyses (and assuming the conservative estimate of our galaxy's planetary population), Kepler scientists estimate that the *Milky Way alone* hosts as many as *40 billion terrestrial planets*.[7]

Meanwhile, at every turn, our understanding of the limits of *"habitability"* and the possible forms of *"living systems"* is being expanded. Recent discoveries by NASA's Cassini probe, for example, have been taken to indicate the *possibility* of microbial life in the geochemically warmed ocean beneath the icy surface of the moon Enceladus, which orbits the *gas giant* Saturn, in a dark, frigid part of our solar system that *not long ago* seemed to us to be far beyond the range in which life could possibly take hold, given the size and energy output distribution range of our Sun, at this stage in its life.[8]

Thus, it is likely that a very substantial proportion of that *huge* number of exoplanets that exist in our galaxy alone are, at least in principle, *habitable*. And if we start with the supposition that the processes that led to the emergence of life on Earth were *not* a *miraculous exception* to general natural laws but, rather, a *normal manifestation* of general natural laws that would lead to similar outcomes in other places where the same basic initial conditions existed, then we are led to the supposition that a reasonably substantial number of exoplanets that are habitable are or have been at some point *inhabited*.

Accordingly, it has begun to look quite possible that "life" could be *common* in the universe. Of course, if even only 0.1% of 40 billion terrestrial planets were, in fact, inhabited (*at any point in their histories*), then that translates to as many as *40 million distinct origins of life in our galaxy alone*. And if even only 0.1% of those origins of life represented *potential* evolutionary arcs toward the emergence of "advanced" life, then that's still *400 hundred thousand potential forms of advanced life in our galaxy alone*—over the course of the galaxy's life. And there are *billions of galaxies*. . . . And ours is *not even* a relatively large one. . . . *Worlds upon worlds upon worlds* . . .

Yet, for all the prescience of the early Atomists' existential assessment of our cosmic situation, this same ongoing expansion of our understanding of what we might call, today, the "*biological universe*" is making it increasingly clear that the mechanistic-physicalist accounts of the nature of the cosmos that modern science and much of Anglo-American philosophy have drawn from, among other sources, the Atomists, are fundamentally mistaken. *Teleology* ("purpose," "intention") has now found its way back into contemporary biological theory, in the form of an emerging consensus recognition of what can very reasonably be termed *neo-Lamarckian*—that is, *nonrandom, self-selective,* or *"directed"*—adaptation and evolution in the context of complex organic systems.

Through recent discoveries regarding microbial phenomena such as "quorum sensing," we now know, for example, that among some colonies of supposedly "simple" microbial organisms that we have observed here on Earth, there are complex, quasi-linguistic systems of chemical signaling at work, in some cases including up to thirty-three distinctly identifiable chemical signature command signals with respectively consistent informational content transmissions, associated with consistent behavioral responses upon reception.

Among other things, such microbial communities use these command-signaling mechanisms in order to *select among themselves* for survival under changing environmental conditions, in order to *force their own evolutionary adaptation* in certain directions, by privileging the transmission of certain DNA and RNA lines over others, *in anticipation of future benefit*. This sometimes involves what may reasonably be described as a form of microbial "altruism," when deselected members of the community self-destruct their own cell membranes in order to relinquish their biomass for the continued sustenance of the members of the selected lines.[9]

OPPORTUNITIES AND CHALLENGES FOR PROCESS THOUGHT

Obviously, this sort of description of even very simple biological organisms producing complex social systems that involve clearly *final causative* (teleological) features, is music to the ears of process philosophers and theologians. According to the process Whiteheadian/neo-Whiteheadian view of reality, the most fundamentally real constituent elements of actuality are not inert particles of "matter" but, rather, *organic unities of experience*. "Experience," of course, is a broadly generic term for Whitehead, which encompasses not only that particularly high-order, complex form of sentient experience we term "consciousness," but also that most primitive, simple unification of elements of vectoral "physical feelings" constituted by a single, momentary, quantum-state-determination.

In other words, according to Whitehead's ontology, a momentary state of a single quantum particle and a moment of human consciousness are to be understood not as fundamentally divergent types of entities but, rather, as qualitatively different exemplifications of the same type of entity: qualitatively different because the latter is vastly more complex than the former in its "prehensive" analysis of the incoming data of its environment (which includes its own past) and the character of its momentary organic self-determination, made in light of this analysis; and, yet, exemplary of the same fundamental ontic type and process, in that both the quantum state and the moment of consciousness are, ultimately, the product of *selective self-construction*, which is to say, they each emerge into being as the outcome of a process of *organic final causation*.

Of course, such final causation does not replace efficient causation in Whitehead's scheme; it complements it. Just as the suddenly reemergent element of neo-Lamarckian selection in our developing understanding of complex organic systems is complementing, rather than displacing, Darwinian natural selection. Moreover, the discovery of a distinct neo-Lamarckian—that is, a *directed evolutionary selective*—dynamic complementing the overarching dynamic of random adaptive selection is far less surprising, and, far more naturally explicable, if one assumes a Whiteheadian or neo-Whiteheadian event-based ontology, rather than a materialistic one. After all, one of the defining differences, if not *the defining* difference, between materialistic and event ontologies is precisely that the explanatory accounts produced by the latter typically involve some form of final causation, whereas those of the former typically reject the very idea of final causation as fundamentally incoherent or conceptually misguided.

There are exceptions, of course. There are some materialist thinkers who try to maintain the notion of final causation in their accounts of "life." Yet even where these attempts to ontologically distinguish between, without dividing, the non-biological and biological worlds are relatively sophisticated, the underlying materialist framework leads to problematic theoretical constructions. Take for example the following illustrative quote from *Matter and Mind*, by the renowned Argentinian-Canadian philosopher of science and physicist Mario Bunge, with whom I'd like to walk here for a moment in conversation. This passage was written toward the end of Bunge's career, in a book intended as a sort of summation of his lifelong pursuit of a rationalized and logically coherent conception of the basic shared elements of scientific and philosophical thought and published only a little more than a decade ago, in 2010 . . .

> I am offered an avocado. *It is* nutritious—an objective statement; and *I like it*—a subjective sentence. Actually only a part of me likes it, namely my brain. *It*—one more thing, of course—is what makes *me* tick. Without *it* there would be no *me*. My brain is a material thing, though a living one, not just a physical one. And its mind, I mean mine, is a subset of my brain's functions, just as my smiles are contractions of my facial muscles—though not automatic but willed by my prefrontal cortex. No organ, no function.[10]

What a difference the last thirteen years of biological science have made! Among numerous other salient developments, the dramatic expansion over the past decade and a half of our understanding of the connections between the brain and other organs, systems, and even *ecologies* within the body makes it unimaginable that any serious philosopher of science would now write a description of the mind-body relation so profoundly simplistic.

How would such a narrative description of this relation go now? Something like the following . . . *setting aside for the moment quibbles with other elements of Bunge's passage, such as his take on the subjective/objective distinctio*n . . .

> I am offered an avocado. *It is nutritious*—an objective statement; and *I like it*—a subjective sentence. Actually *we* like it; after all, I am a collection of beings, including, for example, the *microfaunal* community of my gut biome, which plays a crucial role in how my brain receives and processes elements of my experience like taste. *They*—many more beings—are what make *me* tick. Without *them* there would be no *me*. My brain is an organic component of a larger organic set of systems, which together, in the organic unity of their collective *livingness* (whatever that is), form a whole that transcends the mere collection of their respective parts, and which is the basis of the experiential unity to which I refer when I say "*me.*" No complex system of organic parts—including

> within itself communities of other simpler, and yet still relatively complex, organisms—no organic function of the whole.

For readers of this volume, it probably goes without saying how much more congenial this narrative rings in the ears of process thinkers in the widely Whiteheadian tradition.

It is worth staying with Bunge for a moment. He represents a particular sort of modern materialist with whom, as I hinted above, despite our obvious differences, process thinkers also share some important philosophical intuitions. This is a moment for coalition-building. No particular school of thought should feel overly confident in its own capacity to fully respond to, incorporate, or rationalize the astonishing advancements that are occurring now across scientific fields, along with the new mysteries and questions that are coming with these advances. Mutually critical dialogue in the service of diverse and epistemically humble speculative theoretical development is the desideratum for metaphysics right now. So, it is also worth engaging with Bunge's work because he rejects Whiteheadian and neo-Whiteheadian ontologies on the basis of conceptual critiques that remain powerfully entrenched among current analytic philosophers of science, even as they now move (reluctantly) in certain directions that, again, seem congenial to process thought in relation to the organic-systemic character of both "life" and "consciousness."

Bunge's thought combines elements of what we might now call a somewhat traditional, realist, just-post-Popperian philosophy of science, with an analytic philosophical approach that generally navigates somewhere in between the approaches of W. V. O. Quine and Nicholas Rescher (the influences of both of whom are explicit for Bunge). And these elements of his thought coexist alongside what I find to be a fascinatingly *early-modern-recalling* conception of "matter"—one that also represents an alternative *process philosophical ontology* that emerged out of the Hegelian process tradition and that, for someone like Bunge, traces back, I believe, more or less directly (even if not necessarily consciously), to neo-Hegelian ideas found in transitional modern thinkers like Ludwig Feuerbach and Mikhail Bakunin.

This conception of materialism is certainly not to be confused with Whiteheadian or neo-Whiteheadian forms of post-materialistic, event-based, organic monism, but neither is it quite as distant from such organic-event monism as, perhaps, a thinker like Bunge himself might believe. Here is Bunge, explicating his conception of materialism and contradistinguishing it from "physicalism:"

> I am a materialist but not a physicalist because, as a physicist, I learned that physics can explain neither life nor mind nor society. Physics cannot even

> explain phenomena (appearances), because these occur in brains, which are supraphysical things; nor can it fully explain machines, as these embody ideas, such as those of value, goal, and safety, that are nonphysical. Physics can only account for matter at the lowest level of organization, the only one that existed before the emergence of the earliest organisms some 3,500 million years ago. Hence physicalism, the earliest and simplest version of materialism, cannot cope with chemical reactions, metabolism, color, mentality, sociality, or artifact.
>
> Our contemporary concept of matter is not Democritus', nor even Newton's, which is the one still held by most philosophers, and a reason that most people find it hard to believe that matter can think. They are right: A bunch of marbles cannot think. But brains are made of living tissue, which has peculiar properties that physical matter lacks; and its constituent atoms are far more subtle and complex than the tiny marbles imagined by the ancient atomists. Whence modern materialism is not to be confused with physicalism, let alone mechanism, for it is inclusive rather than eliminative. And yet those confusions are rampant in the philosophical literature.[11]

There is obviously much that could be said about each of these passages from a Whiteheadian or neo-Whiteheadian point of view. What I want to principally call attention to here is the overtly *organic* character of these descriptions of matter as something that is *non-inert* and that *changes in fundamental ways with the infusion of complexity*, as well as the way in which this *vaguely yet fundamentally* organic conception of the nature of matter permits Bunge to cling to his first-order presupposition of teleology—his intractable intuitive conviction that he and other "living" beings *do things for reasons*.

The important point here is that, while such materialist-monist and emergentist ontologies as exemplified here by Bunge (reminiscent also of that of Samuel Alexander) do not go far enough for me as a neo-Whiteheadian in their ontological reimagining; nevertheless, they do embody *allied* ontological intuitions, in ways that are similar to the embodiment of some process-allied intuitions by the tradition of European philosophical Phenomenology with which Whiteheadian and neo-Whiteheadian thinkers are more and more actively engaging in recent years. And, from my point of view, the scientific realism of Bunge is well-juxtaposed as a dialogical counterpoint to the flirtations with antirealism embodied in *some* of the phenomenological sources with which process thinkers are now in dialogue.

That's *not* to endorse Bunge's particular form of scientific realism, which is epistemologically naive in certain respects. But some of his critiques of antirealism and pseudoscientific thinking should be borne in mind as process thinkers continue to assimilate anew the works of thinkers like Husserl and Merleau-Ponty and their contemporary successors. Just as the epistemological perspective of Bunge can equally benefit from corrective contact with the more postmodern epistemologies of the new Phenomenologists, the Speculative Realists, et al. There is a middle ground, I believe. At the same time, I want to say very clearly here that, in my view, process thought desperately needs to avoid giving off even a *whiff* of antirealism, if it is to have any influence at all in the sorts of mainstream discussions in science and philosophy of science in which this volume seeks to be, in part, a productive intervention.[12]

Meanwhile, even as the sorts of developments in biology such as the quorum sensing phenomena discussed above give optimism about the viability of the process perspective as being metaphysically on the right track, I must acknowledge other developments that may represent genuine and fundamental challenges for process thinkers. Regarding the apparent reemergence of teleological explanation as an element of contemporary biology, it is important also to recognize that in physics, some evidence continues to accumulate regarding the possibly accidental fortuity of the evolution of the wider dynamics of reality within which the biological is situated. For example, I mentioned at the very beginning of this chapter the field of planetary chemistry, where theorists now hypothesize, on the basis of expanding real data and highly sophisticated analyses, that, "Geochemical 'paths of least resistance' are the outer constraints that determined core [features] of 'metabolism' [as a necessary condition of Earth biology] and give prebiotic-biotic continuity."[13]

Let us consider for a moment some of the implications of such geochemical serendipity as a foundational basis for the emergence and evolution of terrestrial biology, and the possibility that such foundations might be radically diverse in different environments throughout the universe. Today, we know that, along with volcanism and other geological dynamics, plate tectonics have played an absolutely central role in the development from a geosphere to a biosphere here on Earth. Life, as we know it, needs cyclical processes—change and redistribution of the elements of an environment almost certainly promoted prebiotic forms of production, reproduction, and diversification, leading to the emergence of early biological organisms, what we call "life." Ongoing work in prebiotic chemistry suggests that some combination of tectonic, atmospheric, and tidal dynamics, along with the fact that the Earth rotates on its access, creating a day-night cycle, allowed for the formation of replicable DNA systems before the evolution of the cellular environments in which biological DNA is enabled to reproduce by enzymatic functions.

But what of Enceladus, the moon of Saturn I mentioned at the beginning of this chapter? If there is something closely analogous to Earth life there, in the ice-covered ocean that surrounds its small, warm, rocky core, that life or life-like phenomenon will have necessarily emerged in the absence of both solar cycles (the ice covering the surface of Enceladus reflects away more than 90 percent of the little sunlight that reaches that far out in the solar system) and plate tectonics (there being no "land," only a global ocean covered in ice).

The likely source of warmth and energy fueling such an emergence on an icy planet or moon like Enceladus being serpentinization, the process whereby simple contact between water and rock with very high iron and magnesium content (as is likely found at the core of Enceladus) produces through chemical reaction not only heat but also the whole suite of organic molecules necessary for the production of life.

If we can identify two almost completely different possible geochemical pathways to the production of such an environment here in our own solar system, how much more diverse might such pathways become as we see further and more clearly into the cosmic distance? And if the evolutionary pathways taken by planetary environments might differ radically, how safely can we assume that the definitional aspects of "life" as we know it will be universal requirements for the development of phenomena that manifest characteristics that, on Earth, are solely associated with life—entropic resistance, complex informational transfer through time, intention? We must be exceedingly cautious in making assumptions about the universality of aspects of our experience that are, as a matter of fact, linked to the specific, concrete conditions of the Earth's particular solar environment and planetary history.

Also, in my view, process thinkers have sometimes been too quick to interpolate normative moral, social, and theological claims from factual discoveries in the realm of science. We must be careful where we pitch our tents in this regard. Take, for example, the emerging evidence indicating that cooperative modes of biosocial activity—I gave the specific example of microbial directive communication earlier—are perhaps as pervasive and important in and to the history of biological evolution on this planet as are competitive modes. It is tempting to leap from this observable fact to the claim that what this represents is a manifestation of the normatively framed process understandings of the divine, creativity, and sociality. But cooperation in the relevant biological sense here is a *normatively neutral* concept—after all, even limiting ourselves to human activity, we can point to genocides as triumphs of cooperation.

Furthermore, in light of what we have come to understand so far about both the highly particular and specific routes of evolution on Earth and the vastness and potential differentiation of modes of existence that might be like life in some meaningful way across the cosmic environment, it is now *unambiguously problematic* to continue to simply presume the viability of categories derived from the narrow conceptual confines of past or present human cultural traditions—including religious traditions—in a universalistic way.[14] We must oscillate permanently between the speculative extension of our ideas and the critical reassessment of their adequacy. And our safest metaphysical speculations, in the sense of intellectual grounding, will always be those that can be most completely abstracted from the conditions of our particular niche in the wider cosmic environment and from our particular form of intelligence, while at the same time remaining meaningful in content and concretely associable with our particular forms of existence and experience.

"THEORGANISM:" A SPECULATIVE PHILOSOPHICAL-THEOLOGICAL PROPOSAL

With all of that said, I remain convinced, with Whitehead, that metaphysical speculation is a necessary (and unavoidable) component of any systematic intellectual engagement with reality. And although I am honestly ambivalent about religion as a social-historical phenomenon, I recognize that the spiritual impulse behind this phenomenon is an ineluctable component of human experience—and probably an existentially and psychologically necessary one.

Whitehead's metaphysical explorations eventually led him to a theistic position precisely because his thinking was oriented by the foundational intuition that, instead of using the highly abstracted model of reality encapsulated in mathematical physics to construct our view of the biological world, we ought to use the concretely nuanced model of reality revealed by evolutionary biology and psychology to construct our view of the physical world (and of its metaphysical underpinnings).

I've briefly indicated here some reasons why I think that the Whiteheadian concept of "organism" is better suited than what I take to be the more specific and less abstractable concept of "life" for universalization. I also think the Whiteheadian conception of God can be well-framed in terms of the concept of "organism." That is to say, it seems to me that Whitehead's understanding of God and the God-world relation can consistently be put in these terms: the universe as a whole is (like?) an organism, and "God" is one name that has historically been given to the being whose existence is the ultimate expression of this organic unity, much as the "me" of human selfhood is the ultimate expression of the functional unity of the many beings that make up the human organism, as was discussed at the beginning of this chapter. Thus, the universe may be understood as a society of organisms constituting in their integrative totality a final organism of organisms, or "Theorganism."

The term has the advantage of productive double entendre, implicating both the common grammatical and logical "the" of singularity and unity and the specific "theos" of divine reference. It can function both as a name for the entity it references and as a name for the belief or system of thought involving such an entity. It is a genderless term—the concept of organism being far prior both logically and biologically to the manifestation of gender in complex biological forms. It is a term that conceptually fits with the *non-omnipotence* of the divine in process thought (we do not think of the minds of human or other sentient organisms as having absolute control over the functions of their bodies—I can make my arm go up and down; I cannot order a skin cell on my arm not to become cancerous). And finally, "Theorganism" is a term that *perhaps* could be more congenial to the ears of those in science for whom traditional God-language is irretrievably corrupted. I plan to expand on this conceptual framework for theism and its implications in future works.

NOTES

1. A foundational early paper outlining the mechanism here is found in Bassler, "How Bacteria Talk to Each Other."
2. See Smith and Morowitz, *The Origins and Nature of Life on Earth.*
3. This represents for me a logical extension from my earlier work *Deep Empiricism.*
4. See, Cleland, *The Quest for a Universal Theory of Life.*
5. From 2014 to 2015 I served as Seminar Director for Astrobiology and Society for the Blumberg Program.
6. See Dijksterhuis, *The Mechanization of the World Picture* and Furley, *The Greek Cosmologists.*
7. NASA, "Kepler and K2," https://www.nasa.gov/mission_pages/kepler.
8. NASA, "Cassini," https://solarsystem.nasa.gov/missions/cassini.
9. See Li and Tian, "Quorum Sensing and Bacterial Social Interactions in Biofilms."
10. Bunge, *Matter and Mind,* vii, emphasis original.
11. Ibid., vii–viii. Compare this passage to the following, from Bakunin's *God and the State*: "We may answer that the matter of which materialists speak, matter spontaneously and eternally mobile, active, productive, matter chemically or organically determined and manifested by the properties or forces, mechanical, physical, animal, and intelligent, which necessarily belong to it—that this matter has nothing in common with the *vile matter* of the idealists. . . . The gradual development of the material world, as well as of organic animal life and of the historically progressive intelligence of man, individually or socially, is perfectly conceivable. It is a wholly natural movement from the simple to the complex, from the lower to the higher, from the inferior to the superior; a movement in conformity with all our daily experiences, and consequently in conformity also with our natural logic, with the distinctive laws of our mind, which being formed and developed only by the aid of these same experiences; is, so to speak, but the mental, cerebral reproduction or reflected summary thereof." Bakunin, *God and the State*, ch. 1.
12. As a sidenote, Bunge is also a figure with whom it is worth engaging because of his wide influence over many decades among Spanish-speaking intellectuals, in fields ranging from philosophy of science and technology, to political philosophy and theory, to philosophy of language, to sociology, and so on. See his multivolume, decades-long project: *Tratado de Filosofia.* In earlier decades, the North American academic process community maintained very active interchanges with Spanish-language scholars in philosophy and theology, especially Latin American thinkers, as North American process thinkers sought to export their ontological ideas about God and

human nature and freedom, while they imported moral and epistemological ideas about justice and community and redemption from Liberation Theology and other sources. This is a dynamic of interchange worth replicating in the context of current integrative process philosophical projects like those manifested in a number of the essays included in this volume.

13. Smith and Morowitz, *The Origins and Nature of Life on Earth*, ch. 3.

14. See my essay "Hell is Other Planets."

BIBLIOGRAPHY

Bakunin, M. *God and the State*. New York: Mother Earth Publishing Association, 1916.

Bassler, B. L. "How Bacteria Talk to Each Other: Regulation of Gene Expression by Quorum Sensing." *Current Opinion in Microbiology* 2, no. 6 (1999): 582–87. https://doi.org/10.1016/S1369-5274(99)00025-9.

Bunge, M. *Matter and Mind: A Philosophical Inquiry*. Boston Studies in the Philosophy of Science, Vol. 287. Boston: Springer, 2010.

———. *Tratado de Filosofia*. Vols. I–IV. Barcelona: Gedisa, 2008–2012.

Cleland, C. *The Quest for a Universal Theory of Life: Searching for Life as We Don't Know it*. Cambridge: Cambridge University Press, 2019.

Dijksterhuis, E. J. *The Mechanization of the World Picture*, trans. C. Dikshoorn. Oxford: Oxford University Press, 1961.

Furley, D. J. *The Greek Cosmologists, Vol 1: The Formation of the Atomic Theory and its Earliest Critics*. Cambridge: Cambridge University Press, 1987.

Li, Y-H, and X. Tian, "Quorum Sensing and Bacterial Social Interactions in Biofilms." *Sensors*, Special Issue on *Microbial Quorum Sensing: Linking the Outside World to Microbial Behaviour* 12, no. 3 (2012): 2519–38. https://doi.org/10.3390/s120302519.

Malone-France, D. "Hell is Other Planets: Extraterrestrial Life in the Western Theological Imagination." In *Social and Conceptual Issues in Astrobiology*, edited by K. Smith and C. Mariscol, 21–54. Oxford: Oxford University Press, 2020.

———. *Deep Empiricism: Kant, Whitehead, and the Necessity of Philosophical Theism*. Lanham, MD: Lexington, 2005.

NASA. "Cassini." Accessed April 30, 2023. https://solarsystem.nasa.gov/missions/cassini.

———. "Kepler and K2." Accessed April 30, 2023. https://www.nasa.gov/mission_pages/kepler.

Smith, E., and H. Morowitz. *The Origins and Nature of Life on Earth: The Emergence of the Fourth Geosphere*. Cambridge: Cambridge University Press, 2016.

Chapter 2

The Cosmological Context of the Origin of Life

Process Philosophy and the Hot Spring Hypothesis

Matthew David Segall and Bruce Damer

It is mere rubbish thinking at present of [the] origin of life; one might as well think of [the] origin of matter.

—Charles Darwin[1]

In itself such a material is senseless, valueless, purposeless. It just does what it does do, following a fixed routine imposed by external relations which do not spring from the nature of its being. It is this assumption that I call "Scientific Materialism." Also it is an assumption which I shall challenge as being entirely unsuited to the scientific situation at which we have now arrived. . . . Science is taking on a new aspect which is neither purely physical, nor purely biological. It is becoming the study of organisms.

—Alfred North Whitehead[2]

It is becoming increasingly clear that to understand living systems in any deep sense, we must come to see them not materialistically, as machines, but as (stable) complex, dynamic organization. . . . Our task now is to resynthesize biology; put the organism back into its environment; connect it again to its evolutionary past; and let us feel that complex flow that is organism, evolution, and environment united. . . . Thus, biology is at the

> point where it must choose between two paths: either continue on its current track, . . . or break free of reductionist hegemony, reintegrate itself, and press forward once more as a fundamental science.
>
> —Carl Woese[3]

The authors of this chapter—Segall, a process philosopher, and Damer, an origin of life scientist—are grateful to the editors for the invitation to engage with other philosophers, scientists, and theologians about how best to approach some of the most consequential cosmological questions that biotic minds like ours are capable of imagining. *What is life? How did it originate?* These intimately related questions cannot be answered with metaphysics alone, but nor are they decidable purely on empirical grounds. After all, biologists cannot explain how life originated until they know what they are looking for.[4] These also are not merely academic questions, since the answers we give to them bear directly upon the meaning of our own existence as conscious animals. Ultimately, we want to know more than just the *what* and the *how*. We also want to know *why*.

The authors of this chapter have become convinced that a Kuhnian paradigm shift is afoot, not only in biology, but across the multiple scientific disciplines and methodologies relevant to studying the origin of life. Advances in complex systems science and the study of nonequilibrium thermodynamics have helped narrow the gaps between physics, chemistry, and biology, but many conceptual knots remain to be untangled. Indeed, making progress on the question of life's origin may require a fundamental transformation of traditional conceptions of the relations among the sciences and their varying methods of explanation. While in periods of normal science, master craftspeople and technicians make steady progress on precisely delimited problem-sets, revolutionary science calls upon the aid of scientific seers, people who would have ended up as artists, philosophers, or theologians had they not become scientists.[5] In revolutionary periods, otherwise sharp boundaries between the *how* and the *why* questions begin to blur, thus encouraging closer collaboration between science and philosophy. The how questions constrain the imaginative speculations of philosophers, while the why questions pry scientists out of the shell of specialism, thus allowing, in philosopher of science Sebastian De Haro's terms, "the subject matters and methods of philosophers and of scientists [to] become entangled" such that "the relationship between science and philosophy becomes dynamical."[6]

The authors further affirm that continued progress in the effort to understand the place of life in the cosmos requires a transdisciplinary approach integrating the core insights and methodologies of not only astrobiology and philosophy, but also religious studies and theology. We value the freedom and autonomy of each of the special sciences to invent and test hypotheses unencumbered by the assumptions of other sciences (e.g., molecular biologists operate within a different paradigmatic context compared with evo-devo and systems biologists, etc.). We similarly insist upon the independence of science from theological orthodoxies (e.g., that life was designed and created from scratch by an omnipotent deity, or that the human soul is a supernatural substance existing in causal isolation from the rest of cosmic and biotic evolution). Scientific curiosity is to be checked only by the need for logical coherence and experiential adequacy (including ethical considerations).[7] While metaphysics and theology have been "warned off the premises"[8] of modern experimental laboratories, these ancient disciplines nonetheless retain an essential function in the effort to understand our cosmic origins. For one thing, philosophy and religion inevitably contribute to any final integration of scientific findings into a meaningful and motivating worldview for humanity at large. But even more significantly for natural science, metaphysics has a crucial role to play in shoring up science's own epistemological and cosmological conditions of possibility. Whitehead asks: "What is there in the nature of things which leads there to be any science?"[9] His answer is that trust in science requires a metaphysics explanatory of the insistent rationality of things. For Whitehead, cosmic rationality is a consequence of the inextricable causal entanglement of all things: "there is an essence to the universe which forbids relationships beyond itself, as a violation of its rationality."[10] Natural science thus assumes the universal communicability of the causal nexus across all scales of Nature. Science further presupposes that conscious organisms have arisen within this nexus who are capable of turning back to contemplate and even to explain their own cosmic origins. It is imperative, then, that a way be found for scientific conceptions of physical causation, chemical reaction, and biological origination to hang together with our commonsense experience of conscious awareness and agency. For after all, if our consciousness is a total sham, then so are all our scientific inquiries and religious aspirations. Consciousness must somehow "[have] truck with the totality of things," and it is the job of philosophy to critique and reconstruct the abstractions of the special sciences so as to recover a concrete sense of our connection with the cosmos as a whole.[11] Whitehead invites us to reenvision the modern clash between science and religion as "a sign that there are wider truths and finer perspectives within which a reconciliation of a deeper religion and a more subtle science will be found."[12]

This coauthored chapter aims to contribute to origin of life research by approaching outstanding aporias from two complementary perspectives: 1) *the empirical*—explicating an emerging experimentally grounded hypothesis concerning the "progenitor" of living cells in the setting of wet-dry cycling in fresh water hot springs (written by Damer incorporating feedback from

Segall), and 2) *the metaphysical*—leveraging the conceptual innovations of Whitehead's organic realism to overcome Kantian epistemic quandaries by better defining the cosmological conditions of life's origin (written by Segall incorporating feedback from Damer). These two perspectives may find themselves in dialectical tension on some points. But the ultimate goal is to dynamically integrate the metaphysical and the empirical aspects of this research.

THE EMPIRICAL EVIDENCE FOR A HOT SPRING PROGENITOR OF LIFE

Opportunities for philosophy to connect with science in a mutually respectful and closely collaborative way are not common. Indeed, much of the normal problem solving of day-to-day science finds no need for the big picture ontological reframing proposed by philosophers. Some scientists go further than simply stating the uselessness of philosophy, declaring that "unquestioned philosophical preconceptions have at times been hampering factors of scientific progress."[13] However, when a special branch of science finds itself in the midst of a revolutionary transition or Kuhnian paradigm shift[14], it then becomes apparent that scientific disciplines are themselves founded upon a background of philosophical preconceptions. For this reason, paradigm shifts alert us to the urgent need for collaboration between scientists and philosophers. When scientists discover a new system which is complex or opaque enough that it defies formal treatment and testing with existing tools, a period of fervent creativity ensues, generating new thought experiments, explanatory models, and laboratory instruments. It is easy enough to point to historical examples of philosophy's contributions to new paradigm science, like Kepler's Pythagorean-inspired discovery of planetary harmonics in the seventeenth century or the emergence of quantum mechanics in the twentieth under the influence of Vedic wisdom and Western philosophers as diverse as Plato, Kant, and Schopenhauer.[15] When such shifts are afoot, the specialists and technologists of normal science must endure the wild speculations of their more holistically minded, visionary colleagues. Often those colleagues come in from other disciplines, bringing fresh perspectives but encountering strong resistance until their contribution shines new light upon the obscure questions at hand. These "seer" scientists are open to entertaining and developing ideas beyond the strictly reductionist approaches which had guided the puzzle solving approach of day-to-day science. Scientific seers engage with a wide range of thinkers, going beyond strict disciplinary boundaries to entertain proposals from nonspecialists and from philosophers.

This situation is exactly what we find in the field of origin of life science, which is currently undergoing a very public and controversial paradigm shift from an old scenario to new models. The long-held view that life began in the oceans 4 billion years ago is being supplanted by the approach that life could only plausibly begin in land-based freshwater bodies.[16] One of the leading thinkers in this transition, biochemist David Deamer, discovered that polymers such as RNA and DNA could be self-assembled from their component monomers through a process of repeated hydration and dehydration. Such cycling through wet and dry phases has now been adopted as a common chemical protocol by multiple groups internationally who have repeated and extended the work by demonstrating the formation of polymers from amino acids called peptides. Deamer's approach is based upon five decades of research on membrane biophysics which he also incorporated into the wet-dry cycles to form the cell-sized compartments which can contain and concentrate these polymers. The co-emergence of membranous compartments and their polymer cargo form "protocells," the basic units which can undergo the stepwise transition from non-living to living processes.

A Time Portal to the Hadean

The following story is offered for the layperson and nonspecialist alike to comprehend the proposed scenario of life's origin. The hope is that it captures our latest thinking on how simple biological life can emerge from the background of the sterile but dynamic physics and chemistry of a newly formed world, eventuating in the establishment of an entire microbial biosphere.

> Two scientists, a geologist and a biologist, buy tickets to travel back through a time portal four billion years to the end of the Hadean eon. Their destination is one of the large volcanic land masses rising up through the rusty brown oceans of a turbulent young Earth. After donning environment suits to provide a breathable air supply (free oxygen, a product of life, would not become available for a few more billion years[17]), they grab their kit bags for sampling and step through the portal. In a flash, they tunnel through time and find themselves walking out onto the crunchy black landscape of a lava-filled volcanic caldera. A line of volcanoes roars forth ash as heat is dumped from the mantle of the newborn world. The scientists head for a more quiescent field of geysers jetting forth and filling multi-colored steamy pools interconnected by streams. The orange-brown haze of the atmosphere is lit by a faint young Sun itself surrounded by a brilliant disc of dust and larger rocks from the still-forming solar system. Some of this material enters Earth's atmosphere, painting streaks across the sky and flashing with sonic booms as it hurtles toward the surface. This remarkable scene is conceptually represented by the computer-generated image in figure 2.1.

Figure 2.1. Artist conception of a geyser-fed fluctuating hot spring pool on the Hadean Earth, four billion years ago. (Courtesy of Ryan Norkus and Bruce Damer)

The explorers make their way over to a set of pools and notice that along the edges of some of them a silvery sludge is forming. The geologist has his rock hammer in hand but finds that there is nothing to break. He decides the sludge is not made of minerals and determines that their nature is beyond his disciplinary ken. The biologist leans in to consider the sludge, determining that it is somewhat reminiscent of pond scums formed by bacteria or algae. She scoops up a sample, placing it under a microscope. The view is remarkable: a seething mass of membranes forming ballooning layers which entrap a soup rich in mineral particles and dissolved organics. The biologist looks to the sky, wondering if this material was sourced partly in the dust and meteorites falling all around, accumulating in the hot spring pool at their feet. She determines that the sludge is not merely geological, but also not yet biological. On a hunch, they return to the time portal to turn the dial a few thousand years closer to the present and revisit the same site.

The island is still recognizable. A few more lava fields have formed, and the hot spring system has grown into a larger network of pools and geysers. The scientists can't find their original pool but notice that the sludges now encircle the edges of many more pools in the hydrothermal field. The biologist, fascinated by this apparent growth, takes a knee next to the nearest pool to study the material forming these distinct "bathtub rings." Instead of the uniform silvery-white observed on their first trip, the sludge is now infused with something resembling black ink. She scoops some new samples and inserts them into several instruments to analyze their chemical structure. She notes surprising spikes revealing the presence of polymers, some which are chains dozens of units in length. She then employs a pocket-sized nanopore sequencer to sequence the polymers and is even more astonished to discover that there are whole families of repeating sets of polymers. Some are composed of nucleic acid cousins resembling those making up the RNA and DNA present in all living organisms today. Others are built up from amino acids strung together into families of peptides, the short precursors of proteins. On a whim she subjects a blob of the pond sludge to a powerful UV lamp. A probe records a drop in the pH as the blob becomes more acidic, and resampling shows a surge of polymer production along with a slurry of other molecules, including glucose. She realizes that the sludges have somehow incorporated a means of capturing UV light to turn it into useful products, including repeating sets of polymers and sugars. She suspects that the inky black fluid is a kind of pigment enabling photon absorption, much like chlorophyll in the cyanobacteria that will emerge millions of years in the future. She is not sure if this particular sludge sample is "alive," but the extant evidence of its growth and adaptive capability suggests it is well on its way to evolving into the first living, dividing cells.

Before heading back to the portal to travel home to the terminal phase of the Cenozoic era for a lunch break, the scientists exchange preliminary hypotheses. The geologist remarks that perhaps life organized itself around a liquid form of crystal called a lipid membrane, cycling and growing within small pools formed by the hard geological crystals of silicates, clays, and lava. The biologist wonders in turn what energetic and selective processes may be driving the changes in her soft sludges and what primitive proto-biological feedback loops may be sparking the transition to full-fledged life. She gathers some samples to place in a simulation chamber back in her lab hoping to continue to observe their evolutionary adventure.

The story of this humble sludge and how life can emerge from it is the subject of the rest of part 1.

Foundations of a New Hypothesis

Fifteen years ago, David Deamer began testing his intuition that life could begin in hot springs by visiting bubbling volcanic landscapes in distant Kamchatka, Russia and closer to home at Bumpass Hell in California. These hot springs possess the chemical composition and rhythmic energetic cycling capable of driving *away from equilibrium* chemical interactions. These interactions occur because polymers cycle in what chemists call a "kinetic trap," wherein the rate of their synthesis exceeds that of their hydrolysis (breaking apart).[18] All life today exists in a series of cycles supporting kinetic traps. Geysers and other sources of hot water can fill and refill small pools on a repeating, rhythmic basis, facilitating wet-dry cycling and the emergence of communities of protocells.

Recent analysis of carbonaceous meteorites and missions to asteroids and comets have established that abundant organic molecules for protocell formation would have been delivered to landmasses on the early Earth during the time proposed for life's origin.[19] By focusing on *warm little ponds* on land, where Charles Darwin intuited that life began,[20] the hot spring hypothesis suggests that these organic materials would concentrate in freshwater pools. In contrast, any material landing in the oceans would be diluted and lost to further chemical processes. Furthermore, such geyser-fed hot spring pools not only would have existed on the early Earth, but have been discovered on Mars, and are likely ubiquitous on many rocky exo-planets, so they provide a plausible candidate environment for life's beginnings.

In 2009, while working on a PhD on molecular simulation approaches to the origin of life,[21] coauthor Bruce Damer met Deamer. Coming from a computational and systems background, Damer added a new understanding of how Deamer's wet-dry cycling system could repeat, couple its contents through distinct phases, and generate a protocellular population capable of supporting multiple levels of combinatorial selection, eventuating in the emergence of bonafide biological functions.[22] Deamer and Damer have since engaged in a decade of collaboration on new laboratory science and field trips to hot spring sites at several locations including Bumpass Hell, Lassen Volcanic Park in California, Yellowstone National Park, and Rotorua in New Zealand. They also visited sites in Australia containing fossil evidence of some of the oldest signs of life on Earth, found in preserved hot spring minerals from the Archaean period, 3.5 billion years ago. Their collaboration with colleagues around the world resulted in a number of pivotal publications developing the new scenario and reporting on mounting empirical evidence. This effort culminated in their 2020 *Astrobiology* journal lead article "The Hot Spring Hypothesis for an Origin of Life"[23] which laid out the entire scenario with testable predictions for the next generation of investigators.

Deamer and Damer are now developing an approach to determine which worlds are potentially "urable," a new term denoting zones on rocky planets or moons capable of starting life. While many worlds could be deemed "habitable," i.e., able to support life, fewer would have the proper conditions for initiating it.

Figure 2.2 details the scenario in one dynamic picture starting in the first three steps with sources of organics delivered from space and atmospheric and terrestrial sources to hot spring pool networks on a volcanic island. The fourth stage introduces cycling of these materials which can self-assemble and form polymers as they transition through dry-wet-moist phases. The resulting populations of polymer-containing protocells budding off from the dry into the wet phase are chemically selected for stability with each new test run through the watery medium. As water evaporates, the moist phase allows protocells to nestle together and form collaborative networks of chemical interaction that Carl Woese called "progenotes," which are aggregates of prebiotic processes "in the throes of evolving the genotype-phenotype relationship."[24] These progenote and protocell populations would eventually become

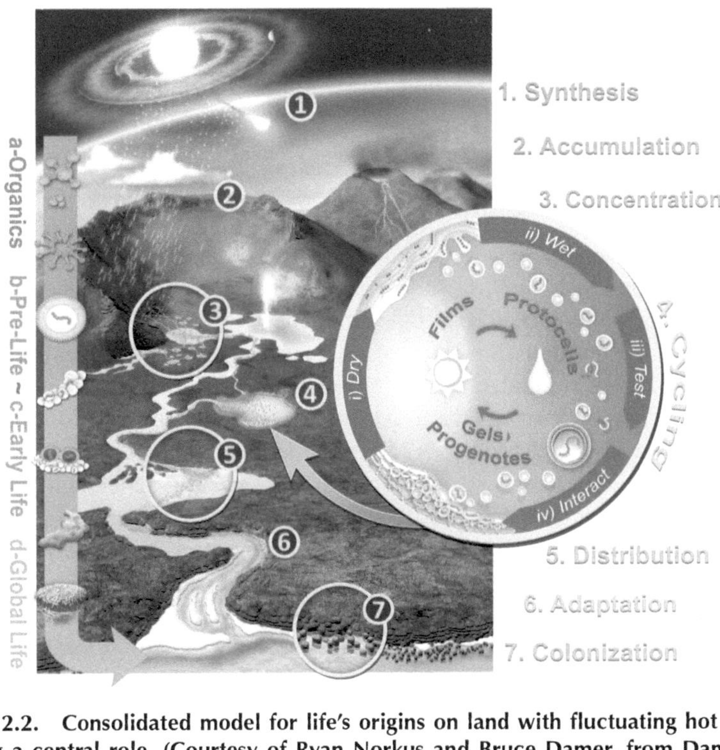

Figure 2.2. Consolidated model for life's origins on land with fluctuating hot springs playing a central role. (Courtesy of Ryan Norkus and Bruce Damer, from Damer and Deamer, "The Hot Spring Hypothesis for an Origin of Life")

robust enough to be distributed across the landscape. Steps five through seven show those aggregates evolving and exchanging material in various settings, flowing downhill and gradually adapting to the more hostile, salty, and disruptive environments of the seashore. On the left side of the figure, the increasing levels of chemical and structural complexity are represented. The assemblage of starting materials which acts as a cradle for the emergence of protocells and progenotes is defined as the "progenitor" and is discussed in detail below.

As this work was emerging and more groups adopted a land-based, wet-dry cycling approach to their prebiotic chemical experiments, controversy erupted in the science media among proponents of the hypothesis that life began in the deep oceans along hydrothermal vents.[25] "Ventists" and those favoring a land-based scenario weighed in, culminating in the December 2020 feature in the journal *Nature*,[26] which was widely considered to be confirmation that the paradigm shift had occurred. This transformative moment in the field has opened the door to new thinking, especially from computer science, complex systems theory, and new speculative cosmological proposals from philosophy. Damer met coauthor, Matt Segall and began collaborating in earnest in late 2019. Segall was intuitively attracted to the process of rhythmic cycling evident in the evolution of protocells in the hot spring setting. Damer was interested in the big picture thinking offered by Segall and Whitehead's process philosophy (see part 2).

The Dawn of the Progenitor

While scientific colleagues debate and test the biochemical minutiae of the stages of various origin of life scenarios, the bigger picture of where life fits into the cosmos as a whole may fade from their view. Stepping back to consider the cosmic context allows us to ask a core metaphysical question: What generative process animated matter? Taking this question apart we can ask what is *animation*, what do we mean by *matter*, and how might we define and describe a *generative process*? This is the problem space that is ripe for renewed twenty-first-century collaboration between science and philosophy. These questions are not new, but for the first time perhaps, they can be informed by an empirically testable artifact, namely a proposed *progenitor* at the origin of life. A dictionary definition of the term progenitor brings up three principal meanings: 1) an ancestor in a direct familial line (e.g., a forefather); 2) a biologically ancestral form; 3) a precursor or originator.[27] The further back we search for a universal, common ancestor the more muddied the waters become. Phylogenetic studies of microbial DNA can only hint at the form, possible lifestyle, and basic functional components of the most ancient cells as there exists no fossil record of their molecular composition.

To further complicate matters, microbial life was fundamentally communal, with cells sharing genes horizontally not only within their own species but across extended ecosystems. Woese's prediction and our recent proposal of a physical model of the progenote, the form which preceded cellular life, could well bring the curtain down on the consensus that life started as simple *protocells*.[28] After tallying up all of these obstacles to understanding, we might be permitted to suggest that the question: *what was the first form of life?* itself breaks down and that there was in fact never a common ancestor. Like the uncertainty of the properties and positions of particles at smallest subatomic scales of matter, the medium and process of life's origination may be stranger than we can yet suppose. If we cease asking ourselves to imagine a distinct, common ancestor of all of life we are liberated to consider an alternative: that cellular life arose through interactions within a complex medium. Drawing from our above dictionary meanings, we could adopt the following definition of an *abiogenesis progenitor* as: *the precursor to and originator of all biological forms.*

Another commonly held belief in origin of life research emerges from the day-to-day practices of *solution* chemists, the majority of workers studying life's origins. Their paradigm is that complex forms can emerge from reactions occurring in relatively simple liquid mixtures called solutions. Such systems work in practice in the laboratory and in industry as long as a continuous supply of reagents and energy are engaged over one or more serial stages. Each stage, if simply left to percolate on its own without further inputs, will run down to what is termed a state of equilibrium, in which bonds forming are balanced with those broken, and no net increase in yield is observed. Applying their skills to origins of life, these same chemists test and publish numerous down-to-equilibrium reactions. However, the bigger picture of life is that cells, bodies and ecosystems continue to produce novel products and continuously move away from equilibrium to a series of ever more complex "steady states." This is achieved by a fiendishly complex set of interactions far beyond the simple serial enrichments of solution chemistry. The interactions are catalyzed by enzymes—gigantic molecules tuned to their tasks of fixing broken links and synthesizing new polymers in an exquisitely choreographed dance. Without this dance of enzymes maintaining a steady state away from equilibrium, Deamer notes that "we would dissolve when taking a shower."[29]

This long-held presumption of a simple, serial start to life, with few inputs and variables, is now yielding to a more complex view that life's origin involved many molecular actors and environmental factors.[30] Leading researchers now consider it plausible that major players resembling DNA and RNA were present together with peptides built out of chains of amino acids, all encapsulated in a membranous housing which facilitated their interaction.

This is a stunning refiguring of the process by which life is thought to have emerged. No longer are we constrained to simple reactions winding down to equilibrium or limited to scenarios of protocells with a handful of polymers and other solutes in interaction. Science thus finds itself at the threshold of novel conceptions and facing new questions. It is here that philosophers might no longer fear to tread and offer their best contributions to a field in flux.

Returning to our original three-part question, the progenitor must be an operating medium capable of supporting a *generative process* which supports the emergence of minimally viable living cells. The current philosophical refiguring of the origin of life is that the progenitor, the medium within and through which life emerged, is plausibly far more complex than the initial emergent biological processes themselves. "Complex" in this context means that the medium possesses a very large number of distinct components in a network of substantial density supporting a great variety of possible encounters and copious means of testing, selecting, and amplifying those components to further densify the network. A related notion emerging from this conjecture is that novel, finely tuned processes like living cells can only emerge and persist within such a richly endowed environmental context. This leads to a conundrum for scientists working on the problem: how can something more complex than the first primitive expressions of life exist before life itself? This seems like a fine exemplar of the age-old "chicken or egg?" question.

We are fortunate to have the time, instruments, and collegial support to pursue the question of life's origin. Following biochemist Albert Eschenmoser, this pursuit is not simply an undertaking to discover life's origin, but an attempt to reinvent it.[31] It is therefore necessary to run numerous experiments in simulated conditions indicative of conditions on the urable early Earth. The most effective way to do this is to start in a carefully designed laboratory setting to get experiments to work, and then take those experiments out to try them again "in the wild" at field sites which are analogues to those which would plausibly have existed on volcanic landscapes four billion years ago. An advantage of fieldwork is that when leaving the clean confines of the laboratory and attempting to perform experiments in much messier natural settings, Mother Nature becomes both teacher and a tough peer reviewer challenging hoped-for outcomes. Damer and Deamer have spent a decade testing their conjectures by traveling to diverse field sites on the flanks of volcanoes or overlying magma plumes where fumarole vents steam and hot spring pools bubble and pulse with geysers.

To justify any argument that the progenitor environment must be extremely complex (and capable), one must simply take a step back and look upon the task this hypothesis is called to perform. This cycling chemical system must, through innumerable trials driven solely by chemical selection, drive systems of polymers across a vast chasm of molecular evolution to the emergence of a minimally viable living cell capable of self-maintenance, growth, and reproduction through fission. The end product is a living organism far simpler and more fragile than your average staphylococcus, but it is still a breathtakingly complicated molecular system. Working backward from first life, the sheer number of finely timed and tuned processes operating in lockstep which must line up to enable the most primitive biotic entity defies our comprehension. To the sixteenth-century alchemists of Prague Castle, a substance capable of transforming base compounds into life would surely be a candidate for their long-sought *philosopher's stone*. Even if a plausible chemical progenitor might be synthesized in twenty-first century science laboratory conditions, it would fall far short of the full creative capacity (and time!) to realize a second genesis. Despite the limitations of our ability to realize Eschenmoser's dictum, it is worth trying our hand to see if we can synthesize candidate progenitors, trying them out both in the lab and in the field.

In late 2021, Deamer and Damer made two site visits to Fly Geyser in northern Nevada. There, while being filmed for science documentaries, they ran experimental panels of wet-dry cycles by setting their instruments down in the hot spring environment and applying drops of hot spring water, allowing the solutions to dry down, and then hydrating them again, for three to four hour long cycles (figure 2.4). As is shown in figure 2.3, dried films became visible on the slides. Microscope studies later revealed that these films were composed of membrane-forming lipids and a mixture of two RNA monomers

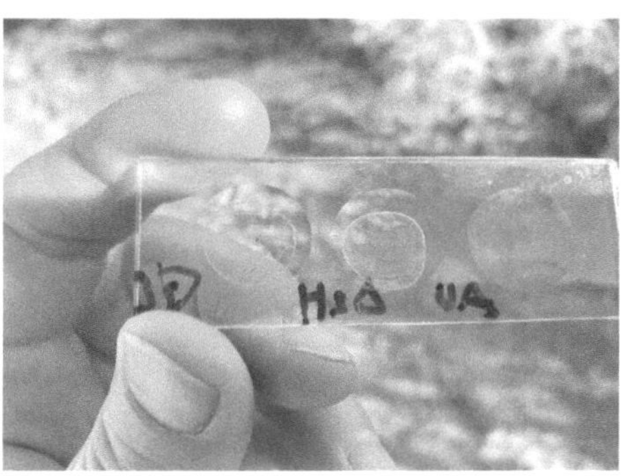

Figure 2.3. Films comprising lipid, dissolved silicates, and RNA monomers in a dried sample at Fly Geyser, November 2021. (Courtesy of Bruce Damer)

Figure 2.4. Wet-dry cycling of repeat and new experiments, Little Pot geyser at Fly Geyser, Nevada, December 2021. (Courtesy of Bruce Damer)

(which were introduced) together with silicate minerals which had crystallized out of the hot spring waters. Further analysis revealed that long chain polymers up to 100 units or more had formed from the RNA monomers. Thus, it was shown that an important biopolymer of life can be linked together from its nucleic acid components, all through a process of dehydration, and in a hot spring setting. The stitching together of some of the biopolymers from which life can emerge must be possible without the need for enzymes which life evolved later to accomplish the same task. This demonstration was only a first step toward life, but a compelling one. As Deamer, Damer, and coresearchers add more ingredients and variables to these tiny primordial soups and cycle them under a variety of conditions, more complex biochemical systems are expected to form and even to begin evolving.

A highly dynamic environment presents itself as the products from the Fly Geyer experiments are viewed under the microscope. Figure 2.5 below shows

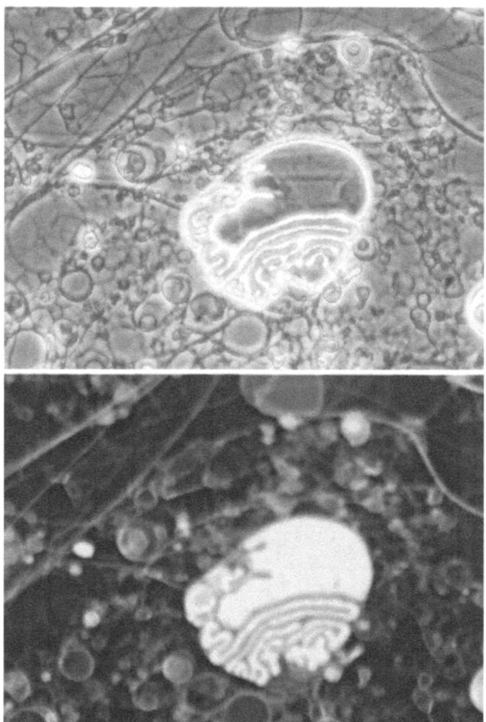

Figure 2.5. Protocell forming from a mixture of RNA components. Top: Phase contrast view of the lipid aggregate with vesicle compartments. Bottom: Fluorescent image of the protocell stained with acridine orange, indicating RNA polymers present within some of the protocell interiors. (Courtesy of David Deamer)

the assemblage of lipid membranes, dissolved silicates, and RNA monomers which assemble into protocell compartments (top). When stained and subject to fluorescence microscopy some of these compartments are observed to have captured polymers of RNA (large volume, figure bottom). As can be clearly seen, complicated spherical compartments, tubes, tendrils, and layers form as lipid membranes self-assemble and start to become animated by the dynamics of the water bath around them. Each microscopic scene unfolds into a complex configuration that arguably will never be exactly repeated anywhere in the entire future of the universe.

At the level below what the microscope can reveal flows a labyrinthine matrix of interacting channels and compartments with concentrated molecules in continuous dynamic interaction. Likely dwelling within such labyrinths would be sets of potentially information-conveying polymers similar to RNA and DNA. When water is present, these move around, coming into contact with other monomer building blocks or polymers, forming bonds and growing new chain links, or sometimes cleaving apart. During dry-down these polymeric tendrils can come into close contact with each other. Other species of polymers, such as peptides made from amino acid chain links, are also players on this stage. As some have polar ends, they can affix themselves to surrounding membranes and ride around on those membranes.[32] Peptides are the precursors to the proteins of our cells, the complex macromolecules essential for the formation of all biological structure and function. These peptide populations, initially formed spontaneously as nonfunctional chance sequences, would travel about in their trillions coming into contact with other riders on the membranes. Significantly, they could also brush against trillions of RNA and DNA-like polymers sandwiched between membrane layers. This dynamic setting might therefore host a stupendously large set of combinatorial possibilities, meshing together life's two primary actors, peptides and oligonucleotides, side to side and head to tail, providing opportunities for creative synthesis. "Creative" here means the emergence of function-forming integrations which benefit the players involved by allowing them to grow and replicate.

Pulling up from our speculative reverie for a moment, we can ask: Is what is being observed in these early micrographs a first scientific view of the proposed progenitor of life? Given the already observed capacity of this system to support the polymerization and lengthening of strands of RNA, DNA, and peptides while organizing them within lipid compartments called protocells, we propose it as a plausible, if greatly simplified, candidate progenitor. Following this first view, let us engage in another extended thought experiment to speculate on the possible properties and processes which might arise within such a progenitor. These imaginings might shed some light on what was going on in the silvery sludges studied by our two-time traveling scientists as they bent over the Hadean hot spring pool, now better informed into which we can peer more deeply:

> The progenitor environment is fed constantly with new organic chemicals and infused with energy from heat in the hot spring, from dehydration, and from the Sun's ultraviolet radiation which can drive important chemical reactions. Flows of electrons and small molecules stream across openings in membranes where momentary dislocations are made by appropriately folded peptides. Meanwhile, the entire system is undergoing continuous transitions between dry, wet, and moist phases. As water is reintroduced onto a pool, the dried layers are contacted and neatly ordered sheets of lipid erupt, budding off into compartments in uncountably large numbers. Some of these compartments contain sets of polymers stitched together during the dry phase, and thereby form protocells. During the flooded, aqueous stage, these protocells engage in a pre-Darwinian "struggle for existence," some holding together through all the stresses they experience in the pond and some falling apart. Each protocell compartment is a distinct chemistry experiment searching for the stability that prevents it from going to pieces. If the protocell pops, then its contents are lost to the dilute watery environment where they break down. If it holds together, the protocell and its polymers have a shot at participating in another cycle. This process may be analogous to the account of evolutionary novelty offered by Whitehead's concrescing "drops of experience," a proposal explored by Segall in Part 2. The physical affinity of polymers and membranes creates a more stable overall package so that both survive. As the pool dries down, the protocells clump together into a moist gel-like aggregate along a mineral surface edge or at the bottom of the pool (the sludge in our time portal story). This gel aggregate of lipid and molecular cargos compresses down as water evaporates. Populations of surviving spherical protocells begin to flatten out into sausage shapes which can spontaneously fuse with each other. The whole system then returns to a layered phase with polymer contents flowing and mixing within vast two-dimensional sheets. During each of these three distinct phases—dry layering, wet protocellular testing, and moist sharing (further unpacked below)—different chemical cycles can be tested and amplified, a very small chance subset of them to become selected as the prototypical components of life.

The progenitor environment would not be very impressive to the naked eye—a shiny thin white slick at the boundary between mineral, water, and air—but it is a powerful catalytic medium capable of generating the first echoes of a life cycle, and ultimately, cellular life itself. The progenitor would be the deepest ancestral form which generates all subsequent lineages of living organisms. This ancestral form is itself not alive but has the capacity to carry a self-assembled system of prebiotic molecules all the way to self-maintaining, reproducing living cells. The progenitor is indivisible from its environment as it arises through processes of concentration and self-assembly within quite specific surrounding conditions.

The exact composition of the actual original progenitor or progenitors aggregating and cycling on the early Earth will never be known. Laboratory and computational simulacra of progenitors will inevitably be missing some critical components and environmental influences. Its primary organizing construct is membranous layers and compartments derived from carboxylic acids delivered by carbonaceous meteorites or dust particles at the time of life's origination. One such meteorite, over four billion years old and known as the *Murchison* for the town it fell near in Australia, has been shown by Deamer and others to contain these membrane-forming molecules.[33] Other exogenously delivered organics such as amino acids and nucleobases have been found on such objects and plausibly generated through atmospheric synthesis[34] are also viable additives to the progenitor soup. The conditions of pools into which the ingredients are assembled can be inferred from current-day volcanic environments, and also from the preserved Archaean rock record as far back as 3.5 billion years.[35] Estimates of atmospheric composition, weather patterns, day-night cycles of radiation under a fainter young Sun, and the underlying geological platform can be pieced together from models and the actual Archaean rock record. Therefore, while pieces of the puzzle are certainly missing, we can assemble a plausible if still rough picture of the likely progenitor conditions on a rocky volcanic planet with fresh water hot springs.

Prebiotic chemists typically explore simpler systems with reduced variables to create clean and publishable outcomes. However, a minimally viable progenitor is going to be extremely complex, its internal processes and products very hard to track and analyze during laboratory trials. Of course, future investigators will have to decide where to set the bar for a minimally viable progenitor, which could simply start as a system which can amplify and diversify polymer populations. If these workers are truly lucky (and patient), they might observe some selected sequences which can perform functions and even self-replicate. From outside, such a system would seem to grow in physical dimensions and self-stabilize in the face of stresses which would otherwise degrade and return the whole structure to an undifferentiated chemical broth.

For future astrobiologists, presented next is an educated guess of the form and properties of progenitor environments. The films on the slides at Fly Geyser provided an early suggestion of how they can emerge. We can attempt to coax them toward growth, adaptation, and even some emergence of function using the best techniques available to chemistry and synthetic biology, all informed by computer models. We can use our best thought experiments and artificial intelligence simulations to predict some of the outputs and behaviors of model progenitors. The epoch of the progenitor and its emergent progeny, protocells and progenote aggregates, may have lasted tens of millions of years. Such a *Progenean* epoch may bound the processes by which biology emerged from physics and chemistry.

SUMMARIZING AND SPECULATING ON THE PROGENITOR HYPOTHESIS

The following list is an initial set of speculative propositions concerning the properties of the progenitor. The first few could be demonstrated through laboratory and field testing in the upcoming decades.

1. Like ecosystems supporting the living world today, the progenitor is an *environment* which provides a *kinetic trap*, in which systems of molecules composed partly of polymers can grow and become complex enough to ultimately accomplish a primitive form of replication. Outside of the supportive environment of the progenitor, these molecules would lose complexity, fragment, and return to a chemical equilibrium of inactive components and short, nonfunctional polymers.
2. Through processes of self-assembly and self-organization, the progenitor supports the arising of primordial versions of the capacities of living cells: capturing energy and incorporating external feedstocks, growing, adapting, and constructing additional progenitors as it is distributed into and colonizes new environments.
3. The progenitor would begin without any biological functions operating within it, but over time it would be gradually taken over (but never completely) by such functions. Even today, some proportion of matter and energy in the construction of ecosystem niches is not caught up in the activities of living cells.
4. The progenitor undergoes a cycling physical metamorphosis between individual protocell units containing sets of polymers and a conjoined indivisible whole formed through an aggregate of protocells termed a *progenote*.
5. In the unit phase, sets of polymers budding off into membrane bounded protocells enter bulk solution and are tested for stability and longevity. In the conjoined moist and dry phases, these protocells and their contents clump together within the progenitor environment wherein network interactions through diffusion, concentration, fluxes of energy, and forms of molecular cooperation and competition can then occur.
6. The progenitor environment consists of an organizing matrix of layered membranous material composed of prebiotically plausible fatty

acids. Within this matrix, polymers can be formed from monomers through dehydration synthesis. These polymers inhabit interior spaces or become affixed to membranes. The biophysics of the environment would then rapidly transport polymers, moving between and on membranous layers. Akin to a highway system running through a city, the vehicles of polymers travel on the high-speed thoroughfares of lipids, or move more slowly along intervening surface streets represented by interior volumes, or concentrate in parking areas represented by vesicular lumens. With this view, the progenitor could be formally modeled as a *directed graph*[36] and bears a resemblance to the living world which emerges from it, from cells to bodies to brains and their progeny: cities and networked computer systems.

7. Ironically, the progenitor would be a much more complex environment than any of the initially prebiotic chemical circuits or protocells emerging within it. Specific events within a progenote aggregate might include: molecule-to-molecule encounters on the membranous highways or between these highways and interior volumes; transiting of molecules and ions carrying energetic potentials across membranes; and budding of new compartments and fusing together of membranous volumes, carrying sets of molecules with them. We hold that the combinatorial potential of these interactive events is large enough to support the arising of biological processes.
8. The progenitor undergoes a down-selection of its own combinatorial expansiveness as more finely tuned and replicating proto-living systems take over from its less efficient self-assembly and random-encounter processes. Protocells and their aggregates, progenotes, "grow" within the supporting matrix of the progenitor and eventually fully colonize it, transforming it into the first living microbial communities, represented in the fossil record as stromatolites.
9. Progenitor substrates would spontaneously form in many watery environments on landscapes on the early Earth and or on similar exoplanets. Captured in the narrative tale at the beginning of this section as an "effusion" or "by-product" of the elemental interactions between air, water, and rock, the progenitor represents an intermediate form between geology and biology. The progenitor is effectively a form of proto-niche construction[37] and becomes the substrate into and from which the super-niche of the biosphere eventually emerges.
10. Progenitor aggregations would assemble with a variety of chemical and structural variations, subsets of which could support polymer accumulation and interaction, combinatorial selection, and stepwise evolution toward cellular life. The evolutionary journey toward life would take many twists and turns within progenitor niches, with many search

pathways petering out as conditions change and environmental stresses become too great. Two factors would support the eventual emergence of life against substantial odds, particularly the forces of degradation on a challenging landscape:

a. progenitor environments would continue to form in abundance supporting new starts toward life, and
b. material from a progenitor in one physical locale could be distributed to nearby or distant environments by flowing water or wind, spreading and recombining molecular innovations with other progenitors.

11. At some point in the progenitor's history, members of its protocellular population could transition from being entirely dependent on external inputs to begin to source some of their essential processes and products in-house. Such a point or points might be referred to as the stage of life's "ignition," i.e., the transition to active work, self-determination, and autopoiesis.[38] Protocells would then begin taking on some of the functions provided by the progenitor surrounds with more efficient, enzyme-driven replacements. Some of the ignition points might include the appearance of the following:
    a. capturing of solar radiation through pigments and coupling it into an energy system represented later by ATP;
    b. the selection and amplification of efficient catalysts which are coupled to the copying of informational templates (the proto-ribosome); and
    c. the collection and division of coherent sets of informational and functional polymers colocalized on membrane surfaces as a first primitive kind of reproduction.
12. The progenitor yields to protocells forming aggregate communities which Woese referred to as a "progenote." Progenitor environments become more "alive" as more efficient biological functions replace earlier stochastically driven and unreliable prebiotic processes. New progenitor aggregates continue to assemble and provide a "feeding" medium as progenote populations colonize them and construct niches.
13. Supported by the surrounding medium of the progenitor and in networked relationships with other protocells occupying a distinct progenote, a key milestone is reached when the first fission of a protocell into two viable daughter compartments occurs. This is another point that some may decide to mark as the event originating life on Earth. These cellular divisions could never happen outside of the protective and nutritive environment of the progenitor. Indeed, many such attempted divisions would fail resulting in the disgorging of protocell contents. This would not present a net loss of that line of polymeric evolution, however, as the contents are reabsorbed into the progenitor to be used again.

14. With cellular fission comes true reproduction, allowing living cells to conserve an increasing proportion of their heritable traits. Vertical descent has begun and with it specialization. The specialization leads to the fundamental division of labor seen in microbial mat communities today: sun-gathering photosynthetic organisms in the top layer, metabolizers below, and detritus digesters toward the bottom. Horizontal sharing of genetic innovations which was completely dominant in the protocellular Progenean epoch continues to be an important factor in microbial communities.
15. We can estimate that the coming of increasingly autotrophic, primitive cellular communities enhances their robustness to distribution, allowing them to escape the confines of hydrothermal pools and, especially after some form of photosynthesis was selected for, to lead lives independent of the chemical feeding zones of hot spring pools and meteoritic organic accumulation.
16. These increasingly robust early microbial communities adapt to a range of settings including rock surfaces and fluid-filled interiors, osmotically challenging saline lakes and marine shores, dilute water bodies such as oceans, and extreme environments like shallow and deep sea hydrothermal vents and the subterranean crust.
17. Progenitor-like environments would continue to arise under similar conditions for billions of years (as has been demonstrated in the above-described Fly Geyser fieldwork). However, as Charles Darwin suspected,[39] competition from hungry extant cellular life prevents these substrates from becoming the cradles of a second genesis. Nonetheless, the synthesis of such conditions in modern laboratory settings could perhaps re-create from sterile ingredients an original "genesis" progenitor, allowing us to observe and test some of the propositions proposed here.

Why the Progenitor Matters

Part 1 concludes with some metaphysical musings. As is witnessed everywhere in the biosphere today, death and extinction play an instrumental role in driving the selection of more efficient functions required for life. Therefore, any system proposed for life's origins must prosecute dead ends for most organisms while a few are provided paths to successful reproduction. The proposed substrate for life's beginning, the *progenitor*, therefore instantiates a "try and try again" molecular search engine that is planetwide across a multitude of environments, all potentially interconnected and sharing their material innovations. Passage through multiple treacherous probability landscapes to the seemingly miraculous emergence of a living, dividing cell can only be undertaken by a very potent and persistent combinatorial selection process. In this view, life's beginning was far from simple. It was nurtured in a spontaneously emergent, self-assembled progenitor system whose capacity for novel interactions vastly exceeded the complexity of the prebiotic geosphere upon which it rested.

Why should the human species go in search of our deepest biogeochemical ancestor? Perhaps the similarity of the progenitor environment to other complex systems arising later in biological and technological evolution suggests that through its study and laboratory re-creation scientists may be able to tease apart the fundamental principles underlying all emergent, adaptive phenomena. This could initiate a Copernican-class revolution, particularly if such knowledge is applied to the development of more powerful artificially intelligent computing learning systems. Further, witnessing a laboratory simulacrum of the enormity of the challenge faced by our biotic ancestors would surely contribute to enriching our cosmological perspective, giving us a new insight into the probabilistic unlikelihood, and therefore the preciousness of all life.

Before we leave the purview of the progenitor, let us take the opportunity to pose some compelling questions, perhaps best taken up by philosophy:

1. Are the matter and energy engaged in the processes of living organisms something truly different from what came before? By discovering a means to use primitive instructional templates acting as the first genes, does life introduce a nonlinear leap in creative potential in the universe? Or is biological life a less remarkable phenomenon, more of a step in a continuum beyond the previous creative assemblages of the prebiotic cosmos?
2. If science can establish that a medium like that proposed for the progenitor is required for life to emerge, then is it a given that life must share some of the same properties of that progenitor? If so, how does this recast our understanding of organisms as an indivisible set of units in densely interconnected relationships?
3. One suggestion of a solidly grounded cosmological context for the progenitor/hot spring origin model is given by the very planetary landscape upon which life would arise. The hot spring pool setting out of which a progenitor can assemble and support the emergence of microbial life lies at the boundary of mineral, water, and atmospheric elements. Such a progenitor is "fed" by chemical and heat energy from the hydrothermal spring system cycling up from below, and also from light energy and organic compounds falling in through the atmosphere above. A good proportion of the infalling material is sourced from the rocky and dusty disc surrounding the young sun which was then accreting to form the planets. The atoms and molecules in this disc in turn formed

from clouds blasted out by generations of stellar explosions. Previous conjecture about the origin of life often places its beginnings in a rare set of molecular interactions, a "pinpoint origin." The novel thinking described here suggests that a landscape full of many thousands of pools and other watery locales across which protocellular material could be exchanged and cycled provided an ideal medium out of which life could emerge. That medium is a "wide field" in which the volcanic landscape is broadly connected to solar system space beyond the boundary of Earth's atmosphere. Therefore, life's emergence and subsequent evolution for billions of years is recast as being in continuous and indivisible connection with the cosmos at large. While the behavior of biological cells is remarkably different from the static grains of ancient dust still orbiting our star, there is a deep underlying and uninterrupted connection between them.

4. Can we gain an understanding of the nature of conscious experience, including our most inexplicable and extraordinary states of mind, by scaling up the fundamental processes operating in the earliest stages of life's origin? For one example of this, let us take it as a given that the progenitor had to be a system for shaping probabilistic outcomes and moving the bar of thermodynamic equilibria. As a result of this inexorable process in subsequent four billion years since life's beginning, a planetwide system has dramatically shaped probability in its favor, at all scales. At the same time, that system has generated a vast store of linear instructions, from genes to memes, from which ever more probabilistic outcomes emerge. It is perhaps this collective power to alter the future in nuanced and highly complex ways that best characterizes life, all the way up to the seemingly miraculous products of the conscious minds of human beings. What can such a view teach us about the nature and potency of the living world? And to conclude, might this understanding of life's origin provide a new explanatory context for our own remarkable capacities, aiding us in our evolutionary bid to become a flourishing and self-sustaining planetary civilization?

THE METAPHYSICAL AND COSMOLOGICAL CONTEXT OF LIFE'S ORIGIN

In part 2, Segall argues that the task of explaining the origin of life is made more tractable by finally laying to rest the metaphysical assumptions of scientific materialism. In place of the now defunct mechanistic world-picture, Whitehead's more general conceptions of both organism and evolution are extended and applied beyond just the biological domain, leading to the idea of an organic cosmology. In order to bring Whitehead's cosmological contributions into relief, part 2 revisits Immanuel Kant's transcendental treatment of biological phenomena. The metaphysical approach Whitehead called "organic realism"[40] is marshaled as a way out of the epistemological quandary that Kant's critical method left any purported science of life. Whitehead's metaphysical generalizations reframe the emergence of biological organization in the course of cosmic and Earth evolution, allowing scientists to reconceptualize the threshold of "life" as a matter of degree rather than of kind.[41] The scientific goals of the interdiscipline of astrobiology, and Damer et al.'s Hot Spring Progenitor Hypothesis in particular, are interpreted as implying such a conceptual shift. Further, the wet-dry cycling scenario described by Damer in Part 1 is interpreted as a special scientific application of Whitehead's more general metaphysical account of concrescence (i.e., the process whereby the universe produces novelty). Such interpretations inevitably raise the controversial question of whether formal and final causation—what Whitehead described in terms of "aims at satisfaction"—play any selective role in the emergence of complexity and the evolution of biological organisms. Relatedly, in conjunction with expanded notions of organism, evolution, and immanent teleology, part 2 unpacks Whitehead's account of the role of environmentality[42] in sheltering the otherwise highly improbable chemical pathways responsible for igniting biological self-organization. Whitehead's conception of environmental sheltering contrasts with the dominant sense of "environment" as fixed and external, opening the door to more processual, relational, and situated analyses of evolutionary novelty.

Since our understanding of abiogenesis bears directly upon the meaning of human existence, the cultural and spiritual implications of process-inflected origin of life science become especially salient. The hope is that a new kind of collaborative transdisciplinary research can contribute to healing the rift between the sciences and the humanities, thus allowing us to better appreciate, in Arran Gare's words, "what it means to be free conscious agents as part of and creative participants within a dynamic, creative nature."[43]

Overcoming the Mechanistic Conception of Nature

It is widely accepted among physicists that "there is nothing in physical law which implies the existence of [biological] organisms."[44] For Whiteheadian or process-inspired researchers, as well as some systems biologists, this means only that the mechanistic metaphysical framework underlying such an interpretation of physical law is not generic enough to account for the reality of life and mind. If there is to be a scientific explanation for the origin of biological cells (not to mention the origin of the scientific minds studying them), then it cannot just be that the more fundamental laws of physics and chemistry *happen to allow for* the emergence of biological organisms as an accidental collocation of molecules. Rather, there must be some set of causal principles of organization and complexity at play—some "all-embracing relations"[45]—linking prebiotic to biotic modes of existence. Contrary to Jacques Monod's claim that living beings on Earth simply got lucky ("Our number came up in the Monte Carlo game"[46]), for the process philosopher random chance is the opposite of a causal scientific explanation. Evan Thompson articulates this point while summarizing Kant's treatment of organisms (unpacked in more detail in the next subsection):

> We cannot explain organisms mechanistically because their organized forms are contingent, not necessary, with respect to the mechanical laws of inorganic nature.[47]

Given classical assumptions about the metaphysical status of matter, the Monodian approach requires viewing the emergence of life as so gratuitously improbable that it borders on the miraculous.[48] To give a scientific explanation is to give a rational account, while an *accident* is said to occur for no reason. It is not that chance plays no role in the unfolding of the universe, nor that biological organization must be understood to be strictly entailed by the laws of physics.[49] The problem is that Monod's influential bifurcation between blind "chance" and iron-clad "necessity" stacks the deck so as to leave no room for life in the universe except as an alien anomaly. A more concrete rendering of the facts of Nature would recognize that, in ontological terms, physical laws are statistical tendencies rather than transcendent impositions, thus softening the dichotomy between randomness and determinism. As in unpacked further below, this allows "laws" to be reformulated as widespread cosmic "habits."[50] Further, Whitehead's panexperiential vision of cosmic creativity allows for the redefinition of chance occurrences in a non-reductionistic way as, to varying degrees depending on the complexity of the environing context, the expression of self-organizing *aims*. Such aims are not imported into Nature from a supernatural beyond but are understood to be intrinsic to self-organizing processes at whatever scale they emerge. At the physical level, "aim" need not imply anything like conscious deliberation: those uncomfortable with such a terminological choice may replace "aim" with "direction," that is, with an affirmation of the irreversible vector of causation and the cumulative character of time.[51] Accepting the reality of cumulative time and immanent telos in the universe allows for a naturalistic treatment of human agency and intelligence, features which might otherwise appear even more absurdly anomalous than the emergence of life.

Mechanistic materialism proposes to explain life as nothing more than an especially complicated, selectively amplified chemical reaction that is ultimately a fluke by-product of physical law under special highly improbable conditions.[52] From a process philosophical perspective, this alleged explanation looks more like a deflationary metaphysical redefinition than a legitimate scientific finding.[53] Life has not been rationally explained, but reductively explained away. An a priori definition of life that excludes the aims and feelings of organisms makes it a good deal easier to explain their arrival in purely mechanistic terms (that is, solely in terms of efficient causal relations between externally related parts). It also leaves a good deal of evident facts entirely unaccounted for, e.g., that as organisms we are ourselves "*directly* conscious of our purposes as *directive* of our actions."[54] Modern science's inability to find aims, feelings, or creativity in Nature follows from its adoption of Descartes's methodologically clarifying but, when ontologically reified, ultimately disastrous dualism separating mind from body, or subject from object. With "mind" thus neatly tucked away outside of the physical world, natural science was free to ignore half the evidence provided by human experience by describing the interaction of mindless bodies according to deterministic rules of succession.[55] As Whitehead put it:

> Scientific reasoning is completely dominated by the presupposition that mental functionings are not properly part of nature. . . . As a method this procedure is entirely justifiable, provided that we recognize the [obvious but undefined] limitations involved. The gradual eliciting of their definition is the hope of philosophy.[56]

The hope of origin of life science, including the Hot Spring Hypothesis, is that a rational account of life's emergence is possible, and that it could be achieved through precisely modeling and experimentally verifying one or more viable chemical pathways leading from molecular motion to biological function, including metabolism,[57] replication, adaptation, anticipation, and ultimately conscious human agency. In summary, the process philosophical response to this revolutionary scientific proposal is twofold:

1. to affirm that an evolutionary continuum bridges any apparent ontological gaps between physical, biological and mental processes such that "the ultimate natures of things lie together in a harmony which excludes mere arbitrariness,"[58] and
2. to question whether the "matter" life is said to have emerged from could be anything like what classical materialism had been imagining.[59]

It follows that the first step in any integration of origin of life science with process philosophy must be to overcome the mechanistic-materialistic substance ontology that continues to inform many biologists' interpretations of the physics and chemistry to which biology is thought to be reducible.[60] Appeals to mechanism may have been justified during the reign of the classical paradigm, when physics was still rooted in well-attested self-consistent physical concepts regarding externally related material particles with simple location in absolute space and time. But following the quantum and relativistic revolutions, no such system of concepts exists.[61] A century later, while complex systems theorists have made important progress toward integrating the special sciences, and while technological applications continue to dramatically reshape human society, the metaphysical foundations of physical knowledge remain in disarray, with dozens of theoretical interpretations of quantum phenomena vying for explanatory priority.[62] Process philosophers are eager to join the growing chorus of quantum physicists, process chemists,[63] and systems biologists to contribute to the development of a post-materialist ontology that may shed more light on otherwise confounding scientific findings.[64]

Whitehead predicted in the mid-1920s that the next great advance in physics would be made in the realm of biology.[65] From Whitehead's perspective, the advance would involve the recognition that "organism" rather than "matter" is key to understanding the regularity of Nature; that mechanism is an abstraction from creative activity; and that enduring entities at every scale from electrons to cells to galaxies are best understood as organisms engaged in the evolution of environments favorable to their persistence and enhancement.[66] Whitehead offered the first systematic sketch of a more generic process-relational ontology that resituates physical and biological entities in a broader cosmic ecology.[67] His scheme allows us to envision a universe of nested self-organizing processes of varying degrees of complexity, with "no absolute gap between 'living' and 'non-living'" systems.[68] His organic realism thus makes good on biochemist Addy Pross' call for an "integration of animate and inanimate matter within a single conceptual framework."[69] In such a context, biology becomes the study of the more complex organisms while physics becomes the study of the simpler organisms.[70] Rather than construing the universe as a structural hierarchy of things, with all causal arrows pointing downward to a base layer of substantial particles, process metaphysics reimagines the cosmos as a dynamic hierarchy of processes, with activities at every level constrained by their internal relations to one another and to the wholes in which they are nested.[71] Process philosophy thus rejects reductive mechanistic accounts of life and joins systems biology in welcoming formal and final causation back into our scientific understanding of Nature.[72]

To be fair, it is undeniable that mechanistic accounts and computer models can provide insights for guiding further research in some limited domains of application, particularly when the kinetic components being investigated are sufficiently stable in the time frames considered.[73] The metaphysical challenge presented by process philosophy does not forbid the machine metaphor as a heuristic device for drawing causal maps of isolated or idealized systems. Computer modeling is obviously an essential instrument in the tool kit of contemporary physics, chemistry, and biology. The process intervention is meant only to check claims to total explanation. As the biologist and creator of General Systems Theory, Ludwig von Bertalanffy, put it:

> We cannot speak of a machine "theory" of the organism, but at most of a machine fiction. . . . We could at most say that organisms can be regarded "as if" they were machines. We do not at all wish to underestimate the value of picturable fictions in science, but we cannot remain satisfied with the one offered in the present case.[74]

Bertalanffy here accepts the machine fiction as a methodological shortcut for aiding research, even though he thoroughly rejects it as a metaphysical explanation of biological organization. His rejection of a machine ontology follows from the fact that the machine metaphor remains "crypto-teleological," i.e., that every machine implies an external engineer whose purposes the machine was designed to fulfill.[75] If modern science rejects the idea of a divine designer imposing purposes on its creation, then it should also reject the mechanistic ontology that continues to imply such a picture. It is important to note that such divinely imposed teleology differs in kind from the sort of *immanent* or *intrinsic* teleology first described by Aristotle[76] and later refined by Kant, whose account of "natural purpose" is discussed below. In the case of the "as if" heuristic deployment of mechanistic or computational models, it is essential to remember that the "designer" in question is the human researcher whose purposes are defined by the parameters of the hypothesis under consideration. Such models at best offer partial descriptions of abstractly demarcated conditions and subcomponents. When it comes to ontology, mechanistic models remain observer dependent in the above sense and are simply too abstract to adequately account for the complex sympoietic[77] dynamism of living organization. This is not because organisms are too *complicated* to accurately model. Regardless of future increases in computational power, Robert Rosen has argued that in principle "organisms cannot be completely formalized" because the *complex* (not merely complicated) "closed causal loops" they instantiate "have nonalgorithmic, noncomputable" self-referential semantic elements that are irreducible to any computational syntax.[78] Further, as Dupré and Nicholson point out, mechanical explanations "are accurate only on the particular timescales of the phenomena they are called upon to explain."[79] When the entire lifecycle of an organism is considered, the seemingly solid components referenced in mechanical accounts (whether genes, proteins, organelles, or organs) dissolve into the continuous stream of activity constituting the organism as an energetically open, ecologically entangled complex dynamic system. While shocking to our substantialist habits of thought, the fact is that "in general, none of the parts of an organism is as old as the organism itself."[80]

It may be argued that the genetic code constitutes a set of instructions older than the organism itself. But this betrays too abstract and decontextualized an idea of genetic "information," particularly in the context of an ontology claiming to be physicalist.[81] As Woese suggested, it is "wrong to consider the codon assignments in cryptographic isolation" since they are just a surface manifestation of something deeper and more fundamental to biological phenomena, i.e., the evolution of the "phenotype-genotype relationship."[82] In some sense, it is true that the *form* of a living organism remains unchanged as its material parts are constantly turned over and replaced, though this form must be understood as a dynamic developmental trajectory or morphogenetic pathway rather than a static shape. In the case of DNA molecules, the nucleic acid sequences remain largely unchanged during the life span of an individual organism thanks to the remarkable fidelity of cellular copying and error correction processes, even if the individual molecules themselves are scattered and replaced during cell division.[83] The popular idea that organisms can be entirely specified by instructions coded in genes is dubious for several reasons. It is here—in what Woese called "the real problem of the gene" and perhaps of biology itself (i.e., how the process of gene translation evolved[84])—that the machine metaphor for life shows itself to be spectacularly inadequate. *Growing an organism is not like booting up a computer*.[85] First of all, the abstraction "gene" (as in a particle-like unit of heritable information for specifying proteins and thus phenotypes) no longer has a single clear biochemical definition.[86] Furthermore, the information required for making an organism cannot be simply located in genetic material but must be distributed throughout the stochastically self-organizing molecular interactions composing the intracellular matrix and indeed the whole historical sequence of environments that organisms have evolved and developed within.[87]

Besides these theoretical limitations to the mechanical method of explanation, further practical and ethical concerns can be raised. Woese, who was well aware of the import of complexity theory for deepening the ontological reach of biology, warned that the tremendous instrumental success of reductionistic molecular biology had now run its course, and that without a new vision of life biology was threatening to become an engineering discipline "that solely does society's bidding" rather than being "society's teacher":

> A society that permits biology to become an engineering discipline, that allows that science to slip into the role of changing the living world without trying to understand it, is a danger to itself. . . . Society cannot tolerate a biology whose metaphysical base is outmoded and misleading: the society desperately needs

to live in harmony with the rest of the living world, not with a biology that is a distorted and incomplete reflection of that world.[88]

Kant's Transcendental Limits and the Self-Organization of Nature

While much progress has been made in physics and biology by means of mechanistic modeling, the modern machine paradigm's overstressing of efficient causes is no less distorting than the medieval Scholastic period's overemphasis on final causes. It remains the task of a sound metaphysics to explain the proper relation between efficient and final causes.[89] Sometimes to move forward, it can be helpful to look back by examining the history of natural philosophy in search of wrong turns and alternative pathways for thinking Nature. The complex causality and intrinsic purposiveness evident in even the simplest organisms was given its earliest modern philosophical articulation by Kant in his *Critique of Judgment* (1790). Kant's claim that there would never be another Newton who might explain how even a mere blade of grass was produced by mechanical causes alone is frequently recounted in histories of scientific progress.[90] Darwin is then triumphantly introduced as precisely the Newton of the grass blade that Kant had philosophically forbidden, with his theory of evolution by natural selection offered as an explanation for how apparently purposive behavior and functional structure are really just products of blind mechanical forces passively amplified by Natural Selection.[91] Unfortunately, this story fundamentally misunderstands the philosophical context of Kant's argument and dramatically overplays Darwin's scientific hand. Darwin's *On the Origin of Species* (1859) offered a theory of speciation, not an account of self-organization and certainly not of biological origination. In fact, his theory must presuppose self-organizing biological individuals that can reproduce before it can do any explanatory work at the level of phylogeny.[92] When it comes to ontogeny, or the development of individual organisms, Kant's skepticism of mechanism remains as valid as ever.

In order to provide the philosophical context missing from most scientific accounts of biological phenomena, this subsection begins with a brief review of Kant's transcendental method.[93] This is followed by a more detailed treatment of his definition of self-organization. The hope is that Kant's critical framing of the biological problem-space brings into sharper relief the important metaphysical advances contributed by Whitehead's organic realism.

In his first critique, the *Critique of Pure Reason* (1781/1787), Kant inaugurated a Copernican Revolution in philosophy by reversing the until then taken for granted relationship between cognition and its objects. Rather than assuming, as dogmatic metaphysicians had, that our knowledge must conform to objects, Kant argued the reverse, that objects must conform to our means of cognition:

> This would be just like the first thoughts of Copernicus, who, when he did not make good progress in the explanation of the celestial motions if he assumed that the entire celestial host revolves around the observer, tried to see if he might not have greater success if he made the observer revolve and left the stars at rest.[94]

Kant thus analogically extends into epistemology what Copernicus had accomplished in astronomy. The upshot of Kant's transcendental maneuver is that it secures the synthetic a priori knowledge claimed by physical science (e.g., mathematical understanding of the laws of Nature). From Kant's point of view, Reason cannot be passively instructed by empirical Nature like a pupil, since (per Hume[95]) accidental observations of particulars provide no inductive basis for the establishment of necessary and universal laws or causal principles. As Kant has it, for scientific knowledge of physical laws to be possible, the scientist must play the role of judge, putting Nature on trial and compelling Her to answer questions as Reason frames them. This is because, according to Kant, "Reason has insight only into what it itself produces according to its own design."[96] Kant would later put it even more starkly: "He who would know the world must first manufacture it."[97] As is evidenced by the mechanistic model-centrism characteristic of so much contemporary scientific materialism[98], Kant's influence is pervasive even among those with no explicit allegiance to his transcendental project. Consider, for example, the words the famous physicist Richard Feynman had written on his blackboard at the time of his death: "What I cannot create, I do not understand."[99]

While Kant's revolution had provided philosophical justification for the knowledge produced by physical science, it came at the cost of forgoing scientific realism. In order to secure the rational necessity and universality of physical laws, Kant had to limit our knowledge of Nature to the sensory domain of phenomena. From his transcendental point of view, "nature is nothing in itself but a sum of appearances . . . merely a multitude of representations of the mind."[100] He construed the phenomenal domain as a kind of cognitive construct, the synthetic product resulting from the application of our innate categories of understanding (e.g., quantity and quality, substance and accident, cause and effect, etc.) to the spatiotemporal display produced a priori by the geometric and arithmetic organization of our sensory intuitions (i.e., geometry is said to be rooted in our pure intuitions of spatial simultaneity, while arithmetic derives number from our pure intuitions of temporal succession[101]). Natural science's synthetic a priori knowledge of the mathematical order and lawfulness of phenomenal Nature is thus grounded in the unity of Reason, rather than in a mind-independent universe,[102] with the nature of the latter remaining hidden behind the dense fog obscuring the noumenal ground of our sensory intuitions.[103]

While Kant resisted Berkeleyan idealism by insisting that *something* exists beyond the rationally organized formal order of the sensory screen, he marked this noumenal realm as a mere "X" of which nothing further can be known. As Whitehead quipped, "According to Kant we never know the real things, but only an *édition de luxe* which has been expurgated into rationality."[104] Kantian quietism as regards ontology resurfaced later in Niels Bohr's "Copenhagen interpretation" of quantum phenomena.[105] Bohr is reported to have responded to a question about whether the quantum theory somehow mirrored an underlying quantum world:

> There is no quantum world. There is only an abstract quantum physical description. It is wrong to think that the task of physics is to find out how nature *is*. Physics concerns what we can *say* about nature.[106]

It is important to note that while Kant's transcendental idealism limited scientific knowledge of Nature to appearances, he nonetheless fiercely upheld the objectivity of this knowledge. All rational knowers share the same set of categories and forms of intuition, thus securing the universality and necessity of the known laws of Nature. But in his final *Critique of Judgment*, focused on the status of teleological judgment in both aesthetics and biology, Kant unwittingly discovered a further limitation to his transcendental approach. While the motion of matter throughout the inorganic world from falling apples to orbiting planets had succumbed to the explanatory power of the new scientific method, Kant realized that living organisms exhibited a form of causality entirely foreign to physical science. Newton's law of gravitation perfectly described the ripened apple's downward trajectory from tree branch to soil, but neither his laws of gravitation nor of motion said anything about the metamorphic process of growth from seed through flower that put the apple up there in the first place.[107] Organic processes of development evidently unfold for the sake of an end, as though the apple were the purpose of the seed (or vice versa). As we've seen, Kant found it absurd even to try to imagine how another Newton, through mechanical causes alone, might explain how even a mere blade of grass could be produced:

> The internal form of a mere blade of grass is sufficient to show that for our human faculty of judgment its origin is possible only according to the rule of purposes.[108]

While he rejected traditional theological explanations, Kant realized that organisms, in that they exemplify "natural purposes," cannot be comprehended rationally without the aid of formal and final causation. Unlike the inorganic or unorganized matter presupposed in mechanistic models of physics, organized beings are *both cause and effect of themselves*, such that *their parts reciprocally produce one another for the sake of the whole to which they belong.*[109] Kant's *Critique of Judgment* thus inaugurates the study of "self-organization" in at least the *living* world if not the whole of Nature.[110] While the causal nexus of the inorganic physical world appeared to be exhaustively determined in mathematico-mechanical terms, the whole-to-part causality of the biological world confounded the Cartesian-Newtonian categories of classical scientific materialism.

But rather than displacing the reign of the mechanical *nexus effectivus* in physics by integrating the formal and final causes evident in living organization, the transcendental strictures of Kant's critical method forced him to confine the explanatory reach of natural purposiveness, not only to the realm of appearances (as was also the case for physical explanations), but also to playing the role of nothing more than a heuristic device regulating our reflective judgments concerning organized beings. While he argued forcefully that even the simplest of organisms is irreducible to mechanism, he was not willing to grant our judgments of their natural purposiveness the status of genuine scientific knowledge. When we think or feel purposes expressed in Nature, whether of a providential or simply individual sort, we are merely analogizing Nature as a whole or in its parts to ourselves, to our own rational faculty. Kant admitted that such analogies are highly suggestive of supersensible possibilities through which our own rational freedom might somehow be brought into harmony with the apparent determinism of inorganic Nature. But he insisted that the analogy remains merely an aesthetic feeling or moral intuition, not a scientific finding. *Aims* are but useful fictions that cannot feature in scientific explanations. As a result, given that Kant defined the domain of life by its purposive self-organization, he insisted that there could be no properly scientific study of living phenomena.

Fortunately, since Kant's day dramatic advances in our scientific understanding of a time-developmental universe have made it easier to understand how self-organization can be physically grounded.[111] But these advances have also dramatically upended the classical scientific understanding of "matter." In the wake of Kant's *Critique of Judgment*, Friedrich Schelling closed the eighteenth century with a series of groundbreaking treatises on natural philosophy that began the double task of resituating mind in Nature and reimagining matter in self-organizing terms.[112] Schelling sought a third way beyond both the Kantian idealistic approach, which begins with the rational subject and explains how an apparently objective Nature is constructed, as well as the materialistic approach, which begins with physical mechanism and tries to explain how apparent subjectivity might become appended to it. He sought instead for a more primordial creative ground from out of which both subjects and objects coemerge. Schelling's procedure was to distill human self-consciousness down to its minimal conceivable experiential potencies, and then to reconstruct the dynamic evolutionary path of Nature upward from these simple experiential beginnings through a series of organizational stages back to mind.[113] Whitehead's organic realism continued this project in the twentieth century greatly aided not only by developments in mathematical logic but by the quantum and relativistic revolutions as well as the progress of physiology and evolutionary theory.[114] He was inspired by these advances to make self-organization constitutive beyond just the biological scale spanning from bacteria to Gaia, *but at every scale* from electrons to galaxies and (as we wait for more empirical data concerning earliest phases of cosmogenesis from the James Webb space telescope) perhaps beyond.

Whitehead credits Kant with initiating a turning point in the progress of metaphysics through his realization that "our trust in science demands a metaphysic which equally supports [the] belief in the coherent rationality of things."[115] Kant attempted to arrive at rational coherence along the Cartesian route by making the measuring, calculating mind the sovereign legislator of a merely phenomenal and entirely mechanical Nature. While Descartes believed that his geometric idea of matter described real things, for Kant the Cartesian method was really only the way the *res cogitans* or thinking substance—itself dreamed up—began to *imagine* matter, reducing Nature to the pure "knowability" of res extensa.[116] Thus, for Kant, an apparently objective world was said to emerge from the constructive activity of the subject (including the activity of the imagination, though always held in check by fidelity to sensory data and to the categorical logic of the understanding). But as we've seen, when confronted with the purposive whole-to-part causality of organisms, which Kant felt was suggestive of analogies to Reason's self-legislating power, he retreated from scientific explanation and inspired a generation of Romantic poets by conjecturing a mysterious noumenal substratum unifying the rational subject with all things from behind the scenes.[117] Whitehead seeks a deeper coherence for scientific explanation across physical and biological domains by further generalizing Kant's transcendental mode of inquiry beyond just the special conditions of human cognition. Rather than a merely apparent world emergent from a transcendental subject, his philosophy of organism "stand[s] Kant on his head"[118] by inverting the process, such that occasional subjects are understood to emerge from the objective conditions of their environments: "the subject emerges from the world—a '*superject*' rather than a subject."[119] In other words, like every organic creature in Nature, we create and come to know ourselves as subject-superjects by actively organizing a real togetherness of formerly alien things.[120] Whitehead's organic realism thus allows for the reintegration of what Kant had torn asunder, "ending the divorce of science from the affirmations of our aesthetic and ethical experiences"[121] by planting the self-conscious power of Reason back into the cosmic context from which it originates:

> The process of cognition is merely one type of relationship between things which occurs in the general becoming of reality . . . [Kant] asked, How is cognizance possible? I suggest to you the more general question, How is any particular entity possible having regard for the relationships which it presupposes?[122]

Every entity that emerges in cosmic history places on the rest of the cosmos the obligation of being patient of it. For example, apart from the patience of the systematic coherence of the electromagnetic field required by its electrons, there can be no living organism even for a billion billionth of a second.[123] Every entity must be studied in the context of an environment providing some systematic character essential to the very nature of the entity in question. The idea of an organism independent of spatiotemporal relations, or independent of electromagnetic and gravitational fields, is meaningless. Any attempt to classify organisms by means of their supposedly isolated properties obscures the fluid togetherness of things, which is to say that the essence of life is inseparable from its cosmic history and habitat. In addition to their systematic or highly ordered character, there are also accidental aspects of any given environment, aspects which are determinate but which cannot be determined by consideration of the environment or the organism in isolation. A particular paramecium could be chemotaxing in a laboratory petri dish, digesting cellulose in a termite's gut, or stowing away on the next SpaceX launch capsule. It is here, in the tension between the givenness of a systematic environment and the constructive activity of individual organisms that Whitehead introduces his controversial panexperientialist postulate: "the togetherness of an entity with the accidental items of its determinate environment is what we mean by the *experience* of the entity," to which he adds the qualification that he does not mean cognitive (i.e., self-conscious) experience.[124] In other words, rather than assuming in advance that all subjectivity is sequestered inside human heads, or at most reserved for highly cephalized animals, Whitehead grants some modicum of experiential potency—some subjective *form*[125] and subjective *aim*[126]—to every self-organizing process in the universe. While the brain and outward facing sense organs are clearly particularly well-organized systems for sensing and responding to accidental features of their environment, it is becoming increasingly well-accepted among biologists that even the simplest forms of cellular life exhibit experiential agency (e.g., decision-making and learning capacities[127]). Some (e.g.,

Michael Levin[128]) are even prepared to entertain Whitehead's adventure in generalization by affirming that *every self-organizing system in Nature enjoys experience to some degree.*

Physical science requires that living organisms be interpreted according to the same principles applying elsewhere in the universe. But this relation of dependence runs in both directions, as it entails that interpretations of other parts of the physical universe must be brought into accord with what we know of living organisms—including both our own first person embodied phenomenological acquaintance with feelings and aims, as well as our third person physiological knowledge of self-organization and auto-/sympoiesis.[129] As Thompson puts it:

> Life is not physical in the standard materialist sense of purely external structure and function. Life realizes a kind of interiority, the interiority of selfhood and sensemaking. We accordingly need an expanded notion of the physical to account for the organism or living being.[130]

In the eyes of Whiteheadian philosopher Bruno Latour, further progress in science requires that researchers come to understand that the "matter" of materialism first conjured into existence by Descartes is really "the most idealist of the products of mind."[131] Even the simpler, supposedly inert processes of the inorganic world actively express themselves (e.g., the vibratory reiteration of electromagnetic waves, the supernova explosions of dying stars, etc.), even if their lack of mental originality severely restricts their expressions to what the causal past allows them to be.[132] The very concept of "force"—which has proven so irreplaceable to physicists in their study of everything from particles to galaxies—emerges from and gains its meaning only by continual reference to *experience*, to our *feelings* of attraction or repulsion, of being pushed or pulled by the insistent presence of others. As Schelling, speaking to Newtonian scientists, wrote in his *Ideas for a Philosophy of Nature* (1797):

> You can in no way make intelligible what a force might be independent of you. For force as such makes itself known only to your *feeling*. Yet feeling alone gives you no objective concepts. At the same time you make objective use of those forces. For you explain the movement of celestial bodies—*universal gravitation*—by forces of attraction and maintain that in this . . . you have [a physical ground of explanation for] these phenomena.[133]

In point of fact, experience can grant us no such physical ground of explanation, if by "physical" is meant the Cartesian idea of res extensa, i.e., a "barren extensive universe" of mute matter in motion set ontologically apart from the organismal experience of our living bodies.[134] As Whitehead put it, echoing Schelling:

> There is nothing in the real world which is merely an inert fact. Every reality is there for feeling: it promotes feeling; and it is felt. Also there is nothing which belongs merely to the privacy of feeling of one individual actuality. All origination is private. But what has been thus originated, publicly pervades the world.[135]

All our scientific knowledge of distant quasars and black holes hits its mark, not because a disembodied mind has correctly represented the formal essences of Nature, but because our organism (equipped with its worldwide network of geometrical notations, telescopes, satellites, computers, and rigorous peer reviewers) has succeeded in translating the *lines of force* at work in the wider universe into the *feelings of life* at work within ourselves. All our knowledge, no matter how abstract or formal, must make its final appeal in the courtroom of experience, since the court of Cartesian-Kantian Reason, having disavowed the facts of feeling involved in all its acts of knowing, has as a result been cut off from its only means of concrete relation to reality. If everything were actually submerged in abstract geometric spacetime, science could never follow the threads of experience, could never arrive at the immanence of a truly de-idealized and naturalistic conception of physicality.[136] Whitehead's organic realism reembodies and concretizes scientific knowledge by attributing unconscious vector feelings to energetic transmission throughout the actual world.[137]

In summary, the emergence and evolution of biological organisms is not a fugitive offshoot from otherwise mechanically determined laws of motion, but evidence of the excessive abstraction of these laws. While mechanical accounts can clearly function as powerful heuristics, closing the explanatory gap between matter, life, and mind requires overcoming the restrictions Kant placed on scientific knowledge by expanding classical conceptions of physical ontology to make room for the formal and final causes (feelings and aims, in Whitehead's terms) characteristic of self-organizing processes at whatever scale they occur, from electrons to E. coli and beyond.

Toward a General Theory of Evolution

Doing justice to the phenomenon of life as we observe it on this planet and as we directly experience it within ourselves requires acknowledging an aim, potency, or directionality intrinsic to prebiotic cosmogenesis that already suggested life even if it did not yet fully achieve it. Far from requiring some sort of extraphysical vital force to explain how life could emerge from dead matter, Whitehead insisted that "there be no jump whatever in principle as between living and inanimate."[138] Accomplishing this explanatory feat not only involves recognizing the self-organizing dynamics constitutive of enduring order across all scales of Nature, but also entails a generalization of evolutionary principles beyond the artificial limits of biology. Such a generalization is a great aid to origin-of-life research, as it closes any ontological gaps between physical vibration, combinatorial chemical selection, and biological function.

Another short historical interlude is in order, both to introduce Whitehead's important influences and to reconsider the road not traveled by twentieth-century biology, despite an appreciation for the limitations of reductionistic methods in the study of emergent complexity in Nature that was evident in the early part of the century. This interlude is also important to correct mistaken assumptions about the influence of evolutionary theory on Whitehead's thought.[139] Upon arriving at Harvard in 1924, Whitehead became acquainted with the work of several philosophically inclined biologists, including the physiologist Lawrence J. Henderson and the entomologist William Morton Wheeler. Whitehead mentions Henderson's work in a section titled "The Order of Nature" in *Process and Reality* as "fundamental for any discussion of this subject."[140] One of the works cited is *The Fitness of the Environment* (1913), the last sentences of which read:

> The properties of matter and the course of cosmic evolution are now seen to be intimately related to the structure of the living being and to its activities; they become, therefore, far more important in biology than has previously been suspected. For the whole evolutionary process, both cosmic and organic, is one, and the biologist may now rightly regard the Universe in its very essence as biocentric.[141]

As for Wheeler, Dennis Sölch has shown that his books on emergent evolution "show a more than general commonality with Whitehead's thought."[142] In his book, aptly titled *Emergent Evolution* (1928), Wheeler writes:

> If the naturalist is to accept both genetic continuity and novelty in evolution, the viable novelty at each emergence must be very small indeed. . . . Novelties such as life and mind, conceived in wholesale fashion, are of such magnitude that we can regard them only as representing the final accumulative stages of a very long series of minimal emergences.[143]

Wheeler builds on Whitehead's new organic conception of physics as laid out in *Science and the Modern World*, characterizing emergent organisms as "intensively manifold spatiotemporal events" rather than static externally related substances.[144] He further suggests that the organic view "resolve[s] the opposition between historicism and naturalism"[145] in biology, a bifurcation that survives today in the form of the two broad approaches to origin of life science, the historical and the universal.[146] Organic realism resolves the bifurcation precisely by interpreting the cosmic evolutionary process as the necessary and sufficient condition for the emergence of biological organisms, whether on Earth or elsewhere. Whitehead's general theory of evolution entails the extension of self-optimizing selection processes beyond just the biological domain. It may be that many possible chemical pathways to life exist, leaving room for historical contingency; but the general conditions of evolution remain the same everywhere, meaning that given the right environmental context, the emergence of biological organisms is all but inevitable. As is unpacked below, Whitehead's generalization of evolutionary theory entails that all order in Nature, including physical particle-fields and laws as much as biological species and instincts, is historically emergent.

Long before much empirical data was available to confirm it, Whitehead mused about earlier cosmic epochs in which "the dominant trend was the formation of protons, electrons, molecules, the stars."[147] How did an entity as stable as the electron arise? Primordial electrons emerging from a magnetic seed field were at first mere flashes in the pan, fitful stretches of enduring organization.[148] At this early stage of the evolutionary process, "reproduction" comes in the simple form of vibratory reiteration. Electrons and other "primate organisms" gradually gained the capacity for reiterative self-propagation by evolving a more favorable environment and by forming stabilizing sympoietic associations among themselves as well as with primates of different species.[149] "Evolution for countless ages stood still until something happened to produce an environment"[150]: evolution thus advanced as organisms produced and transformed their environments for their own purposes (and were themselves transformed in turn).[151] Whitehead offers the example of the associative formation of atomic elements, wherein a positive nucleus merges with negative electrons to produce a neutral atom:

> The neutral atom is thereby shielded from any electric field which would otherwise produce [destructive] changes in the space-time system of the atom.[152]

Just as the theories of niche construction[153] and symbiogenesis[154] have revealed in the biological context, the primate organisms engaged in physical evolution are not isolated systems adapting to a fixed environment, but coevolving agents actively constituting their environments. In Whitehead's terms, primate organisms are vibratory "stream-systems" inseparable from the underlying energy fields from which they emerge.[155] And since the environmental field is largely composed of other organisms, adaptation always means coevolution. Each seemingly independent organism is a distinctive enfoldment of its environment, co-implicated with the entire cosmic ecology of other organisms. In this sense, the problem of evolution is essentially that of procuring a favorable environment, such that organism itself becomes an environmental concept.[156]

> We habitually speak of stones, and planets, and animals as though each individual thing could exist, even for a passing moment, in separation from an environment which is in truth a necessary factor in its own nature. . . . The individual thing is necessarily a modification of its environment, and cannot be understood in disjunction."[157]

Whitehead adds that while thought about discrete "things" is to some degree inescapable, such abstraction always presupposes the systematic background of the environment required for such "things" to exist. Rather than things, Whitehead emphasizes the creative activity of organisms, which are not isolated substances but "[emerge] by reason of the niche which is there for [them] in the Universe" as a potential which they participate in actualizing.[158] The abstractions of science fall into error when they are haphazardly applied across inconsistent backgrounds: environmental context cannot be ignored if we hope to penetrate to the final nature of reality. For example, the internal organization of a living cell shelters greater complexity than the surrounding environment, allowing molecules to engage in functional activities that would be highly improbable outside such a context.

Whitehead's extension of organismic evolution to the physical world offers a coherent, bottom-up account of the emergent stability of organism-environments from particle-fields to biospheres, thus allowing science to reinterpret the order of Nature as a plastic hierarchy of historically achieved stabilities, harmoniously requiring each other.[159] Physical laws can then be understood as enduring habits left in the wake of organismal aims. Whitehead conceived of these stable habits as historically canalized layers of social order. Enveloping physical habits continue to genetically condition without unduly restraining the evolution of organic novelty.

> Thus a society is, for each of its members, an environment with some element of order in it, persisting by reason of the genetic relations between its own members.[160]

Whitehead thus invites us to draw an analogy between the self-organizing molecular societies constitutive of biological organisms and the many examples of physical self-organization across spatiotemporal scales.[161] More recently, the astronomer Eric Chaisson has developed a quantitative model for measuring the "energy rate density" of self-organizing systems across various scales of cosmic evolution, going so far as to attribute simple metabolism to galaxies[162] and primitive forms of variation, adaptation, selection, and even crude replication to stellar societies:

> Much as for biological evolution among living species, the process of selection, generally considered, also seems operative in the physical evolution of

> nonliving systems (although selective pressures for the latter are likely partly internal and autocatalytic). At least as regards energy flow and structural complexity while undergoing stellar evolution, stars have much in common with life—provided that stars are examined broadly, dynamically, and over extremely long periods of time.[163]

Chaisson's proposal for quantifying the evolution of complexity across scales via energy flows through open, nonequilibrium systems is a promising advance lending further credence to Whitehead's speculative generalization of evolution. Stars may not be "alive" in the sense relevant to origin of life scientists, but given the continuity of cosmic evolution they can be placed along the same ascending gradient of self-organization as biological organisms and complex human societies. In Whitehead's terms, such processes of social evolution—from electrons to stars to bacterial and human cultures—constitute "the orderliness whereby a cosmic epoch shelters in itself intensity of satisfaction."[164]

For Whitehead the "electromagnetic society" is among the most dominant environments forming the general background for our cosmic epoch.[165] A bewildering array of more specialized subordinate societies find themselves precipitating out of this wider electromagnetic society, subject to its influence but functioning as vehicles of further evolutionary novelty and more intense experiential satisfaction as they explore the edges of established niches.[166] Whitehead enumerates examples of such societies, beginning with waves of electromagnetic energy and increasing in complexity through atoms, molecules, stars, galaxies, and planets, to living cells, and societies of cells like plants and animals.[167] Each builds on the achievements of the last and is sheltered by the order reproduced by its environing habitats, even as variant organisms scour the instabilities along the edges of established environments in search of new potential niches.[168]

At the base of all this organized activity is the first stirring of life-like motion in the universe, which Whitehead calls "Rhythm":

> The Way of Rhythm pervades all life, and indeed all physical existence. This common principle of Rhythm is one of the reasons for believing that the root principles of life are, in some lowly form, exemplified in all types of physical existence.[169]

Rhythm begins in the vibratory patterns exhibited by physical fields. Gradually these vibrations develop enduring harmonies, experientially valued for their own sake, which then merge into ever higher achievements of harmony.[170] The origination and further evolution of biological cells on Earth is a result of a complexifying series of internal amplifications of the rhythmic energy pervading the cosmic environment.

In *The Function of Reason* (1929), Whitehead expands on his conception of cosmic evolution driven by primitive rhythms into the self-amplifying complexity of the biosphere by developing the general concept of the "cycle," which contemporary biochemists have scientifically elaborated into concepts including *thermodynamic work cycles*[171], *autocatalytic hypercycles*[172], *hypercyclic cooperative networks*[173], and *dehydration/rehydration cycles* (as highlighted in the Hot Spring Hypothesis introduced in Part 1 of this chapter). In Whitehead's words:

> In the Way of Rhythm a round of experiences, forming a determinate sequence of contrasts attainable within a definite method, are codified so that the end of one such cycle is the proper antecedent stage for the beginning of another such cycle. The cycle is such that its own completion provides the conditions for its own mere repetition.[174]

Despite his commitment to a methodologically reductionist interpretation, Pross's theory of "dynamic kinetic stability" in molecular replicating systems (specifically, in RNA oligonucleotides) provides an example of the empirical payoff of a general theory of evolution for origin of life research. Such systems behave analogously to biological systems in that (given the right environmental conditions) they undergo cyclical processes of replication, mutation, and selection. Some molecular systems compete for ecological niches, while still others form cooperative cross-catalytic networks that not only maintain holistic replicative capabilities but accelerate their complexification.[175] While Whitehead and Pross share the goal of integrating the special sciences of physics, chemistry, and biology into a universal science of evolution, Pross's allegiance to the reductionistic method leads him to attempt an explanation of the more complex feelings and aims of life and mind in terms of the simpler kinetic forces of "inanimate" molecules. But Pross's methodological reductionism evidently does not commit him to *metaphysical* reductionism. He refers to his dynamically stable replicators as "molecular fountains," acknowledging Woese's rejection of the machine metaphor in favor of a philosophy of the organism as "patterns in an energy flow."[176] That said, Pross pulls up short of affirming any role for natural purposes in evolution beyond the biological sphere. Purpose (or "teleonomy") only emerges for Pross when thermodynamically constrained chemical replicators cross the threshold into cellular animacy by gaining the metabolic means of freeing themselves, at least locally and temporarily, from entropy.[177]

Whitehead's organic realism is not premised upon ignorance of the laws of thermodynamics. While the rhythmic vibrations pervading the physical world appear highly repetitive—e.g., a proton will endure for countless billions of years—even in these simpler forms probabilistically mapped with mathematical precision, the reiteration of pattern is never perfect. In the cosmic background of our highly organized biosphere, ancient layers of social order give us every indication that they have entered a period of prolonged decay. As physicist Brian Greene puts it, "*the current order is a cosmological relic*."[178] Grandfather galaxies cease to spiral, slowing new star birth as senescent stars burn out, swallowing the planetary spheres probably essential to the sheltering of life. But in earlier cosmic epochs, while still in their creative prime, these astrophysical organisms were expressions of "primordial appetitions," mysterious impulses that under the right conditions cause energy to run upward.[179] The upward impulse evident in the evolutionary expansiveness of our cosmos is only privatively accounted for by a still materialistic science in terms of a postulated extraordinarily "negentropic" or "low entropy" cosmic origin state. While we cannot obtain direct knowledge of the selective striving at work in the early universe, Whitehead's organic realism implies that what appears entropic at present was agentic in the past. Atoms, stars, and galaxies are historically emergent, self-organizing, complex systems—*organisms* nested in a cosmic ecology of evolving processes. The primordial appetitions active in the early formation of the universe could account for the apparent "fine tuning" of physical laws, constants, and other conditions for the emergence of life.[180] It is not that quarks, protons, and electrons somehow conspired with the intention of producing biological cells, but rather that the coevolutionary dynamics of these and other primate organisms assured the universe would unfold in an internally related, organization-enhancing way.[181]

As living organisms, we feel selective impulses directly in the form of our own aims and emotions. The biosphere is full of evidence of activities directed by purposes. Indeed, as Kant so convincingly argued, purposiveness *defines* the biological domain. The reductive dismissal of final causes in Nature explains away biological phenomena, thus making the scientific search for an origin of life redundant. There simply is nothing to explain: the biosphere is just a rare chemical reaction at the extreme edge of physical improbability. Worse, since scientists, too, are organic beings, the elimination of aims as real features of the universe pulls the epistemic rug out from under the entire scientific enterprise. As Whitehead quipped, "Scientists animated by the purpose of proving they are purposeless constitute an interesting subject for study."[182] Purpose may be all but veiled from view in our observations of the present state of the large-scale physical universe, but the challenge of accounting for biological creativity, not to mention our own intellectual capacity to examine the evidence, suggests something more is at play.

While the originating energies that rhythmically coalesced into atoms, stars, and galaxies are now in gradual decay, the more complex progeny of these processes have leaned on the enhanced order and relative stability already achieved to continue their emergent evolution at other scales. As recurrent cycles canalize into enveloping social environments, subordinate societies issue in creative cycles more sensitive to novelty, producing variations which are further elaborated into cycles of cycles as locally evolved complexity ramps up.[183] As in Darwinian evolution, imperfect replication of inherited patterns functions creatively when new patterns are selectively amplified by changing environmental conditions. It follows that a living organism is only "the more highly organized and immediate part of the general environment."[184] But in addition to the chance process of selection described by Darwin, Whiteheadian selection includes the subjective aims of organismal agents seeking deeper intensities of experiential satisfaction. Thereby a gradual transition to more complex types of order is achieved.[185]

While simple enduring physical objects like rocks fail to secure the delicacy of social organization required to favor much intensity of experience, what they lose in subjectivity they gain in survivability. Highly complex animals, on the other hand, have an extremely refined physiological organization capable of sheltering a rich inner life and affording the realization of elaborate goals. But they are also in constant need of food and are susceptible in the extreme to small changes in their usual environment. Increase the temperature a few dozen degrees, alter the chemical composition of the atmosphere just slightly, or withhold water for several days and an animal body will cease to function. Even under the best of conditions, animals age and die. These facts lead Whitehead to deny Herbert Spencer's theory of evolutionary selection based on "survival of the fittest"[186]:

> The fallacy does not consist in believing that in the struggle for existence the fittest to survive eliminate the less fit. The fact is obvious and stares us in the face. The fallacy is the belief that fitness for survival is identical with the best exemplification of the Art of Life. In fact life itself is comparatively deficient in survival value. The art of persistence is to be dead. . . . The problem set by the doctrine of evolution is to explain how complex organisms with such deficient survival power ever evolved. They certainly did not appear because they were better at it than the rocks around them. It may be possible to explain 'the origin of species' by the doctrine of the struggle for existence among such organisms. But certainly this struggle throws no light whatever upon the emergence of such a general type of complex organism, with faint survival value.[187]

Thus the problem for evolution is the production of organic societies which achieve greater complexity without at the same time becoming overly specialized.[188] The selective factor in the evolution of complexity is not mere survivability, but the subjective aims of organisms themselves.[189] What we recognize as full-blown *mind* in ourselves and in the higher animals—the conceptual initiative not only to experience but to *think* about and so anticipate novel experience—is an evolutionarily enhanced expression of thoughtless aesthetic adjustments toward an ideal of harmony swarming already at the subatomic scale (e.g., in the form of the Principle of Least Action).[190] In Whitehead's view, the evolutionary telos of the universe is toward an increase in experiential satisfaction in service to the Art of Life.[191]

The truncated view of evolution implied by scientific materialism ultimately recognizes only the random variation and mindless selection of changes in the external relationships of unchanging material particles. In such a universe, there really is nothing to evolve, there is only purposeless change in the relative position of inert particles. In contrast, in Whitehead's general theory of evolution, everything is thrown into process, such that the only endurances are historically evolved organic structures of activity, each one a unit of emergent value. For Whitehead, the whole point of evolutionary theory is to shed scientific light on the process whereby the more complex organisms arise from simpler antecedents. To deny the evident tendency to increasing complexity and intensity of experience is to deny evolution.[192]

The Place of Feeling and Aim in Nature

Grasping Whitehead's panexperiential proposal requires the wholesale reformulation of classical conceptions of space, time, matter, and mind. The entrenched nature of these abstractions makes it difficult to follow the Whiteheadian proposal, as opponents are apt to import old meanings into Whitehead's novel phraseology. This subsection thus introduces and defines some of the key concepts of Whitehead's process-relational ontology in the hopes of avoiding misunderstanding.

In *Process and Reality*, Whitehead describes two species of process, *macrocosmic transition* and *microcosmic concrescence*, and articulates distinct methods for addressing each. These processes are the systole and diastole of creativity, the metaphysical heartbeat driving emergent evolution from bosons to bacteria to Bach. The macrocosmic process consists in "the transition from attained actuality to actuality in attainment," thus effecting the shift from the actualized past to the real potentiality of an unactualized future. The microcosmic process converts mere potentiality into actual attainment, generating novel experiential value from the materials of the perished past. The macrocosmic process of transition operates as an efficient cause providing the environmental conditions constraining what can be attained in the future, while the microcosmic process of concrescence expresses a final cause by means of which aims are achieved in the present: "The present is the immediacy of teleological process whereby reality becomes actual."[193]

The process of concrescence is ontologically primary, as it constitutes the concreteness of each present moment as it arises and perishes to contribute itself to cosmic evolution. Its phases of growth can be examined using the method of "genetic analysis," which in his Harvard lectures Whitehead also refers to as the "functional point of view" that is essential to higher-order sciences like biology and psychology.[194] This mode of analysis looks at what transpires within each concrescing actual occasion of experience, abstractly dividing occasions into their component "prehensions" (i.e., either physical feelings of perished occasions in their environing past or conceptual feelings of as yet unactualized possibilities), while never forgetting that "the whole function is one fact," with parts and whole in undivided relationship.[195] Genetic analysis provides a "view from within," an *endocosmic* account of the production of novel togetherness in each bud or drop of experience. Whitehead's reimagined account of subject-object relations helps convey the concept of concrescence:

> The word object means an entity which is a potentiality for being a component in feeling; and the word subject means the entity constituted by the process of feeling, and including this process. The feeler is the unity emergent from its own feelings; and feelings are the details of the process intermediary between this unity and its many data.[196]

The other method aimed at the examination of macrocosmic process Whitehead calls "coordinate" or "morphological" analysis. It is the means by which natural science grasps the metrical relations among entities of the already actualized external world. This latter mode of analysis focuses on extensive relations in spacetime, i.e., the "geometrical strains" binding our bodies together into the prehensive solidarity of the physical universe. This method of analysis tempts science to adopt a mechanistic framework in that it foregrounds objectively measurable motion while backgrounding the subjectively motivating feelings and functional aims responsible for driving the evolutionary expansiveness of the world-process. Coordinate analysis of the morphology of extension is Whitehead's way of describing what natural science is doing in all its measurements and modeling of mechanical parts, which are always measurements of *what has already become.* In contrast, the functional, genetic mode of analysis recontextualizes the objective beings of the past by involving them in the eternally recurrent process of concrescence: perished objective beings are prehensively unified into novel subjective becomings. Though it is often modeled as such by physicists, the cosmos is not simply a collection of inert simply located particles nor a continuum of deterministic forces: it is a community of creative participants hurdling themselves beyond the settled past. The universe expands like an embryo grows, as though through cellular division[197], with each concrescent experiential occasion inheriting from its predecessors (=efficient cause) and contributing the novelty it achieves to the process of evolutionary selection (=final cause). In this sense, "the concept of function demands a fusion of efficient and final causes."[198] In all its sophisticated modeling, science must remain cognizant of both finished facts *and* concrescent actualizations of novel facts. No scientific scheme can ever possess the complete set of facts for the simple reason that Nature itself is perpetually perishing, "an incompletion in process of production" that is forever passing beyond itself, caught in creative advance.[199] The incompleteness of Nature must be understood to supervene upon morphological extensiveness.[200] Both genetic/functional and coordinate/morphological modes of analysis contribute to the pursuit of a cosmological scheme adequate for explaining the emergence of life, but Whitehead privileges the former, insisting that "you cannot express a world whose most concrete aspect is function in [merely] morphological terms."[201]

Whitehead uses the phrase "intensity of satisfaction" to describe the subjective aims and feelings constitutive of each process of concrescence, with the relative intensity dependent upon the ordered complexity provided by the environing society from out of which the concrescence arises.[202] Concrescence is the creative process whereby "the many become one and are increased by one,"[203] or the process whereby the perished past is remembered and transitions into to the future with renewed evaluation accruing. The concrescent growth of such complex societies "exemplifies the general purpose pervading nature."[204] The past can pass into the future *only through the concrete duration of the present*: experience is always a function of what William James called the "specious present," which is not a solipsistically frozen instant cut off from its origins and destiny, but the living tension between an inherited past and an anticipated future. For Whitehead, our perception of metrical space and material bodies arises in an abstract present. He calls it "presentational immediacy." It is Descartes's res extensa, granting us clear and distinct perspectival perception through the eyes and other sense organs of the colored surfaces and other relevant features of our immediate surroundings. Though clear and distinct, the sense perceptions afforded by this mode are also prone to error. Time perception, on the other hand, is a function of what Whitehead calls "causal efficacy," which is the feeling of efficient-to-final causal transition from one occasion of experience to the next. We experience causal efficacy in the functioning of our living bodies, not so much via what we see *through* the eyes or hear *through* the ears, but in the blink of the eyes due to a bright flash, or the wince of the ears due to a loud noise, phenomena which clue us into the fact that these organs are themselves immersed within the causal nexus of the universe and not simply neutral windows through which a disembodied mind gazes at a separate reality. Causal efficacy grants us vague but insistent feelings of energetic vectors and physical purposes inherited from the enveloping environment. Concrete experience is a complex hybrid of the two modes of perception, including elements of presentational immediacy and causal efficacy: we distinguish them only for the purposes of intellectual analysis. The two modes are related not through deductive logic or deterministic causality but through analogy and symbolic imagination.[205] Forgetting this epistemic situation leads to the fallacy of misplaced concreteness.

In Whitehead's cosmological scheme, scientific objects abstracted from the "coordinate" or "morphological" domain are mathematizable and Turing-simulable. They belong to what Eastman calls the Boolean domain of geometric extension that can be measured and modeled in bits, rendered exhaustively in binary code as 1s and 0s or yes/no/and/or logical gates.[206] In the "genetic" or "functional" realm of concrescence, by contrast, the classical logical rules of non-contradiction and the excluded middle do not yet apply. The experiential satisfaction of concrescence cannot be measured or digitally modeled, since the prehensions composing it co-exist in organic harmony on their way to final satisfaction as a novel unit of emergent value contributing itself to universal evolution. Concrescence is the process of realization allowing past actualities to grow together and be brought into contrast with "pre-space *potentiae*"[207] hovering in a "poised realm"[208] between quantum coherence and classicality. It is the process whereby the given past environment constrains pure potentiality so as to find a probable pathway to the achievement of final satisfaction in a novel occasion of experience. Before a concrescence is completed, an occasion can be said to be composed of data including many prehensions of its past, some initially in conflict with one another. The process of concrescence integrates contradictory feelings into complex contrasts, sometimes (in more evolutionarily advanced organisms) drawing upon prehensions of novel possibilities not found in its past, transforming clashes into some modicum of aesthetic harmony. These conflicts in the initial data are why the principle of noncontradiction cannot be applied in the genetic analysis of concrescence, since a definite actuality has not yet been achieved. Only once a concrescing subject has achieved its aesthetic aim and perished into objectivity can standard logic and measurements in spacetime be applied.

While abiotic physics may appear reducible to symmetrical dyadic relations of mechanical cause and effect, this is true only of isolated ideal *models* of the physical world. In the real world, dyadic input-output relations inevitably involve *context* and are thus asymmetrical triadic relations in the sense articulated by Eastman.[209] In Whitehead's terms, each concrescent pulse in the vibratory reiteration characterizing the life-history of an electron or other primate organism alters the environmental situation from out of which it emerged. Each novel concrescence inherits and integrates the feelings of the past which it finds before adding itself, a new pulse of emotion and unit of value, to this past: "The many become one, and are increased by one." "It is unnecessary to labor the point," Whitehead says, "that in broad outline certain general states of nature recur, and that our very natures have adapted themselves to such repetitions." Because of Nature's tendency to conform to such repetitive habits, statistical "laws" can be formulated with a good deal of precision, particularly for describing the behavior of the prebiotic world. "But there is a complementary fact," Whitehead continues, "which is equally true and equally obvious: nothing ever really recurs in exact detail."[210] It follows that even the simplest of physical vibrations contributes itself to the irreversibly cumulative character of evolutionary time and thus to the creative advance of Nature.[211]

The intensity of satisfaction of a concrescent actual occasion of experience—its *feelings* and *aims*—cannot be spread out on a coordinate grid to be measured because neither are part of extended spacetime. The realm of intensity or of prehensive feeling is not in extended space and time and has no mass or momentum; rather, measurable spacetime relations are a secondary expression of or emergence from enduring networks (or "societies," in Whitehead's terms) of occasional feelings. Spacetime and other physical fields shaping our cosmic epoch are thus emergent out of the collective decisions of primate organisms, a result of what these experiential occasions have found satisfying, rather than preexistent containers of some kind to which organisms are passively subjected and forced to conform. The extent of conformity to a measurable and predictable spacetime manifold is a function of the stubborn habits accumulated by past environments inherited in the present. The cosmic habits forming the spacetime manifold, the electromagnetic and gravitational societies, etc., set the base notes for further evolution, leaving open the possibility that in the distant future our universe will continue expanding into more dimensions than what at present are detected.

Thus, the very gravitational gradient of spacetime and the energetic dynamics of light are functions of feeling, of *enjoyment*, such that the measurable shapes that the cosmos takes in the extensive domain must be understood to be precipitated products achieved by concrescent activities always *inwardly* underway and so never appearing in the measurable domain. In Whitehead's words: "the creature is extensive, but . . . its act of becoming is not extensive."[212] The concrescent activity of experiential realization does not appear outwardly because it is what does the peering. It is the subjective side of the equation governing cosmogenesis. When Whitehead refers to "intensity of satisfaction," what he means to say is that there is an aesthetic achievement whereby the perished objects of the past are brought together under contrast with one another, "prehensively unified." These processes of unification through contrast find exemplification at all scales in Nature, from the vibratory formation of the first atoms to the cycling of pre-living protocells as described in the Hot Spring Hypothesis (see part 1 and subsection e below). The many objects of the perished past grow together into a new unity, a new whole of some kind, which has an associated experiential vector launching it out of the past, through the present, and into the future. Organismic concrescence is telic, expressing an aim, generating a purposeful unfolding that feels its way toward greater intensity of experience, thus contributing greater organizational complexity to the societies in which it is situated. In thermodynamic terms, we could link these vectors to the many examples of emergent complexity wherein self-organizing systems "fall up" into local minima free energy states.[213] In the case of the Hot Spring Hypothesis, the exploitation of free energy minimization (what Kauffman calls "order for free" emergent throughout the evolutionary process that complements the work of natural selection[214]) is evident in the self-assembly of spherical liposomes and other complex polymers. Such self-assembly and self-organization would be an example of a society of molecules feeling their way to higher-order unities and functions. Talk of affective aims is not meant to conflict with the known laws of thermodynamics or chemistry, but rather is offered as a more generic metaphysical interpretation of increasingly well understood but nonmechanical processes of integrated cell and organismal physiology.[215] The creative advance into what from a mechanistic point of view appear to be more and more improbably complex patterns of activity at every scale of cosmic organization are not unexplained anomalies but expected consequences of Whitehead's organic cosmology. His Philosophy of Organism offers an account of the *why* as a speculative interpretation of the *how*—the latter being a matter of detailed scientific investigation of the efficient causes operating in the morphological domain. The role of metaphysics here is not to propose new testable hypotheses (scientists hardly need the help of philosophers for this), but to interpret the existing findings of the special sciences in light of more generic categories with the ultimate aim of integrating these findings within a broader cosmological context which includes our role as conscious inquirers. As Whitehead explains:

> The useful function of philosophy is to promote the most general systematization of civilized thought. There is a constant reaction between specialism and common sense. It is the part of the special sciences to modify common sense. Philosophy is the welding of imagination and common sense into a restraint upon specialists, and also into an enlargement of their imaginations. By providing the generic notions philosophy should make it easier to conceive the infinite variety of specific instances which rest unrealized in the womb of nature.[216]

The search for a scientific account of *how* life emerged takes on an entirely new philosophical tenor once the metaphysical significance of Whitehead's genetic analysis of concrescence is taken on board. Life is then understood as the realization of experiential purposes, a heightening of mental originality present in germ since the beginning of the universe, and not just an improbable rearrangement of pre-determined material parts.

While material bodies thought to be fully present at an instant remain confined to the abstract morphological domain, energetic activity implies a "time-depth" and thus comes closer to the functional descriptions required in biology.[217] Further, energy can, in Whitehead's terms, be translated into the intensive or functional domain as *experience* or *emotion*—not conscious deliberation, imagination, or self-reflective thought of the sort that human beings enjoy—but a more basic form of feeling, a "vector feeling," in Whitehead's terms. At the most primitive level of physical process, these vector feelings are just gravitational gradients, or the inheritance of the vibratory frequency of a helium atom from moment to moment of its life-history, the repetition and enjoyment of the feeling of that particular frequency, that particular mode of togetherness of protons, neutrons, and electrons. What begin as extremely simple and relatively habitual feeling vectors self-amplify as they cycle and grow together in increasingly organized cosmic environments. After billions of years of accreting value-experience through various stages of expansion and contraction, the already self-organizing order achieved by the physiochemical world sheltered the further cycling of molecular products in suitable planetary environments. No miracles or ontological ruptures are required for hot spring cycling of appropriate ingredients to give birth to auto-/sympoietic metabolism and reliable molecular replication. Life springs naturally from cosmic creativity.

Still, it might be asked: How can molecules within not yet quite living protocells be said to realize aims or achieve experiential satisfaction? These subjective qualities are essential elements in Whitehead's organic realism, which is why a translation is here being attempted. In the case of molecular societies, the embodiment of experiential satisfaction would be expressed by the harmonic resonance of each molecule's vibratory frequency. In Whitehead's process-relational scheme, particles are no longer to be conceived of as solid substances fully present at an instant and requiring nothing but themselves in order to exist. Instead, their being consists in their becoming, such that the classical notion of inert or "dead" matter is replaced with the notion of a nexus of cycling vibratory patterns. The vibratory resonances among the atoms composing a molecule signifies a primitive feeling of "sympathy," i.e., "feeling the feeling in another and feeling conformally with another."[218] What is the source of enduring order in the universe? Whitehead's wager is that material morphologies originally spring from vibratory resonance, and that evolution selects for more intense aesthetic harmonies. Biological organization dramatically enhances the selection process, allowing extremely complex molecular patterns to emerge which would be highly improbable in abiotic environments.[219] Whitehead elsewhere refers to the sympathetic resonance among molecular components as a form of "experiential togetherness," denying that anything can be said to be "together" without presupposing the experiential meaning.[220] Thus, what appear as wave-lengths and vibrations to spectrometers, for the molecular occasion in question is felt as "pulses of emotion."[221] In conscious human experience, emotion is always "interpreted, integrated, and transformed into higher categories of feeling" far more complex than primitive molecular emotions. "But even so," Whitehead continues, "the emotional appetitive elements in our conscious experience are those which most closely resemble the basic elements of all physical experience."[222]

To sum up, there is a creative lure toward more intense relationship operative at every scale of cosmic evolution, but which becomes richer as physical organization complexifies and new means of sheltering otherwise improbable energetic pathways are found (e.g., cooperative chemical networks and genetic memory, symbolic language, digital information technology, etc.). This tendency is an aim toward order that is driven or goaded by the lure of deeper aesthetic satisfaction. It is the great cosmic "counter-agency" to entropy that Whitehead discusses in *The Function of Reason*. His organic realism is an attempt to give physics *animacy* again, not despite the scientific facts, but because these facts themselves (e.g., the causal continuity between physics, chemistry, and biology) cry aloud for such a metaphysical interpretation.[223] This language is not meant to discount the details of physics in the realm of extension. It's just an attempt at reintegrating the far too long neglected domains of creativity and intensity back into the modern scientific understanding of the universe. Whitehead prioritized the realm of intensity as the concrete reality, with the realm of extension being its secondary expression. But you could not have one without the other, an inside without an outside. Both are required for the cosmic engine of evolution to creatively advance. Whitehead's organic realism is a protest against the sort of scientific materialism that tries to explain away subjective interiority by reduction to collisions of externally related objects. Science cannot rationally explain the shapes taken in space by living organisms without giving intensity its due. Intensity is *Natura naturans* (Nature naturing), and without this creative process springing from intensity of satisfaction, then the finished products of *Natura naturata* (Nature natured or already sprung) would not make any sense.

Interpreting the Hot Spring Hypothesis as a Special Application of Whiteheadian Concrescence

Whitehead's speculative method involves using specific notions developed for application to a restricted set of facts for the divination of more generic notions applying to all facts.[224] Following his lead, the environmental rhythms and causal dynamics driving chemical selection at play in the Hot Spring Hypothesis[225] can be interpreted as a special example of Whitehead's more generic account of the production of novelty via the process of concrescence. Following the empirically detailed account of the hot spring progenitor offered in part 1, this final subsection of part 2 draws preliminary connections to Whitehead's metaphysical scheme.

One empirical implication of Whitehead's metaphysical speculations is that the search for the origin of life is really the search for an *environment* suitable for the origin of life. Which early Earth environments provide the rhythms required for something like Pross's dynamic kinetic stability and cooperative cross-catalytic networks to emerge? The burgeoning Hot Spring Hypothesis makes plausible the idea that the wet-dry cycling of geyser-fed thermal ponds would have provided an ideal context for chemical replication-variation-selection cycles to catch fire.

In such a dynamic environment, the chemical selection process can get underway even before genetic templating has been invented. Empirical study has shown that the natural rhythms of the hot spring environment could easily shelter the complex chemistry necessary to initiate and sustain the thermodynamic instability, metabolic stability, and experiential intensity characteristic of even the simplest living organisms. Deamer and Damer's research into life's origins is making clear that "the transition [from physics and chemistry] to [biological] life is a continuum,"[226] supporting the process philosophical intuition that there is no ontological gap between physics and biology, nor between matter and mind, thus alleviating the need for the miraculous emergence of something from nothing.

The geological and astrophysical conditions must be just right for an "urable"[227] planet to ripen into life. Various reliable rhythms in the environment can facilitate the emergence of otherwise improbable kinetic sinks by sheltering and concentrating organic chemicals from a background of relative chaos. The progenitor environment hypothesis shows how sustained cyclical processes of chemical combinatorial selection can generate the complexity required for cellular emergence.

As Damer and Deamer describe it, continued cycling through drying and rewetting phases drives a series of natural experiments that undergo combinatorial selection through three distinct phases:

1. A multilamellar phase characterized by reduced water activity;
2. A hydrated phase in which encapsulated polymer mixtures bud off into protocell compartments whose stability is then tested in the open water; and
3. An intermediate hydrogel phase in which surviving protocells and concentrated solutes form a moist aggregate before drying and fusing into longer polymer chains.[228] As Damer and Deamer describe it:

> Each drying cycle . . . cause[s] lipid membranes of the vesicles to open, allowing polymers and nutrients to mix. On rewetting, the lipid membranes . . . reencapsulate different mixtures of polymers, each mixture representing a kind of natural experiment . . . [P]rotocells would then survive to pass on [their] polymer sets to the next generation, climbing an evolutionary ladder.[229]

If Damer and Deamer are correct, then it was not a single heroic autopoietic cell, but a heroic sympoietic community that gave birth to life. Damer's progenitor hypothesis describes the formation of networks of polymers at the edges of warm little ponds that would be drying out and refilling, drying out and refilling, with a crucial "gel-like" phase in between where complex cities of lipid sheaths would allow for the first gift economy on Earth to emerge as the protocellular survivors of the wet budding phase return to the community during the gel phase to share their battle tested chemical wares. Along the edges of these ponds, continued dehydration cycles would catalyze the synthesis of longer polymers, including nucleic acids and peptides, forming complex chains and molecular worms that gradually begin to manifest the first biological functions on the planet, and perhaps in the universe. In Whiteheadian terms, each wet-dry cycle initiates another throb of experience making its contribution to the evolutionary advance into novelty.

A fruitful application of Whitehead's metaphysically generic account of concrescence to the production of novelty achieved via the chemical selection of polymer-filled protocells is thus possible. Concrescence is Whitehead's description of the iterative, cumulative process whereby novelty emerges over the course of cosmic evolution. Each cycle of concrescence commences as objective beings in the past environment grow together into a new subjective becoming or occasion of experience. The objects of the perished past are felt by the concrescing occasion and brought into harmony with its subjective aim at intensity of satisfaction. Depending on the complexity of the surrounding environment, a given occasion's subjective aim achieves varying grades of experiential intensity. In this way, the formal and final causation operative in concrescence is constrained by the environing context providing the efficient causes out of which it arises.[230] Once subjective satisfaction has been achieved, the concrescent entity perishes into a "superjective" phase, thereby adding whatever novelty it has achieved back to the environment for subsequent cycles of concrescence to inherit.[231] Shifting from Whitehead's generic metaphysical account back to the special case of the progenitor scenario, we can see how as a new wet phase commences, each liposome encased protocell that buds off from the layers formed during the dry phase amounts to a chemical experiment, incorporating whatever polymer chains were achieved in the last cycle for further testing. These buds are like Whitehead's "drops of experience, complex and interdependent,"[232] with each cycle of concrescence achieving greater chemical stability and driving the accumulation of further functionality. While Whitehead insists that some degree of "subjective aim" is expressed in these processes, it is important to note that in the case of protocells or even the first biological cells, the experiential purposes in question are far from intellectual or conscious, but merely "lures for feeling" provocative of the synthesis required for emergent novelty to be generated.[233] While *life* is a matter of degree, Whitehead hints that "an organism [can be considered in the usual sense to be] 'alive' when in some measure its reactions are inexplicable by any tradition of pure physical inheritance."[234] In this sense, while the initial emergence of biological organisms was entirely dependent on suitable environmental conditions, "[biological] life is a bid for freedom"[235] achieving a new level of self-creation relative to its physiochemical components.

Damer likes to say that the universe before the emergence of the biosphere—the atomic, astrophysical, galactic environments—gets a "D" for creativity, in the sense that at these scales relatively few stable forms of organization were found, and for billions of years they have been fixed in place and at present find themselves running down or wasting away. No further evolution can transpire. The abiotic cosmos is thus *ergodic*. It wasn't until the biological realm invented template copying and self-repairing complex adaptive and anticipatory cellular organization that the creativity of the cosmos ratcheted up again to find new, more complex energy cycles to flow into. While it may be true that that the universe before biological life emerged gets a "D" in creativity, the important point here is that it is not an "F." It did just *enough* to pass, expressing enough creativity to achieve the atomic and stellar processes necessary for generating the heavier elements that would later become essential for cellular life. Prebiotic cosmogenesis unfolded at a much slower rate than biological evolution can, with its more potent novelty producing engines, but at least some degree of aim and affective satisfaction was present from the beginning, otherwise atoms, stars, and galaxies could never have formed. These sidereal processes are tremendous organizational achievements in their own right, considering the relative chaos from out of which they came. "Such a change of thought," in Whitehead's terms, "is the shift from materialism to organism, as the basic idea of physical science."[236]

CONCLUSION

As was affirmed in our Introduction, natural science must be granted autonomy to pursue hypotheses concerning the modes of operation of the empirical world independent of the speculative postulates of philosophers and the sacred doctrines of theologians. And yet, in a time of paradigmatic upheaval, if the special sciences are to avoid degenerating into a medley of ad hoc hypotheses then they must themselves become philosophical by engaging in a thorough examination of their metaphysical presuppositions.[237] Without a ground-up re-imagination of entrenched materialistic assumptions, progress on questions like the origin of life (not to mention the origin of matter and mind) will remain stunted. Further, amid an intensifying planetary emergency, philosophy and religion have an essential role to play in the translation of scientific findings into a meaningful and motivating worldview for an increasingly precarious civilization. In large part due to the truly unprecedented scope of our scientific knowledge and the technological power it affords, our species now finds itself on the verge of initiating a major evolutionary transition. The Anthropocene[238] is not the work of a god, but merely that of a conscious animal. Indeed, as we have seen, perhaps humanity's vast endogenous and even vaster technologically augmented information processing capacities are dramatic amplifications of the social networks established by our progenitor ancestors. Whether the present anthropogenic metamorphosis in the Gaian system brings near-term extinction or creative advance for human beings remains to be seen.[239] The coauthors of this paper hope that their transdisciplinary collaboration has contributed some theoretical insight into the origin of life on Earth. As for the destiny of human life on this planet, there remains an urgent practical need to integrate science with philosophy and religion in pursuit of a viable pathway for our species through the great transformation in Earth history that is already underway.

NOTES

1. Darwin, Letter to Joseph Dalton Hooker (March 1863) in Peretó et al., *Charles Darwin and the Origin of Life*, 395. As Peretó et al. point out, Darwin's views on the scientific tractability of abiogenesis changed throughout the course of his life. Less than a decade later, he composed a letter to the same close friend speculating about a "warm little pond" wherein complexifying chemistry could have generated life.
2. Whitehead, *Science and the Modern World*, 17, 105.
3. Woese, "A New Biology for a New Century," 176, 179–80.
4. See Scharf et al., "A Strategy for Origins of Life Research," 1035. See also Pross, "Toward a general theory of evolution,": "A coherent strategy for the synthesis of a

living system is not possible if one does not know what life is, and one cannot know what life is if one does not understand the principles governing its emergence" (1).

5. See Smolin, *The Trouble with Physics*, 310.

6. De Haro, "Science and Philosophy," 25, 310.

7. Whitehead, "First Lecture, 1924" in *Whitehead at Harvard*, 44; Whitehead, *Process and Reality*, 3.

8. As recorded by W. P. Bell in *Harvard Lectures of A.N. Whitehead*, 3.

9. Whitehead, "First Lecture," 43.

10. Whitehead, *Process and Reality*, 4.

11. Ibid., 15.

12. Whitehead, *Science and the Modern World*, 184.

13. See De Haro, "Science and Philosophy," 3.

14. See Kuhn, *The Structure of Scientific Revolutions*.

15. De Haro, "Science and Philosophy," 308. See also Bitbol, "Schrödinger and Indian Philosophy."

16. See Russell, "The 'Water Problem,'" 429.

17. See Luo et al., "Rapid oxygenation of Earth's atmosphere."

18. See Deamer, *First Life*, 100.

19. See Pearce et al., "Constraining the Time Interval for the Origin of Life on Earth," 343–64.

20. Darwin, Letter to Joseph Dalton Hooker, February 1, 1871. See Introduction note 1.

21. Damer, "The Evogrid."

22. Damer and Deamer, "Coupled phases and combinatorial selection in fluctuating hydrothermal pools."

23. Damer and Deamer, "The hot spring hypothesis for an origin of life."

24. Woese and Fox, "The concept of cellular evolution," 3.

25. See Wade, "Meet Luca, the Ancestor of All Living Things."

26. Marshall, "The Water Paradox and the Origin of Life."

27. Merriam-Webster, https://www.merriam-webster.com/dictionary/progenitor. Accessed March 23, 2022.

28. Damer, "A Field Trip to the Archean."

29. Deamer, *First Life*, 100.

30. See Marshall, *The Genesis Quest*.

31. Eschenmoser, "The Search for the Chemistry of Life's Origin."

32. Cornell et al., "Prebiotic amino acids bind to and stabilize prebiotic fatty acid membranes."

33. Deamer, "Boundary structures are formed by organic components."

34. Pearce et al., "CRAHCN-O."

35. See Kranendonk et al., *Earth's Oldest Rocks*.

36. See for example Pavlopoulos et al., "Using graph theory to analyze biological networks."

37. See Odling-Smee et al., *Niche Construction*. See also Damer, "The Hot Spring Hypothesis for the Origin of Life."

38. See Varela et al., "Autopoiesis."

39. Darwin, “Letter to Joseph Dalton Hooker.”

40. Whitehead, *Process and Reality*, 309.

41. Ibid., 102.

42. See Formosinho et al., “Environmentality in biomedicine.” The term “environmentality” has recently been used in several contexts, including environmental governance and ecocriticism. In this case the reference connects with new research by Formosinho, et al., which though focused on the biomedical implications of the microbiome and limited to epistemic or operational use rather than ontology is none the less of great relevance to Whitehead’s environmental conception of cosmic evolution. Formosinho et al. draw on work by Whiteheadian philosopher of science Isabelle Stengers to define environmentality as “the locally described state or quality of being a causal context for something else . . . over time and across scales, from micro to macro” (149, 152).

43. Gare, “Approaches to the Question,” 55.

44. As the theoretical biologist Robert Rosen put it. See “Relational Biology and the Origin of Life,” 421.

45. Whitehead, *The Concept of Nature*, 32–33.

46. Monod, *Chance and Necessity*, 146.

47. Thompson, *Mind in Life*, 132.

48. See Duve, “Life as a cosmic imperative,” 622.

49. See Anderson, “More Is Different”: “The ability to reduce everything to simple fundamental laws does not imply the ability to start from those laws and reconstruct the universe. . . . Instead, at each level of complexity, entirely new properties appear, and the understanding of the new behaviors requires research which . . . is as fundamental in its nature as any other” (339).

50. See also Van Dijk, “Process Physics.” In addition to reinterpreting physical laws as cosmic habits, Van Dijk argues that “these ‘laws’ should better be seen as *measurement phenomenologies*—data-compliant algorithms capable of closely following the changing states of measurement instruments, *not* the changes in nature itself” (28).

51. Whitehead, *Process and Reality*, 237. Thanks is due to Timothy Jackson for this helpful terminological suggestion (personal communication, May 2, 2022).

52. Woese, “A New Biology for a New Century,” 185.

53. See Gare, “Approaches to the Question,” 59.

54. Whitehead, *Modes of Thought*, 156.

55. Ibid., 154.

56. Ibid., 156.

57. The Hot Spring Hypothesis may help overcome the division between proponents of “genetics-first” and “metabolism-first” scenarios for life’s origin (see Eschenmoser, “The Search for the Chemistry of Life’s Origin”), as the “progenitor” environment discussed in part 1 leaves ample room for nucleic and amino acids to complexify in parallel.

58. Whitehead, *Science and the Modern World*, 18. The affirmation of a cosmic evolutionary continuum aligns with astronomer Eric Chaisson’s argument that a philosophically informed “inclusive scientific worldview can rationally explain the

origin, evolution, and complexity of all structured systems in the known Universe . . . [as] more than a subjective, qualitative narrative. See Chaisson, "Energy Rate Density," 60.

59. See physicist and philosopher Timothy Eastman's *Untying the Gordian Knot*. Eastman criticizes materialism for upholding the dogma of "actualism" by denying any role for "*potentiae*" in Nature, a denial that he argues makes the findings of quantum physics (not to mention human agency) unintelligible. See also Kastner et al., "Taking Heisenberg's Potentia Seriously" wherein they define "actualism" as "the doctrine that only actual things exist" despite the fact that a coherent quantum ontology requires making reference to potentialities (5–6).

60. See Dupré and Nicholson, *Everything Flows*: "Although physics emancipated itself from the mechanicist worldview at the turn of the twentieth century (which is partly what led Whitehead to embrace process metaphysics . . .), mechanicism never really lost its grip on biology" (28).

61. Whitehead, *Science in the Modern World*, 97. While classical mechanistic models may still offer predictive value when deployed within narrowly defined conditions, they have been obsolete as ontology for over a century.

62. In *Philosophy of Physics*, philosopher of science Tim Maudlin puts it starkly: "No consensus at all exists among physicists about how to understand quantum theory. . . . Instead, there is raging controversy" (2). He adds that there really cannot even be said to be a "quantum theory" at all; rather, there is "a recipe or prescription . . . for making predictions about data" (5). See also the quantum physicist Carlo Rovelli, who goes so far as to say that "a lot of current work in theoretical physics is a map without a territory." See "Carlo Rovelli on Consciousness," YouTube, timestamp: 37:55.

63. See for example Stein, "Toward a Process Philosophy of Chemistry": "molecules . . . are not static, unchanging substances, but rather are dynamic entities that negotiate complex energy landscapes. Process thought, and its powerful metaphor of molecule-as-ecosystem, gives us the language to express . . . molecular change" (19).

64. See Segall, *Physics of the World-Soul* for a detailed explication of Whitehead's contributions to contemporary new paradigm natural sciences, including quantum, relativity, evolutionary, and complexity theories.

65. Whitehead, *Harvard Lectures of ANW*, 12.

66. Ibid., 141, 156. See also Whitehead, *Process and Reality*, Part II, Chapter III: "Organism and Environment."

67. As Jesse Bettinger and Timothy Eastman put it, physics and biology thus become "two pillars of an even more general discipline: complex systems. See "Foundations of Anticipatory Logic in Biology and Physics," 118.

68. Whitehead, *Process and Reality*, 102.

69. Pross, "Toward a general theory of evolution," 2. While Pross notes the remarkable prescience of Whitehead's ontological shift from substance to process for bridging the gap between chemistry and biology (12), it remains to be seen whether he is willing to replace classical conceptions of "matter" with Whitehead's generalized concept of organism.

70. Whitehead, *Science and the Modern World*, 129.

71. Dupré and Nicholson, *Everything Flows*: "The complex web of causal dependencies between the various levels means that we cannot fully specify the nature of an entity merely by listing the properties of its constituents and their spatial relations. It also means that we cannot pick out any level in the hierarchy as ontologically or causally primary" (27).

72. See Mariusz Tabaczek's article "The Metaphysics of Downward Causation" for a helpful conceptual history of emergence and an insightful analysis of downward causation in terms of formal causation. Tabaczek ultimately defends a substantialist Aristotelean formal cause, distancing himself from processual accounts because of their panpsychist implications. For further helpful discussions of the place of formal and final causality in the biological world, see: Thompson's *Mind in Life*, 129–33; Juarrero's *Dynamics in Action*, 46–48, 126–28; and Deacon, *Incomplete Nature*.

73. Dupré and Nicholson, *Everything Flows*, 29.

74. Bertalanffy, *Modern Theories of Development*, 38.

75. Ibid., 37.

76. See Shields, *Aristotle*, 84.

77. In *Staying with the Trouble*, Whiteheadian philosopher of science and technology Donna Haraway defines "sympoiesis": "Sympoiesis is a simple word; it means 'making-with.' Nothing makes itself; nothing is really autopoietic or self-organizing. . . . Sympoiesis is a word proper to complex, dynamic, responsive, situated, historical systems. It is a word for worlding-with, in company. Sympoiesis enfolds autopoiesis and generatively unfurls and extends it" (58). See also Meincke, "Autopoiesis" for an argument that the autopoietic view of life implies a process ontology.

78. Rosen, *Essays on Life Itself*, 54, 136. See also Gare, "Approaches to the Question,": 61–67: "As Kepler and Newton freed science from the assumption of earlier astronomers that all planetary motion is in circles, Rosen has freed science, and biology in particular, from assumptions about mathematical modeling which effectively made life itself unintelligible" (66).

79. Dupré and Nicholson, *Everything Flows*, 30.

80. Ibid., 17.

81. See Walker and Davies, "The algorithmic origins of life." Walker and Davies understand genetic information and functionality is context dependent, distributed throughout the environment and emergent from the whole history of the organism, rather than being "a local property of a molecule" (2). Their critique of digital-first conceptions of biological information, which neglect the analog format of the proteome, is well taken (3). Yet by defining life in terms of "information control" ("information manipulates the matter it is instantiated in" [6]) and insisting its emergence represents a fundamental shift in the causal structure of Nature, they beg many metaphysical questions about physical ontology, which they hint at but do not directly address at least in the cited paper. It is not at all clear that the "top-down" informational causality they point to is compatible with materialism.

82. Woese, "A New Biology for a New Century," 176.

83. Dupré and Nicholson, *Everything Flows*, 17.

84. Woese, "A New Biology for a New Century," 176. Woese adds that while the complex process of gene expression known as translation "is describable in

reductionist terms," it is nonetheless "neither predictable nor fully explainable therein" (180).

85. For a comprehensive critique of the literalization of computational metaphors in molecular biology, see Evan Thompson, *Mind in Life*, 179–87: "whereas software and hardware are independent in a computer—the hardware has to be there before the program can be run, and hardware and software do not produce each other autopoietically—DNA replication and gene activation are entirely dependent on the autopoiesis of the cell. They contribute enormously to this process, but they also owe their existence to it" (180).

86. For a helpful discussion of the dissolution of the classical conception of the gene, see Godfrey-Smith, *Philosophy of Biology*, 81–85. See also Bellazzi, "The Emergence of the Postgenomic Gene."

87. For another thorough treatment of the inadequacy of the computer metaphor, and the role of environmental context (including intracellular context) in specifying phenotypic development, see Lewontin, *The Triple Helix*, 17–18.

88. Carl Woese, "A New Biology for a New Century," 173, 185.

89. Whitehead, *Process and Reality*, 84.

90. Kant, *Critique of Judgment*, Sec. 75.

91. Godfrey-Smith, *The Philosophy of Biology*, 60.

92. See Thompson, *Mind in Life*, 131.

93. As Steven J. Dick has pointed out, Kant's relevance to astrobiology was established well before his development of transcendental philosophy in the final two decades of his life. His first major work, "Universal Natural History and Theory of the Heavens" (1755) explored the possibility of alien life. Dick wonders if this imaginative exercise "affected Kant's unusually generalized yet deeply critical philosophy." See Dick, "Critical Issues in the History, Philosophy, and Sociology of Astrobiology," 917–18.

94. Kant, *Critique of Pure Reason*, 110/Bxvi.

95. See Hume, *An Enquiry Concerning Human Understanding*, 843, 856, 899.

96. Kant, *Critique of Pure Reason*, 109/Bxiii.

97. Kant, *Opus Postumum*, 240.

98. See Auxier and Herstein, *The Quantum of Explanation*, 199, 233.

99. See the CalTech Archives (February 1988): https://archives.caltech.edu/pictures/1.10-29.jpg

100. Kant, *Critique of Pure Reason*, 236/A114.

101. Kant, *Kant's Prolegomena to any Future Metaphysics*, 36/Sec. 10.

102. Kant's transcendental framework could thus be understood to provide a rational answer to the question raised by Eugene Wigner in his famous paper "The Unreasonable Effectiveness of Mathematics in the Natural Sciences."

103. Kant, *Critique of Pure Reason*, 354/B295.

104. Whitehead, "First Lecture," 46.

105. For more on this connection, see Bitbol and Osnaghi, "Bohr's Complementarity and Kant's Epistemology." See also Bell and Iyengar, "Whitehead and Kant at Copenhagen" in *Whitehead at Harvard.*

106. Petersen, "The Philosophy of Niels Bohr," 12.

107. While Newton conceived of his laws of motion in terms of a direct exchange of forces between impenetrable bodies (i.e., efficient causes), his law of gravitation could not be so construed, as it implied action at a distance with no known medium to propagate efficient causal forces. This led Newton to theologically speculate that the space between masses is divinely occupied, i.e., is the *sensorium of God.* See Henry "Newton, the sensorium of God, and the cause of gravity."

108. Kant, *Critique of Judgment*, 167/Sec.67. Whitehead would add that there is nothing in the materialistic doctrine of evolution that might explain how a man like Newton could come to be. See Price, *Dialogues of Alfred North Whitehead*, 283.

109. Kant, *Critique of Judgment*, 162–63/Sec. 64–65.

110. Evan Thompson argues that Kant's conception of self-organization is a precursor to the contemporary concept of "autopoiesis." See Thompson, *Mind in Life*, 138.

111. Thompson, *Mind in Life*: "Our conception of matter as essentially equivalent to energy and as having the potential for self-organization at numerous spatiotemporal scales is far from the classical Newtonian worldview" (140). See also Swimme and Evelyn Tucker, Ch. 10: "Rethinking Matter and Time" in *Journey of the Universe*.

112. These treatises include: in 1797, *Ideas for a Philosophy of Nature*; in 1798, *On the world soul, a hypothesis of the higher physics for the clarification of universal organicity*, selections of which have been translated by Grant in *Collapse*, 58–95; and in 1799, *First Outline of a System of the Philosophy of Nature*. See also Gare, "Approaches to the Question," 68–72.

113. See Esposito, *Schelling's Idealism and Philosophy of Nature*, 80ff. See also Gare, "Approaches to the Question," 68–69.

114. In *Science and the Modern World*, Whitehead mistakenly suggests that "the whole of the German idealistic movement [remained] out of touch with its contemporary science so far as reciprocal modification of concepts is concerned" (139). Schelling in particular had an important influence on the leading-edge scientists of his day. See Esposito, *Schelling's Idealism and Philosophy of Nature*, 125ff and Gare, "Approaches to the Question, 70. In *The Romantic Conception of Life*, Robert Richards characterizes Schelling's natural philosophical works as "[groaning] with the weight of citations of the most recent, up-to-date experimental work in the sciences" (128). What can be said is that Whitehead philosophically amplified twentieth century advances in physics and biology to breathe new life into Schelling's non-mechanistic naturalism, which had languished for over a century following the rise of materialistic positivism. For more on the Schellingian resonances in Whitehead's philosophy, see Maciel, "Outlines of a Speculative Cosmology," 100–2.

115. Whitehead, "First Lecture," 46.

116. See Latour, *An Inquiry Into Modes of Existence*, 110, 112, 121.

117. Kant, *Critique of Judgment*, 188–93/Sec. 77.

118. As recorded by Bell in *Harvard Lectures of A.N. Whitehead*, 113.

119. Whitehead, *Process and Reality*, 88.

120. Whitehead, *Science and the Modern World*, 151.

121. Ibid., 157.

122. Whitehead, "First Lecture," 46–47.

123. Ibid., 49.

124. Ibid., 50 (italics added). See also *Process and Reality*: "The principle that I am adopting is that consciousness presupposes experience, and not experience consciousness. It is a special element in the subjective forms of some feelings. Thus an actual entity may, or may not, be conscious of some part of its experience" (53).

125. In *Process and Reality*, Whitehead defines "subjective forms" as "private matters of fact" (22), including "emotions, valuations, purposes, adversions, aversions, consciousness, etc." (24). "[A]n actual entity, on its subjective side, is nothing else than what the universe is for it, including its own reactions. The reactions are the subjective forms of the feelings" (154).

126. "In its self-creation the actual entity is guided by its ideal of itself as individual satisfaction and as transcendent creator. The enjoyment of this ideal is the 'subjective aim,' by reason of which the actual entity is a determinate process. This subjective aim is not primarily intellectual; it is the lure for feeling. This lure for feeling is the germ of mind." Whitehead, *Process and Reality*, 85.

127. See Offord, "Can Single Cells Learn?"

128. While Whitehead uses the term "cognition" as a near synonym for human thought, Levin defines it more broadly in cybernetic informational feedback terms as part of an empirically grounded effort to establish an electrically mediated experiential continuum from chemical networks to human minds. Levin suggests his "scale-free" conception of cognitive agency "can perhaps be seen as a form of panpsychism." See Levin, "The Computational Boundary of a 'Self,'" 18.

129. Whitehead, *Process and Reality*, 119. As Haraway notes, in the living world, nothing really makes itself: sympoietic relationships enfold and extend autopoiesis. See Haraway, *Staying With the Trouble*, 58.

130. Thompson, *Mind in Life*, 238.

131. Latour, *An Inquiry Into Modes of Existence*, 106.

132. Whitehead, *Process and Reality*, 177.

133. Schelling, *Ideas for a Philosophy of Nature*, 18.

134. Whitehead, *Process and Reality*, 122.

135. Ibid., 310.

136. Latour, *Modes of Existence*, 106.

137. Whitehead, *Process and Reality*, 177.

138. As recorded by Bell in *Harvard Lectures of A. N. Whitehead*, 172.

139. See Lucas, "Evolutionist Theories and Whitehead's Philosophy" wherein Lucas argues that evolution played no significant role in Whitehead's cosmology. Newly published lecture notes from Whitehead's students at Harvard have shown, on the contrary, that Whitehead had much to say about Darwin and evolutionary theory more generally. Lucas has since recanted his earlier position, which was based on a partial reading of Whitehead's published works. See Lucas, "Uncovering a 'New' Whitehead" in *Whitehead at Harvard*, 330.

140. *Process and Reality*, 89.

141. Henderson, *The Fitness of the Environment*, 312.

142. Sölch, "Wheeler and Whitehead," 491.

143. Wheeler, *Emergent Evolution*,18.

144. Ibid., 17.

145. Wheeler, "Present Tendencies in Biological Theory," 108.

146. See Scharf, et al., "A Strategy for Origins of Life Research." Scharf et al. add a third approach, the synthetic, but for our purposes it can be folded into the other two.

147. Whitehead, *The Function of Reason*, 24.

148. As recorded by Bell, *Harvard Lectures of A. N. Whitehead*, 135. See also more recent research on this early phase in cosmic evolution by Subramanian, "The origin, evolution and signatures of primordial magnetic fields" and Ichiki et al., "Cosmological magnetic field."

149. Whitehead, *Science and the Modern World*, 112, 133.

150. As recorded by Bell, *Harvard Lectures of A. N. Whitehead*, 141.

151. Whitehead, *The Function of Reason*, 7. See also Levins and Lewontin, *The Dialectical Biologist*: "It is impossible to avoid the conclusion that organisms construct every aspect of their environment themselves. They are not passive objects of external forces, but the creators and modulators of these forces" (104).

152. Whitehead, *Science and the Modern World*, 134.

153. See Odling-Smee et al., *Niche Construction*.

154. See Aanen and Eggleton, "Symbiogenesis."

155. Whitehead, *Science and the Modern World*, 37.

156. As recorded by Bell, *Harvard Lectures of A. N. Whitehead*, 155.

157. Whitehead, *Adventures of Ideas*, 154.

158. As recorded by Bell, *Harvard Lectures of A. N. Whitehead*, 356.

159. Whitehead, *Process and Reality*, 93.

160. Ibid., 90. See also 92: "This doctrine, that order is a social product, appears in modern science as the statistical theory of the laws of nature, and in the emphasis on genetic relation."

161. Whitehead, *The Function of Reason*, 24.

162. Chaisson, "Energy Rate Density," 59.

163. Ibid., 37–38.

164. Whitehead, *Process and Reality*, 119.

165. Ibid., 98.

166. As recorded by Bell, *Harvard Lectures of A. N. Whitehead*, 141.

167. Whitehead, *Process and Reality*, 98.

168. See Allan, "Diagrams and Myths" in *Whitehead at Harvard*: "The relationship of two different kinds of stable environment with their differing kinds of stable entities creates a minor instability along their adjacent edges, a niche chipped away from the continuity of their shared boundary, a slight breach creating an openness that is neither one environment nor the other. A crack of some sort in the established order of things can offer an opportunity to escape the repression of unsuitable possibilities, to be free to achieve something different, to create a novel variant that until then had been impossible or even inconceivable" (294).

169. *Function of Reason*, 21.

170. Whitehead, *Science and the Modern World*, 96.

171. Kauffman, *A World Beyond Physics*, xi.

172. Juerraro, *Dynamics in Action*, 121.

173. Pross, "Toward a general theory of evolution," 3.

174. Whitehead, *Function of Reason*, 21.
175. Pross, "Toward a general theory of evolution," 3–4.
176. Ibid., 6, 8.
177. Ibid., 11.
178. Greene, *The Fabric of the Cosmos*, 174.
179. *Whitehead, Function of Reason*, 24; *Process and Reality*, 105.
180. See Simon, "Fine-Tuning."
181. See Grandpierre, "Extending Whiteheadian Organic Cosmology," 70ff.
182. Whitehead, *Function of Reason*, 16.
183. Ibid., 22.
184. Whitehead, *Process and Reality*, 119.
185. Ibid., 91.
186. See Spencer, *Principles of Biology* (1864).
187. Whitehead, *The Function of Reason*, 4.
188. Whitehead, *Process and Reality*, 100.
189. Whitehead's approach to evolutionary selection can thus be fruitfully compared with James Mark Baldwin's theory of "Organic Selection." See Scarfe, "James Mark Baldwin with Alfred North Whitehead on Organic Selectivity."
190. Whitehead, *Process and Reality*, 102.
191. Whitehead, *The Function of Reason*, 8.
192. Whitehead, *Science and the Modern World*, 109–10. See also Pross, "Toward a general theory of evolution": "Within the biological world there is no doubting that a definite process of complexification over the extended evolutionary time frame has taken place . . . That evolutionary drive toward greater complexity cannot be denied" (3).
193. Whitehead, *Process and Reality*, 214–15.
194. As recorded by W. E. Hocking, *The Harvard Lectures of A. N. Whitehead*, 77.
195. Ibid.
196. Whitehead, *Process and Reality*, 88.
197. See Van Dijk, "Process Physics," 44–45.
198. As recorded by Bell, *The Harvard Lectures of A. N. Whitehead*, 56.
199. Whitehead, *Process and Reality*, 215.
200. As recorded by Bell, *The Harvard Lectures of A. N. Whitehead*, 301.
201. As recorded by Hocking, *The Harvard Lectures of A. N. Whitehead*, 77.
202. Whitehead, *Process and Reality*, 100.
203. Ibid., 21.
204. Ibid., 100.
205. See Whitehead, *Process and Reality*, 121ff.
206. See Eastman, *Untying the Gordian Knot*.
207. Ibid., 46–47.
208. Vattay et al., "Quantum Biology on the Edge of Quantum Chaos."
209. See Eastman, *Untying the Gordian Knot*, 27, 49.
210. Whitehead, *Science and the Modern World*, 5.
211. Whitehead, *Process and Reality*, 237.
212. Ibid., 69.

213. See Eastman, *Untying the Gordian Knot*, 134. See also Kirschner et al., "Molecular 'Vitalism.'"

214. Kauffman, *The Origins of Order*, 30.

215. See Kirschner et al., "Molecular 'Vitalism'" for more on the chemical principles at play in biological self-assembly, as well as the limitations of machine analogies in biology: "biological systems, which might seem machine-like, on closer examination operate on very different principles" (79).

216. Whitehead, *Process and Reality*, 17.

217. As recorded by Bell, *The Harvard Lectures of A. N. Whitehead*, 76–77.

218. Whitehead, *Process and Reality*, 162.

219. See Marshall, et al., "Identifying molecules as biosignatures." Marshall et al. have devised a means of experimentally identifying biosignatures by measuring molecular complexity using mass spectrometry.

220. Whitehead, *Process and Reality*, 189.

221. Ibid., 163.

222. Ibid.,163

223. Whitehead, *Science and the Modern World*, 110.

224. Whitehead, *Process and Reality*, 5.

225. Damer and Deamer, "The Hot Spring Hypothesis," 432.

226. Damer, "A Field Trip to the Archean," 6.

227. To reiterate, "urability" is a new concept Damer is developing with Deamer to refer to the thermodynamic and chemical conditions necessary for life to emerge on a planet. We are used to thinking of the "habitability" of planets, but "urability" has to do with establishing not just habitability for existing life but the conditions for the origin (ur-) of life. For more on urability, see part 1. See also Deamer et al., "Urability."

228. Bruce Damer and David Deamer, "The Hot Spring Hypothesis for an Origin of Life," 436–37. See also See Damer and Deamer, "Coupled Phases and Combinatorial Selection in Fluctuating Hydrothermal Pools."

229. "Life Springs" by Martin J. Van Kranendonk, David W. Deamer, and Tara Djokic in *Scientific American*, August 2017.

230. Whitehead, *Process and Reality*, 87.

231. Ibid., 84.

232. Ibid., 18.

233. Ibid., 85.

234. Ibid., 104.

235. Ibid.

236. Ibid., 309.

237. Whitehead, *Science and the Modern World*, 18.

238. See Meera, "Humans versus Earth."

239. See Kelly, *Becoming Gaia*.

BIBLIOGRAPHY

Aanen, D. K., and P. Eggleton. "Symbiogenesis: Beyond the Endosymbiosis Theory?" *Journal of Theoretical Biology* 434 (2017): 99–103.

Anderson, P. W. "More Is Different." *Science* 177, no. 4047 (1972): 393–96.

Auxier, R., and G. Herstein. *The Quantum of Explanation: Whitehead's Radical Empiricism*. New York: Routledge, 2017.

Bellazzi, F. "The Emergence of the Postgenomic Gene." *European Journal for Philosophy of Science* 12, no. 17 (2022). https://doi.org/10.1007/s13194-022-00446-0.

Bertalanffy, L. von. *Modern Theories of Development: An Introduction to Theoretical Biology*. Oxford: Oxford University Press, 1933.

Bettinger, J., and T. Eastman. "Foundations of Anticipatory Logic in Biology and Physics." *Progress in Biophysics and Molecular Biology 131* (December 2017): 108–20. https://doi.org/10.1016/j.pbiomolbio.2017.09.009.

Bitbol, M. "Schrödinger and Indian Philosophy." In *Cahiers du service culturel de l'ambassade de France en Inde*. Allahabad, 1999.

Bitbol, M., and S. Osnaghi. "Bohr's Complementarity and Kant's Epistemology." In *Bohr, 1913–2013, Séminaire Poincaré XVII*, edited by O. Darrigol et al., 145–66. Cham: Springer: 2013.

Bogaard, P. A., and J. Bell, eds. *The Harvard Lectures of Alfred North Whitehead, 1924–1925: Philosophical Presuppositions of Science*. Edinburgh: Edinburgh University Press, 2017.

Chaisson, E. "Energy Rate Density as a Complexity Metric and Evolutionary Driver." *Complexity* 16 (2010): 27–40.

———. "Energy Rate Density. II. Probing Further a New Complexity Metric." *Complexity* 17, no. 1 (2011): 44–63.

Cornell, C. E., et al. "Prebiotic amino acids bind to and stabilize prebiotic fatty acid membranes." *PNAS*, 201900275 (2019): 17239–44. https://doi.org/10.1073/pnas.1900275116.

Damer, B. F. "The Evogrid: An Approach to Computational Origins of Life Endeavours." PhD Diss., University College Dublin, 2011. https://doi.org/10.13140/RG.2.1.4800.7206.

———. "A Field Trip to the Archean in Search of Darwin's Warm Little Pond." *Life*, No. 6, 2016.

———. "The Hot Spring Hypothesis for the Origin of Life and the Extended Evolutionary Synthesis" (2019). Essay for Extended Evolutionary Synthesis project. Accessed 31 December 2021. http://extendedevolutionarysynthesis.com/the-hot-spring-hypothesis-for-the-origin-of-life-and-the-extended-evolutionary-synthesis/.

Damer, B. F., and D. W. Deamer. "Coupled Phases and Combinatorial Selection in Fluctuating Hydrothermal Pools: A Scenario to Guide Experimental Approaches to the Origin of Cellular Life." *Life* 5, no. 1 (2015): 872–87. https://doi.org/10.3390/life5010872.

———. "The Hot Spring Hypothesis for an Origin of Life." *Astrobiology* 20, no. 4 (2020): 420–52. https://doi.org/10.1089/ast.2019.2045.

Darwin, C. Letter to Joseph Dalton Hooker, February 1, 1871. Darwin Correspondence Project, "Letter no. 4065," accessed on 7 March 2023, https://www.darwinproject.ac.uk/letter/?docId=letters/DCP-LETT-4065.xml

Davis, A. M. et al., eds. *Process Cosmology: New Integrations in Science and Philosophy*. Cham: Palgrave: 2022.

De Haro, S. "Science and Philosophy: A Love-Hate Relationship." *Foundations of Science* 20 (2020): 297–314. https://doi.org/10.1007/s10699-019-09619-2.

Deacon, T. *Incomplete Nature: How Mind Emerged from Matter*. New York: W. W. Norton, 2011.

Deamer, D. "Boundary Structures are Formed by Organic Components of the Murchison Carbonaceous Chondrite." *Nature* 317 (1985): 792–94. https://doi.org/10.1038/317792a0.

Deamer, D. *First Life: Discovering the Connections Between Stars, Cells, and How Life Began*. Berkely: University of California Press, 2011.

Deamer, D. et al. "Urability: A Property of Planetary Bodies That Can Support an Origin of Life." *Astrobiology* 22, no. 7 (2022): 889–900. http://doi.org/10.1089/ast.2021.0173.

Dick, S. J. "Critical Issues in the History, Philosophy, and Sociology of Astrobiology." *Astrobiology* 12, no. 10 (2012): 906–27. https://doi.org/10.1089/ast.2011.0786.

Dupré, J., and D. J. Nicholson. *Everything Flows: Toward a Processual Philosophy of Biology*. Oxford: Oxford University, 2018.

Duve, C. "Life as a Cosmic Imperative?" *Philosophical Transactions of the Royal Society A*, 369 (2011): https://doi.org/10.1098/rsta.2010.0312.

Eastman, T. E. *Untying the Gordian Knot: Process, Reality, and Context*. Lanham, MD: Lexington, 2020.

Eschenmoser, A. "The Search for the Chemistry of Life's Origin." *Tetrahedron* 63 (2007): 12821–44. https://doi.org/10.1016/J.TET.2007.10.012.

Esposito, J. L. *Schelling's Idealism and Philosophy of Nature*. Lewisburg, PA: Bucknell University, 1977.

Formosinho, J. et al. "Environmentality in biomedicine: microbiome research and the perspectival body." *Studies in History and Philosophy of Science*, 91 (2022): 148–58. https://doi.org/10.1016/j.shpsa.2021.11.005.

Friederich, S. "Fine-Tuning." *The Stanford Encyclopedia of Philosophy* (Winter 2021), edited by Edward N. Zalta, https://plato.stanford.edu/archives/win2021/entries/fine-tuning/).

Gare, A. "Approaches to the Question 'What Is Life?': Reconciling Theoretical Biology with Philosophical Biology." *Cosmos and History: The Journal of Natural and Social Philosophy* 4, no. 1–2 (2008): 53–77.

Godfrey-Smith, P. *Philosophy of Biology*. Princeton: Princeton University, 2014.

Greene, B. *The Fabric of the Cosmos: Space, Time, and the Texture of Reality*. New York: Knopf Doubleday, 2007.

Haraway, D. J. *Staying with the Trouble: Making Kin in the Chthulucene*. New York: Duke University Press, 2016.

Henderson, L. J. *The Fitness of the Environment: An Inquiry into the Biological Significance of the Properties of Matter*. New York: Macmillan, 1913.

Henning, B. and J. Petek, eds. *Whitehead at Harvard, 1924–1925*. Edinburgh: Edinburgh University Press, 2021.

———. *The Harvard Lectures of Alfred North Whitehead, 1925–1927: General Metaphysical Problems of Science*. Edinburgh: Edinburgh University Press, 2021.

Henry, J. "Newton, the Sensorium of God, and the Cause of Gravity." *Science in Context* 33, no. 3 (2020): 329–51. https://doi.org/10.1017/S0269889721000077.

Hume, D. *An Enquiry Concerning Human Understanding*. In Steven M. Cahn, ed. *Classics in Western Philosophy*. Indianapolis: Hackett Publishing Company, 2012.

Ichiki, K et al. "Cosmological Magnetic Field: A Fossil of Density Perturbations in the Early Universe." *Science*311, no. 5762 (2006): 827–29. https://doi.org/10.1126/science.1120690.

Juarrero, A. *Dynamics in Action: Intentional Behavior as a Complex System*. Cambridge: Bradford Books, 1999.

Kant, I. *Critique of Judgment*. Mineola, NY: Dover, 2005.

———. *Critique of Pure Reason*. New York: Cambridge University Press, 1998.

———. *Kant's Prolegomena to any Future Metaphysics*. Chicago: Open Court, 1912.

———. *Opus Postumum*. Cambridge: Cambridge University Press, 1993.

Kastner, R. et al. "Taking Heisenberg's Potentia Seriously." *International Journal of Quantum Foundations* 4, no. 2 (2018): 158–72. https://doi.org/10.48550/arXiv.1709.03595.

Kauffman, S. *A World Beyond Physics: The Emergence and Evolution of Life*. Oxford University Press, 2019.

———. *The Origins of Order: Self-Organization and Selection in Evolution*. Oxford University Press, 1993.

Kelly, S. *Becoming Gaia: On the Threshold of Planetary Initiation*. Tumwater: Revelore Press, 2020.

Kirschner, M. et al. "Molecular 'Vitalism.'" *Cell* 100, no. 1 (2000): 79–88. https://doi.org/10.1016/S0092-8674(00)81685-2.

Kranendonk, M. et al., eds. *Earth's Oldest Rocks* (2nd Edition). Amsterdam: Elsevier, 2018.

———. "Life Springs." *Scientific American* 317, no. 2 (2017): 28–35. https://doi.org/10.1038/scientificamerican0817-28.

Kuhn, T. *The Structure of Scientific Revolutions*. Chicago: Chicago University, 2012.

Latour, B. *An Inquiry into Modes of Existence: An Anthropology of the Moderns*. Harvard: Harvard University, 2013.

Levin, M. "The Computational Boundary of a 'Self': Developmental Bioelectricity Drives Multicellularity and Scale-Free Cognition." *Frontiers in Psychology*10 (2019): https://doi.org/10.3389/fpsyg.2019.02688.

Levins, R., and R. Lewontin. *The Dialectical Biologist*. Cambridge, MA: Harvard University Press, 1985.

Lewontin, R. *The Triple Helix: Gene, Organism, and Environment. Cambridge, MA*: Harvard University Press, 2000.

Lucas, G. R. "Evolutionist Theories and Whitehead's Philosophy." *Process Studies* 14, no. 4 (1985): 287–300.

Luo, G. et al. "Rapid Oxygenation of Earth's Atmosphere 2.33 Billion Years Ago." *Science Advances* 2, no. 5 (2016): https://doi.org/10.1126/sciadv.1600134.

Marshall, M. *The Genesis Quest: The Geniuses and Eccentrics on a Journey to Uncover the Origin of Life on Earth*. Chicago: University of Chicago Press, 2020.

Marshall, M. "The Water Paradox and the Origin of Life." *Nature* 588 (2020): 210–13. https://media.nature.com/original/magazine-assets/d41586-020-03461-4/d41586-020-03461-4.pdf

Marshall, S. M., et al. "Identifying molecules as biosignatures with assembly theory and mass spectrometry." *Nature Communications* 12, no. 3033 (2021): https://doi.org/10.1038/s41467-021-23258-x.

Maudlin, T. *Philosophy of Physics: Quantum Theory*. Princeton: Princeton University Press, 2019.

Meincke, A. S. "Autopoiesis, Biological Autonomy and the Process View of Life." *The European Journal for Philosophy of Science* 9, no. 5 (2019): https://doi.org/10.1007/s13194-018-0228-2.

Monod, J. *Chance and Necessity*. New York: Vintage Press, 1972.

Odling-Smee, F. et al. W. *Niche Construction: The Neglected Process in Evolution*. Princeton: Princeton University Press, 2000.

Offord, C. "Can Single Cells Learn?" *The Scientist* (May 2021). Accessed December 21, 2021. https://www.the-scientist.com/features/can-single-cells-learn-68694

Pavlopoulos, G. A., et al. "Using Graph Theory to Analyze Biological Networks." *BioData mining* 4, no. 10 (2011): https://doi.org/10.1186/1756-0381-4-10.

Pearce, B. K. D., et al. "CRAHCN-O: A Consistent Reduced Atmospheric Hybrid Chemical Network Oxygen Extension for Hydrogen Cyanide and Formaldehyde Chemistry in CO2-, N2-, H2O-, CH4-, and H2-Dominated Atmospheres." *J. Phys. Chem. A* 124, no. 41 (2020): 8594–8606. https://doi.org/10.1021/acs.jpca.0c06804.

———. "Constraining the Time Interval for the Origin of Life on Earth." *Astrobiology* 18, no. 3 (2018): 343–64. https://doi.org/10.48550/arXiv.1808.09460.

Peretó J. et al. "Charles Darwin and the Origin of Life." *Origins of Life and Evolution of Biospheres* 39, no. 5 (2009): 395–406. https://doi.org/10.1007/s11084-009-9172-7.

Petersen, A. "The Philosophy of Niels Bohr." *Bulletin of the Atomic Scientists* 19, no. 7 (1963): 8–14. https://doi.org/10.1080/00963402.1963.11454520.

Price, L. *Dialogues of Alfred North Whitehead: As Recorded by Lucien Price*. Westport, CT: Greenwood Press, 1977.

Pross, A. "Toward a General Theory of Evolution: Extending Darwinian Theory to Inanimate Matter." *Journal of Systems Chemistry* 2, no. 1 (2011): https://doi.org/10.1186/1759-2208-2-1.

Richards, R. *The Romantic Conception of Life: Science and Philosophy in the Age of Goethe*. Chicago: Chicago University Press, 2002.

Rosen, R. "Relational Biology and the Origin of Life." In *Molecular Evolution and Protobiology*, edited by Matsuno K., Dose K., Harada K., Rohlfing D.L. Boston: Springer, 1984, 421–32.

———. *Essays on Life Itself*. New York: Columbia University Press, 2000.

Rovelli, C. "Carlo Rovelli on Consciousness, the Illusion of Time, and Philosophy of Relational Quantum Mechanics." On the "Theories of Everything" YouTube Channel, hosted by Curt Jaimungal. December 17, 2021. https://www.youtube.com/watch?v=r_fUPbBNmBw.

Russell, M. J. "The 'Water Problem' (sic), The Illusory Pond and Life's Submarine Emergence—A Review." *Life* 11 (2021): https://doi.org/10.3390/life11050429.

Scarfe, A. C. "James Mark Baldwin with Alfred North Whitehead on Organic Selectivity: The 'Novel' Factor in Evolution." *Cosmos and History: The Journal of Natural and Social Philosophy* 5, no. 2 (2009): 40–107. https://cosmosandhistory.org/index.php/journal/article/view/136

Scharf, C. et al. "A Strategy for Origins of Life Research." *Astrobiology*15, no. 12 (2015): 1031–42. https://doi.org/10.1089/ast.2015.1113.

Schelling, F. W. J. *Ideas for a Philosophy of Nature*. Cambridge: Cambridge University Press, 1988.

———. *On the World Soul, A Hypothesis of the Higher Physics for the Clarification of Universal Organicity*. In *Collapse: Philosophical Research and Development*, Vol. 6. Falmouth: Urbanomic, 2010, 58–95.

———. *First Outline of a System of the Philosophy of Nature*. Albany: SUNY Press, 2004.

Segall, M. D. *Physics of the World-Soul: Whitehead's Adventure in Cosmology*. Grasmere: SacraSage, 2021.

Shields, C. *Aristotle*. New York: Routledge, 2007.

Smolin, L. *The Trouble with Physics: The Rise of String Theory, the Fall of Science, and What Comes Next*. New York: Houghton Mifflin Harcourt: 2006.

Sölch, D. "Wheeler and Whitehead: Process Biology and Process Philosophy in the Early Twentieth Century." *Journal of the History of Ideas* 77, no. 3 (2016): 489–507. https://doi.org/10.1353/jhi.2016.0021.

Spencer, H. *Principles of Biology*. New York: D. Appleton and Co., 1864.

Stein, R. "Toward a Process Philosophy of Chemistry." *Hyle* 10, no. 1 (2004): 5–22.

Subramanian, K. "The Origin, Evolution and Signatures of Primordial Magnetic Fields." *Reports on Progress in Physics* 79, no. 7 (July 2016): https://doi.org/10.1088/0034–4885/79/7/076901.

Subramanian, M. "Humans Versus Earth: The Quest to Define the Anthropocene." *Nature* 572 (2019): https://doi.org/10.1038/d41586-019-02381-2.

Swimme, B., and M. E. Tucker. *Journey of the Universe*. Yale University Press, 2011.

Tabaczek, M. "The Metaphysics of Downward Causation: Rediscovering the Formal Cause." *Zygon* 48, no. 2 (June 2013): 280–404. https://doi.org/10.1111/zygo.12012.

Thompson, E. *Mind in Life: Biology, Phenomenology, and the Sciences of Mind*. Cambridge, MA: Harvard University Press, 2007.

Vattay, G., et al. "Quantum Biology on the Edge of Quantum Chaos." *PLOS ONE* 9. no. 3 (2014): https://doi.org/10.1371/journal.pone.0089017.

Varela, F. G., et al. "Autopoiesis: The Organization of Living Systems, its Characterization and a Model." *Biosystems* 5, no. 4 (1974): 187–96. https://doi.org/10.1016/0303–2647(74)90031–8.

Wade, N. "Meet Luca, the Ancestor of All Living Things." *The New York Times* (July 25, 2016): https://www.nytimes.com/2016/07/26/science/last-universal-ancestor.html

Walker, S. I., and P. C. W. Davies. "The Algorithmic Origins of Life." *Journal of the Royal Society Interface* 10 (2012): https://doi.org/10.1098/rsif.2012.0869.

Wheeler, W. M. "Present Tendencies in Biological Theory." *Scientific Monthly* 28, no. 2 (1929): 97–109. https://www.jstor.org/stable/14574.

———. *Emergent Evolution and the Development of Societies*. London: Kegan Paul, Trench, Trubner and Co., Ltd., 1927.

Whitehead, A. N. *The Concept of Nature*. Cambridge: Cambridge University Press, 1920.

———. *Science and the Modern World.* New York: The Free Press, 1925.

———. *The Function of Reason.* New York: The Free Press, 1929.

———. *Process and Reality.* Corrected edition, D.R. Griffin and D. Sherburne. New York: The Free Press, 1929.

———. *Adventures of Ideas*. New York: The Free Press, 1933.

———. *Modes of Thought.* New York: The Free Press, 1938.

Wigner, E. "The Unreasonable Effectiveness of Mathematics in the Natural Sciences." *Communications in Pure and Applied Mathematics* 13, no. 1 (1959): 1–9. https://doi.org/10.1002/cpa.3160130102.

Woese, C. "A New Biology for a New Century." *Microbiology and Molecular Biology Reviews* 68, no. 2 (June 2004): https://doi.org/10.1128/MMBR.68.2.173-186.2004.

Woese, C. R., and G. E. Fox. "The Concept of Cellular Evolution." *J Mol Evol* 10, no. 1–6 (1977): https://doi.org/10.1007/BF01796132.

Chapter 3

The Organic Universe and Otherworldly Lives

Bergson and Sagan

Wahida Khandker

In his 1907 work *Creative Evolution*, Henri Bergson's image of the élan vital arises less in response to the question, "What is life?," than in the effort to articulate what life *does*. An overview of the many orders of life and the dynamics of organic evolution as debated in the decades after the publication of Darwin's *Origin of Species* suggests, for Bergson, only two key movements. The first is a gradual accumulation of energy; the second, "an elastic canalization of this energy in variable and indeterminable directions, at the end of which are free acts."[1] Following this identification of two "necessities" in what is otherwise a plethora of contingencies in evolution, Bergson speculates on what the outcome of these movements might have been on other planets, with other elements at their disposal. What might life look like that was not based on the specific configurations and processes grounded on carbon, nitrogen, and oxygen that define the forms and processes of bodies on Earth? Indeed, the "fitness" of living processes would be determined differently in each solar system, and even further, Bergson continues, "it is not even necessary that life should be concentrated and determined in organisms properly so called, that is, in definite bodies presenting to the flow of energy ready-made through elastic canals."[2] What might life look like if its impetus ran across "a matter not yet solidified"? There would be little difference, Bergson concludes, "between this vitality, vague and formless, and the definite vitality we know, than there is, in our psychical life, between the state of dream and the state of waking."[3] At this point, there is a tantalizing glance forward to the kind of speculations envisioned in works of science fiction, such as Stanislaw Lem's *Solaris*,[4] in which an ocean-covered planet seems to manifest an unfathomable form of consciousness able to create and respond, but not communicate, conventionally at least, with its human explorers.

Yet we need only turn to the work of physicist Carl Sagan, who revels in the kind of inquisitiveness we see here in Bergson's text, and indeed who himself wrote a speculative novel, *Contact*,[5] about the possibilities of detecting other, more advanced, life-forms among the stars. Reflecting on the search for more rudimentary signs of life, Sagan, in his 1985 Gifford Lectures on Natural Theology at the University of Glasgow (published posthumously as *The Varieties of Scientific Experience*[6]), writes of the spectroscopic analysis of comets that reveals that what is "out there" (a range of organic molecules in the tails of passing comets) is very close to what is "in here" (the organic molecules in living organisms on Earth). He finds in this molecular affinity between us and the stars a reason for humility and a greater sense of interconnectedness with other living organisms, both on this planet and in the potential forms they may take on other planets. The first part of this chapter will be devoted to such questions in Sagan's work. As I will return to in the second part of the chapter, Bergson's own speculations will terminate in a more cautious stance on the tendency toward reductionism in the sciences, tempered by an optimism in philosophical creativity to enable the translation from observations of inert materiality to visions of a vital impulsion. I conclude with a consideration of features of life that have evolved in the deepest parts of our oceans, and that have only recently become accessible with the development of technologies able to navigate ecosystems at these otherwise inhospitable depths. There, we discover forms of life capable of evolving and flourishing in the absence of sunlight, sustained by chemosynthesis, mirroring obliquely the development on land of ecosystems founded upon photosynthesizing plants. In summary, this chapter explores varied perspectives on the challenges of *traversing vast distances* in space, in evolutionary time, between land and the deep sea, and indeed between the living and the nonliving, and how all such endeavors might enhance our understanding and appreciation of our own unique habitat.

UNEARTHLY LIFE

In a flyby of the Earth in December 1990, the Galileo spacecraft, it is reported, served as a "control experiment" for the detection of life on other planets. Its findings of abundant water in all of its states (vapor, solid ice, and snow and ocean) indicates the probability of water-based life, and high concentrations of methane not generated on other physically similar planets might signal to passing explorers from other worlds that something unique, perhaps life, was present here. Finally, radio emissions generated by human technology are sufficiently powerful to be detected over interstellar distances, advertising the presence of intelligent life quite regardless of whether we intend to be seen or heard by others.[7] The experiment broaches two main features in Sagan's writings on the origins of life: an interest in the conditions necessary for the production of organic molecules that may result in the emergence of life, as well as the conditions that might limit or destroy life, such as the proliferation of nuclear arms and anthropogenic climate change; and the search for extra-terrestrial intelligence focused on the detection of anomalous radio emissions as well as active attempts to communicate with other civilizations.

Sagan's speculations on the possibilities of detecting intelligent life on other planets rest on the relatively high probability that the chemical conditions on the early Earth made the emergence of life reasonably likely on other planets or moons at similar stages of formation. In a 1963 paper, Sagan discusses the equation $N = R_* f_p n_e f_l f_i L$,[8] which cites, among other factors, the combination of the vast number of star systems hosting planets with environmental conditions favourable to life, and the fact that only a tiny fraction of those worlds need generate life for it to be relatively plentiful. "Given sufficient time," Sagan writes, "and an environment which is not entirely static, the evolution of complex organisms is apparently inevitable. In our own solar system, the origin of life has probably occurred at least twice."[9] He goes on to outline the unique difficulties of interstellar communication, which would require the span of a human lifetime just to hold a conversation, whereas interstellar travel might "only" take a few decades, well within the lifetime of a healthy human explorer willing to make the trip.[10] What they might find at the end of such a journey is likely, it is argued, to be a benign civilization. Furthermore, a civilization, argue Sagan and Newman, "thousands or millions of years more advanced than we are, beneficiaries of an exponentiating technology over all those intervening millennia, will not be engaged in a simple extrapolation of our activities, or be driven by our motives."[11] On the other hand, a society driven by the need for aggressive colonization is more than likely to destroy itself before it has the opportunity to spread its malignancy to other worlds. In the meantime, while terrestrial billionaires today have their sights set on space tourism for the elite, a keystone of Sagan's work on the detection of extraterrestrial intelligence, the observatory at Aricebo, has rather spectacularly succumbed to significant damage (on 10th August 2020, "an 8-centimeter-thick steel cable, one of 18 suspending a 900-ton instrument platform high above the dish, had pulled out of its socket at one end and fallen, slicing into the dish. By December, more cables snapped resulting in even more damage to the dish").[12] Yet the more modest endeavor to discern the origins and possibilities of life still ring a more hopeful note in Sagan's work, reflected perhaps in the recent missions to explore signs of ancient life on Mars.[13]

The origins of life on our planet are traced against a physical clock marking the lifetimes of the early Earth and particularly of the sun as it reached its present intensity and, speculatively, as it continues to evolve. Eventually, with increasing intensity of radiation from the sun, will come increased temperatures on Earth such that it will resemble the atmosphere of Venus, while Mars will start to look more like Earth, at least in terms of temperature.[14] "If there are any organisms left on our planet in that remote epoch," Sagan and Mullen write, "they may wish to take advantage of this coincidence."[15] In our present epoch, and as projected in a paper in 1979, Sagan et al. find such inevitable changes in the life of the sun preempted by "human alteration of the environment," and that such modifications have not only been effected by modern humans, but that "it is more likely that the human species has made a substantial and continuing impact on climate since the invention of fire."[16]

Casting eyes farther back in time, Sagan's writings on the conditions necessary to produce the "prebiological" constituents of life, amino acids, note the role of ultraviolet radiation in their emergence[17], a nod to the benevolent influence of the sun on the primitive Earth, as well as "shocks" from the entry of meteorites into the Earth's atmosphere.[18] Such "shocks" may have been necessary, alongside tendencies toward a greenhouse effect on the early Earth, to generate enough energy required for the synthesis of organic compounds, otherwise unlikely given the relative paucity of radiation emitting from the early "faint sun."[19] As J. B. S. Haldane in a 1944 paper notes, the emergence and evolutionary paths of early forms of life would have been constrained at each turn by fluctuations in the availability of energy. At these early stages, Haldane explains,

> a living organism would have been unable to provide even the small energy needed for cell division or amoeboid movements. At a later time, life of a simple sort would have been possible, but locomotion would have been very difficult, and large swimming or crawling animals could not have existed.[20]

Underpinning such radical differences between these most ancient of living organisms and the complexity and variety of life so abundant today is a fundamental affinity between all living things found in their common basis in organic or carbon-based molecules. Since our most abundant source of organic molecules on Earth is already biological in origin, our best bet to discern their origin, Sagan explains, is to search for them on other worlds.[21] If organic molecules are found almost exclusively on Earth, then the likelihood of life arising elsewhere diminishes. Here, Sagan relays the story of the astronomer William Huggins, an early spectroscopist who analyzed in the late nineteenth century the "spectra" of Comets Brorsen and Winnecke II in 1868, and whose results helped to precipitate mass panic across several nations in 1910. The analysis showed "that a molecular fragment, diatomic carbon or C_2—two carbon atoms attached together—is present when you look at the spectrum of the comets and also when you look at natural gas and the vapor from heated olive oil."[22] Such experiments, Sagan continues, also discovered the presence of CN or the cyanide molecule, a single grain of which can kill a human being instantly.[23] On the news that the Earth was due to pass through the tail of Halley's Comet in 1910,

> Astronomers tried to reassure people. They said it wasn't clear that the Earth would pass through the tail, and even if the Earth did pass through the tail, the density of CN molecules was so low that it would be perfectly alright. But nobody believed the astronomers.[24]

The panic was unfounded, Sagan explains, because evidence of organic molecules can be found right across our solar system, for example, on the outer moons of Saturn and, a little closer to home, on moons of Mars.[25] Even more tantalizing is the example of the largest of Saturn's moons, Titan. Like the Earth it holds a thick atmosphere, loaded with organic molecules. There, "the land is covered with this organic muck that falls from the skies. There is a submarine deposit underneath this ocean of liquid ethane and methane of more of this complex stuff, and then down deep is frozen methane and frozen water and so on."[26] One could imagine, in the light of the speculations of Sagan and Mullen mentioned earlier, that at some point in the progressive intensification of the sun over its lifetime that conditions may become warm enough to spark and sustain life on Titan, though it would be far beyond the lifetime of our own species.[27] If the evidence for the emergence of life on Earth is anything to go by, forming somewhere between 4000 million and 3500 million years ago, when algal stromatolites evolved, then such speculations on the creation of life in places such as Titan might even seem rather conservative. Yet, as I discuss in the next section, Henri Bergson argues that our ability to imagine the emergence of life on other worlds is hindered by our all-too-human tendency to create a conceptual gulf between living and nonliving processes. How are we to bridge this gulf in a way that might also promote a better appreciation of our home, the terrestrial ecosystem?

THE LIMITS OF ANALYSIS

In a short paper published in *Nature* in 1933, the physicist Niels Bohr compares the problem of the analysis of atoms in their quantum state to the problem of analyzing the functions of living organisms. A certain "renunciation" of knowledge is necessary in the case of quantum states, due to the complementary status of two otherwise theoretically incompatible modes of being, expressed here in the properties of light:

> The spatial continuity of light propagation, on one hand, and the atomicity of the light effects, on the other hand, must, therefore, be considered as complementary aspects of one reality, in the sense that each expresses an important feature of the phenomena of light, which, although irreconcilable from a mechanical point of view, can never be in direct contradiction, since a closer analysis of one or the other feature in mechanical terms would demand mutually exclusive experimental arrangements.[28]

The "ingrained idea," Bohr goes on to explain, of the separation of objects from our modes of observation is supplanted by this recourse to a certain "complementarity." In the realm of physics, there is an acknowledgement of the disturbance caused by measuring instruments on the object under scrutiny. For example, "any imaginable procedure aiming at the coordination in space and time of the electrons in an atom will unavoidably involve an essentially uncontrollable exchange of momentum and energy between the atom and the measuring agencies, entirely annihilating the remarkable regularities of atomic stability for which the quantum of action is responsible."[29] We are confronted here with a combined theoretical and methodological problem that compares readily with the problems encountered in the biological analysis of life. On one hand, the sheer magnitude of the complexity of life seems to be irreducible to conventional methods of measurement; on the other hand, invasive experimental methods betray their inadequacy in the study of life in their tendency to extinguish the very functions they seek to analyse. As Bohr underlines, any conventional attempt to discern, down to the atomic level, the natural functions of a living organism would succeed only in killing it.[30]

Compare Henri Bergson's diagnosis of the limits of the human intellect, that while "so skilful in dealing with the inert, is awkward the moment it touches the living."[31] The intellect "proceeds with the rigour, the stiffness and the brutality of an instrument not designed for such use."[32] The problem at hand, of course, is that of physical reductionism, or the assumption that the function of a thing can be discerned by breaking it down into its component parts and reconstructing those parts from the ground up. The analogy that Bohr draws between quantum physics and biology speculates on the possibilities for revising the principles of biological analysis, in the light of the incorporation of once contradictory processes (e.g., the wave and corpuscle theories of light) into a single, physical scheme. This resonates with Bergson's aspiration, in his analysis of the function of the intellect in *Creative*

Evolution, to foster new ways of thinking about life, time, or process in general. Bohr and Bergson alike summarize the challenges faced by the biological sciences to find concepts that are adequate to the complexity of organic life, and to theorists across disciplines who are interested in challenging the continued use and "disposability" of animals for scientific research.

Bergson's anti-reductionist stance spans his most famous works, commencing with his discussions of the division between homogeneity and heterogeneity in *Time and Free Will* arising from a "phenomenological" consideration of the simple act of recollection. In *Matter and Memory*, the differences between intensities of attention to life, and the exercise of will in a free action, are grounded upon the dynamics of expansion and contraction of matter. Finally, the characterization of living matter in *Creative Evolution* from the perspective of the evolution of life is developed in the terms given to us in *Matter and Memory* when we consider the operation of consciousness, now applied to the contracting and expanding creative impetus, simply expressed as a process of "making and unmaking." The movement between each of these texts is far from uncomplicated, but we can identify an overarching concern with the inability of mechanistic models to encapsulate the processes of life.

We saw, in the previous section, how Sagan's work extrapolates *in time* the physical conditions on distant planets toward those favorable for life (at least as indicated by the geological record leading up to the environment presently on Earth), in a way that bridges *spatial* distances between organic molecules on Earth and those on other worlds. Astrobiology's microscopic focus on the fundamental units of life suggests their final expression, or reduction, into atomic structures. The endeavor to bridge this gap between the living and the nonliving has proven, thus far, to be much more elusive. For Bergson, the "mechanical utility," or the fundamental connection constantly enacted between the physical and the biological, enjoyed by organisms shows itself to be both a necessity of life and a critique of the limitations of the function of the intellect, motivating the entire project of *Creative Evolution*. Bergson sets out to identify the structures that inform our everyday knowledge of the world, including life both internal and external to ourselves, and to speculate on the conditions of their formation, not in a priori structures of the intellect, but rather in the evolved conditions, in a fully biological sense, of our physiological interactions with our environment. What Bergson names "the double form of consciousness,"[33] or the intuitive and intelligent aspects of conscious life reflects the material and vital aspects of reality. At this point, it is easy to become immersed in the speculative cosmology implied in such statements, but we need not treat it as much more than the affirmation that the long history of the study of nature (*phusis*) has tended to divide itself into the studies of physics and biology (nonliving, material nature on the one hand, and the living, natural world on the other). The history of physics is, it could be said, plagued by the resistance of living beings to its principles. In parallel to this history, we might add the divergence of epistemology from metaphysics. One seeks the structures of human knowledge, the other the principles of their formation, and yet:

> each of these two lines of thought leads to the other; they form a circle, and there can be no other centre to the circle but the empirical study of evolution. It is only in seeing consciousness run through matter, lose itself there and find itself there again, divide and reconstitute itself, that we shall form an idea of the mutual opposition of the two terms, as also, perhaps, of their common origin.[34]

The "mutual opposition" of consciousness and matter (indeed, of life and matter) and the discovery of their "common origin" is none other than the unification of biology and physics. Where Erwin Schrödinger, for example, talks of the recourse to statistical models for taking account of the massive numbers of atomic interactions at play at the macroscopic scale of organic processes,[35] Bergson's epistemological approach seeks to identify the dynamic relationships between opposing manifestations of living matter: whether in the taxonomical generalizations imposed upon organisms such that they can be divided (with Georges Cuvier) into the plant and animal kingdoms, or in the "double direction" of the evolution of consciousness along the lines of instinct and intelligence.[36] Ultimately, this epistemological approach will lead to a different kind of "renunciation" of the specificity of classical mechanical description in favour of an affirmation of "only a current of existence and the opposing current."[37] Thus the tools available to us for thinking, measuring, interacting with our physical environment must themselves be subjected to critique. As Deleuze underlines succinctly, the "problematizing, differentiating, and temporalizing" method of Bergsonism often concludes with the resolution of opposing terms in a common source.[38] In this case, intellect and matter might be revealed in their making as the products of an underlying ontological process, one that requires a revision of our method of enquiry, "for psychology, cosmology and metaphysics take intelligence, in all that is essential to it, as given, instead of, as we now propose, engendering it in its form and in its matter."[39]

What, then, of the place of the science of physics in this evolutionary scheme? When Bergson talks of the opposition between space and geometry on one side, and organic life on the other, the object of critique is not simply a caricature of instrumental science that seeks to submit nature to its laws. In accordance with the "differentiating" method cited above, the aim is to expose the relationship between physical analysis and the ordinary operation of intelligence. Intelligent structures are carried in a particular direction by the sinuosity of matter, "but, the impulsion once received, mind continues its course. The idea that it forms of pure space is only the schema of the limit at which this movement would end."[40] Compare, here, with Kant's transcendental framework, in particular the role of the intuitions of space and time. Rather than space being the form of intuition, or the field within which sensations are organized, Bergson makes of space the completion of a process. Our intellect has been formed after the contours of material objects (themselves, insofar as any "thing" is an object for us, products of a petrification and selection of a fundamentally processive reality). The idea of space that we have is, then, the relative passivity of our minimal conscious engagement with the world (perception with the barest intervention of memory), taken to its farthest limit. This terminal point becomes our concept of abstract geometrical space, and

> once in possession of the form of space, mind uses it like a net with meshes that can be made and unmade at will, which, thrown over matter, divides it as the needs of our action demand. Thus, the space of our geometry and the spatiality of things are mutually engendered by the reciprocal action and reaction of two terms which are essentially the same, but which move each in the direction inverse of the other.[41]

For Bergson, physics "understands its role when it pushes matter in the direction of spatiality,"[42] but metaphysics ought to serve, it seems, as a tempering force that seeks knowledge against the grain of the relatively passive, perceiving consciousness. Classical physics, characterized in its alignment with mathematics and geometry, is concerned only with matter, but not *living* matter.[43] By contrast, the role of metaphysics is to attempt to scrutinize its own operation not just in the formation of distinct objects, but also in its "creative" or free acts, in the direction of intuition. In this endeavor, metaphysics constitutes a change of perspective, characterized by Bergson as a "twisting on itself, [where] the faculty of seeing should be made to be one with the act of willing,—a painful effort which we can make suddenly, doing violence to our nature, but cannot sustain more than a few moments."[44] The metaphysician's intuitive "glance" might, then, shed light on matters concerning the biologist such as the "simple fact" of vision, belied by the accompanying structural complexity of the organ of sight:

> As soon as the eye opens, the visual act is effected. Just because the act is simple, the slightest negligence on the part of nature in the building of the infinitely complex machine would have made vision impossible. This contrast between the complexity of the organ and the unity of the function is what gives us pause.[45]

P. A. Y. Gunter has noted Bergson's use of a variety of images or metaphors to describe the movement of life that tends to elude mechanistic articulation (such as the hand moving through iron filings or the explosive motion of a firework), and he updates the list of metaphors to include the wave function in quantum mechanics.[46] This analogy is particularly instructive in helping to demystify the image of the élan vital:

> In both cases there is passage from a broad, or indefinite, thoroughly dynamic state to one that is narrower, more definite, more stable. In both cases, perfect predictability is not possible. Quantum measurement eliminates dynamism and uncertainty, replacing them with the actual and the definite. While Bergson stresses the opposition of life and matter, he also asserts that matter causes the division and dissociation of the original thrust of life. The analog of quantum measurement would be the action of physical matter on the vital impetus, causing the "collapse" of the original impetus into kingdoms, genera, and species.[47]

In summary, the epistemological goal of *Creative Evolution* is broadly to describe a process (life) once it has already passed: the élan vital is a retrograde reconstruction of a directly experienced process that is either typically mistaken for a real thing (a vitalist or animating principle), or, viewed with true philosophical insight (intuition), it is acknowledged to merely indicate the afterimage of a fleeting process.

A commentator on the intersections of physics and biology in the work of both Niels Bohr and, briefly, on Henri Bergson, the physicist Walter Elsasser helps to emphasize some of the interdisciplinary overlaps highlighted so far in the above sections. Elsasser and Bergson both seek to overcome physical reductionism, but Elsasser, with a view to establishing solid scientific principles for "biological theory," makes a considerable leap forward to describe the physical processes that underpin organic synthesis. His attempt to avoid physical reductionism is based in the problem of immense numbers. For the purposes of calculating the various permutations of the elements that make up organic compounds (carbon, hydrogen, oxygen, nitrogen), the sheer immensity of the possible combinations of these elements makes the verification of the behaviour of classes of biological entities simply impossible to achieve. In short, the experimenter would run out of specimens before verification could be achieved. This points to the unfathomable complexity of living processes.[48] In this reconciliation of quantum and biological theory, Elsasser suggests one way of naturalizing Bergson's formulation of the problem of memory without reducing it to the structure of a storage device such as the mammalian brain. Where, then, does this leave Bergson's account of the dynamic that defines life, and that functions in place of the otherwise unbridgeable gulf between life and matter, as set out in *Creative Evolution*?

We saw above that Bergson's anti-reductionist stance was motivated and defined by a combination of epistemological and specific biological problems. The function of intelligent life is to carve up its environment into relatively homogeneous entities that facilitate greater ease of action; and this function is then reflected in mechanical description. Nonliving matter lends itself to organization into such homogeneous classes, to use Elsasser's term, but living matter presents too high a degree of complexity to be subjected to the same process. The often-cited example of the evolution of the eye (not to mention the convergent evolution of the organ of sight) indicates the presence of both complexity and a certain element of direction (and unity) in evolution. Where Elsasser "resolved" this problem of extreme complexity by proposing a basis for theoretical biology on a principle of holism and an acknowledgement of the "unfathomable complexity" of organic processes, Bergson's metaphysical approach dwells on the *imprecision* of language as a means of expressing the process of the creation of possibilities. There is, Bergson writes, "a reality of quite another order, which contrasts with the atom as the thought of the poet with the letters of the alphabet."[49] This complementary feature of material reality is expressed in the second law of thermodynamics "the most metaphysical of the laws of physics, since it points out without interposed symbols, without artificial devices of measurement, the direction in which the world is going."[50] Bergson's "unacceptable" metaphysical solution to the problem of information transmission in the case of cerebral memory, the allusion to an underlying Process (i.e., *durée*), is reiterated in the case of life in general in terms of the second law of thermodynamics. Borrowing this law's own imprecision, Bergson's description of the interaction between life and matter moves from a general expression of a process to its more specific manifestations, in what are assumed to be the two most dominant kingdoms of life: the plant and the animal. Life is

> an effort to remount the incline that matter descends. In that, they reveal to us the possibility, the necessity even of a process the inverse of materiality, creative of matter by its interruption alone. The life that evolves on the surface of our planet is indeed attached to matter. If it were pure consciousness, a fortiori if it were supraconsciousness, it would be pure creative activity. In fact, it is riveted to an organism that subjects it to the general laws of inert matter. But everything happens as if it were doing its utmost to set itself free from these laws. It has not the power to reverse the direction of physical changes, such as the principle of Carnot determines it. It does, however, behave absolutely as a force would behave which, left to itself, would work in the inverse direction. Incapable of stopping the course of material changes downwards, it succeeds in retarding it. The evolution of life really continues, as we have shown, an initial impulsion: this impulsion, which has determined the development of the chlorophyllian function in the plant and of the sensori-motor system in the animal, brings

> life to more and more efficient acts by the fabrication and use of more and more powerful explosives. Now, what do these explosives represent if not a storing-up of the solar energy, the degradation of which energy is thus provisionally suspended on some of the points where it was being poured forth?[51]

The image of an impulsion thus encompasses and expresses a number of elements of life. The maintenance of a low level of entropy, as Schrödinger might word it, is presented by Bergson in the form of the "release of energy" in animal movement being offset by the storage of starches in green ("chlorophyllian") plants.[52] This work of "offsetting" or delaying split across the plant and animal kingdoms demonstrates the applicability of the same principle at the level of interspecific interactions as well as that of the basic operation of an individual organism (with which Schrödinger's account is concerned). The application of this same principle at different levels is therefore best indicated in Bergson's use of metaphors such as the action of raising one's arm and letting it fall again, or ultimately in the "psychological" image of an élan or impetus, recalling the theses of both *Time and Free Will* and *Matter and Memory* in which the multiplicity of mental states is reconciled with the unity of free, conscious action. What remains is an image of a movement (rather than, say, a subject moving an object), such that when we speak of "vital activity," we can speak of its "afterimage" or "that which subsists of the direct movement in the inverted movement, a reality which is making itself in a reality which is unmaking itself."[53]

EXTREMOPHILES

"Despite the utter mediocrity of our position in space and time," write Carl Sagan and William Newman, "it is occasionally asserted, with no sense of irony, that our intelligence and technology are unparalleled in the history of the cosmos. It seems to us more likely that this is merely the latest in the long series of anthropocentric and self-congratulatory pronouncements on scientific issues that dates back well before the time of Claudius Ptolemy."[54] They cite the work of Frank J. Tipler for his "solipsist world view" according to which the lack of obvious evidence of extraterrestrial intelligence indicates that there is none. Furthermore, the promise of the development of self-replicating machines by human ingenuity indicates that any such intelligence elsewhere will have already succeeded at building machines of this complexity. Their absence might suggest that human civilization has reached a technological peak unmatched elsewhere in the universe. However, as Sagan and Newman contend, "absence of evidence is not evidence of absence."[55]

To conclude this chapter, I want to turn to studies of some of the extraordinary and unlikely manifestations of life on our own planet that point to the paucity of our knowledge of what makes life what it is and, indeed, what might ground and constitute intelligent life on the many forking paths it has, and could, take over the course of evolutionary development. Recall Bergson's speculations on alternative possible forms of life that may use elements in combinations quite different to the ones that make up our own bodies. Merlin Sheldrake, in his book, *Entangled Life*, provides his own otherworldly intervention into debates about the evolution of intelligent life on Earth, albeit indirectly by illuminating the complex "entanglement" of all life, including that of human beings, with the many and often enigmatic species of fungi whose processes appear to underpin every other ecosystem on the planet. Sheldrake notes the significance of lichens, "living riddles" that, in their constitution from an intricate union of fungal and algal partners, or the mycobiont and photobiont respectively, confound our traditional conceptions of biological identity. He describes how, from a taxonomical perspective, lichens have long posed a problem for the traditional Darwinian image of the divergence of lines of descent from a common ancestor. That lichens appeared to merge two (or more) lineages would mean that "evolution could no longer be thought of solely in terms of competition and conflict. Lichens had become a type of inter-kingdom collaboration."[56] Their significance, explains Sheldrake, also lies in their status as a "geological force" engaged in a process of "weathering" enacted by their growth on rock surfaces, their use of "an arsenal of powerful acids and mineral-binding compounds to digest the rock," and finally their death that generates soils from which new ecosystems can emerge.[57] Furthermore, some lichens can be classed as "extremophiles," able to survive in space exposed to cosmic radiation, and in some instances to recover their metabolic activities or to repair damage sustained in their exposure to space.[58]

Thus, in research toward the origins of life, lichens have provided a glimpse into the possibilities of evolution more akin to the model of endosymbiosis proposed by Lynn Margulis, which contends that the different parts of eukaryotic cells that make up plant and animal bodies were originally separate, free-living organisms (typically bacteria) that eventually came to form a range of symbiotic relationships. Lichens are instructive in this regard, Sheldrake writes, not because they "re-enact" this original symbiosis, but rather because:

> They are cosmopolitan bodies, a place where lives meet. A fungus can't photosynthesise by itself, but by partnering with an alga or photosynthetic bacterium it can acquire this ability horizontally. Similarly, an alga or photosynthetic

> bacterium can't grow tough layers of protective tissue or digest rock, but by partnering with a fungus it gains access to these capabilities—suddenly.[59]

The plasticity afforded by the ability of bacteria to "exchange" genes horizontally has enabled the emergence of communities of organisms in the most inhospitable sites on Earth, such as cold seeps and deep-sea hydrothermal vents. These communities are built upon the labors of microorganisms that can harness the chemical energy of elements such as hydrogen sulfide and methane that seep from geological faults.[60] The process of "chemosynthesis" was first described by Sergei Vinogradskii (or Winogradsky) in 1890, in studies of the energy requirements of plants. Vinogradskii found that the particular species of bacterium under investigation, Beggiatoa, was able to convert hydrogen sulfide into food.[61] His experiments, extended to other kinds of bacteria able to effect similar conversions of ferrous oxide and nitrates, indicated that

> organisms could oxidize inorganic matter unassisted by the energy of either light or organic material. For him, these autotrophic organisms were the biological agents that transferred matter between nature's inorganic and organic realms—that is, his new 'physiological types' were essentially direct links between nature's biological and mineral realms.[62]

The ability of certain microorganisms to extract and use energy to sustain life from decidedly lifeless places points to the likelihood that living organisms could emerge and develop on other worlds, echoing (or perhaps preempting) the emergence of life on Earth more than three thousand million years ago. However, studies of the prevalence of chemosynthesis have noted that it does not wholly replace reliance upon photosynthetic organisms. Rather, "almost all of the observable biomass associated with deep-sea communities such as hydrothermal vents, cold seeps, whale falls and wood falls depends on the action of aerobic chemosynthetic organisms which could not survive in complete isolation from the sun due to their oxygen requirement."[63]

We have seen how Sagan's extrapolations *in time* of the physical conditions on distant planets toward the formation of the rudiments of life and, eventually, of intelligent life, depend upon insights that are always, if we might say, *too late* for us to grasp. Messages sent to other worlds might elicit responses only for unknown people belonging to future generations, while insights gleaned about the origins of life on Earth form ultimately unknowable connections with emergent life on other planets. Furthermore, as we saw from a Bergsonian perspective, the history of efforts to discern the link between matter and even the "simplest" of living organisms is bedevilled by a human perception that is utility-driven and, therefore, always only finding processes after their completion (again, *too late*). Bergson shows how knowledge of life, even if only fleeting, requires us to relinquish claims to possess, reduce or replicate life, to better appreciate the creative possibilities of the evolution of life on other worlds as well as our own. Examples of extremophilic and chemosynthetic organisms that can subsist in extreme environments, with less, if not completely without, oxygen dependency, show us that life can exist decidedly without us, and far from the reach of our specifically calibrated sense-organs. It emphasizes, in the end, that it is very much our privilege to inhabit and, if we heed that it is a gift, to care for this Earth.

NOTES

1. Bergson, *Creative Evolution*, 255.
2. Ibid., 256.
3. Ibid.
4. Lem, *Solaris*.
5. Sagan, *Contact*.
6. Sagan, *Varieties of Scientific Experience*.
7. Sagan et al., "A search for life on Earth from the Galileo spacecraft," 720.
8. The equation expresses, as Sagan explains, "The number of extant advanced technical civilizations possessing both the interest and the capability for interstellar communication." Sagan, "Direct contact among galactic civilizations," 486.
9. Ibid., 487.
10. Ibid., 489–91.
11. Sagan and Newman, "The Solipsist Approach to Extraterrestrial Intelligence," 118–19.
12. Clery, "Fallen Giant," accessed October 9, 2021, https://www.science.org/news/2021/01/how-famed-arecibo-telescope-fell-and-how-it-might-rise-again.
13. Farley et al., "Mars 2020 Mission Overview."
14. Sagan and Mullen, "Earth and Mars," 55.
15. Ibid.
16. Sagan et al., "Anthropogenic albedo changes and the earth's climate," 1367.
17. Sagan and Khare, "Long-wavelength ultraviolet photoproduction of amino acids," 417.
18. Bar-Nun et al., "Shock synthesis of amino acids," 472.
19. Sagan and Chyba, "The early faint sun paradox," 1217.
20. Haldane, "Radioactivity and the Origin of Life in Milne's Cosmology," 555.
21. Sagan, *Varieties of Scientific Experience*, 69.
22. Ibid., 72.
23. Ibid., 75.
24. Ibid., 76.
25. Ibid., 83–84; 87–88.

26. Ibid. 95.

27. Indeed, Arthur C. Clarke's *Space Odyssey* novels set their drama of the past and futures of human and extraterrestrial evolution against the backdrop of Saturn's moons (and, by contrast, on moons of Jupiter, in Stanley Kubrick's film, *2001: A Space Odyssey*).

28. Bohr, "Light and life," 422.

29. Bohr, *Atomic Physics and Human Knowledge*, 19.

30. Ibid., 5.

31. Bergson, *Creative Evolution*, 165.

32. Ibid.

33. Ibid., 178.

34. Ibid.

35. Schrödinger, *What is Life?*, 10.

36. Bergson, *Creative Evolution*, 184.

37. Ibid.

38. Deleuze, *Bergsonism*, 34–35.

39. Bergson, *Creative Evolution*, 186–87.

40. Ibid., 202.

41. Ibid.

42. Ibid., 208.

43. Ibid., 219.

44. Ibid., 237.

45. Ibid., 93.

46. Gunter, "Bergson's Creation of the Possible," 36.

47. Ibid., 37.

48. Elsasser, *Reflections on a Theory of Organisms*, 37.

49. Bergson, *Creative Evolution*, 240.

50. Ibid., 242–43.

51. Ibid., 245–46.

52. Ibid., 246.

53. Ibid., 248.

54. Sagan and Newman, "The Solipsist Approach to Extraterrestrial Intelligence," 113.

55. Ibid, 114.

56. Sheldrake, *Entangled Life*, 82.

57. Ibid., 85.

58. Ibid., 89.

59. Ibid., 92–93.

60. Fujikura et al., "The deepest chemosynthesis-based community yet discovered," 17.

61. Ackert, "The role of microbes in agriculture," 381.

62. Ibid., 382.

63. Smith, "Chemosynthesis in the deep-sea," 17045.

BIBLIOGRAPHY

Ackert, L. T. "The Role of Microbes in Agriculture: Sergei Vinogradskii's Discovery and Investigation of Chemosynthesis, 1880–1910." *Journal of the History of Biology* 39, no. 2 (2006): 373–406. https://doi.org/10.1007/s10739-006-0008-2.

Bar-Nun, A., et al. "Shock Synthesis of Amino Acids in Simulated Primitive Environments." *Science* 168, no. 3930 (1970): 470–73. https://doi.org/10.1126/science.168.3930.470.

Bergson, H. *Matter and Memory*, trans. N. M. Paul and W. S. Palmer. New York: Zone Books, 1991.

———. *Creative Evolution*, trans. Arthur Mitchell. New York: Dover, 1998.

———. 1999. *Duration and Simultaneity*, trans. Leon Jacobson, ed. Robin Durie. London: Clinamen, 1999.

Bohr, N. "Light and Life." *Nature* 131 (1933): 421–23. https://doi.org/10.1038/131457a0.

———. *Atomic Physics and Human Knowledge*. New York: Dover, 2010.

———. *Atomic Theory and the Description of Nature*. Vol. 1. Cambridge: Cambridge University Press, 1961.

Clery, D. "Fallen Giant." Accessed October 9, 2021. https://www.science.org/news/2021/01/how-famed-arecibo-telescope-fell-and-how-it-might-rise-again.

Deleuze, G. *Bergsonism*. Translated by H. Tomlinson and B. Habberjam. New York: Zone, 1988.

Elsasser, W. M. *Reflections on a Theory of Organisms: Holism in Biology*. Baltimore: Johns Hopkins University Press, 1987.

Farley, K. A., et al. "Mars 2020 Mission Overview." *Space Science Review* 216, no. 142 (2020). https://doi.org/10.1007/s11214-020-00762-y.

Fujikura, K. et al. "The Deepest Chemosynthesis-Based Community yet Discovered from the Hadal Zone, 7326 m deep, in the Japan Trench" *Marine Ecology Progress Series* 190 (1999): 17–26.

Gunter, P. A. Y. "Bergson's Creation of the Possible." *SubStance* 36, no. 3 (2007): 33–41. https://doi.org/10.1353/sub.2007.0043.

Haldane, J. B. S., "Radioactivity and the Origin of Life in Milne's Cosmology." *Nature* 153, no. 555 (1944): https://doi.org/10.1038/153555a0.

Lem, S. *Solaris*. London: Faber and Faber, 1970.

Mayr, E. "Cause and Effect in Biology." *Science* 134, no. 3489 (1961): 1501–6

Robbins, S. E. "Bergson and the Holographic Theory of Mind." *Phenomenology and the Cognitive Sciences* 5, no. 3–4 (2006): 365–94. https://doi.org/10.1007/s11097-006-9023-1.

Sagan, C. *The Varieties of Scientific Experience*. New York: Penguin, 2006.

———. *Contact*. London: Orbit, 1997.

———. "Direct Contact among Galactic Civilizations by Relativistic Interstellar Spaceflight." *Planetary and Space Science* 11, no. 5 (1963): 485–98. https://doi.org/10.1016/0032–0633(63)90072–2.

Sagan, C., and C. Chyba. "The Early Faint Sun Paradox: Organic Shielding of Ultraviolet-Labile Greenhouse Gases." *Science* 276, no. 5316 (1997): 1217–21. https://doi.org/10.1126/science.276.5316.1217.
Sagan, C., and F. Drake. 1975. "The Search for Extraterrestrial Intelligence." *Scientific American* 232, no. 5 (1975): 80–89.
Sagan, C., and B. N. Khare. "Long-Wavelength Ultraviolet Photoproduction of Amino Acids on the Primitive Earth." *Science* 173, no. 3995 (1971): 417–20. https://doi.org/10.1126/science.173.3995.417.
Sagan, C., and G. Mullen. "Earth and Mars: Evolution of Atmospheres and Surface Temperatures." *Science* 177, no. 4043 (1972): 52–56. https://doi.org/ 10.1126/science.177.4043.52.
Sagan, C., and W. I. Newman. "The Solipsist Approach to Extraterrestrial Intelligence." *Quarterly Journal of the Royal Astronomical Society* 24 (1983): 113–21.
Sagan, C., et al. "A Search for Life on Earth from the Galileo Spacecraft." *Nature* 365, no. 6448 (1993): 715–21. https://doi.org/10.1038/365715a0.
Sagan, C., O. B. Toon, and J. B. Pollack. "Anthropogenic Albedo Changes and the Earth's Climate." *Science* 206, no. 4425 (1979): 1363–68. https://doi.org/10.1126/science.206.4425.1363.
Schrödinger, E. *What is Life?: With Mind and Matter and Autobiographical Sketches*. Cambridge: Cambridge University Press, 1992.
Sheldrake, M. *Entangled Life: How Fungi Make Our Worlds, Change Our Minds, and Shape Our Futures*. London: The Bodley Head, 2020.
Smith, C. "Chemosynthesis in the Deep-Sea: Life without the Sun." *Biogeosciences Discussions* 9, no. 12 (2012): 17037–52. https://doi.org/10.5194/bgd-9-17037-2012.

Chapter 4

The Connection-Action Principle

A Basis for Process Philosophy, Cosmic Creativity, and Value?

Mark Lupisella

THE CONNECTION-ACTION PRINCIPLE

Creativity, Laws, Action

A key concept in Whitehead's original articulation of process philosophy is *creativity*.[1] The science of cosmic evolution seems to suggest a highly "creative"—perhaps *increasingly* creative—evolving universe. "Creative" is in quotes to convey the concept of creativity in a very broad sense and to start by allowing for the possibility that the universe may not be creative in the sense of highly intentional and specific forms of creativity that human beings often engage in, but rather that the universe gives rise to forms and objects as a realization of the unfolding of particular dynamics over the course of cosmic evolution.

Indeed, we can always appeal to a scientific brute-fact kind of explanation that suggests the laws of physics and/or initial conditions of the universe are fundamental facts to accept without cause or explanation, and those laws and conditions explain why the universe is creative. Specifically, an initial high energy and low entropy state naturally gave rise to expansion,[2] with cooling and clumping emerging over time—consistent with the second law of thermodynamics, gravity, and other physical forces—causing the aggregation, or "creation," of objects like atoms, stars, galaxies, planets, and life.

But we can still ask, as many have, why those initial conditions? Why these "laws"? Why any order or laws at all? Why an origin, and was it truly from "nothing" as some suggest?[3] Merely being able to ask these questions does not mean they are well-posed or that they have answers, let alone scientific answers, but there have been a wide range of intriguing and compelling suggestions ranging from: (1) design by a god or gods, or some kind of entity or beings, including the possibility that our universe is a simulation of sorts,[4] (2) anthropic principles,[5] (3) an eternally oscillating universe, going back to the Greeks and forms of Eastern worldviews such as Hinduism, and now by some in modern cosmology,[6] (4) cosmological natural selection,[7] (5) a multiverse that suggests the possibility of many universes[8] and (6) even more provocative versions of anthropocentric thinking that suggest conscious beings in some sense create the universe and possibly even its laws via extreme interpretations of quantum theory.[9]

Regardless of the kind of explanation for our universe's origin and its particular laws and initial conditions, many views seem to rest on, assume, or at least imply, that our universe (or multiverse) is dynamic—that it is an "action-laden" reality. Indeed, it appears we live in a universe of *action*, and action is central to our understanding of our world.[10] Even contemplations of a cosmic origin as a quantum fluctuation from a quantum vacuum state or quantum "foam" (a realm of virtual particles which are wavelike fluctuations in the quantum vacuum at "absolute zero") seem to imply there is still some kind of dynamic at work—that there is at least some kind of action. The quantum vacuum state appears to be a dynamic realm of action, or at least a realm of potential action. We seem to live in a fundamentally dynamic physical reality—in a universe full of action. But we can still ask: Why should there be any action at all? Indeed, asking about how cosmic creativity arises seems to be very close to, if not the same as asking why there are "laws," or physical dynamics, or more generally, *action*?

Connectedness, Relationality, Action

> You cannot evade accounting for the relevance of a particular fact A to another particular fact B. The answer which I would give in outline is that the becomingness of reality is a process of exhibiting the togetherness of things, and that this togetherness is essential.
>
> —Alfred North Whitehead[11]

In wondering why there is action in the universe, we can appeal to a kind of scientific brute-fact explanation that suggests our particular laws of physics cause, or facilitate, or at least permit, forms of action, without any other explanation. But presumably there can be action without particular laws.[12] A conceptual way to think about a source of action is to appeal to the oft-cited "connectedness" of the universe (perhaps very similar to Whitehead's "togetherness" noted in his quote above and explored further in his *Process and Reality*). The connected unity of the universe is a theme often explicitly embraced by many perspectives such as various interpretations of quantum mechanics, including explicitly "holistic" views of quantum theory,[13] environmental movements.[14] and cosmic evolution more generally.[15]

As is often the case with many terms and concepts, it can be difficult to precisely define "connectedness"—particularly in this context, partly because its use is intentionally broad. Minimally, connectedness would seem to imply some kind of relationship. If there is a connection, regardless of the kind of connection, and however abstract it may be, it seems a relationship of some form would be required to make that connection relevant or meaningful, to realize and instantiate it—to ultimately make it in some sense "actionable." We may be able to use connectedness and relationality almost interchangeably since they are so close in this articulation. However, connectedness might be better seen as a somewhat broader, more fundamental, or slightly different construct—e.g., as a fundamental general property that can perhaps be both abstract and concrete—for which relationality is one of what may be multiple manifestations of connectedness. The basic suggestion here is that the property of connectedness is realized as relations and relationality more generally.

Perhaps more importantly, it could be argued that something needs to *happen* to realize and instantiate a relation. A specific dynamic, an "*event*," or perhaps more generally, an *action* of some kind, seems to be needed to manifest a relationship—to actualize a relation and arguably relationality more generally. Actions realizing and manifesting relations seems consistent with most forms of process philosophy in the sense that actions can also be thought of as dynamic events or processes more generally. Indeed, a central construct for Whitehead's "event" metaphysics is "*actual* occasions."[16]

Our present understanding of the universe can be interpreted to suggest that there does not appear to be any truly "static" state of affairs—including the "emptiness" of a putative quantum vacuum as noted previously. Could this be because relationality, as a manifestation of the basic property of connectedness, drives toward some form of action in order to realize relationships and ultimately a general fundamental property of connectedness? Otherwise, how would relations, or relationality more generally, be realized without "action" or "events" or "process" to actualize relations?[17]

Connectedness giving rise to action might be called the "connection-action principle." The most basic, or minimalist, version of the connection-action principle could be summed up as: *Connectedness is manifested as action.* Put into the parlance of philosophical property talk, it might be stated as: *The universe's property of connectedness is instantiated as action.*

The universe's property of connectedness can also be thought of as a sort of latent "relational potential," and actions are a manifestation of that potential.[18] In this sense "relationality" is also foundational[19] because connectedness is relational and actions are relational. Relations are more specific than connectedness but less concrete than actions. This makes "relationality," or "relationalism," a useful bridge and framework in which to understand the connection-action principle—as suggested by the simple graphic in Figure 4.1 and contextualized further in my recent work.[20] For example, a slightly more specific and/or perhaps slightly stronger version of the connection-action principle might be something like this: *Connectedness gives rise to relations which are realized as actions.*

This basic "minimalist" version of the connection-action principle does not necessarily claim that the property of connectedness is manifested *only* as action. It suggests primarily that action is an important, perhaps critical, manifestation of connectedness and relationality, allowing for the possibility of other forms of instantiations and realizations of connectedness. However, the basic version of the connection-action principle explored here suggests that action is a significant and robust manifestation of connectedness.

Degree and Necessity: Stronger Versions of the Connection-Action Principle

If we allow for the general possibility that possessing a property can lead to increasing degrees of its instantiation,[21] a stronger version of the connection-action principle can be interpreted to suggest that the universe's property of connectedness could be manifested as *increasing degrees* of action, perhaps *ever*-increasing degrees of action. Increasing degrees of action could be manifested in many ways and is arguably consistent with many forms of process philosophy. As suggested prior, process, or at least "actual occasions," can be seen as a form of, or possibly synonymous with, *action.* Increasing realization of connectedness via relations and actions can give rise to more and novel "processes" in the form of increasing creativity, increasing

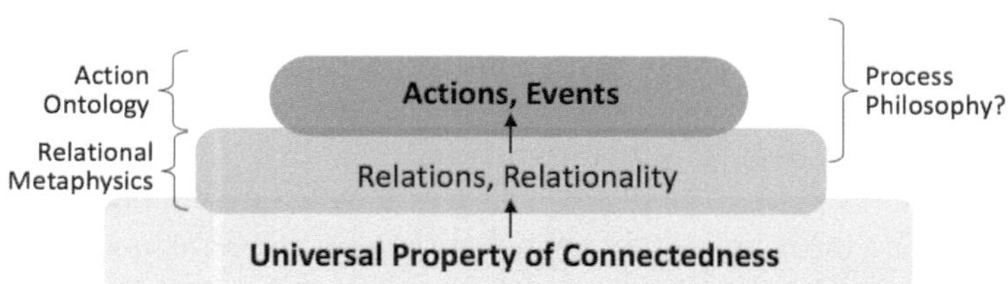

Figure 4.1. Graphical representation of the connection-action principle. (Courtesy of Mark Lupisella)

diversity, increasing complexity—continually driving the emergence of new relations, new actions, new dynamics, new phenomena.[22] On this interpretation, increasing degrees of relationality and action (for example, increasing degrees of diversity and novelty) are a part of the universe increasingly realizing its nature—its fundamental property of connectedness.

The more the universe produces "action," the more the universe's property of connectedness is realized—the more its potential is realized. This arguably implies a soft form of teleology, suggesting the universe is, at least in part, about realizing connectedness and manifesting relations, and in so doing, is evolving in many diverse ways, always bringing forth new creations, new relations, and increasing novelty and diversity as suggested by certain forms of process philosophy—and perhaps as implied by the "temporalized" version of the principle of plenitude.[23] So a broader, stronger version of the connection-action principle might suggest that the universe's property of connectedness is realized through relations as *increasing* degrees of action. The greater the degrees of relationship and action, the greater the degree to which the universe's connectedness and "relational potential" is realized.

But why "connectedness"? One start at supporting connectedness as a fundamental property of the universe is to ask how we might conceive of the universe as being otherwise, that is, as being fundamentally disconnected. What would that look like? How would such disconnectedness obtain or originate? Can we give an account of how even very distant parts of the universe are fundamentally disconnected or unrelated? The Big Bang model (including the observed homogeneity of the early universe) along with physical phenomena such as energy, gravity, and aspects of quantum mechanics/quantum field theory and quantum entanglement[24] seem to ultimately imply a deep fundamental physical connectedness, relationality, and dynamism in the universe as a whole. For example, gravity and other fundamental forces such as the strong nuclear force literally connect physical material together.[25]

So it appears it would be extremely difficult, if not impossible, to describe how the universe could be fundamentally internally disconnected despite what might have been a highly inflationary phase that could have caused different parts of the early universe to move away from each other at extremely high speeds. Such an inflationary phase still permits the most distant parts of the universe to be connected. Any attempt to describe disconnectedness would seem to ultimately involve, or be traceable as, a chain of causality giving rise to what is incorrectly seen as fundamentally disconnected parts of the universe that were once at least causally connected. For example, although "baby universes" being "pinched off" from their parent universe might seem truly disconnected, there is still at least a "temporally" connected physical chain of actions giving rise to the newly created universe—which, notably, may have similar properties to its parent universe.[26] In any cyclical models of the universe or universes, connectedness would also seem to apply across the various temporally connected cycles.

Even if there have been multiple cosmic origins, or an infinite number of separate origins from a more fundamental state of reality such as a quantum vacuum state or quantum field, such universes would apparently still arise from some common base of reality, as opposed to a true "state of nothingness" in which universes come into existence without any connection to any other reality of any kind. Indeed, independently arising universes still would appear to be connected or related in the broad sense of having originated from the same fundamental base of reality, perhaps even with a possibility that there could even be interactions between coexisting universes[27] or perhaps between "branes."[28]

As a general point then, it doesn't seem like we can provide a compelling account of fundamental disconnectedness in our universe, or any universe, or any collection of universes.[29] It may be that it is logically necessary for the whole of reality to be completely connected—even if it is connected in an unusual or tenuous way. It may be that it is physically, or even conceptually impossible for universes, or any kind of fundamental reality in general, to be anything other than what is ultimately a fundamentally deeply connected unity of some kind—however nebulous or unclear that connectedness may be.

Given what appears to be an empirical and perhaps conceptual or logical basis for deep connectivity of our universe and reality in general, and given that it may not be possible to provide a sufficient account of fundamental disconnectedness, can we suggest that connectedness is a *necessary* property of the universe? If this reasoning holds, then a stronger version of the connection-action principle could also suggest that connectedness is a *necessary* property of the universe. If relations and actions are ultimately required or necessary to realize connectedness, then a stronger, fuller version of the connection-action principle might be: *The universe's necessary property of connectedness requires relations which are necessarily manifested as increasing degrees of action.*

So, the greater the degree of action, the greater the degree to which the universe's fundamental necessary nature, or potential, is realized.[30] Whether a basic characteristic or property of the universe could, should, or must be realized as ever-increasing degrees is not necessarily required by weaker versions of the connection-action principle, but stronger versions would imply that kind of necessity and trend. The logical possibility that the property of connectedness could be realized as a trend toward increasing degrees of actualizing connectedness through actions may be enough for our general purpose here to articulate plausible stronger versions of the connection-action principle and the potential trend-like implications. However, there are likely to be more specific ways to infer a trend toward increasing degrees of action from a universal property of connectedness.

One possible approach for the connection-action principle to more specifically imply a trend is to consider that increasingly robust forms of relations and actions result from building upon the old to create the new,[31] since otherwise, there may only be mere repetition. This construction or creativity built upon what has come before can lead to emergence and the emergence of increasingly complex systems, which can then be continually built upon to endlessly create novel and increasingly robust forms of relationships and action.

Hints of Intra-action and Relational Metaphysics

The connection-action principle appears to be consistent with process philosophy in the sense that it suggests a conceptual or "metaphysical foundation" that gives rise to processes, events, actions. An even stronger version of the connection-action principle can be interpreted to make an ontological claim that relations are all that really exists.[32] Interpretations of this kind can be seen to support the idea that many conceptions of philosophical substantivism may be a kind of illusion.

However, in its stripped-down minimalist form, the connection-action principle can be interpreted to be agnostic regarding whether relations and interactions are between material objects (materialism/substantivism) or *intra*-actions[33] within a highly blended web of relations that may not require any particular material objects at all. Nevertheless, an interpretation that only relations and process are ultimately required, e.g., relational interpretations of relativity and quantum mechanics,[34] combined with a consideration that material objects do not appear to be ontologically necessary, suggests it is reasonable to interpret the connection-action principle as incorporating, or at least implying, a form of *intra*-action as opposed to interaction among discrete, concrete, material, irreducible, objects.[35]

It could also be argued that *intra*-action allows for a more robust realization of the connection-action principle in the sense that it is ultimately relations and actions "all the way down"[36] (and "all the way across"). There is no discrete, material, concrete "atomic" end point at which relationality and action are ultimately constrained. There is no material limit, no "point" at which relations and actions stop in a single instance of irreducible actionless matter. A relational metaphysics of this kind (or "relation-action" metaphysics more specifically) arguably allows for more "depth" of opportunity for more relations and actions to be realized, perhaps of infinite depth. Indeed, the history of particle physics seems to reflect a kind of endless "depth" to "particles"—including in the sense that quantum field theory sees particles primarily as transient waves in seamless fields. So, under stronger interpretations of the connection-action principle, *intra*-action arguably implies more potential for realizing the nature of the universe.

Interaction, or intra-action, and a relational metaphysics more broadly, could be further explored in more detail by leveraging fields such as systems theory, cybernetics, complexity theory, modern network theory and action ontology[37]—all of which have overlapping and interrelated subsets that could allow for more detailed exploration (and perhaps some degree of formalization) of the kind of metaphysics implied by the connection-action principle. The connection-action principle arguably provides a conceptual foundation for an action-based ontology altogether (as shown in Figure 4.1). More detailed applications leveraging the above areas are beyond the scope explored here, but could be examined more thoroughly in follow-on treatments. Related, what follows is an attempt to at least briefly consider how information and complexity relate to the connection-action principle.

Emergence, Biology, Information, Complexity

Theories of emergence and complexity may ultimately have much to say about how increasing degrees of action can be more precisely measured and how complexity might be linked to ideas such as the connection-action principle. A trend of increasing complexity can at least conceptually follow from increasing degrees of action in part because complexity, with its diverse and large number of connections and relationships, as well as high information content, can be thought to correspond to a high degree of inter/intra-action. The importance of connections and inter/intra-action is also realized by considering the following view regarding the origin of life from Stuart Kauffman:

> As the diversity of molecules in our system increases, the ratio of reaction to chemicals, or edges to nodes becomes ever higher. In other words, the reaction graph has ever more lines *connecting* (italics added) the chemical dots. The molecules in the system are themselves candidates to be able to catalyze the reactions by which the molecules themselves are formed. As the ratio of reaction to chemicals increases, the number of reactions that are catalyzed by the molecules in the system increases. When the number of catalyzed reactions is about equal to the number of chemical dots, a giant catalyzed reaction web forms, and a collectively autocatalytic system snaps into existence. A living metabolism crystallizes. Life emerges as a phase transition.[38]

The conceptual key to this account is that a relatively high ratio of *connections* (in this case, interactions in the form of chemical reactions) to other elements in the system (in this case molecules) is what causes a phase change to "emerge." In an important sense, the system's degree of connectedness is fundamental to the system's evolution and ultimately gives rise to the emergence of new dynamics.

The connection-action principle is consistent with this kind of account in the sense that it implies that what is important are connections in a system which ultimately drive internal relations and actions of the system. It might even be that we can see this connectedness and relationality of a prebiotic system as a beginning for how "software" might emerge from "hardware"—something Paul Davies suggests is an important transition for the emergence of life.[39] Connections and the associated relations and interactions within a system arguably represent some relative informational state within the system,[40] so a high degree of connectedness within a system can also be seen as representing a high degree of information content.

On this "informational" view, we may not have the fundamental conceptual problem of needing a transition from hardware to software since the universe is in some important sense already primarily "software," or information, or perhaps computation in the general forms of connections, relations, and inter/intra-action.[41] One interpretation of this view, consistent with the past discussion, is that there are no concrete "things" necessarily, just information, or more generally, *relations*.[42] In this context, life would be a matter of degree of actualization of the universe's informational/relational nature, as opposed to an unlikely or radical emergent phase transition that appears to be dramatically different from the rest of the reality from which it arose. Nevertheless, something like phase transitions (which again, may ultimately be matters of degree, albeit importantly, large jumps in degrees)[43] would still be a way for highly novel emergent phenomena to occur, helping to realize yet more novel action—a way for the universe to explore a very large, if not infinite, possibility space of its potential driven by connectedness realized as relations and action.

As noted previously, complexity is another way to think about the kinds of biological and emergent dynamics touched on above. Numerous connections and any associated actions can be seen as being similar to complexity. Complexity can be a problematic subject,[44] but a bottom line here is that in a universe that is realizing connectedness in the form of actions and increasing degrees thereof, complexity is arguably "predicted" in the sense that complexity can be a way for the universe to increasingly manifest additional novel, robust, "advanced" relations and actions.[45]

Holism and the Unending Process of Movement

> I would say that in my scientific and philosophical work, my main concern has been with understanding the nature of reality in general and of consciousness in particular as a coherent whole, which is never static or complete, but which is in an unending process of movement and unfoldment. Thus, when I look back, I see that even as a child I was fascinated by the puzzle, indeed the mystery, of what is the nature of movement.
>
> —David Bohm[46]

The suggestions in the previous section point to the possibility of thinking about the universe as a kind of deeply connected holistic ecosystem, perhaps a kind of information ecosystem that describes relations and actions. The connection-action principle arguably suggests why the universe can be seen as a holistic integrated collection of relations and actions. Bohm's reflection seems consistent with the motivation of asking why there is action and also seems consistent with some implications of the connection-action principle. *Movement* and *action* are very close in meaning, so Bohm's reference to the universe as an *unending process* of *movement* and *unfoldment* is very similar to the implication of the connection-action principle that it gives rise not only to action but, in stronger interpretations, to ever-increasing or unending degrees of action (or movement) over time.

Laws, Time, Plenitude, Multiverse?

In the context of the connection-action principle, we might see laws (or more generally, anything that acts as a fairly consistent and stable physical framework), as one way for connectedness and action to be robustly manifested because physical consistency can allow for the relatively stable "construction" of new, perhaps increasingly complex, creations. Laws can permit the new to be built upon the old, and hence new and more robust forms of action can be realized (similar to precepts of many articulations of process philosophy). Otherwise, random chaotic action may persist and presumably not allow for the creation of new, increasingly diverse and complex forms. This is arguably a similar conceptual foundation of an evolutionary developmental ("evo devo") model (Smart 2009)[47]—albeit broader and less constrained. That is, some form of developmental constraints (e.g., "laws") are needed to create relatively stable complexity, leveraging and building upon the less constrained dynamics of more random and open-ended evolutionary processes.

If connectedness is realized over time as ever-increasing degrees of action and hence new relations (e.g., increasing diversity and increasing complexity, etc.), does this provide support for the idea of the temporalized version of the principle of plenitude that suggests diversity increases (or, "anything that can happen, will happen"[48]), perhaps to infinity, over time? Could "time" be a way for the universe to increasingly realize connectedness through ongoing, perhaps eternal, creativity and emergence? Even if time's manifestation, or its "arrow," comes from the second law of thermodynamics and an initial low entropy state, time can still be seen as a way to give the universe many, maybe infinite, opportunities to realize connectedness through action since there is some form of "duration" that allows for continuous creativity.

Related, if time's arrow can ultimately be traced back to the Big Bang when there was a low entropy state, we can wonder why the universe might have had that initial low entropy condition (Carroll 2010a, b).[49] One possibility is that our particular universe is not a closed system and that the initial low entropy state came from a previous universal cycle and/or a manifestation of a larger multiverse from which universes of initial low entropy emerge—either as statistical brute facts or perhaps for some other reason. For example, a cyclical universe (or multiverse?) could be a robust manifestation of the most generalized version of the connection-action principle—a strong realization or "instantiation" of a generalized property of connectedness across all of reality, giving rise to more action across of all of reality, across all cosmological cycles, across all universes.

So, a multiverse may be a way to realize very high levels of action, particularly when considered in the aggregate. Presumably more universes gives rise to more action, which contributes to more creativity, diversity, complexity, etc. The postulated "ground" of reality from which many, possibly infinite, universes may emerge (e.g., a quantum vacuum state or quantum foam), while still a realm of dynamism and action, can also be interpreted to suggest it is nevertheless a low-action state. Such a minimal state may be seen as a lower or weaker manifestation of connectedness in the sense that it is a relatively homogenous and minimally dynamic state—with, nevertheless, high potential. A homogenous state could be considered a kind of "low connection-action state" since there is low diversity of action and hence less action overall. These lower states of homogenous action may give rise to "higher" states of action with more differentiation and diversity (e.g., the emergence particles or universes) and arguably then more robust manifestations of connectedness as more action, diversity, and emergent creativity.

The connection-action principle may be a conceptual/metaphysical foundation for why the base of physical reality "moves off" its "low" base state. Something like a quantum vacuum state, or a low energy quantum field, could be the lowest state of action possible, from which higher levels of action spring into existence, perhaps necessarily, "automatically," so that connectedness is better realized as increasing action relative to that base state. Perhaps we can interpret the instability of a quantum vacuum state as a manifestation of the need to push out of the lowest action state into some new, or more robust action state.[50] If the lowest energy quantum field state (e.g., a "quantum vacuum") is the "floor" of reality, maybe it's only "up" and "out" from there?

VALUE THEORY AND THE CONNECTION-ACTION PRINCIPLE

Intrinsic Value

G. E. Moore says the following about intrinsic value: "To say that a kind of value is 'intrinsic' means merely that the question whether a thing possesses it, and in what degree it possess it, depends solely on the intrinsic nature of the thing in question."[51] Or, as Lemos puts it: "The intrinsic value of a state of affairs is absolute, in the sense that it is determined completely by the nature of the state of affairs in question taken completely in abstraction from any consideration of the value of other states of affairs that are not constituents, elements, or parts of it."[52]

There are two very important ideas expressed in these conceptions of intrinsic value: independence and nature-dependence. Moore uses the word "solely" to capture the idea that intrinsic value is internal to a thing in that its value is not dependent on anything else but its own nature. Intrinsic value is independent of anything else. Lemos uses the word "absolute" to capture the same idea. Both Lemos and Moore emphasize the importance of grounding value in the nature of a thing, so that it is in virtue of how a thing is, or what a thing is, that gives it its intrinsic value. In describing intrinsic value, it seems we would want to give an account of the nature of some general state of affairs and why such a state of affairs is intrinsically valuable. So we need a description attempting to capture some nature, and more importantly, some account as to why that nature should be intrinsically valuable.

If, as the connection-action principle suggests, the universe's property of connectedness is realizable through relations and action, then we might think of this as a kind of cosmic grounding of value, or "cosmological *intrinsic* value," in the sense that the property of connectedness and its manifestation and implications are intrinsic to the nature of the universe as a whole and its evolution. This suggests that the universe may "be about something"—albeit broad and highly open-ended. Nevertheless, if connectedness is a fundamental property of the universe, and if connectedness gives rise to relationships, action, creativity, diversity, complexity, etc., then the universe is realizing its nature via these characteristics and dynamics. Can the universe realizing its nature be a sufficient grounding for intrinsic value? If the universe (or multiverse/metaverse) is all that is, all that was, and all that ever will be, then perhaps it can serve as a compelling "objective" framework for grounding and assessing intrinsic value.

If the universe is realizing its nature, then that realization is intrinsic to the universe and hence can be argued as a basis for value that goes well beyond the value created by valuing agents—it is arguably "of value" to the universe for it to realize its nature (or at least part of its nature). But even if this bridging from a "fact" to a value is insufficient to truly ground intrinsic value, we can at least argue that it is a reasonable philosophical choice to use the nature of the universe and any associated trend to ground and justify intrinsic value. Even if truly intrinsic value in its purest strictest sense cannot be completely justified with this approach, it is at least arguably a very strong form of value—it is a value associated with something significant and compelling and seemingly "objective," namely the whole of our universe (or multiverse).

The connection-action principle can be interpreted to suggest that relationality in general, and specific relationships and associated actions in particular, are valuable. In this sense, connection-action principle suggests a rough conceptual basis, if not a solid metaphysical foundation, for some kind of intrinsic value in the universe. The connection-action principle suggests what is arguably an objective ontological or cosmological grounding for the value of relations and resulting actions that emerge—a grounding that does not depend on valuing agents per se, but that can be recognized by valuing agents nonetheless. On this view, all forms of relations and intra/interactions are in some sense valuable—which can obviously be problematic. One way to address some challenges of this implication is to explore degrees of value or degrees of intrinsic value. Indeed, the connection-action principle can be interpreted to provide at least a theoretical framework for exploring *degrees* of value.

Degrees of Value

Regardless of the ontological commitments as to whether there are ultimately irreducible concrete "things" vs. solely relations, what is valuable in the context of the connection-action principle is realizing connectedness through relations and action in the many forms that may take. What is valuable is an unending unfolding of creativity, perhaps leading to ever-increasing degrees of creativity. Can we say then that ever-increasing degrees of connectedness, relations, and action, and the creativity that results, represent forms or expressions of *degrees* of intrinsic value?

Most attempts to deal with the issue of value measurement have focused on value as it pertains to human ends.[53] Philosopher, Alan Marshall, in exploring how extraterrestrial life might be viewed, stresses that intrinsic value is not imposed by humans but is merely recognized by humans.[54] As touched on above, this polar tension is indicative of how value has been viewed in such fundamentally different ways and reflects the difference between instrumental value and intrinsic value.

Treating connectedness, relationality, and action as intrinsically valuable provides what could serve as a broader objective frame of reference against which to assess value. In theory, the connection-action principle could provide a reasonably objective, possibly even "scientific," way to assess value. The degree to which connectedness is realized via action (e.g., degrees of complexity as one example) could be seen to correlate with degrees of value or possibly intrinsic value—or at least *relative* degrees of value. A basic utilitarian approach could still be used in the sense that a kind of "measurement" of what is valuable can be attempted, but the measurement could be informed by something other than solely anthropocentric or biocentric motivations. Assessing and measuring the value of action would presumably be very complex and perhaps even ultimately untenable for a variety of reasons. What precisely is meant by relationality, action, interaction, or intra-action so that it could be accurately, precisely and fairly assessed or even measured? As was asked previously, what constitutes lesser or greater degrees of relationality and action?

Organized complexity, as a possible way to recognize robust forms of action, is consistent with the idea of a self-organizing universe trending toward ever-increasing degrees of organized complexity. Scientists are attempting to measure complexity, and some contemporary philosophers have been evaluating it to inform ethical thinking.[55] Organized complexity is arguably a robust actualization of connectedness in the sense that complexity usually involves a large number of diverse relationships and interactions with high information content useful for driving novel and complex interactions, creations, and emergent phenomena. But the assessment of value as action would involve much more than just complexity since connectedness could presumably be manifested in many other ways. Indeed, connectedness and action are meant here in the very broadest sense. We are not necessarily limiting the notion of action to the kind of physical action we normally experience. It does not mean the universe and everything in it must always be buzzing as a sea of physical action as the only way to manifest connectedness and relationships and hence value. Connectedness, relationality, and action can presumably be realized in many different, unusual, and unknown ways. In effect, the connection-action principle arguably implies the potential realization of a very large, if not infinite possibility space of value.

Diversity and Freedom

The relationship between the connection-action principle and freedom and diversity is something like the relationship between teleological (utilitarian) and deontological (virtue or duty-oriented) ethical theories. Deontological theories can be thought of as shorthand rules of thumb for realizing the greatest good.[56] Similarly, diversity and freedom can be seen as rough "rules of thumb" for generating relationships and action since they give rise to the production of novel action in the world. Freedom, with its reduced restrictions on human agency (and other life-forms as well), can be seen as an effective "mechanism" for moving through a very large possibility space of action to find increasingly rich and complex relationships, interactions, and diversity. Unique or rare creations might be worth preserving because of how they contribute to diversity.

The connection-action principle, like the "principle of the sanctity of existence," also implies that existence is intrinsically valuable[57] in the sense that the more the universe produces, the more that exists, the more the universe's potential is realized. In its barest form, we again see the consistency with the principle of plenitude, versions of which have been proposed since Plato's Demi-urge—a "God" who wanted the world to lack nothing. Giordano Bruno, Spinoza, and Leibniz (in the form of the principle of sufficient reason) have also proposed ideas similar to the principle of plenitude and are the Western parents of Lovejoy's more modern articulation. Nozick also explores the idea of "fecundity" in way that is similar to the principle of plenitude.[58] The connection-action principle arguably provides a conceptual foundation for why the universe appears to be one of high fecundity and arguably high diversity. This kind of account is also consistent with those who cite diversity as a goal worthy in and of itself such as Freeman Dyson, who writes:

> Diversity is the great gift which life has brought to our planet and may one day bring to the rest of the universe. The preservation and fostering of diversity is the great goal which I would like to see embodied in our ethical principles and in our political actions.[59]

The connection-action principle differs from Dyson's first sentence by suggesting that the *universe may bring diversity to itself*, with the biosphere of Earth being an important example, perhaps one of many that may exist throughout the universe.[60] Indeed, in making choices consistent with this view, humanity may help encourage diversity here on Earth and throughout the universe. This kind of view might also assign a significant degree of value to unique nonliving objects such as the Grand Canyon or Valles Marinaris on Mars. We recognize these valuable actualizations to have "formed integrity"—to use Holmes Rolston's phrase[61]—and they contribute significantly to the diversity of the universe. If the continued existence of such forms does not prevent degrees of action from being realized in other ways, their preservation is presumably justifiable.

Freedom provides many paths of action. However, freedom is also distinct from diversity in an important sense. Freedom can be seen as a natural uninhibited realization of action—a kind of unfettered path to action. Freedom is in some sense a path of least resistance to creating relationships and actions and as such contributes as a kind of efficient diversity generator. Biological and intelligent agents appear to exhibit and strongly pursue freedom of action, so we might say that life and intelligence can be robust sources of cosmic value—but not at the exclusion of all else. This kind of perspective transcends human and biological biases and extends value to the broadest sphere, while also acknowledging high degrees of value for living things that benefit from freedom of action and as a result may bring additional diversity to the universe. In general, thinking about the universe and value this way would arguably encourage us to be careful about how we interact with the rest of the universe. A commonsense consequentialist approach would encourage us to exercise thoughtful caution about possible consequences of our actions before we act. It would require us to be responsible and respectful to more than just the human species—again, something many humans are increasingly recognizing intuitively or by exposure to ideas regarding the value of nonhuman life.[62]

Aesthetic Value and Organic Unity

In the spirit of seeking underlying unifying principles, it can help to have a comprehensive account of value so that it might tie together ethical value theory and aesthetics. We can argue that to some extent ethical value has a strong, if not critical, "aesthetic" component—at least in the sense that perceived ethical value might arise from internal emotional guidance that pulls us toward things we find "appealing" or "attractive" for any variety of reasons, including evolutionary reasons. Pursuing this kind of approach is potentially a way to find common ground between emotive and realist views of values and ethics if those aesthetic principles can be shown to "supervene" on or emerge from mental states which are generally shared by all humans or all beings capable of the relevant experiences.

Robert Nozick draws on the concept of "organic unity" from aesthetics and uses it to help ground intrinsic value. He writes: "The more diverse the material that gets unified (to a certain degree), the greater the value."[63] This unity in diversity is what Nozick suggests can be equated with intrinsic value.

Importantly, he suggests that organic unity might be the best approximation to value because our experience may be limited regarding what is valuable.

Although Nozick gives a helpful account of how organic unity fits with our general perception of value, a more objective justification for why organic unity should be considered valuable might be worth pursuing in the context of the connection-action principle. Nozick is careful to distinguish between demonstrating the value of organic unity from demonstrating why there should be value at all. Without going into all the details of Nozick's criteria, the first question as to why organic unity would be valuable is not addressed in a fundamentally objective approach because his criteria for being valuable are ultimately connected to a valuing agent. Nozick admits this dependence when he acknowledges that all we can do is approximate value because, as noted above, our experience is likely to be incomplete. Regarding objective value, Nozick writes:

> The philosopher's quest for a basis for (and a theory of) objective values, to ground them and understand them, itself involves a value judgment: it is better that there be objective values.[64]

Regarding why there should be any value at all, Nozick writes:

> What is needed to bring value to our universe is our reflexive choice that there be value, our reflexive imputation of the existence of value. . . . The fundamental choice open to us is whether or not there will be value.[65]

Andres Altman responds with:

> The basic problem with Nozick's explanation of how value arises is that it presupposes the very phenomenon which it is supposed to explain. To suppose that the fundamental value choice can be reflexive in the way Nozick describes is to presuppose that human choice can create objective value. But that is the mystery that needs explanation. How can human choice bestow on any fact (including facts about organic unity) prescribe authority? How can we, by a simple act of will, bestow upon anything the normative power to determine what our preferences and choices ought to be? Nozick's theory does not answer these questions, and by failing to do so, it fails in explaining how organic unity, or any other factual feature of the world, could have the status of objective value.[66]

Altman may underestimate the possibility for reflexive choice to produce something that is, in some important sense, "objective." For example, perhaps in the sense that "emergent" social conventions, relations, and properties can arise and become independent of any one person's views, there can be a certain kind of objectivity—perhaps at least some limited or partial objectivity of sorts. Altman himself arguably comes close to this suggestion by indicating that social agreement be required for what is objective. However, Altman rightly brings light to the more fundamental challenge of how true objectivity can be fundamentally linked with value.

Regardless of whether organic unity can be considered "objective" value and/or intrinsically valuable, organic unity arguably follows from the connection-action principle in the sense that organized complexity is similar to organic unity because organized complexity involves the organization, or "unity," of diverse interactions. Indeed, as Nozick points out, biologists often refer to organisms as having organic unity. It may even be that organic unity is redundant in that the term "organic" captures both diversity (the diversity enabled by the organic world) as well as unity, in the form of "organization"—a word apparently related to "organic." Notably, Whitehead called his philosophical framework a "philosophy of organism."[67] The connectedness of the universe corresponds to unity, and action creates diversity. The connection-action principle says that diversity (e.g., novel action) comes from the connectedness (e.g., unity) of the universe as it realizes its nature. The connection-action principle appears to at least be consistent with Nozick's use of organic unity, if not an underlying theoretical foundation for it.

CHALLENGES

We might ask why the connection-action principle rises to the level of being called a principle. The connection-action principle makes a fundamental claim with varying degrees of strength and consequence depending on the version and the associated implications—as touched on previously. The fundamental claim proposes what can be considered a conceptual or metaphysical foundation for why the universe is a relational, dynamic, "creative" entity. The idea does not stipulate a relational metaphysics as a first principle but instead provides a potential conceptual grounding for why there might be relationality to begin with—namely that the fundamental property of connectedness gives rise to relationality, which is then instantiated as many forms of action—perhaps necessarily and perhaps in ever-increasing degrees in stronger versions touched on previously.

The connection-action principle then is a specific fundamental theoretical proposal about the nature of reality that is arguably consistent with, if not a foundational explanation of, the relational nature of fundamental science such as relativity and quantum theory,[68] and perhaps even informational models of reality that are receiving increasing attention. So, the connection-action principle is simultaneously a specific form of a relational metaphysics as well as an explanation for why a relational metaphysics might exist at all. The principle also has potential for broad applicability—including applicability to value theory as briefly explored above.

But of course, there are challenges. For example: (a) incompleteness of the science and philosophy that attempts to defend the idea of connectedness as a fundamental property of the universe, (b) questioning the linkage that relations and actions follow from connectedness, (c) questioning the need to invoke a conceptual construct or abstraction to explain metaphysical "why" questions that may be misguided, unanswerable or untestable, (and for which much science may suffice anyway), (d) inferring value, and possibly intrinsic value, from what may be tenuous conceptual speculations that could also have problematic consequences and be misused, and (e) a high degree of generality or variability and a lack of specific "measurable" practical utility. Many of these concerns have been at least partially addressed previously in this chapter and much further in *Cosmological Theories of Value*, but not all of these challenges will be addressed further here. Instead, we will focus primarily on the potential for the connection-action principle to have some practical utility.

Admittedly, it is not clear how to make an idea like the connection-action principle sufficiently scientifically operational and concrete enough to produce precise measurable predictions (although some general predictions are suggested in *Cosmological Theories of Value*). One significant "epistemological" challenge with the connection-action principle is that it can be interpreted to be consistent with almost any observations and predictions involving relationships, action, interaction and/or intra-action, creativity, diversity, complexity and increasing degrees thereof—which is arguably applicable to "everything" and too general to be specifically practical and useful. Indeed, one way to think about a good explanation or theory is that it can't be varied too much.[69] Also, as touched on previously, how would we "measure" implications of the principle in a consistent and credible way that could inform questions of value and ethics?

Practical Ethical Challenges

When it comes to making practical decisions, either personally or collectively, that's when the ethical rubber hits the road and when many general principles can become insufficient or even counterproductive. The obvious difficult question here is how the connection-action principle could help with more practical ethical questions, as well as more metaethical guidance that might help inform specific ethical deliberations.

Perhaps the biggest challenge of applying the connection-action principle to practical ethics is that even if the universe is about realizing its property of connectedness as action, can human beings, or intelligent beings more generally, sufficiently identify with the universe or use it for ethical analyses in any sensible way?[70] Further, why *should* we? The universe (or multiverse?) may be the biggest, most compelling absolute we are aware of (other than suggestions of "spirit" or God), but it does not necessarily follow that we must then use the universe to obligate ourselves to something like the connection-action principle and any associated ethical implications. This is what the naturalistic fallacy and the "is–ought" distinction properly sensitize us to. However, the kind of reasoning that takes us from a perceived reality of the universe to more value-oriented judgments based on that reality can also arguably be a philosophical choice that sufficiently aware beings can make if it is thought to be warranted and carefully reasoned.

While there might be aspects of this principle that are consistent with human predispositions, presumably there are some that are not. For example, human beings very often value simplicity and a lack of diversity and a lack of action in many cases. If evolution can explain these predispositions, maybe we don't need more explanation or further basis for human values—we simply accept "evolutionary values" and make messy practical choices that attempt to create relatively stable, balanced, and rewarding individual and social lives. Maybe that's all that can and should be attempted in practical ethics—and perhaps in our worldviews more generally. Even if we were to accept something like the connection-action principle as a way to assess practical ethical questions, how would it work in terms of the details for any given ethical challenge? For example, how would we weigh or measure the best "action" in any situation? Is it the *amount* of action, or *kind* of action, or both, or something else? And, how could it be assessed with satisfying precision, consistency and credibility? While there may be ways to ultimately attempt these kinds of assessments, it is also reasonable to suppose that it will be essentially impossible to make these assessments with enough consistency or credibility to satisfy enough people. There would presumably be many open-ended and highly contradictory value judgments about how best to realize connectedness through relationships and action—and that raises a flag about practical applications.

Nevertheless, the idea of "value measurement" (in this case "measuring" degrees of relationality and action in all the forms it may take), while possibly having some undesirable effects such as being used improperly (as "survival of the fittest" or "social Darwinism" has been), could, in theory, be carefully pursued and based on careful assessments of what constitutes degrees of relationality and action in a way that is cautious and thoughtful. As noted previously, this could be extremely complicated but may ultimately have a helpful basis in science to the extent that things like interactions and *intra*-action, diversity and complexity, etc., are sufficiently definable, measurable, and perhaps inter-relatable. Such "measurements" could in principle include many complicated factors including a wide range of human emotions, perceptions, experiences, desires, and aspirations.

Cosmic Destruction and Cosmic Perpetuation

The fact that there is much destruction in the world—a primary form being the basic need for life to destroy and consume life—can be interpreted to go against, or at least be inconsistent with, the connection-action principle. However, we might consider that this is just how the universe has evolved in our little corner of the cosmos, and that it need not necessarily be this way everywhere. Or maybe cultural evolution can lead to significant reductions, if not elimination, of unnecessary destruction.

Despite the powerful logic of natural selection, it might be justified to consider sufficiently different forms of it elsewhere, or perhaps it may not be necessary at all for the evolution of certain kinds of life and intelligence elsewhere—or at least for sufficiently complex "self-organizing" and self-replicating systems elsewhere.[71] Natural selection may be completely transcended by many forms of cultural evolution, including the evolution of artificial intelligence, for which reduced destruction might be an organizing ethic. Artificial intelligence, or "super-intelligences" may be able to more easily transcend destructive tendencies since presumably such intelligence will not be controlled by biologically driven tendencies and will have a broad base of knowledge and experience to draw from to help reduce destruction.

We might also consider that in some sense the universe, or multiverse, is not capable of perfectly or optimally manifesting connectedness, so that there are limitations of sorts to what the universe can do (e.g., limitations to complexity?).[72] So, what we observe as destruction, including the second law of thermodynamics, may be partly indicative of some kind of cosmological or "universal" limitation. Related to that, we might also consider the oft-cited notion that there must be destruction in order to have creativity—what is often referred to as "creative destruction." This may be indicative of an inherent limitation to the universe. Conservation laws may require it. Regardless, the connection-action principle does not necessarily require that there be an "optimal" universe.[73]

Additionally, it may be that part of what we can do as intelligent beings is to help the universe actualize itself to greater extents—we may be a means by which unnecessary destruction, or destruction more generally, can be minimized or maybe even eliminated entirely. The problem of destruction is more powerfully realized by considering that in an oscillating universe model, it is often suggested that no form of life or any kind of information could survive an event such as big crunch or big bounce. Such complete destruction would seem to oppose the idea that the universe trends toward ever-increasing actualizations of connectedness—particularly increasing organized complexity like life and intelligence—if ultimately it is to only be completely destroyed. We might deal with this concern by again considering the notion that the destruction of what is created may ultimately be necessary for more new creations to subsequently emerge. This may simply be a fundamental limitation to the universe or any given universe. The universe might just get to a point where it must "start over" or "recycle" itself because there are certain limitations to the universe and its evolution. Perhaps this "recycling" is "good" in that it allows for fundamentally new creations to emerge, for new actualizations of the universe's nature to be realized.[74]

In a cosmic "heat death," or what is sometimes referred to as the "big freeze" or "big chill," it appears action would essentially come to an end.[75] However, if there is a multiverse, then the destruction that may occur in one universe, or the destruction of any particular universe as whole, may not be a significant problem for realizing the connection-action principle since there would be many other universes—especially if universes give birth to more universes before they themselves come to an end. Similarly, the "many-worlds" interpretation of quantum mechanics may help realize the connection-action principle in the sense that constantly branching universes are essentially alternate realities in which destruction in one universal timeline may be avoided in another.

Worldviews that suggest the intentional creation of endless universes as a fundamental ethic (e.g., consistent with ideas from Vidal[76]) would be consistent with the connection-action principle. It may very well be that life and intelligence are a vehicle by which the nature of the universe can be robustly realized as we evolve and add to the creations of the universe. We may very well be a way for the universe to realize its greatest potential, perhaps for an eternity. So perhaps the big crunch will not be the end. Maybe it will be possible to somehow survive a big crunch? Could it be avoided? Could we send information "through" such an event? Perhaps there will be a way for intelligent beings to survive it. For example, could intelligent beings become so embedded in the fabric of the universe (perhaps as forms of information embedded in the universe) that a big crunch will just become a "rebooting" opportunity with some of the same, yet somewhat different, operating "software" in place?

It is impossible of course to predict what life and the universe will be like at long-term epochs, but one general possibility is that intelligence and the universe will become so interconnected, so *intra*-connected, that the universe will become a kind of cosmic "mind" or universal "consciousness" with value theory and ethical pursuits as powerful motivations (Teilhard De Chardin 1955, Tipler 1994).[77] Indeed, it may be that for intelligent beings to survive and evolve, such a "merging" might not be as fanciful as it sounds. If that kind of cosmic merging between mind and matter were to occur, consciousness would no longer be a subset of the universe and could be less vulnerable to destruction at a big crunch or other cosmic critical events that might occur, including a heat death. Instead, "we" would be completely and truly "one with the universe" in some new unimaginable form, forever existing, forever morally creating.

COSMIC MEANING AND ETHICS

The meaning of meaning can be slippery. It can be broad, vague, diverse, and very subjective—even more so than ethics which often tries to seek more practical common ground than meaning-seeking. The intent here is not to cover the subject of meaning or worldviews in general, for which there is already much academic and popular literature.[78] Instead the intent here is to briefly touch on some potential implications of the connection-action principle for meaning and ethic.[79]

The Lens of Evolutionary Psychology

Here we will touch briefly on the "psychology of philosophy," noting how some psychological predispositions can find affinity with certain kinds of philosophical orientations—particularly as it might apply to the connection-action principle. For example, we previously noted Robert Nozick's observation that the pursuit of an objective basis for values presupposes, and/or at least implies, the value judgment that it is indeed better to have objective values as opposed to subjective values. This kind of perspective could be influenced heavily by particular psychological tendencies or predispositions for which evolutionary psychology may be able to shed light.[80]

For example, if group members can be convinced that there is an objective basis for values, then those values may carry more force and take hold better within the group. "Objective" values may be perceived to have a source that is not restricted solely to what other group members think or want for themselves, i.e., objective values can be seen to be less selfish or less beneficial to subsets of groups. Objective values could be perceived to be associated with something much more powerful or all-encompassing—something more deeply true and highly consequential. If those values help integrate a group to act cohesively (including aggressively toward out-groups), to act as good resource managers, and to learn about what is independently real about the environment, then such tendencies for seeking objectivity could persist in subsequent generations of groups via group-level selection.

Similarly, and more specifically with respect to the connection-action principle, certain psychological architectures may be predisposed to be highly sensitive to connections and relationships in order to navigate and survive in a complex natural and social world. A sensitivity to relationships and how things are "connected," can help individuals and groups feel better and understand the world better. A sensitivity to connectedness and relationships in one's group can make an individual a good group member and help increase personal and group fitness.

A sensitivity to relationships is consistent with how the idea of a "relational self" can stem from child-rearing.[81] Related to this, love, and how it may motivate familial and group connectedness, may be appealing because of its power to motivate a sense of belonging and commitment to children and one's group. Additionally, a general sensitivity to action in the world, perhaps in the form of being "active," e.g., in motion, producing, accomplishing tasks, etc., presumably can also increase the fitness of individuals and groups.

Diversity may be appealing because it may help a group be prepared for different challenges and perhaps also help "outlier" individuals promote values that would allow them to be better tolerated by a group.[82] Extreme diversity, perhaps even in the form of the principle of plenitude, may also have a certain appeal in the sense that it allows for all things, all actions, all possibilities to be realized—perhaps creating the impression that constraints can be overcome and "anything is possible." Related to this, putting ourselves in a privileged position of being powerful creators, and/or cocreators and arbiters of value and reality can also be psychologically comforting and empowering in the sense that it confers great privilege and power to ourselves—a tendency for which there is an obvious problematic history.

Highly active minds, or unusually hyperactive minds, may be drawn to action (particularly as action relates to productivity), diversity, creativity and complexity and find those pursuits appealing because it can keep people active, engaged, and cognitively challenged. This kind of appeal would merely be a result of how such brains work, perhaps as a result of selective pressures, not necessarily because there is anything more objective or independently true or valuable about such tendencies. Indeed, perhaps the simple process of natural selection can predispose us to see more connectedness and complexity than there really is—to see patterns, meaning and purpose where there may otherwise be none. We should not underestimate how much the simple mechanism of natural selection can produce so much complexity, including unnecessarily complex mental states and behaviors.

So, perhaps behind much philosophical musings and meaning-making are important, but sometimes misleading, psychological motivations, particularly evolutionary motivations driven by individual and group selection—at least as "by-products" of such selective pressures, if not direct fitness-enhancing selective pressures. But exploring these kinds of psychological factors and sensitivities does not need to fundamentally undermine philosophical pursuits or philosophical psychological predispositions. It is touched on briefly here to heighten our sensitivity and encourage a certain amount of humility and open-minded caution in our philosophical explorations—particularly those of a highly speculative and sweeping nature such as those involving the universe as a whole.

Cosmocentric Ethics: A Martian Scenario

Having suggested foundations of value theory that relate to cosmological considerations, we can now apply those views to an unusual ethical challenge. Here we will emphasize stronger versions of cosmocentric ethics, based primarily on the connection-action principle which arguably provides a foundational metaphysics as a way to more strongly support the idea of a "cosmocentric ethic" since it provides something foundational and consequential in the form of a proposed fundamental characteristic of the universe, i.e., connectedness and its implied "derivatives" such as relationality, action, diversity and complexity. Stronger versions of cosmocentric ethics could be characterized with criteria such as (1) establishing the universe as a priority in a value system, (2) appealing to something characteristic of the universe, which might then (3) provide a justification of value, perhaps intrinsic value, and perhaps (4) allow for some kind of assessment of degrees of value.[83]

At first glance, talk of a cosmocentric ethic might seem paradoxical. How can an ethical view be centered or focused on "all that is"? From egocentrism to geocentrism, we are able to more easily "centralize," focus, and prioritize value. Nevertheless, we can explore the possibility that such a cosmic ethic may be helpful in dealing with value-based questions of many kinds, including those involving extraterrestrial issues such as interaction with extraterrestrial life.

The Potential Value of Martian Life

Underlying many of the above noted challenges are questions of value. How much might we value the preservation of "primitive" extraterrestrial life such as microorganisms and why—and what actions might be needed to preserve that value?[84] Certainly, there is instrumental value, or more specifically, scientific value associated with Martian life. Masking the existence of such life and/or destroying it beyond recognition would be a scientific loss of immense proportion. Biology would benefit from learning about a "second data point" (another kind of life), and there are many important questions that need consideration if we are to ensure the benefits associated with this scientific value. However, as history has painfully demonstrated, the momentum of accomplishing certain goals, particularly big goals, often overshadows contemplation of consequences and any associated policy needs. The exploration and exploitation of the Americas are a poignant example of the harm our species is capable of (intentionally and unintentionally). Looking ahead, we can wonder how we will guide our actions before and after the scientific utility and novelty wears off.

Sanctity of Existence

Don MacNiven has suggested that a central tenet of a cosmocentric ethic would be the "principle of the sanctity of existence," which he notes would make it difficult to justify the significant modification or destruction of indigenous life.[85] In a minimal sense, the principle of the sanctity of existence seems to satisfy criteria one and two for a cosmocentric ethic noted above because the universe, and all therein, *exists*. MacNiven additionally suggests appealing to a "selective concept of uniqueness" as we sometimes do in considering terrestrial matters (e.g., preserving the Grand Canyon). With MacNiven's principle of the sanctity of existence we see the challenges of not directly satisfying criterion three of the previously noted cosmocentric ethic criteria, as well as the additional challenge associated with criterion four of "measuring" value—or more specifically, the relative value of human activity compared with other forms of value such as the preservation of an extraterrestrial life-form.

Somewhat related to the "sanctity of existence," others have applied Eastern views such as Buddhism and Hinduism to suggest that ideas such as (a) the principle of non-duality (emphasizing the "unity" of all existence and the de-emphasis of self), (b) "dependent co-arising," and (c) the common origin, interrelatedness, and interdependence of all beings for their existence, to suggest a certain degree of sensitivity and moral consideration to that which is seen as "totally other."[86] Here too we see broad suggestions seemingly implying some kind of intrinsic value and deep sensitivity for other life-forms, but without specific suggestions about how to assess what might be relative degrees of value to help guide practical behavioral actions and policies.

Complexity

As explored previously, complexity—possibly *increasing* complexity—may be one of many manifestations of the connection-action principle.This suggests we may want to consider the emerging science of complexity as a basis for a cosmocentric ethic, particularly when thinking about extraterrestrial life. A science of complexity could not only provide a scientific basis for value, but it could conceivably allow for relatively precise assessments, possibly even quantification, of degrees of value.

Kelly Smith ties rationality to a triad that also includes sociality and culture which ultimately gives rise to immense complexity, where complexity appears to be a trend of biological evolution.[87] Seen in totality, the biocultural evolutionary trajectory here on Earth can be interpreted to exhibit a trend from relative simplicity to increasing complexity. That complexity could be theoretically contingent on the particular twists and turns that natural selection took here on Earth, but as Smith suggests, it may also be indicative of "universal" trends for all life in the universe, and hence, could serve as a foundational ethic for all life and intelligence in the universe. Delahaye and Vidal go further and suggest that organized complexity might have intrinsic value and hence provide a "measurable" basis for universal ethics.[88]

As explored previously, the connection-action principle goes further in the sense that it goes beyond biology and organized complexity and suggests a conceptual theoretical underpinning for why the universe might trend in directions of organized complexity (among other possible trends that manifest connectedness via increasing degrees of relationships and actions), and hence may provide a basis for something like "intrinsic cosmological value" (intrinsic to cosmic evolution) for which complexity may be one of many forms.

Diversity

For those who would suggest that Martian life has "rights," a peaceful coexistence compromise might not be satisfactory. Only a noninterference policy would be acceptable if there is a concern that our presence would compromise the existence of that life.[89] However, we might also consider Chris McKay's view that the rights of Martian life "confer upon us the obligation to assist it in obtaining global diversity and stability."[90] This kind of view might also be consistent with the connection-action principle and its implications for favoring diversity throughout the universe—assuming we were indeed capable of assisting life on another planet (something that would require great caution of course). So if we interpret the connection-action principle to suggest that diversity is a manifestation of a form of cosmic value, then perhaps we can justify the preservation and "assistance" of other extraterrestrial life-forms based on a contribution to cosmic creativity and cosmic diversity?

A Morally Creative Cosmos?

The idea that the universe may be realizing its fundamental property of connectedness through relationships and actions might help inform our perceptions of value and meaning of some ethical and moral sentiments by seeking out and creating diversity, novelty, dynamism, complexity, and creativity more generally. This kind of view, while potentially problematic for a number of reasons touched on previously, might still help guide our ethical perspectives by recognizing a broader base of value beyond the subjective value of our individual and collective predispositions and deliberations.

The connection-action principle suggests not only that we can recognize and perhaps revere a highly connected, relational, dynamic, and diverse universe, but that we can also encourage those characteristics in our contemplations and actions and act as cocreators with the universe to help give rise to a "morally creative cosmos"—perhaps with other extraterrestrial beings that may exist who we may see as robust creations of the nature of the universe.[91] We might generalize further, noting that some ethical interpretations of the connection-action principle could suggest that our connectedness and relationships to each other, human or otherwise, suggest the importance of taking action on behalf of each other.

NOTES

1. Whitehead, *Process and Reality*.

2. The initial expansion may have included an extremely rapid and unusually accelerated "inflationary" expansion phase suggested by inflationary theory. See Guth and Steinhardt, "The Inflationary Universe"; Linde, "The Self-Reproducing Inflationary Universe"; and Guth, *The Inflationary Universe*.

3. See Krauss, *A Universe from Nothing* for a recent scientific exploration of an origin from nothing, but which nevertheless seems to fall short of explaining the emergence of the universe from truly nothing—at least in the traditional philosophical sense of "nothing" (itself a premise Krauss appears to challenge).

4. Bostrom, "Are You Living in a Simulation?"

5. Barrow and Tipler, *The Anthropic Cosmological Principle*.

6. Steinhardt and Turok, "A Cyclic Model of the Universe" and "Cosmic Evolution in Cyclic Universe."

7. Smolin, *The Life of the Cosmos*.

8. Tegmark, "Parallel Universes."

9. Von Neumann, *The Mathematical Foundations of Quantum Mechanics*; Wheeler, "Information, Physics, Quantum"; Davies, "Life, Mind, and Culture as Fundamental Properties."

10. Turchin explicitly links the epistemological criticality of action with an action ontology. See Turchin, "The Cybernetic Ontology of Action." Mermin emphasizes action when he writes of QBism: "in QBism, on the other hand, a measurement can be *any* action taken by any user on her external world. The outcome of the measurement is the experience the world induces back in that particular user, through its response

to her action." Mermin, "Why QBism Is Not the Copenhagen Interpretation," 89. This is also arguably similar to relational quantum mechanics (RQM) in the sense that it emphasizes the tight relationship between observer and the "external world" and generalizes to all observers and so could be seen as a subset of RQM which generalizes relational dependencies beyond observers, or "agents" or "users," to *all* objects/systems in the world. See Rovelli, "Relational Quantum Mechanics" and Vidotto, "Relational Quantum Cosmology." We can also note here an apparent consistency with Paul Dirac's "*Interaction* Picture" of quantum mechanics.

11. Whitehead, "First Lecture," 12. Thanks to Andrew M. Davis for bringing this recently discovered First Lecture by Alfred North Whitehead to my attention (Dec 20, 2021).

12. Indeed, the very early universe may have been highly chaotic, without much, if any, "normal" lawlike behavior. See Linde, "Eternally Existing." Wheeler questions the traditional notion of physical laws in general. See Wheeler, "Law Without Law."

13. Bohm, *Wholeness and the Implicate Order*; Bohm and Hiley, *Undivided Universe*.

14. Leopold, *A Sand County Almanac*.

15. Sciama, *The Unity of the Universe*.

16. See McHenry, *Event Universe*.

17. A similar sentiment seems to be expressed by Whitehead when he writes: "In what sense can unrealized abstract form be relevant? What is its basis of relevance? 'Relevance' must express some real fact of togetherness among forms. All real togetherness is togetherness in the formal constitution of an actuality." Whitehead, *Process and Reality*, 32.

18. Quante and Engelhard provide a variety of perspectives regarding "potentiality" and ways to use words such as "manifest," "instantiate," "actualize," and "realize"—all of which are essentially used interchangeably in this context. See Quante and Engelhard, *Handbook of Potentiality*.

19. See Davis, *Mind, Value and Cosmos* for a recent exploration of the "relational nature of ultimacy."

20. See Lupisella, *Cosmological Theories of Value*.

21. This is an important consideration that will not be explored in detail here. Instead, it is suggested that it is reasonable to consider that to possess a property, can, at least in some cases, imply some form of *increasing degree* of realizing that property, perhaps over time, but also perhaps in ways that are independent of time as well.

22. Whitehead proffers a similar observation with stronger, and arguably narrower, language when he writes in *Process and Reality*: "'Creativity' is the principle of *novelty*" (21).

23. Lovejoy, *The Great Chain of Being*.

24. Esfeld, "The Reality of Relations."

25. For example, stars are highly active, highly dynamic entities, essentially created by the "connecting" force of gravity. Stars are literally "turned on," turned into high-energy states as matter is connected together by gravity, which then often leads to the creation of new elements and their distribution through highly dynamic explosions (or supernovae) which then contribute further to creative dynamics such

as providing energy to nearby systems, causing mutations in replicating molecular systems, ultimately becoming the machinery of our bodies and leading to intelligence. Indeed, it seems we are, in many critical ways, the direct result of star dynamics, or "star stuff," as Carl Sagan and many others have poignantly observed.

26. Smolin, *The Life of the Cosmos*.

27. Kashlinsky et al. provide observational data that some have interpreted as another universe interacting with ours, e.g., "tilting" it. See Kashlinsky et al., "A Measurement of Large-Scale Peculiar Velocities." See also Mersini-Houghton and Holman, "'Tilting' the Universe"; Mackenzie et al., "Evidence Against a Supervoid"; Alonso-Serrano et al., "Interacting Universes."

28. See Khoury et al., "Ekpyrotic Universe" and Webb, *Out of This World*. We can also make note of the "connectedness" of potentially interacting parallel universes under the "many-worlds" interpretation which is a kind of connected branching of universes. See Hall et al., "Quantum Phenomena Modeled by Interactions."

29. See Sciama, *The Unity of the Universe*.

30. What constitutes "greater" action has been briefly touched on in the very general sense of suggesting characteristics and qualities such as emergence, creativity, complexity, self-organization, diversity, etc. But this is a very open-ended question, and "greater" action could presumably take many forms, which could be explored in subsequent work, perhaps even with quantified analyses.

31. Volk has used the term "combogenesis" to emphasize how simpler smaller entities have continuously been brought together throughout cosmic evolution to create new larger forms with new sets of relations. See Volk, *Quarks to Culture*.

32. See Oliver, *A Relational Metaphysics*.

33. Barad, *Quantum Physics*.

34. See Rovelli, "Relational Quantum Mechanics;" Vidotto, "Relational Quantum Cosmology."

35. This is also arguably consistent with interpretations of quantum mechanics such as David Mermin's suggestion that only correlations are ultimately real, that reality is constituted by "correlations without correlate." See Mermin, "What Is Quantum Mechanics Trying to Tell Us?"

36. For a contemporary refutation, see Briceño and Mumford, "Relations All the Way Down?"

37. See Turchin, The Cybernetic Ontology of Action"; Heylighen and Beigi, *Mind Outside Brain*.

38. Kauffman, *At Home in the Universe*, 62.

39. Davies, *The Fifth Miracle*.

40. This is similar to claims of relational quantum mechanics. See Rovelli, "Relational Quantum Mechanics."

41. See Wolfram, *A New Kind of Science*; Lloyd, *Programming the Universe*; Davies and Gregersen, *Information and the Nature of Reality*; and Bokulich and Jaeger, *Philosophy of Quantum Information and Entanglement*.

42. For a balanced and cautious view on using information is a fundamental construct, see Timpson, "Information, Immaterialism, Instrumentalism."

43. Turchin explored the idea of a "metasystem transition" to help explain significant transitions. See Turchin, *The Phenomenon of Science*.
44. See Rescher, *Complexity*.
45. Rescher, a proponent of process philosophy, sees nature as continuously evolving process toward increasing complexity. See Rescher, *Process Metaphysics* and *Metaphysics: Key Issues*.
46. Bohm, *Wholeness and the Implicate Order*, viii.
47. Smart, "Evo Devo Universe?"
48. Deutsch argues that much of what we think of as fiction is likely to be fact somewhere in the multiverse. See Deutsch, *The Beginning of Infinity*.
49. See Carroll, *From Eternity to Here*.
50. Krauss suggests that the unstable nature of the quantum vacuum state postulated by quantum field theory is effectively a state of nothingness from which the universe arose via the activities of virtual particles that emerge from that quantum vacuum state. However, there are disagreements as to whether a quantum vacuum state is truly "nothing" since there appears to be activity associated with such a state. However, it is not clear that this distinction is significant for our purposes here, with the possible exception that a true state of nothingness might be "unstable" and/or logically impossible if connectedness needs to be manifested by action (or in some other way). See Krauss, *A Universe From Nothing*; Albert, "On the Origin of Everything."
51. Moore, *Principia Ethica*, 286.
52. Lemos, *The Nature of Value*, 34.
53. See Handy, *The Measurement of Values*; Zenzen and Hammer, "Value Measurement and Existential Wholeness."
54. Marshal, "Ethics and the Extraterrestrial Environment."
55. See Smith, "Manifest complexity"; Delahaye and Vidal, "Universal Ethics."
56. Altman, "Nozick's Theory of Value"
57. See MacNiven, *Creative Morality*.
58. Nozick, *Philosophical Explanations*.
59. Dyson, *Infinite in All Directions*.
60. The large number of exoplanets that have been discovered points to an increasing probability of "Earth-like" planets beyond our solar system.
61. See Rolston, "The Preservation of Natural Value"; and *A New Environmental Ethics*.
62. See Singer, *The Expanding Circle*; Rollin, *Animal Rights and Human Morality*.
63. Nozick, *Philosophical Explanations*, 416.
64. Ibid., 434.
65. Ibid., 536.
66. Altman, "Nozick's Theory of Value," 150.
67. Whitehead, *Process and Reality*.
68. Lupisella, *Cosmological Theories of Value*.
69. Deutsch cautions against theories that are too easy to vary. See Deutsch, *The Beginning of Infinity*.
70. Lupisella, "Is the Universe Enough?"

71. See Lupisella, "Caring Capacity and Cosmocultural evolution," for speculation regarding a biological non-Darwinian intelligence.

72. For example, John Smart explores a model that "would predict the necessity of a branching tree of self-organizing complexity underlying our universe, and an abundance of very simple proto-universes coexisting in the multiverse with a comparatively tiny number of complex universes such as ours." Smart, "Evo Devo Universe?," 231.

73. Freeman Dyson notes his brand of Socinian theology as one where "God is neither omniscient nor omnipotent. God learns and grows as the universe unfolds." Dyson, *Infinite in All Directions*, 119. We can also take note of a related suggestion by John Corliss that the evolution of the universe is a learning process. See Corliss, "The Flow of Energy."

74. Stapledon, in his classic science fiction novel *Star Maker* (1937) explores the theme of endless universal cycles as part of an "infinite spirit's" (the Star Maker) attempts at perfect, but failing, creative activity.

75. Related, Freeman Dyson was one of the first to explore the possibility of life and communication continuing forever even as the universe approached a potential heat death. See Dyson, "Time Without End."

76. Vidal, *The Beginning and the End*.

77. Teilhard de Chardin, *The Phenomenon of Man*; Tipler, *Physics of Immortality*.

78. See for example Alerts et al., *Worldviews*; *World Views and the Problem of Synthesis*; Babbage and Ronan, "Philosophical Worldview and Personality Factors"; Wilson, *Darwin's Cathedral*; Shermer, *Why People Believe Weird Things*; Koltko-Rivera, *The Psychology of Worldviews*; Gershenson et al., *Worldviews, Science and Us*; Bulbulia et al., *The Evolution of Religion*; Johnson et al., "Integrating the Study of Culture and Religion"; Henriques, *A New Unified Theory of Psychology*; Vidal, "Metaphilosophical Criteria" and *The Beginning and the End*; Nilsson, "The Psychology of Worldviews"; Hedlund-de Witt et al., "Exploring Worldviews"; Saucier, "Isms Dimensions"; Saucier et al., "Cross-Cultural Differences."

79. For additional treatment, see Lupisella, *Cosmological Theories of Value*.

80. Related, Rescher makes a useful distinction between epistemic values and other kinds of values, where epistemic values are values that help inform how we learn about the world. In this context, seeking an objective basis for values could be considered an epistemic value, and arguably has an evolutionary source. See Resher, *The Strife of Systems*.

81. Barlow, *Green Space, Green Time*.

82. Haidt discusses how group selection can produce different levels of psychological predispositions toward diversity in order to satisfy a variety of group functions and also enhance individual fitness. See Haidt, *The Righteous Mind*.

83. Lupisella and Logsdon, "Do We Need a Cosmocentric Ethic?"; Lupisella, "Is the Universe Enough?"

84. Callicott, "Moral Considerability and Extraterrestrial Life"; "On the Intrinsic Value of Nonhuman Species"; Haynes, "Ecopoiesis"; Marshall, "Ethics and the Extraterrestrial Environment"; Lupisella, "The Rights of Martians?", "Ensuring the Scientific Integrity of Possible Martian Life," and "The Search for Extraterrestrial

Life"; Cockell, "The Value of Microorganisms" and "The Ethical Status of Microbial Life"; Rummel et al., "Ethical Considerations for Planetary Protection"; Impey et al., *Encountering Life in the Universe*; Race, "Space Exploration and Searches for Extraterrestrial Life"; Ehrenfreund, *COSPAR Workshop*; Wilson and Cleland, "The Moral Subject of Astrobiology"; Smith, "The Trouble with Intrinsic Value" and "The Curious Case of the Martian Microbes"; Schwartz and Milligan, *The Ethics of Space Exploration*; Dick, *Astrobiology, Discovery, and Societal Impact*; Peters, "Does Extraterrestrial Life Have Intrinsic Value?"; Green, "Convergences in the Ethics"; Smith and Mariscal, *Social and Conceptual Issues in Astrobiology*.

85. MacNiven, "Environmental Ethics and Planetary Engineering."

86. Irudayadason, "The Wonder Called Cosmic Oneness."

87. Smith, "Manifest Complexity."

88. Delahaye and Vidal, "Universal Ethics."

89. Marshall, "Ethics and the Extraterrestrial Environment."

90. McKay, "Does Mars Have Rights?," 194.

91. Related to the connection-action principle and ideas of a moral universe, Burkhart points out how Native American thought often invokes principles, or "ways of being," he calls the "meaning-shaping principle of action" and "the moral universe principle," which, taken together can be interpreted to suggest that through our actions we can help realize and enhance a kind of "cosmic" morality that may already exist or that we can help create and bring to the universe. See Burkhart, "What Coyote and Thales Can Teach Us."

BIBLIOGRAPHY

Aerts, D. et al. *Worldviews: From Fragmentation to Integration*. Brussels: VUBPRESS, 1994.

———. *World Views and the Problem of Synthesis*. Dordrecht: Kluwer Academic, 1999.

Albert, D. "On the Origin of Everything." *New York Times Book Review* (2012). https://www.nytimes.com/2012/03/25/books/review/a-universe-from-nothing-by-lawrence-m-krauss.html.

Alonso-Serrano, A et al. "Interacting Universes and the Cosmological Constant." *Physical Letters B* 719, no. 1–3 (2013): 200–5. https://doi.org/10.1016/j.physletb.2013.01.013.

Altman, A. "Nozick's Theory of Value and its Implications." *Southern Journal of Philosophy* 22, no. 1 (1984): 139–53. https://doi.org/10.1111/j.2041–6962.1984.tb00331.x.

Babbage, D. R., and K. R. Ronan. "Philosophical Worldview and Personality Factors: Studying the World in Our Own Image." *Personality and Individual Differences* 28, no. 2 (2000): 405–20. https://doi.org/10.1016/S0191-8869(99)00117–8.

Barad, K. *Quantum Physics and the Entanglement of Matter and Meaning*. Durham and London: Duke University Press, 2007.

Barlow, C. *Green Space, Green Time: The Way of Science*. New York: Springer, 1997.

Barrow, J. D., and F. J. Tipler. *The Anthropic Cosmological Principle*. Oxford University Press, 1986.

Bohm, D. *Wholeness and the Implicate Order*. London: Routledge & Keegan Paul, 1980.

Bohm, D., and B. J. Hiley. *Undivided Universe: An Ontological Interpretation of Quantum Theory*. New York: Routledge, 1993.

Bokulich, A., and G. Jaeger. *Philosophy of Quantum Information and Entanglement*. Cambridge: Cambridge University Press, 2010.

Bostrom, N. "Are You Living in a Simulation? *Philosophical Quarterly* 53, no. 211 (2003): 243–55.

Briceño, S., and S. Mumford. "Relations All the Way Down? Against Ontic Structural Realism." In *The Metaphysics of Relations*, edited by A. Marmodoro and D. Yates, 98–217. Oxford: Oxford University Press, 2016.

Bulbulia, J. et al., eds. *The Evolution of Religion: Studies, Theories, and Critiques*. Santa Margarita: Collins Foundation, 2008.

Burkhart, B.Y . "What Coyote and Thales Can Teach Us: An Outline of American Indian Epistemology." In *American Indian thought*, edited by A. Walters, 15–26. Mauldin: Blackwell, 2004.

Callicott, J. B. "Moral Considerability and Extraterrestrial Life." In *Beyond Spaceship Earth: Environmental Ethics and the Solar System*, edited by E. C. Hargrove, 227–59. San Francisco: Sierra Club Books, 1986.

———. "On the Intrinsic Value of Nonhuman Species." In *The Preservation of Species*, edited by B. Norton, 138–72. Princeton, NJ: Princeton University Press, 1986.

Carroll, S. *From Eternity to Here: The Quest for the Ultimate Theory of Time*. New York: Dutton, 2010.

Cockell, C. S. "Originism: Ethics and Extraterrestrial Life." *JBIS* 60 (2007): 147–53.

———. "The Ethical Status of Microbial Life on Earth and Elsewhere: In Defense of Intrinsic Value." In *The Ethics of Space Exploration: Space and Society*, edited by J. Schwartz and T. Milligan, 167–80. Cham: Springer, 2016.

———. "The Value of Microorganisms." *Environ Ethics* 27, no. 4 (2005): 375–90.

Corliss, J. "The Flow of Energy, Natural Learning Systems, and the Creation of Life on Earth." *Acta Astronautica,* 19, no. 11 (1989): 869–73. https://doi.org/10.1016/0094–5765(89)90078–7.

Davies, P. *The Fifth Miracle: The Search for the Origin and Meaning of Life*. New York: Simon & Schuster, 1999.

———. "Life, Mind, and Culture as Fundamental Properties of the Universe." In *Cosmos and Culture: Cultural Evolution in a Cosmic Context*, edited by S. J. Dick and M. L. Lupisella, 383–98. Washington, DC: NASA History Series, 2009.

Davies, P., and N. Gregersen. *Information and the Nature of Reality: From Physics to Metaphysics*. Cambridge: Cambridge University Press, 2010.

Davis, A. M. *Mind, Value, and Cosmos: On the Relational Nature of Ultimacy*. Lanham, MD: Lexington, 2020.

Delahaye, J. P., and C. Vidal. "Universal Ethics: Organized Complexity as an Intrinsic Value." In *Evolution, Development and Complexity: Multiscale Evolutionary*

Models of Complex Adaptive Systems, edited by G. Y. Georgiev, J. M. Smart, C. L. Martinez, and M. E. Price, 135–54. Berlin: Springer, 2018.

Deutsch, D. *The Beginning of Infinity: Explanations that Transform the World*. London: Penguin Books, 2011.

Dick, S. J. *Astrobiology, Discovery, and Societal Impact*. Cambridge: Cambridge University Press, 2018.

Dyson, F. *Infinite in All Directions*. New York: Harper & Row, 1988.

———. "Time Without End: Physics and Biology in an Open Universe." *Reviews of Modern Physics* 51, no. 3 (1979): 447–60. https://doi.org/10.1103/RevModPhys.51.447.

Ehrenfreund, P., et al. *COSPAR Workshop on Developing a Responsible Environmental Regime for Celestial Bodies*. Committee on Space Research of the International Council for Science and The Space Policy Institute. George Washington University, 2013. https://www.gwu.edu/~spi/assets/ docs/PEX_WorkshopReport_ES_March14_Web%20(1).pdf.

Esfeld, M. "The Reality of Relations: The Case from Quantum Physics." In *The Metaphysics of Relations*, edited by A. Marmodoro and D. Yates, 218–34. Oxford: Oxford University Press, 2016.

Gershenson, C. et al., eds. *Worldviews, Science and Us: Philosophy and Complexity*. Singapore: World Scientific Publishing, 2007.

Green, B. P. "Convergences in the Ethics of Space Exploration." In *Social and Conceptual Issues in Astrobiology*, edited by K. C. Smith and C. Mariscal. New York: Oxford University Press, 2020.

Guth, A. *The Inflationary Universe: The Quest for a New Theory of Cosmic Origins*. New York: Perseus Books, 1998.

Guth, A., and P. Steinhardt. "The Inflationary Universe." *Scientific American* (1984). https://www.scientificamerican.com/article/the-inflationary-universe/

Haidt, J. *The Righteous Mind*. New York: Random, 2012.

Hall, M. J. W. et al. "Quantum Phenomena Modeled by Interactions between Many Classical Worlds." *Physical Review* 4, no. 4 (2014): 041013. https://doi.org/10.1103/PhysRevX.4.041013.

Handy, R. *The Measurement of Values*. St. Louis: Warren H. Green Inc., 1970.

Haynes, R. "Ecopoiesis: Playing God on Mars." In *Moral Expertise*, edited by D. MacNiven, 161–83. London: Routledge, 1990.

Hedlund-de Witt, A. et al. "Exploring Worldviews and their Relationships to Sustainable Lifestyles: Towards a New Conceptual and Methodological Approach." *Ecological Economics* 84 (2014): 74–83. https://doi.org/10.1016/j.ecolecon.2012.09.009.

Henriques, G. *A New Unified Theory of Psychology*. Berlin: Springer, 2011.

Heylighen, F., and S. Beigi. "Mind Outside Brain: A Radically Non-Dualist Foundation for Distributed Cognition." In *Socially Extended Epistemology*, edited by A. Clark, S. O. Palermos, and D. Pritchard, 59–86. Oxford: Oxford University Press, 2018.

Impey, C. et al., eds. *Encountering Life in the Universe: Ethical Foundations and Social Implications of Astrobiology*. Tucson: University of Arizona Press, 2013.

Irudayadason, N. "The Wonder Called Cosmic Oneness: Toward Astroethics from Hindu and Buddhist Wisdom and Worldviews." In *Encountering Life in the Universe: Ethical Foundations and Social Implications of Astrobiology*, edited by C. Impey, A. Spitz, and W. Stoeger, 94–119. Tucson: University of Arizona Press, 2013.

Johnson, K. A., et al. "Integrating the Study of Culture and Religion: Toward a Psychology of Worldview." *Social and Personality Psychology Compass* 5, no. 3 (2011): 137–52. https://doi.org/10.1111/j.1751–9004.2010.00339.x.

Kashlinsky, A. et al. "A Measurement of Large-Scale Peculiar Velocities of Clusters of Galaxies: Results and Cosmological Implications." *The Astrophysical Journal Letters* 686, no. 2 (2008): L49–L52. https://doi.org/10.1086/592947.

Kauffman, S. *At Home in the Universe*. New York: Oxford University Press, 1995.

Khoury, J. et al. "Ekpyrotic Universe: Colliding Branes and the Origin of the Hot Big Bang." *Physical Review D* 64, no. 12 (2001). https://doi.org/10.1103/PhysRevD.64.123522.

Koltko-Rivera, M. E. "The Psychology of Worldviews." *Review of General Psychology* 8, no. 1 (2004): 3–58. https://doi.org/10.1037/1089–2680.8.1.3.

Krauss, L. *A Universe from Nothing: Why Is There Something Rather Than Nothing?* New York: Free Press, 2012.

Lemos, R. *The Nature of Value: Axiological Investigations*. Gainesville: University Press of Florida, 1995.

Leopold, A. *A Sand County Almanac*. New York: Oxford University Press, 1966.

Linde, A. "Eternally Existing Self-Reproducing Chaotic Inflationary Universe." *Physical Letters B* 175, no. 4 (1986): 395–400. http://dx.doi.org/10.1016/0370–2693(86)90611–8.

———. "The Self-Reproducing Inflationary Universe." *Scientific American* (1994).

Lloyd, S. *Programming the Universe: A Quantum Computer Scientist Takes on the Cosmos*. New York: Random House, 2006.

Lovejoy, A. *The Great Chain of Being*. Cambridge, MA: Harvard University Press, 1936.

Lupisella, M. "The Rights of Martians?" *Space Policy* 13, no. 2 (1997): 89–94. https://doi.org/10.1016/S0265-9646(97)00009-X.

———. "Caring Capacity and Cosmocultural Evolution: Potential Mechanisms for Advanced Altruism." In *Extraterrestrial Altruism*, edited by D. A. Vakoch, 93–109. Berlin: Springer, 2013. https://doi.org/10.1007/978-3-642-37750-1_7.

———. "Ensuring the Scientific Integrity of Possible Martian Life." In *Paper IAA-99-IAA.13.1.08 Presented at the International Astronautical Federation Congress*. Amsterdam: American Institute of Aeronautics and Astronautics, 1999.

———. "Is the Universe Enough? Can It Suffice as a Basis for Worldviews?" *Journal of Big History* 3, no. 3 (2019): 123–40. https://doi.org/10.22339/jbh.v3i3.3370.

———. "The Search for Extraterrestrial Life: Epistemology, Ethics, and Worldviews." In *Exploring the Origin, Extent, and Future of Life: Philosophical, Ethical and Theological Perspectives*, edited by C. Bertka, 261–80. Cambridge: Cambridge University Press, 2009.

———. *Cosmological Theories of Value: Science, Philosophy, and Meaning in Cosmic Evolution*. Cham: Springer, 2020.

Lupisella, M., and J. Logsdon. "Do We Need a Cosmocentric Ethic?" In *Paper IAA-97-IAA.9.2.09 Presented at the International Astronautical Federation Congress*. Turin: American Institute of Aeronautics and Astronautics, 1997.

Mackenzie, R., et al. "Evidence Against a Supervoid Causing the CMB Cold Spot." *Monthly Notices of the Royal Astronomical Society* 470, no. 2 (2017): 2328–38. https://doi.org/10.48550/arXiv.1704.03814.

MacNiven, D. "Environmental Ethics and Planetary Engineering." *Journal of the British Interplanetary Society* 48 (1995): 441–43.

———. *Creative Morality*. New York: Routledge, 1993.

Marshall, A. "Ethics and the Extraterrestrial Environment." *Journal of Applied Philosophy* 10, no. 2 (1993): 227–36. https://doi.org/10.1111/j.1468-5930.1993.tb00078.x.

McHenry, L. B. *Event Universe: The Revisionary Metaphysics of Alfred North Whitehead*. Edinburgh: Edinburgh University, 2015.

McKay, C. "Does Mars Have Rights?" In *Moral Expertise*, edited by D. MacNiven. London: Routledge, 1990.

Mermin, N. D. "What Is Quantum Mechanics Trying to Tell Us?" *American Journal of Physics* 66, no. 9 (1998): 753–67. https://doi.org/10.48550/arXiv.quant-ph/9801057.

———. "Why QBism Is Not the Copenhagen Interpretation and What John Bell Might Have Thought of It." In *Quantum [Un]speakables II: Half a Century of Bell's Theorem*, edited by R. Bertlmann and A. Zeilinger, 83–94. Springer International, 2017.

Mersini-Houghton, L., and R. Holman. "'Tilting' the universe with the landscape multiverse: the 'Dark' flow." *Journal of Cosmology and Astroparticle Physics* (2009). https://doi.org/10.1088/1475–7516/2009/02/006.

Moore, G. E. *Principia Ethica*. Cambridge: Cambridge University Press, 1903.

Murphy, N., and G. Ellis. *On the Moral Nature of the Universe: Theology, Cosmology, and Ethics*. Minneapolis: Fortress Press, 1996.

Nilsson, A. "The Psychology of Worldviews: Toward a Non-Reductive Science of Personality." PhD diss., Lund University, 2013.

Nozick, R. *Philosophical Explanations*. Cambridge, MA: Harvard University Press, 1981.

Oliver, H. *A Relational Metaphysics*. Boston: Martinus Nijhoff Publishers, 1981.

Peters, T. "Does Extraterrestrial Life Have Intrinsic Value? An Exploration in Responsibility Ethics." *International Journal of Astrobiology* 17, no. 2 (2018), 1–7. https://doi.org/10.1017/S147355041700057X.

Quante, M., and K. Engelhard, eds. *Handbook of Potentiality*. Netherlands: Springer Netherlands, 2018.

Race, M. S. "Space Exploration and Searches for Extraterrestrial Life: Decision Making and Societal Issues." In *Encountering Life in the Universe: Ethical Foundations and Social Implications for Astrobiology*, edited by C. Impey, A. Spitz, and W. Stoeger. Tucson: The University of Arizona Press, 2013.

Rescher, N. *Complexity: A Philosophical Overview*. New Brunswick, NJ: Transaction Publishers, 1998.

———. *Metaphysics: The Key Issues from a Realistic Perspective*. Buffalo: Prometheus, 2006.

———. *Process Metaphysics: An Introduction to Process Philosophy*. Albany: SUNY Press, 1996.

———. *The Strife of Systems: An Essay on the Grounds and Implications of Philosophical Diversity*. Pittsburgh: University of Pittsburgh Press, 1985.

Rollin, B. *Animal Rights and Human Morality*. Buffalo: Prometheus Books, 1981.

Rolston, H. *A New Environmental Ethics: The Next Millennium for Life on Earth*. New York: Routledge, 2012.

———. "The Preservation of Natural Value in the Solar System." In *Beyond Spaceship Earth: Environmental Ethics and the Solar System*, edited by E. C. Hargrove, 140–82. San Francisco: Sierra Club Books, 1986.

Rovelli, C. "Relational Quantum Mechanics." *International Journal of Theoretical Physics* 35 (1996): 1637–78. https://doi.org/10.1007/BF02302261.

Rummel, J. D. et al. "Ethical Considerations for Planetary Protection in Space Exploration: A Workshop." *Astrobiology* 12, no. 11 (2012): 1017–23. https://doi.org/10.1089/ast.2012.0891.

Saucier, G. "Isms Dimensions: Toward a More Comprehensive and Integrative Model of Belief-System Components." *Journal of Personality and Social Psychology* 104, no. 5 (2013): 921–39. https://doi.org/10.1037/a0031968.

Saucier, G., et al. "Cross-Cultural Differences in a Global 'Survey of World Views.'" *Journal of Cross-Cultural Psychology* 46, no. 1 (2015): 53–70. https://doi.org/10.1177/0022022114551791.

Schwartz, J. S. J., and T. Milligan, eds. *The Ethics of Space Exploration.* Space and policy series. Berlin: Springer, 2016.

Sciama, D. W. *The Unity of the Universe*. Garden City: Doubleday & Company, 1959.

Shermer, M. *Why People Believe Weird Things*. New York: Henry Holt and Company, 2002.

Singer, P. *The Expanding Circle: Ethics and Sociobiology*. New York: Farrar, Straus & Giroux, 1981.

Smart, J. "Evo Devo Universe? A Framework for Speculations on Cosmic Culture." In *Cosmos and Culture: Cultural Evolution in a Cosmic Context*, edited by S. J. Dick and M. Lupisella, 201–95. Washington, DC: NASA History Series, 2009.

Smith, K. C. "Manifest Complexity: A Foundational Ethic for Astrobiology?" *Space Policy* 30, no. 4 (2014): 209–14. https://doi.org/10.1016/j.spacepol.2014.10.004.

———. "The Trouble with Intrinsic Value: A Primer for Astrobiology." In *Exploring the Origin, Extent, and Future of Life: Philosophical, Ethical and Theological perspectives*, edited by C. Bertka, 261–80. Cambridge: Cambridge University, 2009.

———. "The Curious Case of the Martian Microbes: Mariomania, Intrinsic Value and the Prime Directive." In *The Ethics of Space Exploration, Space and Society*, edited by J. Schwartz and T. Milligan, 195–208. Berlin: Springer, 2016.

Smith, K. C., and C. Mariscal, eds. *Social and Conceptual Issues in Astrobiology*. New York: Oxford University Press, 2020.

Smolin, L. *The Life of the Cosmos*. New York: Oxford University Press, 1997.

Sontag, S. *At the Same Time: Essays and Speeches*, edited by P. Dilonardo and A. Jump. New York: Farrar, Straus & Giroux, 2007.

Stapledon, O. *Star Maker*. Mineola, NY: Dover Publications, 1937.

Steinhardt, P. J. and N. Turok. "A Cyclic Model of the Universe." *Science* 296, no. 5572 (2002): 1436–39. https://doi.org/10.1126/science.1070462.

———. "Cosmic Evolution in Cyclic Universe." *Physical Review D*, 65 (2002): 126003. https://doi.org/10.1103/PhysRevD.65.126003.

Tegmark, M. "Parallel Universes. Not Just a Staple of Science Fiction, Other Universes Are a Direct Implication of Cosmological Observations." *Scientific American* 288, no. 5 (2003): 40–51.

Teilhard de Chardin, P. *The Phenomenon of Man*. New York: Harper & Row, 1959.

Timpson, C. G. "Information, Immaterialism, Instrumentalism: Old and New in Quantum Information." In *Philosophy of Quantum Information and Entanglement*, edited by A. Bokulich and G. Jaeger, 208–28. Cambridge: Cambridge University Press, 2010.

Tipler, F. *Physics of Immortality*. New York: Doubleday, 1994.

Turchin, V. F. "The Cybernetic Ontology of Action." *Kybernetes* 22, no. 2 (1993), 10–30. https://doi.org/10.1108/eb005960.

———. *The Phenomenon of Science: A Cybernetic Approach to Human Evolution.* New York: Columbia University Press, 1977.

Vidal, C. "Metaphilosophical Criteria for Worldview Comparison." *Metaphilosophy* 43, no. 3 (2012): 306–47. https://doi.org/10.1111/j.1467–9973.2012.01749.x.

———. *The Beginning and the End: The Meaning of Life in a Cosmological Perspective*. Berlin: Springer, 2014.

Vidotto, F. "Relational Quantum Cosmology." In *The Philosophy of Cosmology*, edited by K. Chamcham, J. Silk, J. Barrow, and S. Saunders, 297–316. Cambridge: Cambridge University Press.

Volk, T. *Quarks to Culture: How We Came to Be*. Columbia University Press, 2017.

Von Neumann, J. *The Mathematical Foundations of Quantum Mechanics*. Princeton, NJ: Princeton University Press, 1996.

Webb, S. *Out of This World: Colliding Universes, Branes, Strings, and Other Wild Ideas of Modern Physics*. New York: Copernicus Books, 2004.

Wheeler, J. A. "Information, Physics, Quantum: The Search for Links." In *Complexity, Entropy, and the Physics of Information*, edited by W. Zurek, 309–36. Redwood City: Addison-Wesley, 1990.

———. "Law Without Law." In *Quantum Theory and Measurement*, edited by J. A. Wheeler and W. H. Zurek, 182–213. Princeton, NJ: Princeton University Press, 1983.

Whitehead, A. N. "First Lecture: September, 1924." *Process Studies* 48, no. 2 (2019): 159–81.

———. *Process and Reality. An Essay in Cosmology. Gifford Lectures Delivered in the University of Edinburgh During the Session 1927–1928*. New York: Macmillan, 1929.

Wilson, D. S. *Darwin's Cathedral: Evolution, Religion, and the Nature of Society*. Chicago: The University of Chicago Press, 2002.

Wilson, E., and C. Cleland. "The Moral Subject of Astrobiology: Guideposts for Exploring Our Ethical and Political Responsibilities Towards Extraterrestrial Life." In *The Impact of Discovering Life Beyond Earth*, edited by S. J. Dick, 207–21. Cambridge: Cambridge University Press, 2015.

Wolfram, S. *A New Kind of Science*. Wolfram Media, Inc., 2002.

Zenzen M. J., and L. Hammer. "Value Measurement and Existential Wholeness: A Critique of the Rokeachean Approach to Value Research." *The Journal of Value Inquiry 12,* no. 2 (1978): 142–56.

Chapter 5

Astrobiological Searches for Life and Shared Knowledge

Chelsea Haramia

Humans are currently searching for extraterrestrial life in the cosmos, and some are engaged in the search for technologically advanced life—also known at the Search for Extraterrestrial Intelligence (SETI). These searches are growing in scope and intensity, and they raise pervasive philosophical questions about the assumptions under which astrobiologists, SETI practitioners, and others are working. These include biological questions (What is life?), ethical questions (What grounds claims of ethical value?), and, as I will show, metaphysical questions (Are numbers real? Do non-physical entities exist?).

By their nature, philosophical questions resist empirical analysis. While empirical study may inform philosophical debate, philosophical conclusions typically cannot be reached through empirical study alone. Of course, it is unreasonable to require philosophical certainty for most projects that are undergirded by crucial philosophical assumptions. My goal here is not merely to uncover uncertainty and skepticism with respect to the search for extraterrestrial life in the cosmos. Nor is it to reject or dismiss the relevant skeptical accounts. Instead, my goal is to provide good reason not only to proceed with the search for extraterrestrial life and technology despite legitimate skepticism, but also to expand that search to include a quest for shared ethical knowledge among humans and extraterrestrial others, even in the face of legitimate skepticism about that possibility as well. The result will be a defense of the claim that humans and extraterrestrial intelligence (ETI) will have shared axiological, biological and mathematical knowledge. While the ontological status of entities such as ethical properties and mathematical equations is well-trodden philosophical territory, my argument leads to a novel account regarding the ontological status of life itself and has the potential to fruitfully inform certain approaches in the search for extraterrestrial life in accordance with these considerations.

THE SEARCH

Astrobiological searches are ongoing scientific endeavors with the potential for great significance. Some researchers explore the cosmos for evidence of life-supporting conditions beyond Earth, including conditions that support unicellular and microbial life. Others, such as SETI practitioners, scan the cosmos for evidence of technosignatures—evidence of extraterrestrial technology such as repeated laser pulses or powerful modulated narrowband signals. These could be the first recognizable signs of advanced technology, and therefore of technologically intelligent life, elsewhere in the universe.[1]

At this time, we humans cannot be certain that we will find extraterrestrial life or technology, but we believe it is worth looking for. Since its beginnings in 1984, the SETI Institute has been central in the search for extraterrestrial technology.[2] The express mission of the SETI Institute is "to explore, understand and explain the origin and nature of life in the universe and the evolution of intelligence."[3] However, absent from the SETI mission is an axiological dimension. There is no mention of the search for and exploration of extraterrestrial values. But intelligent life may have ethical capabilities as well as technological capabilities. The intermediate aim of this chapter is to provide justification for the search for universal, and therefore extraterrestrial, values. It is not surprising that the SETI mission omits a search for values, so let us begin by addressing predictable concerns involving skepticism about extraterrestrial values.

EARTHLY EVIDENCE

One prominent concern about the possibility of universal, extraterrestrial values stems from the idea that human ethics is a product of evolution and is therefore Earthbound.

Consider the following two claims:

> Claim (1): Values are the kind of thing that we can hope to identify elsewhere in the cosmos, because values and ethics are not things that humans just made up.

> Claim (2): Life is the kind of thing that we can hope to identify elsewhere in the cosmos, because life is not something that humans just made up.

There are many people who will affirm (2) who are at the same time deeply skeptical of (1). Michael Ruse joins biologist E.O. Wilson in framing the problem with (1) in the following way

> Morality, or more strictly, our belief in morality, is merely an adaptation put in place to further our reproductive ends. Hence the basis of ethics does not lie in God's will . . . or any other part of the framework of the Universe. In an important sense, ethics . . . is an illusion fobbed off on us by our genes to get us to cooperate.[4]

Many scientists, laypeople, and others assume that our ethics and values are mere human constructs beholden only to personal bias or cultural influence. These people doubt that universal, objective ethical values could exist independently of human beliefs, and so they'd doubt that we could have knowledge regarding extraterrestrial ethical value. Furthermore, as justification for this skepticism, they may point to the fact that ethical debate is full of disagreement and does not produce clear results. While such skepticism has merit, a similar skepticism about extraterrestrial life is also warranted, though not as commonly voiced. Let us turn now to a defense of this scientific skepticism.

Why, on the one hand, might we be *skeptical* of the truth of (2)? As Carol Cleland and Elspeth Wilson note, "our current understanding of life as a natural phenomenon—which is based solely on our experience with life here on Earth—is myopically Earth-centric."[5] Our Earth-centric biases may make it difficult to generalize from our available evidence to a universal, objective account of life—a point I will return to later. So, both (1) and (2) appear to engender some legitimate skepticism.

On the other hand, why might we assume the *truth* of (2)? One answer is that Earth and its inhabitants provide some evidence for what could exist elsewhere. Perhaps, because life evolved in our corner of the universe, it is the kind of thing that can evolve elsewhere in the universe. This evidence is typically enough to engender credence in (2). In turn, I contend that similar evidence should be enough to engender credence in (1).

Consider one obvious challenge to the existence of extraterrestrial ethical values. Arguably, our values are a product of evolution. Thus, one might argue, what we think of when we think of, say, *goodness*, is not the kind of thing we can expect to find beyond Earth where things may have evolved differently. But one cannot consistently believe this and not apply the same standards to the search for life beyond Earth.[6] It is possible that *what we think of* when we think of life is not the kind of thing we can expect to find beyond Earth where things may have evolved differently, given that these organisms are products of evolution.

The alternative is to assert that the search for life—including technologically advanced life—is nonetheless justified because its existence here is some evidence that it evolved elsewhere as well. And the same can be said of values. Because ethical values evolved here on Earth, they are the kind of thing that could evolve elsewhere. This, then, is reason to think that we may share ethical understandings with ETI, and shared beliefs about values will be *prima facie* evidence that values themselves independently exist.

THE ROLE OF EVOLUTION

It is, however, possible that we may detect extraterrestrial values that are mere evolutionary adaptations. The above appeal to evolutionary explanations leaves open the role that evolution is playing. There are two potential explanations of this role. Either we evolved to be capable of recognizing certain objective facts about the world, or we merely have certain subjective beliefs about the world and our evolving the way we did generated these beliefs. I call the former view "Accurate Perception" and the latter "Mere Adaption."

i. *Accurate Perception:* We evolved to have certain capabilities that enable us to recognize some objective facts about the world. Evolution is the cause of these specific capabilities—capabilities that have been evolutionarily advantageous—and they generate true beliefs about the world.

e.g., Our evolving the capacity for sight caused us to be able to correctly recognize objects as near or far.

i. *Mere Adaptation:* Some beliefs are merely subjective beliefs about the world, and our evolving the way we did generated these beliefs. Evolution is the cause of certain beliefs, and these beliefs are held only because they are evolutionarily advantageous and not because they are true.

e.g., We believe that fats and sugars are tasty because it was once evolutionarily advantageous for us to eat such foods.

So, even though life and values are products of evolution, perhaps evolution played a different role in our knowledge and understandings of these distinct phenomena. Stating that evolution played a vital *causal* role in our beliefs does not entail that evolution is the explanation for the *truth* of those beliefs. The former explanation posits a non-human source. Presumably, someone who denies the conjunction of (1) and (2) above may deny this because they assume that human beliefs are *not* the ultimate source of our concepts of life, whereas they assume that human beliefs *are* the ultimate source of our concepts of value.[7]

But why make these assumptions? Let's consider the respective contrarian positions. On the one hand, we have beliefs about the conditions required for life and intelligence. Yet it is possible that the source of these beliefs is merely our having evolved to interpret the world in the way that we do. We call things "alive" but this may just be an Earthly custom. On the other hand, it's also possible that our evolving the way that we did caused us to be capable of recognizing true ethical facts about the world that exist independently of human beliefs about them. We call things "good" and "bad," and they may be genuinely good or bad.

Consider the following mathematical analogy:

We evolved to be capable of higher-order thinking skills, which allow us to understand mathematical truths. I will return to this point later, but it is fairly uncontroversial to assume that these mathematical truths were true before any human evolved to have beliefs about them (though anti-realist philosophers and mathematicians will remain skeptical, of course). At least, the study of mathematics is well-motivated, and it would be hard to motivate this discipline as we know it if there were a widespread belief that studying equations in math is just like studying, say, cadence in poetry—something that humans did make up. While human evolution was needed to generate the capability to understand and hold these beliefs, common accounts of mathematics entail that evolution is not the source of mathematical truths. If this is the case, one may claim that evolution played a vital *causal* role in our ability to have accurate mathematical beliefs, but these mathematical truths exist outside of, and prior to, our beliefs and evolutionary history.

In the same way, one viable account of axiological value is that truths about value exist independently of our ability to comprehend those truths. Our evolving into higher-order thinkers allowed us the *capability* to understand ethical values. We may have incorrect or incomplete understandings of ethics, just as we may have incorrect or incomplete understandings of mathematics and science. If this doesn't weaken our confidence in the existence of real mathematical or scientific truths, it likewise should not weaken our confidence in the existence of real ethical truths. So, insofar as this does not weaken the astronomer's confidence in the objective status of, say, the mathematics used in the search for extraterrestrial life or ETI, she should be equally confident that extraterrestrial values are real as well. That is, if objective values are out there, then other intelligent beings are likely to discover and understand them as well, in the same way that we often assume that, if the mathematical truths are out there, then other intelligent beings are likely to discover and understand them. In sum, recognizing that evolution plays a role in the content of our beliefs does not entail that those beliefs must be mere adaptations.

And, of course, this mere-adaptation concern is equally serious for biological theories of life as it is for axiological theories of value. Within biological and scientific frameworks, we are able to categorize easy cases of life quite well. Pigs and bees and jellyfish are alive. Rocks and water and chairs are not. The same can be said for intelligence. Rocks and jellyfish are not intelligent, and as SETI practitioner Seth Shostak says, "[y]ou're intelligent if you can build a powerful laser or a thumping radio transmitter."[8] At the same time, we struggle with the borderline cases in these scientific frameworks and lack a perfect, complete account of the relevant concepts. Are pigs intelligent? Are guinea pigs or bees? Are crystals alive? Are protein viruses, computer viruses or prions alive? These questions underscore just some of the challenges we face when studying life and intelligence in all its variability. We do not have a perfect and complete understanding of the boundaries of biology or intelligence. If the search for extraterrestrial life is justified, then, it must be justified despite these lacunae in our understandings of the concepts. In turn, recognizing conceptual shortcomings regarding our beliefs about, say, ethical goodness and badness, is not sufficient to dismiss these beliefs as mere adaptations and thereby not sufficient to dismiss the search for extraterrestrial values.

Consider another parallel:

Gratuitous suffering is bad. Joy is good. We have clear concepts of goodness and badness that capture these easy cases, even though we struggle with borderline cases in ethical debates. Are lies bad when they produce a better outcome? Is it good to give aid to some when others need it more? We lack a perfect and complete account of ethical value that provides us with clear answers to every ethical question. This should not prevent us from searching for extraterrestrial ethical values any more than our lacking a perfect and complete account of life or intelligence or mathematics prevents us from searching for extraterrestrial others.

It is, arguably, possible that life, intelligence, mathematics, and values are all mere human constructs and do not exist independently of human beliefs. But if we are content assuming that evolution here on Earth is not the source of all concepts of life, intelligence or math, then, on pain of consistency, we should also not assume that evolution on Earth is the source of all concepts of ethical value. Thus, these source-based concerns are no more challenging for the ethicist advocating for the search for extraterrestrial values than they are for the scientist advocating for the search for extraterrestrial life or intelligence.

CAPABILITY AND COMPREHENSIBILITY

Notably, humans and human evolution are nonetheless the source of the specific tools and capabilities we use to analyze the concepts in question. Our symbols, languages and technologies are highly contingent on humans having evolved in the specific ways that we did. This does not prevent us from believing that we can use these tools to comprehend universal objective truths. But neither does it guarantee that we can access such truths.

The arbitrary symbols we use to represent mathematical equations are meant to represent non-arbitrary, universal truths. Of course, no one assumes that if we discover ETI, that they will, say, represent the number 4 with the symbol "4," or with any other symbol that represents that number here on Earth. It is much more likely, however, that ETI will have a concept of the number 4 that they represent differently.[9]

We evolved to interact with the world in ways that inform the content of our beliefs. What we understand to be good, or to be true, or to be alive is likely determined to some extent by our human capabilities. These capabilities arose from our evolving in the way we did. Some might narrowly worry that humans evolved in a very specific cosmic setting that informs our value-concepts. Because ETI likely evolved in a much different setting, we cannot assume that we will share any values with ETI. This is a legitimate worry, even if *Accurate Perception* is the correct explanation of our beliefs about values. But it is an *equally* legitimate worry for other projects, even if *Accurate Perception* is the correct explanation of our beliefs about, say, mathematics, life and intelligence. ETI may have evolved to look different enough from what we call "life" and "intelligence" that we are unable to recognize ETI when we come across them, even if we have certain objectively correct beliefs about these concepts. The key point here is that we are taking on the search for life and intelligence despite these epistemic challenges. Therefore, if recognizing the possibility that we may have evolved to have certain capabilities and concepts distinct from the capabilities and concepts of ETI—and that ours may prevent us from being able to identify ETI—does not cause us to abandon the search for extraterrestrial life or intelligence, neither should it prevent us from searching for extraterrestrial values in general and shared values in particular.

PERSONAL, CULTURAL, AND ANTHROPOCENTRIC BIASES

It is possible that we have arrived, or can arrive, at objectively true beliefs about the world, but concerns about the processes of our belief-formation might remain. Even our best-faith efforts to form true beliefs are not immune to charges of bias. One might worry that, even if ethical values really exist, we are likely to have a distorted understanding of them, given that our concepts of, say, goodness and badness are clouded by our personal, cultural and anthropocentric biases. But the same can be said of our understandings of life and intelligence. Intelligence as a concept is particularly fraught. Our assignations of intelligence, of degrees of intelligence, and of who is presumed to possess intelligence are likely to be, and have historically been, distorted by anthropocentric, cultural and personal biases—and these biases have led to oppressive hierarchies across human history.[10] If this history of bias with respect to intelligence does not cause us to abandon the search for intelligent life in the universe,[11] then neither should evidence of our biases block our search for values. This does not mean that we shouldn't try to overcome or mitigate these biases and oppressive circumstances in the search for truth. It does mean that we should be careful to control for biases and harm to the extent that we can, whether we're searching for extraterrestrial life, extraterrestrial technology, extraterrestrial mathematics, or extraterrestrial values.

This problem is connected to common worries that undergird some accounts of ethical relativism. We humans seem to encounter a lot of interpersonal and cross-cultural disagreement about ethical values—disagreement that many chalk up to mere bias. I have argued elsewhere that presuming similar interstellar disagreement about basic values is not as justified as many assume it to be.[12] This is because there are in fact basic human value claims about which there is little to no disagreement—recall the claims from above that joy is good and gratuitous suffering is bad. Furthermore, we ought to consistently apply objections from disagreement, but in practice we don't. Disagreements among mathematicians do not typically cause us to abandon belief in mathematics.[13] And disagreement is rampant in scientific research, a point that Renford Bambrough highlighted in defense of ethical realism.[14] For example, physicists currently disagree about things such as dark matter and the correct interpretation of quantum physics. Furthermore, claim (2) above relies on scientific analyses of life. But science does not guarantee perfect knowledge of this phenomenon—and *many* other phenomena—any more than ethical analysis guarantees perfect moral knowledge. As many acknowledge, and as Erika Offerdahl states, "science is uncertain, subject to sociocultural influence, and often unable to produce clear results."[15] And there remains genuine, persistent scientific and philosophical debate about what constitutes life and technology and whether we might be able to identify those things.[16] Yet, we typically believe that the pursuit of real biological and mathematical knowledge is nonetheless justified. Because disagreement is common in scientific, mathematical, and axiological debates, and because we nonetheless have good reason to pursue biological, mathematical, and ethical knowledge, then the search for ethical values remains on par with the search for extraterrestrial life and intelligence.

Thus, engaging in the search for ETI justifies abandoning skepticism regarding extraterrestrial values. Furthermore, if we detect evidence of extraterrestrial intelligence, we will have good reason to explore not only their scientific and mathematical knowledge but also their ethical knowledge. As Holmes Rolston puts it

> Scientists have what Michael Polanyi (1962, 65) called "universal intent"; they hope for theories true at all times and places, true for all peoples everywhere. If this is true in science, we can also ask whether the human mind can reach ethical principles that may transcend our somatic embodiment. It might be that in the search for extraterrestrial intelligence, the question to ask is about the value of pi, or the atomic number of carbon. Another, more revealing test of their intelligence might be to ask whether one should tell the truth, keep promises, or be just.[17]

LIFE AND NON-OBSERVABLES

I have shown that, from an epistemic perspective, the search for shared extraterrestrial values is as justified as the search for extraterrestrial life and intelligence. I have also argued that the assumption that we will have shared axiological understandings of what is truly good, bad and valuable is as justified as the assumption that we will share mathematical or scientific understandings with extraterrestrial others. This means that we have good reason to believe that ethical value is as real as life, intelligence and math. Part of the justification for this partners-in-innocence argument comes from highlighting the difficulties of defining something that we typically agree exists and is worth searching for: life. So, let us look more carefully at this debate.

What is life? The answer simply isn't clear, despite years of scientific and philosophical exploration of the question. Cleland and Wilson write,

> In recent years, the question "What is life?" has taken on increasing scientific urgency. Molecular biologists who are investigating the origin of life or trying to synthesize life in the laboratory from basic building blocks want to know at what stage an ensemble of non-living molecules becomes a primitive living

> thing. The answer is not clear. Likewise, astrobiologists charged with designing remote or *in situ* robotic experiments for detecting extraterrestrial life on other planets and moons worry that we might not recognize life if it differs significantly from familiar life on Earth.[18]

They go on to explain that metabolic, Darwinian, thermodynamic, biochemical, cybernetic and autopoietic attempts to define life "all face serious counterexamples."[19] They then use these difficulties and others to underscore the challenges involved in providing a justified account of life's fundamental nature as a natural kind. "Life" appears to resist both empirical and ideal definitions. Cleland ultimately argues that the fundamental problem with defining "life" can be chalked up to the fact that life is a natural kind and all natural kinds resist ideal definitions.[20] Indeed, as Hilary Putnam famously argued, a natural kind's referent is not fully determined by human concepts (1979).[21] Cleland builds on this to argue that the definitional approach is not only inherently misguided, attempts to define life will primarily serve to reinforce misconceptions of what life is—misconceptions that arise from our inadequate empirical evidence.[22] When analyzing life, we have a sample size of 1: Earthly life. For our purposes here, the key takeaway is that "life" consistently resists definitions or understandings that operate under the assumption that it refers to a *natural* kind.[23]

But what if life is not a natural kind? There are two alternatives to consider if this is the case. Either life is a mere human construct that can be defined entirely by human concepts, or life is real and independent of human minds, but it is also not a natural kind that is amenable to empirical analysis. The former flies in the face of our fundamental assumptions about life in the universe. The latter is rarely considered, but it is the option I wish to consider here.

Natural kinds are amenable to scientific analysis. With enough empirical investigation, we may discover what is needed to understand what certain natural kinds *really* are (not merely to understand what our concepts determine of them). But we may worry that empirical research might not be the right solution to understanding the boundaries of "life," given the apparent intractability of the corresponding scientific investigation. The problem is that life has evaded our attempts to understand its boundaries despite sustained and advanced empirical investigation. This evasiveness is not as prominent in other natural kinds that have undergone advanced empirical investigation, such as water. If, on another planet we find clear, odorless, tasteless, potable liquid that is not composed of H_2O, we will have found something other than water. But finding entities that exhibit some of the characteristics of life on Earth but lack the chemical or molecular foundations of Earthly life is *not* reason to conclude that we have found something other than life.[24] Furthermore, the challenges and counterexamples to candidate categorizations of "life" are frustrating in part because it seems intuitively obvious to us that some things in the world are clearly alive whereas others are clearly non-living. An account of what separates life from non-life remains the elusive goal. Again, while accurately categorizing natural kinds can be challenging, Cleland and Wilson maintain that this project is especially difficult for the project of categorizing life due to the paucity of our available evidence. They write

> What we need in order to provide a scientifically compelling answer to the question "What is life?" is not a definition (an analysis of human concepts of life) but rather a general, empirically-grounded theory of living systems. Unfortunately . . . contemporary scientists are in no position to formulate such a theory. Familiar life on Earth represents a single—and quite possibly unrepresentative—example of life.[25]

Now, something similar can be said for the project of categorizing ethical value. To date we have evidence only of ethical value that is tethered to Earth. We do not know how value manifests elsewhere in the universe. Earthly ethical value represents a single—and quite possibly unrepresentative—example of axiological value.[26] Of course, contemporary scientists are also in no position to formulate a theory or definition of value because the study of value is not a scientific project. Values, such as goodness and badness, are not ostensible entities. They do not have quantifiable measurements, extension, shape or any other physical qualities that would enable us to empirically single these things out and study them in the physical world as we know it. So, if they really exist, they are clearly not observable in the way that, say, form and size are observable. They are, as I will call them, "non-observables." In the same way that we see hands on a clock move but do not literally perceive *time* itself, I claim that we may perceive the concrete actions of genocidal agents but not *badness* itself, or, likewise, we may perceive the various processes of plants and animals but not *life* itself.

And yet, we seem to be able to identify value in the world, and the apparent truth of value-claims. Genocide is bad. Arguably and intuitively, this is objectively true. We do not discover this truth by recognizing any physical properties of badness, and of course science cannot prove that it is bad. Rather, we recognize that genocide is bad by understanding both badness and the act of genocide. Recognizing that genocide is bad does not give us perfect knowledge of all value claims, but there is no reason to think that recognizing an instantiation of a non-observable will provide us with perfect knowledge of the phenomenon itself. Numbers, too, are non-observables if they really exist. Numbers themselves do not have literal sizes, extension, shape or any other physical qualities. Yet it is arguably and intuitively objectively true that

2+2=4. And of course, recognizing this truth does not somehow garner perfect mathematical knowledge.

While anti-realists maintain that these things are mere constructs and not an independent part of our universe, realists maintain that these things are instead real and independent of human beliefs. This means that they are real but not accessible through direct empirical apprehension. Andrew M. Davis explains such entities in the following way

> Mathematical entities, for example, are not the kind of realities grasped with our physical senses. [David Ray] Griffin provocatively inquires, "Do we perceive numbers and mathematical truths with our eyes? Do we hear them with our ears? Do we touch them with our hands? Do we smell them with our palates" (Griffin, 2016, p. 145). The answer is no. Neither are objective moral or ethical truths or infinite "possibilities of beauty" grasped with our physical senses.[27]

Davis, following A.N. Whitehead, refers to our acquisition of such knowledge as "prehension," as opposed to the apprehension involved in our empirical observations. If correct, this means that empirical science cannot provide good definitions or categorizations of such non-observables, but they are targets of analysis precisely because they nonetheless seem to really exist and to affect the world in important ways.

The project of defining things such as numbers and values is therefore a project of prehension and not strictly an empirical or scientific project. I contend that, likewise, defining "life" is not strictly an empirical or scientific project. We learn about life through prehension. If this is true, it is because *aliveness* is non-observable. Many of the difficulties of defining life and distinguishing it from non-life therefore come not from the incorrect assumption that life is a natural kind of which we have only Earthly examples, but rather they come from the difficulties inherent to studying and theorizing about non-observable universals.

This sets life apart from genuine natural kinds such as water. Again, if on another planet we find water-like liquid that is not composed of H_2O, we will have found something other than water. But finding entities that exhibit some of the characteristics of life on Earth but lack the chemical or molecular foundations of Earthly life is *not* reason to conclude that we have found something other than life. As noted before, chemical and molecular attempts to differentiate life from non-life fail. One viable explanation for this failure is that non-observables are not amenable to molecular analysis, and life is a non-observable.

We are nonetheless able to empirically observe a multitude of things about entities that are alive. We may observe movement, growth, reproduction, metabolic processes, etc. But, again, we don't observe life per se; we merely observe the organisms that are alive. Likewise, we don't observe badness per se or time per se, even though we can observe individual actions that are bad, or the clock's hands ticking by. Technological advances and the scientific method cannot therefore be the means of identifying life per se any more than they could be the means of identifying value per se. Likewise, we don't observe numbers per se. We observe highly-contingent and artificial representations of numbers, such as "2." And we also observe naturally occurring entities that possess *twoness*. For example, we observe distinct moons orbiting Mars along with their physical dimensions, locations, etc., and we then understand that there are two of them, even though we do not and cannot empirically observe twoness itself in this collection of moons.

While I am sympathetic to a realist account of entities such as numbers, it is not the aim of this paper to prove whether and how these things are fundamentally real. Instead, I wish to highlight the fact that accounts of life reliably and fundamentally evade purely empirical understandings in much the same way that accounts of axiological and mathematical phenomena fundamentally evade purely empirical understandings. Because of this, there is reason to believe that we cannot proceed with the search for extraterrestrial life with scientific knowledge about what exactly life is, and, if I am right, we cannot hope for further empirical study to provide us with an answer. In sum, if biological, axiological and mathematical entities are real, then they may all exist in the same non-observable manner.

AN OBJECTION FROM UNCERTAINTY

One may raise the following concern in light of this reasoning: Perhaps the search for extraterrestrial life is unfounded or fundamentally misguided and therefore should be abandoned, given that we are crucially unsure of what we are looking for. But there is not sufficient reason to jump to this kind of conclusion. There is, instead, excellent reason to search for what we do understand. We understand, for example, that the obvious cases of life on Earth are alive. We also understand the difference between even numbers and odd numbers, and we understand that gratuitous suffering is bad. We may assign a very high degree of confidence to all of these understandings, even as we acknowledge the incompleteness of our understandings of every entity, equation or claim in biological, mathematical and axiological domains. In sum, these concerns do not justify abandoning the search for life because they do not undermine our confidence in the easy cases. Instead, these concerns justify using our powers of prehension, a healthy dose of epistemic humility as we search (a humility that is already present in many searches), and a recognition of this necessary uncertainty alongside the limits of our current knowledge.

So, the proper response to these concerns about lacunae in our cosmic understandings is to search for what we do understand while simultaneously acknowledging what we do not understand and continuing to pursue greater understandings amidst uncertainty. These projects are already in process, in both scientific and philosophical fields.[28]

CONCLUSION

To return to the first half of this paper, we are taking on the search for extraterrestrial intelligent life not because we are certain that we know what exactly constitutes intelligence or aliveness. We are searching because we believe it is possible that we can successfully recognize at least some such life. If this search is justified, which I maintain it is, so too is the search for evidence of universal value. We may hope to find evidence of extraterrestrial ethical capacities in the same way that we hope to find evidence of extraterrestrial mathematical prowess. Because of this, we are justified in pursuing evidence of shared knowledge between humans and extraterrestrial others—specifically, understandings of real non-observables such as life, value and math.

This view has further practical import. Successful searches for ETI will create the potential for communication. Given the presumably vast distances involved, we will likely need to be judicious about what information we attempt to convey and about what information we attempt to solicit. Because the possibility for miscommunication is so high, there is reason to focus our communication on topics about which we believe we may possess shared understandings. If we lack information regarding ETI's physical capabilities, we will be in a position to assume very little about their empirical understandings. However, if they are alive and communicating and we are able to recognize that, then we have good reason to begin with a focus on non-observables, and we have good reason to hope that our basic understandings of these non-observables are shared by extraterrestrial others.

NOTES

1. Special thanks to the participants at the 2019 Making Contact Workshop, sponsored by the Breakthrough Listen Initiative, and to members of the astronomy department at UC Berkeley and the Berkeley SETI Institute, for their valuable feedback on portions of this work.

2. The SETI Institute paved the way for many vibrant and active SETI programs, such as The Berkeley SETI Research Center, The Penn State Extraterrestrial Intelligence Center, etc., and these other programs have taken up a more prominent role in SETI searches today.

3. https://www.seti.org/about-us/mission

4. Ruse and Wilson, "The Evolution of Ethics," 51–52.

5. Cleland and Wilson, "Lessons from Earth," 27.

6. Renowned SETI practitioner Jill Tarter acknowledges the possibility that we may be searching for that which we cannot even imagine, but she does not conclude that this possibility justifies abandoning searches. See Tarter, "Contact."

7. For example, Sharon Street denies the conjunction on similar grounds, though the scope of this paper chapter not directly address her dilemma. See Street, "A Darwinian Dilemma for Realist Theories of Value."

8. Shostak, "Are We the Galaxy's Dumbest Civilization?," 221.

9. It should be noted that we may not be justified in simply assuming even this degree of overlap in our interstellar understandings. We evolved to have certain sensory capabilities, and these capabilities dictate how we apprehend and interact with the world. An alien civilization might have evolved to have *very* different sensory capabilities, and they may interact with and apprehend the world in a much different manner. For example, as John Traphagan notes, some imagine what might be involved in the mathematical understandings of an intelligent species who evolved to apprehend the world through primarily olfactory means—perhaps understanding math in terms of gradations rather than discrete numerals. See Traphagan, *Extraterrestrial Intelligence and Human Imagination*. Our perceptual capabilities inform our conceptual framework, and ETI might have evolved to have utterly different perceptual capabilities and an utterly different conceptual framework. This does not entail that, say, mathematical or scientific truths are non-objective or non-universal. It simply means that we might have access to universal truths that ETI do not have access to, and vice versa. And this does not make it the case that mathematical or scientific truths must be mere relative truths. We might both be accessing universal truths that are inaccessible to the other. While this cuts against my thesis regarding the *probability* of shared understandings, it does not challenge the claim that shared understandings are *possible*, nor does it provide sufficient reason not to search for shared understandings about objective value.

10. See Shorter, "On the Frontier of Redefining 'Intelligent Life'" for a radical critique of the colonialist attitudes inherent to certain concepts and assumptions in "settler" science in general and inherent to SETI and the search for "intelligence" in particular.

11. While such a critique has not caused a general retreat from searching, it has raised legitimate calls to revise and improve the beliefs and conditions that undergird searches, which is appropriate given the history of bias and oppression in scientific study and in concert with my conclusion here.

12. Haramia and DeMarines, "Ethically-Informed METI Analysis." See also Rachels, "The Challenge of Cultural Relativism," for a similar argument.

13. For but one of the many examples of unsolved problems in mathematics, see Filippov et al., *Dniester Notebook* which collects hundreds of unresolved problems in algebra.

14. Bambrough, "A Proof of the Objectivity of Morals."

15. Offerdahl, "A Scientifically Minded Citizenry," 225.

16. See, for example, Mix, "Defending Definitions of Life" for challenges to defining "life." And SETI practitioners currently have but one criterion for qualifying as extraterrestrial "intelligence"—the capability to produce detectable technology. Many readily admit that this condition is merely functional based on what we may hope to find. See for example, Wright and Oman, "Visions of Human Futures in Space and SETI." They typically present technology as a sufficient condition for intelligence, not a necessary condition. Furthermore, they do not claim to be working under a complete account of intelligence, which shows that SETI practitioners are prepared to search within the constraints of our abilities and limited understanding of the target of the search. Insofar as our understanding of ethical value is also incomplete, this likewise does not count against the search for extraterrestrial values.

17. Rolston, "Terrestrial and Extraterrestrial Altruism," 214.

18. Cleland and Wilson, "Lessons from Earth,"19–20.

19. Ibid., 21.

20. Cleland, *The Quest for a Universal Theory of Life*.

21. Putman, *Mind, Language, and Reality*.

22. Cleland, *The Quest for a Universal Theory of Life*.

23. The Stanford Encyclopedia of Philosophy defines "natural kind" in the following way: "To say that a kind is *natural* is to say that it corresponds to a grouping that reflects the structure of the natural world rather than the interests and actions of human beings." https://plato.stanford.edu/entries/natural-kinds/

24. Though there is room for contention with the following, one potential avenue for challenging this is to claim that Earthly life has a single origin whereas Earthly water does not, so the study of life has an N=1 problem that the study of water lacks. Thanks to Carol Cleland for pointing this out in a Technosignatures Seminar talk on May 11, 2022.

25. Cleland and Wilson, "Lessons from Earth," 25.

26. Assuming universal value exists, which I have shown we may assume insofar as we assume life exists.

27. Davis, "Whiteheadian Cosmotheology," 439.

28. For an example of astrobiological efforts, see Walker et al. "Exoplanet Biosignatures." Furthermore, continuing to pursue greater understandings amidst uncertainty is arguably the central project of philosophy, and philosophers continue to grapple with the realism/anti-realism debate discussed here. See, for example, Davis, "Whiteheadian Cosmotheology."

BIBLIOGRAPHY

Bambrough, R. "A Proof of the Objectivity of Morals." *The American Journal of Jurisprudence*, Vol. 14, No. 1 (1969): 37–53. https://doi.org/10.1093/ajj/14.1.37.

Cleland, C.E. and E.M. Wilson. "Lessons from Earth: Toward and Ethics of Astrobiology." In *Encountering Life in the Universe*, edited by C. Impey, A. H. Spitz, and W. Stoeger, 17–55. Tucson: The University of Arizona Press, 2013.

Cleland, C.E. *The Quest for a Universal Theory of Life: Searching for Life as We Don't Know It.* Cambridge: Cambridge University Press: Cambridge, 2019.

Davis, A.M. "Whiteheadean Cosmotheology: Platonic Entities, Divine Realities and Shared Extraterrestrial Values." In *Process Cosmology: New Integrations in Science and Philosophy*, edited by A.M. Davis et al., 423–452. Cham: Palgrave, 2022.

Filippov, V. et al. *Dniester Notebook: Unsolved Problems in the Theory of Rings and Modules*. Fourth Edition, Mathematics Institute, Russian Academy of Sciences Siberian Branch: Novosibirsk, 1993.

Haramia, C. and J. DeMarines. "The Imperative to develop an Ethically-Informed METI Analysis." *Theology and Science*, Vol. 17, No. 1 (2019): 38–48. https://doi.org/10.1080/14746700.2019.1557800.

Mix, L. "Defending Definitions of Life." *Astrobiology*. Vol. 15, No. 1 (2015): 15–19. https://doi.org/10.1089/ast.2014.1191.

Offerdahl, E. "A Scientifically Minded Citizenry: The Ethical Responsibility of All Scientists." In *Encountering Life in the Universe*, edited by C. Impey, A. H. Spitz, and W. Stoeger, 222–235. Tuscon: The University of Arizona Press, 2013.

Putnam, H. *Mind, Language, and Reality*. Cambridge: Cambridge University Press, 1979.

Rachels, J. "The Challenge of Cultural Relativism." In *The Elements of Moral Philosophy*, 14–32. McGraw Hill: New York, 1999.

Rolston, H. "Terrestrial and Extraterrestrial Altruism." *Extraterrestrial Altruism: Evolution and Ethics in the Cosmos*, edited by D. Vakoch, 211–222. Heidelberg: Springer, 2013.

Ruse, M. and E.O. Wilson. "The Evolution of Ethics." *New Scientist*, Vol. 108, No. 17 (1985): 50–52.

Shorter, D.D. "On the Frontier of Redefining 'Intelligent Life' in Settler Science." *American Indian Culture and Research Journal*, Vol. 45, No. 1 (2021): 19–44. https://doi.org/10.17953/aicrj.45.1.shorter

Shostak, S. "Are We the Galaxy's Dumbest Civilization?" (2002). Accessed at www.seti.org/epo/news/features/are-we-the-galaxys-dumbest.php.

Smith, K.C. "The Curious Case of the Martian Microbes: Mariomania, Intrinsic Value and the Prime Directive." In *The Ethics of Space Exploration, Space and Society*, edited by J. Schwartz and T. Milligan, 195–208. Berlin: Springer, 2016.

Street, S. "A Darwinian Dilemma for Realist Theories of Value." *Philosophical Studies: An International Journal for Philosophy in the Analytic Tradition*, Vol. 127, No. 1 (2015): 109–166. https://doi.org/10.1007/s11098-005-1726-6.

Tarter, J.C. "Contact: Who Will Speak for Earth and Should They?" In *Encountering Life in the Universe*, edited by C. Impey, A. H. Spitz, and W. Stoeger, 178–199. Tuscon: The University of Arizona Press, 2013.

Traphagan, J.W. *Extraterrestrial Intelligence and Human Imagination: SETI at the Intersection of Science, Religion, and Culture*, Cham: Springer International, 2015.

Walker, Sara I. et al. "Exoplanet Biosignatures: Future Detections." *Astrobiology*, Vol. 18, No. 6 (2018): 779–824. https://doi.org/10.1089/ast.2017.1738.

Wright, J.T. and M.P. Oman-Regan. "Visions of Human Futures in Space and SETI." *International Journal of Astrobiology* Vol. 17, No. 2 (2017): 1–12. https://doi.org/10.48550/arXiv.1708.05318.

PART II

Exotheology

Chapter 6

Evolution Connected to Theory of Value by Process-Relational Astrotheology and Cosmotheology

Theodore Walker Jr.

BIOLOGICAL EVOLUTION

Process-relational theologian John B. Cobb Jr. called for a "richer account of evolution" in *Back to Darwin: A Richer Account of Evolution* (2008). In this collection of twenty-three essays, fifteen contributors,[1] all of whom affirm biological evolution, sought to identify and solve selected problems. All fifteen "accept and admire [Darwin's] immense achievement in showing that all living forms, certainly including human beings, evolved from common ancestors in a process in which natural selection has played a central role."[2] The problems are problems with theories developed by followers of Darwin, not by Darwin. Cobb states:

> The problems highlighted in this book are with certain assumptions and overstatements in the post-Darwin development of evolutionary theory. The book implicitly proposes that we go behind those developments—back to Darwin—and evaluate the later evolutionary theories more critically.[3]

In part, Cobb's call for "a richer account of evolution" is a call to correct the mistaken post-Darwin or Neo-Darwinian restriction of "natural selection" to selections from among random-chance genetic mutations. Correcting this mistaken restriction was helped by going "back to Darwin" and observing that Darwin's account of evolution was richer than that of his followers.[4] Beyond appreciating Darwin and depreciating Neo-Darwinians, Cobb and others added various supplements, including theological supplements, to advance a still richer account of evolution. Enrichment continues.

BIOLOGICAL EVOLUTION INCLUDES BEHAVIORAL EVOLUTION

Starting with evolution in biology, in *The Biology of the Cell Surface*, Ernest Everett Just (1883–1941),[5] who pioneered the study of marine egg cell fertilization and early embryo development, concluded his descriptions of cell-surface-mediated interactions with sperm and environment by adding a richer account of biological evolution. According to Just, from egg cells to humans, biological evolution includes *both* evolution of physical structures and functions *and* evolution of behaviors, including competitive behaviors and, more importantly, *cooperative behaviors*.

In emphasizing cooperative behaviors over competitive behaviors, Just was following Peter Kropotkin (1842–1921). Contrary to the prevailing idea that evolution is driven exclusively by competitive struggles against others and against the environment (what Just says is a misrepresentation of Darwin's account of "struggle for existence"), Kropotkin argued that "co-operation" or "mutual aid" is an essential factor in evolution.[6] To be sure, field observations by Kropotkin and others, including Darwin, did include observations of animals cooperating. For the sake of establishing Kropotkin's thesis, however, zoological field observations had to be supplemented by microscope observations of cells cooperating. Just theorized that observations of embryo development from an egg cell would be decisively important to theory of biological evolution because embryo development from a single egg cell resembles evolution from our common single cell ancestor, the very first link in the evolutionary chain.

More so than competitive behaviors, cooperative behaviors are essential to egg cell development, and essential to biological evolution. More attention to the evolution of cooperative behaviors (in tandem with the evolution of physical structures) enriches the otherwise fatally impoverished post-Darwin exclusive emphasis upon competitive behaviors. Starting from single cells, attending to the development and evolution of both physiology and behavior (both physical structures and inspired behaviors, especially cooperative behaviors) enriches evolutionary accounts of the natural origins of ethical behavior.[7]

COSMIC EVOLUTION

There are accounts of evolution that transcend biology. These include behavioral evolution,[8] "cultural evolution,"[9] "stellar evolution,"[10] and "cosmic evolution."[11] Eric J. Chaisson describes our cosmic context in terms of participation in "the cosmic-evolutionary narrative," and "energy flow" is essential to the narrative. As Chaisson states,

> More than any other single factor, energy flow would seem to be the principal means whereby all of nature's diverse systems have naturally spawned complexity in an expanding universe, in fact some of them evolving impressive degrees of order characteristic of life, mind, and civilization. Energy, specifically humanity's use of it optimally and wisely, will also likely guide our fate along the future arrow of time. For we, too, partake in the cosmic-evolutionary narrative, an epic-class story of rich natural history for the new millennium.[12]

Here, with attention to energy flow and cosmic context, Chaisson forwards the idea that evolution extends to "all spatial and temporal scales,"[13] even to cosmic scales and cosmic evolution.

Cosmic evolution can be implied by evolutionary cell biology. For instance, Just described the egg cell as "a universe." Egg cell biology is connected to astronomy and cosmology such that Just states,

The universe, however much we fragment it, abstract it, ever retains its unity.

> The egg cell also is a universe. And if we could but know it we would feel in its minute confines the majesty and beauty which match the vast wonder of the world.[14]

Just's idea of seeing the universe in an egg cell reminds me of a Mandelbrot fractal image (a universe) containing deep within it an identical image (also a universe). According to Just, the experience of seeing "majesty and beauty" through a microscope lens matches the experience of seeing "the vast wonder of the world" through a telescope lens. Hence, the colorfully illustrated biography of Just—*The Vast Wonder of the World: Biologist Ernest Everett Just*—by Melina Mangal—has a very appropriate title. Telescope viewing of "the vast wonder" of the universe is connected to microscope viewing of "the majesty and beauty" of egg-cell fertilization. Just again:

> The lone watcher of the sky who in some distant high tower suddenly saw a new planet floating before his lens could not have been more enthralled than the first student who saw the spermatozoon preceded by a streaming bubble moving toward the egg-centre. And as every novitiate in astronomy must thrill at his first

glance into the world of stars, so does the student to-day who first beholds this microcosm, the egg-cell.[15]

Given the unity of the universe, telescopes and microscopes reveal connections between astronomy and biology. Evolution applies to both.

Intuitively and imaginatively, before there was astronomical evidence of stellar and cosmic evolution, by describing the egg cell as a universe, Just was implicitly connecting astronomy and cosmology to evolutionary biology, especially to developmental and evolutionary processes revealed by studying egg cells. Starting in 1946, Just's 1939 intuition (connecting evolutionary biology to astronomy) would come to be justified by astronomical evidence demonstrating that planet Earth and living creatures are made from evolved stardust.

STELLAR EVOLUTION

We and our planet are made from heavy elements originally synthesized in previous generations of stars. The technical scientific literature on stellar synthesis of elements heavier than hydrogen includes a succession of works by Fred Hoyle and his colleagues.[16] The article "Synthesis of the Elements in Stars," famously signified by B2FH (for coauthors Burbidge, Burbidge, Fowler, and Hoyle), is so widely known among astronomers and astrophysicists because it was "a turning point in our knowledge of how the universe works."[17] According to astronomy and astrophysics, biological evolution was made possible by "stellar evolution."

THEORY OF VALUE, ETHICS, AND THEOLOGY

Another enrichment to theory of evolution is the idea of connecting cosmic evolution to theory of value and ethics. In *Cosmological Theories of Value*, Mark Lupisella presents three cosmological theories of value: (1) unidirectional "cosmological reverence," (2) bidirectional "cosmocultural evolution," and (3) "connection-action principle" with "cosmological intrinsic value." He holds "that reality is action-laden," that "connectedness requires relations which are realized as actions" and that "the more the universe produces action . . . the more its potential is realized."[18] In agreement with process philosophy, Lupisella states:

> This [connection-action principle] arguably implies a soft form of teleology, suggesting the universe is, at least in part, about realizing connectedness, and

> in so doing, it is evolving in many highly dynamic ways, always . . . new creations . . . and increasing novelty and diversity as suggested by certain forms of process philosophy . . .[19]

From among forms of process philosophy attentive to creativity, increasing novelty, and diversity, there are forms instructed by Alfred North Whitehead's magnum opus *Process and Reality*.

Whitehead's cosmology (*Process and Reality* is subtitled "an essay in cosmology") is *theologically* classified as "panentheism" by Charles Hartshorne and William L. Reese.[20] And contrary to nontheological renderings of process philosophy, the small-print index of *Process and Reality* includes references to the word "God" that run for two-thirds of the page. Obviously, a form of theology is essential to Whitehead's cosmology.

Not obviously, but logically, theology is *necessarily* essential to Whitehead's cosmology. This is because Whitehead's cosmology is a "philosophy of organism" that recognizes (from biology and from experience) that a creaturely individual is greater than the sum of its parts, and that the "universal individual,"[21] the "all-inclusive whole of reality,"[22] is greater than the sum of all included parts (all parts of reality). Logically, this all-inclusive reality is the only reality that is, and cannot fail to be, "that than which none greater can be conceived" (Anselm), the "religious idea of God" (Hartshorne)[23] and the prayerful meaning of "God is great, God is good."

In cosmic biology, we emphasize that Whitehead describes his "essay in cosmology" as a "philosophy of organism."[24] From cosmology to atomic events, science is, Whitehead states, "becoming the study of organisms," and biology is "study of the larger organisms; whereas physics is the study of the smaller organisms."[25] And "reality" is a creative "process," a creative synthetic process.[26] Whiteheadian-Hartshornean forms of process philosophy thus universalize creativity. In agreement with preclassical religious thought, the categories of Creator, creatures, and creations (procreations and recreations) embrace all that is actual. And with universal creativity comes a universal process of becoming. According to Hartshorne's theology, and according to Hoyle's astrobiology and cosmology, creation and evolution go together. Universal creativity yields a fundamentally evolving universe.[27]

The general idea of connecting cosmology to theory of value is consistent with Whitehead's cosmology. And some of the specific ideas described by Lupisella (such as bidirectionality, realization by actualization, and "cosmological intrinsic value") are also consistent with Whiteheadian-Hartshornean cosmology. For example, bidirectionality is consistent with the Whiteheadian-Ogdenian theological idea that God is both the one who makes a partly determinative difference to all things, and the one to whom all things make a partly determinative difference.[28] Metaphorically put, God transmits a signal that reaches (and influences) every receiver, and God receives (and is influenced by) every transmission. According to Hartshorne, everything that happens also happens to God, and God always responds in the most loving way.

METAPHYSICS

Process metaphysics (called neoclassical metaphysics by Hartshorne and Ogden) is a form of metaphysics that can help with connecting cosmology to theory of value and ethics. In cosmology, and in science generally, many fact-value and naturalistic fallacy discussions are disabled by the failure to recognize that science includes both contingent-factual truths *and* necessary truths. Necessary truths include necessary truths of abstract logic and mathematics, *and* logically necessary truths about concreate reality, including mereologically necessary truths about distinguishing the whole of reality from all and some parts of reality.

Consider the contrast between factual claims and logically necessary claims. The "sky is blue" is a contingent factual claim that is true if weather and other conditions obtain (and factually false if weather and other contingencies do not obtain). Given such factual contingencies, we need falsifiable observations for discerning if the sky is, or is not, blue. By contrast, "something is happening now" is necessarily true and never falsifiable.

The study of logically necessary truths about reality (metaphysics) is sharply separate from and opposite to popular-spooky-unnatural paranormal "metaphysics," and sometimes the former is called "transcendental metaphysics." Transcendental metaphysics includes a metaphysics of nature and a metaphysics of value and morals/ethics. Where natural science correctly appreciates transcendental metaphysics, natural sciences can correctly appreciate natural scientific theology (natural theology) and natural scientific theological ethics (natural law).

Modern science (including modern cosmology) characteristically fails to appreciate the necessary reality of the all-inclusive divine whole of reality. Likewise, modern moral theory characteristically fails to appreciate the necessity of references to the all-inclusive/comprehensive good—the "divine good." Both characteristic failures are caused by committing the "partialist fallacy,"[29] which is failing to conceive that all parts of reality are parts of the whole of reality.

Referencing only parts of reality (absent the all-inclusive whole of reality) can only yield a seriously inadequate, or wrong, or incoherent, account of reality. For example, *humanism* characteristically commits the partialist fallacy by valuing only human parts of no whole.[30] In *Beyond Humanism*, Charles Hartshorne argues that theology serves theory of value by providing "explicit recognition" of "the whole of which all lesser values are parts.[31] Hartshorne thus argues for the logical necessity of reference to God in moral theory.[32]

Following Hartshorne and Ogden,[33] Franklin Gamwell holds that the "necessary ground" for "any moral claim" is "the divine good," which is the "comprehensive variable that identifies the good as such."[34] An adequately grounded moral theory must refer to "the divine good" (the divine whole of reality) and to more than one part of reality, because "greater or lesser good can be identified only by the concrete comparisons within the divine relativity."[35] And because reference to God (whose necessary existence can be demonstrated by valid metaphysical arguments) is a necessity for adequate moral theory, modern moral theory, on account of its nontheistic ground (its partialist fallacy), is a necessary failure. Increasingly, however, the metaphysical necessity of God is coming to be appreciated in constructive postmodern science (or reconstructive postmodern science).[36]

EVOLUTION AND GOD

Fred Hoyle's article "Stellar Evolution and the Expanding Universe" indicates that the cosmic context of stellar evolution is "the expanding universe." Hoyle, who coined the term "big bang" to describe the expanding universe, had been famous for rejecting big bang cosmology in favor of steady-state cosmology; but the observational evidence changed his mind.[37] Moreover, while Hoyle had been famous for rejecting theism, again, he changed his mind. Hoyle's atheism was changed to a form of theism by the outcomes of his own mathematical calculations concerning stellar evolution and the coupling constants (of the expanding universe that accommodates stellar evolution) and by the outcome of mathematical calculations by Hoyle and Chandra Wickramasinghe concerning the astronomically improbable assembly of a living cell in some warm little pond by random chance.

Atheist Fred Hoyle became theist Fred Hoyle.[38] Hoyle's nonsupernatural theism is consistent with the neoclassical/process-relational conception of God. Hoyle recognized the mathematical necessity of an "all embracing intelligence."[39] The "all embracing" intelligent universe (described by Hoyle) is the "all-inclusive whole of reality" (described by Ogden).[40]

Concerning random cell assembly, in their "theory of cosmic creationism" Hoyle and Wickramasinghe calculated that the chance of correctly arranging amino acids into around two thousand enzymes crucial for life by random assembly is one in $10^{40,000}$! Hoyle illustrated the point by saying, "The chance that life forms might have emerged in this way is comparable with the chance that a tornado sweeping through a junk-yard might assemble a Boeing 747 from the material therein."[41] By comparing the random chance assembly of a Boeing 747 to the random chance assembly of two thousand enzymes, Hoyle illustrated the astronomical improbability of both. Hoyle and Wickramasinghe thus wrote:

> life cannot have had a random beginning. Troops of monkeys thundering away at random on typewriters could not produce the works of Shakespeare, for the practical reason that the whole observable universe is not large enough to contain the necessary monkey hordes, the necessary typewriters, and certainly the waste paper baskets required for the deposition of wrong attempts. The same is true for living material.[42]

The mathematical evidence from biological evolution converges with mathematical evidence from finely tuned cosmic and stellar evolution. Hoyle and Wickramasinghe saw the convergence of cosmic, stellar, and biological evolution as a "convergence to God."[43]

CULTURAL EVOLUTION AND COCREATING WITH GOD

In addition to Whiteheadian forms of process philosophy and theology, as Lupisella correctly indicates, there are other forms of process-relational theology that value creativity. For example, consider the relational "womanist perspective" that inspires the metaphor of "dancing with God" advanced by Karen Baker-Fletcher.[44] And consider Søren Kierkegaard's conception of "the self as a process of becoming" and Natalia Marandiuc's claim that "human and divine love dance together to cocreate the self."[45] Consider too the inspired bodily "walking with God" and "dancing with God" during worship by mostly Zulu followers of Isaiah Shembe (Isaiah Mloyiswa Mdliwamafa Shembe) from Kwa-Zulu-Natal, South Africa. Spiritual and physical "walking and dancing with God" is cocreating a theo-cultural evolution in congregational church worship.[46]

EASTERN PERSPECTIVES

In their Introduction to *Cosmos and Culture*, Steven J. Dick and Mark Lupisella lament that "Eastern perspectives are not properly represented, but perhaps will be in the future."[47] In small measure (not properly represented), Eastern perspectives are present.

There is an Eastern perspective on Just's *Biology of the Cell Surface*. In a thirty-eight-page article on "The Significance of the Cell Surface" in the

Journal of the Zoological Society of India, Cedric Dover says, in India, "we have preserved in biology a view of life as process, as interrelatedness, as a totality of subtle harmonies rather than a Darwinian war."[48] Dover compares the work of Calcutta zoologists Nelson Annandale, Sunder Lal Hora, and other "Asian biologists" to the work of Just; and he argues that Just "belongs to their company" because "his philosophy, like theirs, was a unitary one."[49] Certainly, Dover is correct in appreciating Just's unitary philosophy. And, insofar as Just "belongs in their company" (belongs in the company of Indian biologists), Asian perspectives are not entirely absent. Furthermore, preserving "a view of life as process, as interrelatedness" agrees with process-relational thinking.

There is a substantial body of literature indicating that Whiteheadian panentheism is resonant with Eastern thought. For example, in *Philosophers Speak of God*, Charles Hartshorne and William Reese place Sri Jiva, Hindu scriptures, and Lao-tse among ancient quasi-panentheists; and modern panentheists include Muslim scholar Muhammad Iqbal and Hindu scholar Radhakrishnan.[50] Furthermore, in the discussion below, there is considerable attention to work by Chandra Wickramasinghe, who grew up among Buddhists in Sri Lanka, and Gensuke Tokoro from Japan.

PARADIGM SHIFT?

According to Chandra Wickramasinghe and co-authors—Kamala Wickramasinghe and Gensuke Tokoro, recent enthusiasm for deliberating about cosmic context, energy flow, cosmic creativity, and cosmic evolution indicates the coming of a "long-overdue paradigm shift" in biology, a shift from "geocentric life to cosmic life," and with "far-reaching and profound" moral implications.[51] This long-overdue paradigm shift had been predicted in *The Relation of Biology to Astronomy*, where Fred Hoyle said:

> I suspect that the cosmic quality of microbiology will seem as obvious to future generations as the Sun being the centre of our solar system seems obvious to the present generation.[52]

The "future generations" that Hoyle spoke of in 1980 are now nearly our contemporaries. When this future becomes present, the paradigm for relating biology to astronomy will shift from exclusively Earth-centered biology to recognizing stellar evolution, cosmic evolution, cosmic life, and "our cosmic ancestry in the stars."[53] Echoing Hoyle's 1980 prediction, in 2019 Wickramasinghe and coauthors predicted that "ten years from now our cosmic origin will be deemed as obvious as the sun being the center of the solar system is considered obvious today"[54] (= paradigm shift by the year 2029!).

Panspermia (theory affirming that seeds of possible life are panoramically distributed, not restricted to planet Earth) was advanced from the early 1980s by Wickramasinghe and Hoyle. It was and remains provocative. Along with continuing advances in astrobiology, however, panspermia is becoming less provocative and perhaps nearly mainstream. David Darling has reported that unlike previous times, when panspermia was ridiculed and oppressed, by the year 2001 panspermia was coming to be tolerated. Darling wrote, "No longer does the mere mention of the word jeopardize a promising career or invite ridicule by one's scientific peers."[55] Only four years later, Steven J. Dick and James E. Strick described and advanced the development of astrobiology and the idea of a living universe.[56]

Hence, due to advances in the science of astrobiology, panspermia is not the most provocative feature of Wickramasinghe's and coauthors' *Our Cosmic Ancestry in the Stars*. Instead, among numerous other provocative features, the most provocative feature is the very far-reaching idea that shifting from geocentric energy flow to extraterrestrial and stellar energy flows is a "cosmic imperative" driven by "the acquisition of cosmic viruses" that can sustain and nourish the evolution of "unbridled greed."[57]

TWO CATEGORIES: LE AND LU

According to Wickramasinghe and co-authors, when considering the "use and expenditure of energy," we should consider the contrast between conservative and "unbridled," between "frugal (life-economical)" and "profligate (life-uneconomical)" expressed in the contrast between *Le* and *Lu*. Le and Lu are "two categories of presumed virally acquired characteristics: the former (Le/frugal/conservative) that subsists in harmony with a planetary ecology, and the latter (Lu/profligate/unbridled) that does not."[58]

> We came from space, we are an assembly of cosmic viruses, and ultimately we must return to the cosmos.[59]

Life has cosmic origins. Some of what comes from space is life-bringing.[60] For preeminent example, our earliest microbial ancestors came from space. In this sense, we came from space. And some of what comes from space is life-threatening, including "diseases from space" and viral pandemics from space.[61] Ultimately, "we must return."[62] While we are presently earthbound, we require more Le than Lu. In the future, for shifting to extraterrestrial energy flows, we will require more Lu than Le. Both Le and Lu were acquired via viruses with extraterrestrial and cosmic origins.

PANENTHEISM

There is a noteworthy theological difference between *The Big Bang and God: An Astro-Theology* (2015) by Walker and Wickramasinghe and *Our Cosmic Ancestry in the Stars* (2019) by Wickramasinghe and others. This theological difference resembles the important difference between panentheism and pantheism.

Panentheism, instructed by Charles Hartshorne, extrapolates from experience and biology, and recognizes mereological distinctions between whole and parts, and between all parts and some parts of reality. Accordingly, a whole living individual creature, such as a human creature, includes and exceeds (transcends) the sum of its living and nonliving bodily parts (its living cells and other parts). Similarly, the all-inclusive living individual Creator is the "one all-inclusive [divine] whole of reality" that includes and exceeds (transcends) the sum of all living and nonliving parts of reality.[63] Hence, taking the mere nontranscendent sum of all living and nonliving parts of reality to be "God" is a theological mistake; and this mistaken theology is called "pantheism" (pan = theos [instead of pan <theos]).

In *Big Bang and God*, "God" is understood as "the one all-inclusive whole of reality" that includes and exceeds (transcends) all parts of reality. This is explicit panentheism. In *Our Cosmic Ancestry in the Stars*, the conclusion of chapter 3 (titled "A Cosmic Creator?") is that Sir James Jean was correct in concluding that "God = universe."[64] Jean's equation is pantheistic. Nevertheless, the "purely logical case" for "intelligent design" and the idea of a universal "superintelligence" and "Cosmic Creator" finely tuning "the coupling constants of physics" reaches beyond the equation (= universe) in ways that clearly imply panentheism.[65] This is explicitly pantheistic, and implicitly panentheistic.

UNIVOCAL REFERENCING

In astronomy, cosmology, and theology, important mereological distinctions, such as the distinction between pantheism and panentheism, are sometimes easily lost because the word "cosmos" is at least bivalent, the word "universe" is at least trivalent, and the word "God" is multivalent. For instances:

Sometimes *cosmos* refers to the transcendent whole of reality, and

sometimes cosmos refers to the nontranscendent sum of all parts of reality.
Sometimes *universe* refers to the transcendent whole of reality, and
sometimes universe refers to the nontranscendent sum of all parts of reality, and
sometimes universe refers to some very large part(s) of reality (as with the idea of multiple universes).

Similarly, the word God:
sometimes refers to the transcendent whole of reality (panentheism),
sometimes refers to the nontranscendent sum of all parts of reality (pantheism),
sometimes refers to some part or parts of reality (hence, possible dualism and possible polytheism), and
sometimes refers to no part of reality, or to imaginary parts of reality (atheism).

Such variable referencing is problematic. Univocal references are needed for advancing astronomy, cosmology, natural scientific theology (natural theology, philosophical theology), and natural scientific theories of value, aesthetics, and natural law ethics. Variable referencing can be especially problematic for interdisciplinary approaches to philosophy (such as astro-philosophy/astrophilosophy), theology (such as astro-theology/astrotheology and cosmo-theology/cosmotheology), and theological ethics (such as astro-liberation theology, "cosmo-liberation theology,"[66] and "black natural law"[67]). For the sake of univocal referencing within and across natural scientific disciplines, unless explicitly qualified in ways indicating transcendence (such as "intelligent universe,"[68] "living universe,"[69] and "cosmic creator."[70]), the words "cosmos" and "universe" should refer to the nontranscendent sum of all parts of reality, and the word "God" should refer to the transcendent "all-inclusive whole of reality,"[71]

NOTES

1. Francisco J. Ayala, Ian Barbour, L. Charles Birch, Philip Clayton, John B. Cobb Jr., John C. Greene, David Ray Griffin, Pete A. Y. Gunter, John F. Haught, Lynn Margulis, Reg Morrison, Dorion Sagan, Jeffrey P. Schloss, Robert J. Valenza, and Howard J. Van Till.
2. Cobb, *Back to Darwin*, viii.
3. Ibid.
4. Similarly, see Loye, *Rediscovering Darwin*.
5. See Manning's award-winning biography of Just in *Black Apollo of Science*. See also Jenkins, "Black Apollo of Science."
6. Kropotkin, *Mutual Aid*.

7. Accordingly, *The Biology of the Cell Surface* was followed by *The Origin of Man's Ethical Behavior*. Also: Walker, "The Bioethical Significance of 'The Origin of Man's Ethical Behavior'"

8. Ibid.

9. Dick and Lupisella, *Cosmos and Culture*.

10. Hoyle, "Stellar Evolution and the Expanding Universe."

11. Hartshorne, *Beyond Humanism*, 140; "Creation through Evolution," 62; Tyson and Goldsmith, *Origins*; Chaisson "Cosmic Evolution."

12. Chaisson "Cosmic Evolution," 25.

13. Ibid.

14. Just, *The Biology of the Cell Surface*, 368.

15. Ibid.

16. Chronologically, see Hoyle, "The Synthesis of the Elements from Hydrogen," "On the Condensation of the Planets," "The Chemical Composition of the Stars," "On the Formation of Heavy Elements in Stars," "A New Model for the Expanding Universe," "Stellar Evolution and the Expanding Universe," and "On Nuclear Reactions Occurring in Very Hot Stars." See also Hoyle et al., "Origin of the Elements in Stars" and Burbidge et al., "Synthesis of the Elements in Stars." For this and related work on nuclear reactions in stars, William A. Fowler and Subrahmanyan Chandrasekhar shared the Nobel Prize in Physics in 1983.

17. Tyson and Goldsmith, *Origins*, 165.

18. Lupisella, *Cosmological Theories of Value*, 233, 235, 237, 238.

19. Ibid., 238.

20. Hartshorne and Reese, *Philosophers Speak of God*.

21. Hartshorne, *Reality as Social Process*, 176.

22. Ogden, "Process Theology and the Wesleyan Witness," 21.

23. Sia, *God in Process Thought*, 9.

24. Whitehead, *Process and Reality*, xi, 18.

25. Whitehead, *Science and the Modern World*, 103.

26. Whitehead, *Process and Reality*; Hartshorne, *Reality as Social Process*; *Creative Synthesis and Philosophical Method*; *Creativity in American Philosophy*.

27. Hartshorne, "Creation through Evolution"; Hoyle, "a new view of creation and evolution" (subtitle of *Intelligent Universe*); both published in 1984.

28. Ogden, "The Concept of God in Process Metaphysics," 80, 85, 88.

29. Gamwell, *Beyond Preference*; *The Divine Good*.

30. A unique form of humanism called "humanocentric theism" avoids committing the partialist fallacy by advancing a nonclassical and antiracist theism. Jones, *Is God a White Racist?*

31. Hartshorne, *Beyond Humanism*, 25.

32. See also Hartshorne, *Man's Vision of God and the Logic of Theism*, ch. 4.

33. See Hartshorne, *Beyond Humanism*; "God and Righteousness" in *Man's Vision of God and the Logic of Theism*; *Creative Synthesis and Philosophical Method*. See also, Ogden, *The Reality of God and Other Essays*, "The Criterion of Metaphysical Truth and the Senses of Metaphysics," "The Concept of God in Process

Metaphysics," "Process Theology and the Wesleyan Witness," and "The Metaphysics of Faith and Justice."

34. Gamwell, *The Divine Good*, 18, 178. See also *Beyond Preference*; *Existence and the Good*; and *On Metaphysical Necessity*.

35. Gamwell, *The Divine Good*, 182.

36. See for example Griffin, *The Reenchantment of Science*; *Religion and Scientific Naturalism*; and *Reenchantment Without Supernaturalism*.

37. See Hoyle, "A New Model for the Expanding Universe."

38. Walker and Wickramasinghe, *The Big Bang and God*.

39. Hoyle, *The Intelligent Universe*, 225.

40. Ogden, "Process Theology and the Wesleyan Witness," 21.

41. Hoyle and Wickramasinghe, *Evolution from Space*, 105.

42. Ibid., 148.

43. "Convergence to God" is chapter 9 in Hoyle and Wickramasinghe, *Evolution from Space*. Furthermore, they advanced an evolutionary "Case for Life as a Cosmic Phenomenon." Wickramasinghe, *Journey with Fred Hoyle: The Search for Cosmic Life*, *Search for Our Cosmic Ancestry*, and *Vindication of Cosmic Biology*. And, according to "Creation from Chaos and from the Cosmic Egg" by Charles H. Long, contemporary scientific creation narratives resemble long-ago mythological creation narratives.

44. Baker-Fletcher, *Dancing with God*.

45. Marandiuc, *The Goodness of Home*, 6, 19.

46. Vilakazi, *Shembe: The Revitalization of African Society*. For an ethical prescription for theo-cultural evolution in church worship, see Hermans-Webster, "The Eucharist and Planetary Wellbeing."

47. Dick and Lupisella, *Cosmos and Culture*, 5.

48. Dover, "The Significance of the Cell Surface," 3.

49. Ibid., 3–4. Also, concerning Dover, see *The Prism of Race: W. E. B. Du Bois, Langston Hughes, Paul Robeson, and the Colored World of Cedric Dover* by Nico Slate.

50. See Hartshorne and Reese, *Philosophers Speak of God*, ch. 7. For more recent Eastern applications of panentheism, see Göcke and Medhananda, *Panentheism in Indian and Western Thought*; Clayton and Biernacki, *Panentheism across the World's Traditions*; Biernacki, *The Matter of Wonder*; Cooper, *Panentheism*, ch. 9.

51. Wickramasinghe et al., *Our Cosmic Ancestry in the Stars*, 1.

52. Hoyle, *The Relation of Biology to Astronomy*, 24–25.

53. Wickramasinghe et al., *Our Cosmic Ancestry in the Stars*. Similarly, Wickramasinghe et al., "The Transition from Earth-Centred Biology to Cosmic Life."

54. Wickramasinghe et al., *Our Cosmic Ancestry in the Stars*, 1.

55. Darlin, *Life Everywhere*, 47.

56. Dick and Strick, *The Living Universe*.

57. Wickramasinghe et al., *Our Cosmic Ancestry in the Stars*, 49–58.

58. Ibid., 52.

59. Ibid., 1.

60. Hoyle and Wickramasinghe, "Comets—a Vehicle for Panspermia," *Evolution from Space*, ch. 9; *Space Travellers*.

61. Hoyle and Wickramasinghe, *Diseases from Space*.

62. Wickramasinghe et al., *Our Cosmic Ancestry in the Stars*, 1.

63. Ogden, "Process Theology and the Wesleyan Witness," 21.

64. Wickramasinghe et al., *Our Cosmic Ancestry in the Stars*, 3.

65. Ibid., 27, 31.

66. Graff, *Undisciplining Knowledge*; Walker, "Interdisciplinary Ethics: From Astro-Theology to Cosmo-Liberation Theology." Walker is instructed by Barbara A. Holmes, who argues that cosmology is important for liberation theology in her books *Liberation and the Cosmos*, and *Race and the Cosmos*, and by Karen Baker-Fletcher's *Womanist Wordings on God and Creation*.

67. Lloyd, *Black Natural Law*.

68. Hoyle, *The Intelligent Universe*.

69. Dick and Strick, *The Living Universe*.

70. Wickramasinghe et al., *Our Cosmic Ancestry in the Stars*, 3.

71. Ogden, "Process Theology and the Wesleyan Witness," 21.

BIBLIOGRAPHY

Baker-Fletcher, K. *Dancing with God: Trinity from a Womanist Perspective*. St. Louis: Chalice Press, 2006.

———. *Sisters of Dust, Sisters of Spirit: Womanist Wordings on God and Creation*. Minneapolis: Fortress Press, 1998.

Biernacki, L. *The Matter of Wonder: Abhinavagupta's Panentheism and the New Materialism*. Oxford: Oxford University Press, 2023.

Burbidge, E. M. et al. "Synthesis of the Elements in Stars." *Rev. Mod. Phys.* 29, no. 547 (1957): 547–654. https://doi.org/10.1103/RevModPhys.29.547.

Chaisson, E. J. "Cosmic Evolution: State of the Science." In *Cosmos and Culture: Cultural Evolution in a Cosmic Context*, edited by Steven J. Dick and Mark L. Lupisella, 7–26. Washington, DC: National Aeronautics and Space Administration, 2009.

Clayton, P., and L. Biernacki. *Panentheism across the World's Traditions*. Oxford University Press, 2013.

Cobb, J. B. Jr., ed. *Back to Darwin: A Richer Account of Evolution*. Grand Rapids, MI: William B. Eerdmans Publishing, 2008.

Cooper, J. W. *Panentheism: The Other God of the Philosophers*. Grand Rapids, MI: Backer Academic, 2006.

Darling, D. *Life Everywhere: The Maverick Science of Astrobiology*. New York: Basic Books, 2001.

Dick, S. J., and J. E. Strick. *The Living Universe: NASA and the Development of Astrobiology*. New Brunswick, NJ: Rutgers University Press, 2005.

Dick, S. J. and M. Lupisella, eds. *Cosmos and Culture: Cultural Evolution in a Cosmic Context*. Washington, DC: National Aeronautics and Space Administration, 2009.

Dover, C. "The Significance of the Cell Surface (The Work of E. E. Just)." *Journal of the Zoological Society of India* 6, no. 1 (1954): 3–42.

Gamwell, F. I. *Beyond Preference: Liberal Theories of Independent Associations*. Chicago: University of Chicago Press, 1984.

———. *Existence and the Good: Metaphysical Necessity in Morals and Politics*. Albany: SUNY Press, 2011.

———. *On Metaphysical Necessity: Essays on God, the World, Morality, and Democracy*. Albany: SUNY Press, 2020.

———. *The Divine Good: Modern Moral Theory and the Necessity of God*. Dallas: Southern Methodist University Press, 1996.

Göcke, B. P., and W. Medhananda, *Panentheism in Indian and Western Thought: Cosmopolitan Interventions*. New York: Routledge, 2024.

Graff, H. J. *Undisciplining Knowledge: Interdisciplinarity in the Twentieth Century*. Baltimore: John Hopkins University Press, 2015.

Griffin, D. R. *Reenchantment Without Supernaturalism: A Process Philosophy of Religion*. Ithaca: Cornell University Press, 2001.

———. *Religion and Scientific Naturalism*. Albany: SUNY Press, 2000.

———. *The Reenchantment of Science*. Albany: SUNY Press, 1988.

Hartshorne, C. "Creation through Evolution." In *Omnipotence and Other Theological Mistakes*, 65–95. Albany: SUNY Press, 1984.

———. *A Natural Theology for Our Time*. La Salle, IL: Open Court Publishing, 1967.

———. *Beyond Humanism: Essays in the Philosophy of Nature*. Gloucester: Peter Smith Books, 1975.

———. *Creative Synthesis and Philosophical Method*. La Salle, IL: Open Court Publishing, 1970.

———. *Creativity in American Philosophy*. Albany: SUNY Press, 1984.

———. *Man's Vision of God and the Logic of Theism*. Hamden, CT: Archon Books, 1964.

———. *Reality as Social Process: Studies in Metaphysics and Religion*. Boston: Beacon Press, 1953.

Hartshorne, C., and W. L. Reese. *Philosophers Speak of God*. Chicago: University of Chicago Press, 1953.

Hermans-Webster, T. "The Eucharist and Planetary Wellbeing: Norman Pittenger's Process Theology of the Eucharist for a Sacramental Ecotheology." Phd diss., Boston University School of Theology, 2022.

Holmes, B. A. *Liberation and the Cosmos: Conversations with the Elders*. Minneapolis: Fortress Press, 2008.

———. *Race and the Cosmos: An Invitation to View the World Differently*. London: Bloomsbury T&T Clark, 2002.

Hoyle, F. "A New Model for the Expanding Universe." *Monthly Notices of the Royal Astronomical Society* 108, no. 1748, (1948): 372–82. https://doi.org/10.1093/mnras/108.5.372.

———. "On Nuclear Reactions Occurring in Very Hot Stars. I. The Synthesis of Elements from Carbon to Nickel." *Astrophysical Journal Supplement* 1 (1954): 21. https://doi.org/10.1086/190005.

———. "On the Condensation of the Planets." *Monthly Notices of the Royal Astronomical Society* 106, no. 5 (1946): 406–22. https://doi.org/10.1093/mnras/106.5.406.

———. "On the Formation of Heavy Elements in Stars." *Proc. Phys. Soc.* 59, no. 6 (1947): 972. https://doi.org/10.1088/0959–5309/59/6/306.

———. "Stellar Evolution and the Expanding Universe." *Nature* 163 (1949): 196–98. https://doi.org/10.1038/163196a0.

———. "The Chemical Composition of the Stars." *Monthly Notices of the Royal Astronomical Society* 106, no. 4 (1946): 255–59. https://doi.org/10.1093/mnras/106.4.255.

———. "The Synthesis of the Elements from Hydrogen." *Monthly Notices of the Royal Astronomical Society* 106, no. 5 (1946). https://doi.org/10.1093/mnras/106.5.343.

———. *The Intelligent Universe: A New View of Creation and Evolution*. New York: Holt, Rinehart, and Winston Publishing, 1984.

———. *The Relation of Biology to Astronomy*. Cardiff: University College Cardiff Press, 1980.

Hoyle, F., and C. Wickramasinghe. "The Case for Life as a Cosmic Phenomenon." *Nature* 322, no. 6079 (1986): 509–11. https://doi.org/10.1038/322509a0.

———. "Comets—a Vehicle for Panspermia." In *Comets and the Origin of Life: Proceedings of the Fifth College Park Colloquium on Chemical Evolution*, edited by C. Ponnamperuma, 227–39. Dordrecht: Springer, 1981. https://doi.org/10.1007/978-94-009-8528-5_15.

———. *Diseases from Space*. London: J. M. Dent & Sons, 1979.

———. *Evolution from Space: A Theory of Cosmic Creationism*. New York: Simon & Schuster, 1981.

———. *Lifecloud: The Origin of Life in the Universe*. London: J. M. Dent & Sons, 1978.

———. *Space Travellers: The Bringers of Life*. Cardiff: University College Cardiff Press, 1981.

Hoyle, F., et al. "Origin of the Elements in Stars." *Science* (new series)124, no. 3223 (1956): 611–14. https://doi.org/10.1126/science.124.3223.6.

Jenkins, L. R. "Black Apollo of Science: The Life of Ernest Everett Just—Summarizing Timeline, Sumitography, and Concept Poster" (2021): 1–31. Perkins Faculty Research and Special Events. 27. https://scholar.smu.edu/theology_research/27.

Jones, W. R. *Is God a White Racist?: A Preamble to Black Theology*. Boston: Beacon Press, 1998, originally 1973.

Just, E. E., *The Biology of the Cell Surface*. Philadelphia: P. Blakiston's Son & Co., 1939.

Keller, C., and M. Rubenstein, eds. *Entangled Worlds: Religion, Science, and New Materialisms*. New York: Fordham University Press, 2017.

Kropotkin, P. *Mutual Aid: A Factor of Evolution*, edited by C. Aldarondo. Kindle Edition: Amazon Digital Services. CreateSpace Independent Publishing Platform, 2014, originally 1902.

Lloyd, V. *Black Natural Law*. New York: Cambridge University Press, 2016.
Long, C. H. *Alpha: The Myths of Creation*. Chico, California: Scholars Press, 1963.
Loye, D. *Rediscovering Darwin: The Rest of Darwin's Theory and Why We Need It Today*. Pacific Grove: Romanes Press, 2018.
Lupisella, M. *Cosmological Theories of Value: Science, Philosophy, and Meaning in Cosmic Evolution*. Cham: Springer, 2020.
Mangal, M. *The Vast Wonder of the World: Biologist Ernest Everett Just*. Minneapolis: Millbrook Press, 2018.
Manning, K. R. *Black Apollo of Science: The Life of Ernest Everett Just*. Oxford: Oxford University Press, 1983.
Marandiuc, N. *The Goodness of Home: Human and Divine Love and the Making of the Self*. Oxford: Oxford University Press, 2018.
Ogden, S. M. "The Concept of God in Process Metaphysics." In *Faith and Freedom: Toward a Theology of Liberation*, 75–95. Nashville: Abingdon Press, 1979.
———. "The Criterion of Metaphysical Truth and the Senses of Metaphysics." *Process Studies* 5, no. 1 (1975): 47–48.
———. "Process Theology and the Wesleyan Witness." *Perkins School of Theology Journal* 37, no. 3 (1984): 18–33.
———. "The Metaphysics of Faith and Justice." *Process Studies* 14, no. 2 (1985): 87–101.
———. *The Reality of God and Other Essays*. New York: Harper & Row, 1966.
Sia, S. *God in Process Thought: A Study of Charles Hartshorne's Concept of God*. Boston: Martinus Nijhoff, 1985.
Slate, N. *The Prism of Race: W. E. B. Du Bois, Langston Hughes, Paul Robeson, and the Colored World of Cedric Dover*. New York: Palgrave Macmillan, 2014.
Tyson, N., and D. Goldsmith. *Origins: Fourteen Billion Years of Cosmic Evolution*. New York; London: W. W. Norton, 2004.
Vilakazi, A., with B. Mthethwa and M. Mpanza. *Shembe: The Revitalization of African Society*. Johannesburg: Skotaville, 1986.
Walker, T., Jr. "The Bioethical Significance of 'The Origin of Man's Ethical Behavior' (October 1941, unpublished) by Ernest Everett Just and Hedwig Anna Schnetzler Just," *Journal of the South Carolina Academy of Science* 18, no. 1, Article 4 (January 2020).
———. "Interdisciplinary Ethics: From Astro-Theology to Cosmo-Liberation Theology." In *Entangled Worlds: Religion, Science, and New Materialisms*, edited by C. Keller and M. J. Rubenstein, 221–27, Bronx: Fordham University Press, 2016.
Walker, T.. Jr. and C. Wickramasinghe. *The Big Bang and God: An Astro-Theology*. New York: Palgrave Macmillan, 2015.
Whitehead, A. N. *Adventures of Ideas*. New York: Macmillan, 1933.
———. *Process and Reality: An Essay in Cosmology*. Corrected Edition, edited by D. R. Griffin and D. W. Sherburne. New York: Free Press, 1978.
———. *Science and the Modern World*. New York: Macmillan; Free Press, 1967.

Wickramasinghe, C. *Our Cosmic Ancestry in the Stars: The Panspermia Revolution and the Origins of Humanity*. Rochester: Bear & Company, 2019.

———. *A Journey with Fred Hoyle: The Search for Cosmic Life*, edited by Kamala Wickramasinghe. London: World Scientific Books, 2005.

———. *The Search for Our Cosmic Ancestry*. Hackensack: World Scientific Publishing, 2014.

Wickramasinghe, C., ed. *Vindication of Cosmic Biology: Tribute to Sir Fred Hoyle (1915–2001)*. Hackensack: World Scientific, 2015.

Wickramasinghe, C.. et al. "The Transition from Earth-Centred Biology to Cosmic Life." *Journal of Astrobiology & Outreach* 3, no. 1 (2015): 1–6. https://www.walshmedicalmedia.com/open-access/the-transition-from-earthcentred-biology-to-cosmic-life-2332-2519-1000122.pdf.

Chapter 7

Astrobiology, Astrotheology, and Cosmic Consciousness

Ted Peters

As a science, astrobiology—accompanied by sister disciplines such as astronomy, astrophysics, and cosmology—is almost inherently religious. Looking at the stars, said Plato, "compels the soul to look upwards and leads us from this world to another."[1] Two and a half millennia after Plato, Harvard astronomer Owen Gingerich adds, "Cosmology is a voyage of the human spirit."[2]

Why is the science of astrobiology almost religious? Because it plucks the strings of spiritual sensibilities. More specifically, our built-in psychological and cultural sense of *the Beyond* begins to vibrate at thoughts of astronomical distances and the possibility of meeting extraterrestrial neighbors.

For our archaic ancestors, it was the sky that attuned itself to our innate sense of the Beyond. Power comes from the sky. The life-giving sun is in the sky. Shooting stars bring the sky down to earth. Thunderbolts in the sky dazzle and even destroy. The sky became personalized as Aman-Re in Egypt, Indra in India, Thor in Scandinavia, Zeus in Greece, and Jupiter in Rome.

Phenomenologist of religion Mircea Eliade recovers our primal experience with the sky. "It is a total awareness on man's part: beholding the sky, he simultaneously discovers the divine incommensurability and his own situation in the cosmos. For the sky, *by its own mode of being,* reveals transcendence, force, eternity. It *exists absolutely,* because it is *high, infinite, eternal, powerful.*"[3] The sky's transcendence calls to the quest for transcendence within the human soul.

But our human awe for the sky disappeared in the modern world when we secularized the sky, when we rid the sky of Jupiter and all the other sky-gods. Our sky loses its enchantment when scientized, technologized, and demythicized. The sky loses its awe when we daily consult weather reports on our Apple Watch.

But right behind the secularized sky we discovered a second Beyond. That second Beyond is outer space. Outer space is deep, dark, unfathomable. Be there gods there? Our spiritual sensibilities are struggling to attune themselves to every new scientific discovery about that Beyond. Might we ask our astrobiologists to dazzle us with prophecies, revelations, and awe?

No. Why not? Because subjectivity has been expunged. Awe has been expunged. Reverence has been expunged. Transcendence has been expunged.

The excitement and meaning and sense of grandeur has been expunged in order to limit science to the data. Like other scientists, astrobiologists gather data about exoplanets, habitability, and biota. Data gathering is objective. Strictly objective. There is no room in this science or any other science for attunement to spiritual sensibilities.

What an astrobiologist knows about the planet Jupiter today must be denuded of the sky-god Jupiter known and worshipped by our Roman ancestors. To discover and store data about the largest planet in our solar system, the astrobiologist is obligated to rid Jupiter of any and all spiritual resonance.

Jupiter is religious. Astrobiology, which studies Jupiter, is not. This makes the science of astrobiology almost, but not quite, religious.

This observation should interest the astrothologian. On the one hand, we have so much to learn about outer space from the astrobiologist's discoveries. On the other hand, the scientific denudation of Jupiter's previous religious meaning must be taken into account. Astrobiological knowledge cannot count as theological knowledge without considerable interpretation.

In what follows we will look at the mission of astrobiology, the almost but not quite religious science. We will look also at the task of astrotheology, namely, to bridge the gulf dividing the strictly objectivist epistemology of astrobiology from our human experience with built-in spiritual sensibilities.

We will then ask: might this bridge be built from materials drawn from cosmology, ontology, physics, and metaphysics? Might the physics of David Bohm or the metaphysics of Alfred North Whitehead help in bridge construction? The Whiteheadian and Bohmian cosmological schemes include both objective science and subjective human experience in their respective worldviews. Might the astrotheologian take advantage of one or both?

ASTROBIOLOGY: THE ALMOST RELIGIOUS SCIENCE

In the 1990s, the term *astrobiology* quickly replaced the term previously used by Carl Sagan, Frank Drake, SETI, NASA, and others, *exobiology.* SETI's former director, Jill Tarter, provides a definition shared today with

NASA: "Astrobiology is the science that deals with the origin, evolution, distribution, and future of life in the Universe."[4] Hybrid astrobiologist-theologian Lucas Mix expands this slightly. "Astrobiology is the scientific study of life in space. It happens when you put together what astronomy, physics, planetary science, geology, chemistry, biology, and a host of other disciplines have to say about life and try to make a single narrative."[5]

Within our solar system, astrobiologists are looking for biota, for chemical processes that suggest biological processes. They expect to find microbial life on Mars and the moons of Saturn. Beyond our solar system yet still within the Milky Way Galaxy, astrobiologists are looking at exoplanets in the habitable zone (H_z) for techno-signatures. They expect to find signs of intelligent civilizations, especially more technologically advanced civilizations.

This research carries a spiritual valence, despite the sober objectivity of the research methods employed. Questions such as meaning or purpose cling to astrobiology like barnacles to a ship's hull. But, for methodological reasons, the astrobiologist is obligated to scrape off those barnacles of meaning. Meaning dare not contaminate the raw data that researchers log in their computers. Regardless, the published data incite questions of meaning to those of us in the extrascientific culture who watch the astrobiologists with crossed fingers and hopeful prayers.

The astrotheologian asks: but what counts as explanatory adequacy? Is materialist reductionism enough? No. "The issue is precisely," says process theologian David Ray Griffin, "*which* set of concepts and principles is most adequate for interpreting all events."[6] When a science such as astrobiology at the level of method eliminates all subjectivity, then the results of its research can be only partial, devoid of meaning, and incomplete.

Astrobiologists, like Big Bang astrophysicists, find it difficult to avoid questions of origin, destiny, and meaning.[7] NASA's percipient astrophilosopher Mark Lupisella tries to pave a road to maximum comprehensiveness. But, alas, he runs out of road and hits a dead end.

> If our worldviews need to be comprehensive and include specific guidance for human behavior and address most of our complex subjective needs, then the universe is probably not enough for most people most of the time . . . there are broader questions such as why the universe exists at all, or more generally, why there is something rather than nothing, that modern cosmology arguably does not provide satisfying answers for.[8]

On the one hand, astrobiology invites us to a comprehensive worldview replete with ultimate meaning. On the other hand, crippled by a restrictive scientific epistemology, the space sciences are unable on their own to do the work only a theologian can.[9]

What is important here is that the natural world interpreted by astrobiology cannot help but emit vibrations of ultimacy that are attuned to our spiritual sensibilities. This leads Unitarian Universalist theologian Dawn Cooley to make an observation. “It is our human imperative to look for and make meaning—this is the foundation for the establishment of religions. It is also the impetus behind the creation of the field of astrobiology.”[10] Whether the astrobiologist admits it or not, this branch of science sprays spiritual meaning like a lawn sprinkler sprays water.

ASTROTHEOLOGY: A SCIENTIFICALLY INFORMED THEOLOGY OF NATURE

The attunement between the subjective and the objective—especially the attunement of the soul with the Beyond—belongs inherently to human experience. It belongs as well to theological reflection on that experience. Michael Fishbane makes this clear. “Theology is . . . situated at the border of the known and unknown, of the manifest and the concealed. It is at this nexus that the self seeks God . . . [Seeking God extends] to the utmost depths of Being and beyond (beyond the Beyond of all conception).”[11]

The systematic theologian relies upon this attunement of our spiritual sensibilities when conceptualizing the divine. “God,” according to Katherine Sonderegger, “is the Shattering Object, the Uncontained and Unconstrained Reality, the One Who Is There. God is not coordinated with our salvation or our hunger to know Him because He is Beyond, always Beyond such pairing.”[12] With this in mind, the theologian will be drawn to astrobiology like filings drawn to a magnet.

Astrotheology, then, is the systematic theologian’s response to astrobiology. *Christian Astrotheology* is that branch of theology which provides a critical analysis of the contemporary space sciences combined with an explication of classic doctrines such as creation and Christology for the purpose of constructing a comprehensive and meaningful understanding of our human situation within an astonishingly immense cosmos.[13]

Astrotheology fits within a slightly more inclusive category, namely, *Theology of Nature*. A theology of nature, according to Ian G. Barbour, “must take the findings of science into account when it considers the relation of God and [humanity] to nature, even though it derives its fundamental ideas elsewhere.”[14] Elsewhere? Yes, special revelation. This indicates that what we learn from science about nature contributes to our understanding of God’s creation, even though what we know about God is extrascientific in origin. A theology of nature requires an epistemology that includes science yet more than science.[15]

In both its classical and contemporary form, systematic theology is a field encompassing field. The theologian attempts to draw the most comprehensive picture of reality that can be conceived. St. Thomas Aquinas describes theology as a "sacred science" with God as the object of this science, along with all other things. "But in sacred science, all things are treated under the aspect of God: either because they are God Himself or because they refer to God as their beginning and end. Hence it follows that God is in very truth the object of this science."[16]

I put it this way: a theology of nature tries to comprehend all things in reality in relationship to the one God of grace. In astrotheology, this includes the galaxies, the stars, the planets, and perhaps our future space neighbors.[17] And, according to Lewis Ford, it most assuredly includes our extraterrestrial neighbors. "God is necessarily operative in the development of every life and in every culture, whether terrestrial or extraterrestrial."[18]

LOSS OF THE BEYOND SENSIBILITY IN CARTESIAN DUALISM

Medieval and Reformation Europeans lived in one world, a single world that included both the objective and the subjective.[19] The human mind, it was assumed, was attuned by reason to both physics and metaphysics. The founders of modern empirical science—Copernicus, Kepler, Galileo, and Newton—marveled in gratitude that God had graced the human mind with the same rational principles that guided the stars in their courses.

The sundering of subject from object began inadvertently with René Descartes (1596–1650). Descartes engaged in a devastating thought experiment about truth. He asked: how can I be certain that the image in my subjectivity corresponds accurately to what I perceive objectively? With this question the inner soul and the outer world became sharply distinguished, defined as separate substances. Despite the substantial difference, Descartes was confident that God would maintain the attunement between mind and matter. What terrorizes us is this question: what might happen if we eliminate God's binding work? Might subject and object undergo a permanent divorce? History tells us that this is just what happened in the late Enlightenment period.

Pierre-Simon LaPlace (1749–1827), while writing his five-volume *Celestial Mechanics*, told Emperor Napoleon that he no longer needed the God hypothesis to hold things together. Nature sufficed on its own. This process of materializing science is referred to as disenchantment or *Entgotterung* by David Ray Griffin. "God was at first stripped of all causal power beyond that of the original creation of the world; later thinkers turned this deism into a complete atheism. . . . The ironic conclusion is that modern science, in disenchanting [*Entgotterung*] nature, began a trajectory that ended by disenchanting science itself. If all human life is meaningless, then science, as one of its activities, must share in this meaninglessness."[20]

Once God became eliminated from the equation, the bridge between mind and matter crumbled and fell into a chasm. A continental divide opened up between the inner world of meaningful subjectivity and meaningless objective knowing. Modern science would now march to the drumbeat of objectivity alone.

But the objective realm is only part of reality! Right? Ouch! The epistemological surgery—the subjectechtomy—left the objective scientific researcher only half alive.

Philosophers of science such as Thomas Nagel have filed a wrongful malpractice suit against modern science. The mind is just as real as the objects the mind perceives, Nagel observes. "The great advances in the physical and biological sciences were made possible by excluding the mind from the physical world. This has permitted a quantitative understanding of that world, expressed in timeless, mathematically formulated physical laws. But at some point it will be necessary to make a new start on a more comprehensive understanding that includes the mind."[21] Objective scientific data excludes the human mind, but the human mind is the only place where this objective data is collected, monitored, and cared for. We need a new start, a start that reconnects the mind with what the mind knows.

That new start—where we work from within a "more comprehensive understanding that includes the mind"—should tempt the astrotheologian like an ice cream cone tempts a child running home after school.

No purchase on reality can be comprehensive if it includes the stars but not the eyes that watch those stars. No purchase on reality can be comprehensive if it points to the physical beyond denuded of the spiritual sensibility that appreciates the Beyond.[22]

Whether the astrobiologist admits it or not, this branch of science exudes spiritual meaning while it excludes spiritual meaning.

RETRIEVING SUBJECTIVITY IN WHITEHEADIAN HOLISM

In this field encompassing field, the astrotheologian will want to incorporate scientific knowledge without necessarily repudiating that knowledge. Scientific knowledge is genuine knowledge, to be sure. But its scope is limited to the natural causal nexus as objectively reported.

In my own methodological musings, I have been attracted to holistic hypotheses that retrieve the connection between subjectivity and objectivity, data and meaning, the mundane and the divine. Alfred North Whitehead's metaphysics and David Bohm's physics have drawn my attention.

Whitehead's speculative philosophy attempts to understand all things real at the highest level of abstraction. But before he abstracts, Whitehead attends to what is concrete, namely, experience. "Speculative philosophy is the endeavour to frame a coherent, logical, necessary system of general ideas in terms of which every element of our experience can be interpreted."[23] Most importantly, this speculative philosophy includes human subjectivity, objective knowing, and the divine. "But what ever else God is," contend Herman Daly and John Cobb, "God is also the inclusive whole."[24]

If our fundamental datum is experience, then we should note that experience unavoidably includes both the objective and subjective poles. Objective data, in effect, is data abstracted from the more primary experience that includes the subjectivity of the scientist gathering the objective data. For the scientist to claim that the abstracted data is more primary than our subjective interpretation of it is to commit the fallacy of misplaced concreteness. "The fallacy of misplaced concreteness" Whitehead states, "consists in neglecting the degrees of abstraction involved when an actual entity is considered merely so far as it exemplifies certain categories of thought."[25] In sum, a domain of natural science will shortchange its purchase on reality if it fails to incorporate the subjective pole. Experience, upon which empirical knowledge is based, includes a primordial bond between subject and object. The process metaphysician never suspends this holistic insight.

RETRIEVING SUBJECTIVITY IN BOHMIAN HOLISM

Switching from Whitehead's metaphysics to Bohm's physics, we see a second variant of holism at work. Physicist David Bohm proposes that all of reality, both objective and subjective, is best described as undivided wholeness. Inseparable quantum interconnectedness of all things in the universe is the fundamental reality.

> The more comprehensive, deeper, and more inward actuality is neither mind nor body but rather a yet higher-dimensional actuality, which is their common ground and which is of a nature beyond both . . . so we do not say that mind and body causally affect each other, but rather that the movements of both are the outcome of related projections of a common higher-dimensional ground.[26]

A prior higher dimensional ground unites the substances sundered by substance dualism.[27] But this is a dynamic wholeness, not a static one. Bohm's term is *holomovement*.

> I propose a view that I have called *unbroken wholeness.* Relativity and quantum physics agree in suggesting unbroken wholeness, although they disagree on everything else. That is, relativity requires strict continuity, strict determinism, and strict locality, while quantum mechanics requires just the opposite—discontinuity, indeterminism, and nonlocality. . . . They both agree, however, on the unbroken wholeness of the universe.[28]

In a manner not unlike Whitehead, Bohm would allow us to attend to our primary experience with the sky and with outer space, an experience with the spiritual sense of awe at the beyond already built in at the concrete level. Once this level of experience is acknowledged, then the astrobiologist is free to measure and mathematize at the level of abstraction.

Might one think of Bohm's holomovement as divine? It looks like something divine. Could it be the divine component to pantheistic metaphysics?

Bohm himself describes the implicate order as *holy*, but not *sacred.* The idea of the sacred makes Bohm nervous because it reminds him of organized religion. The idea of the holy, however, connotes the whole of reality. So, that keeps Bohm in his comfort zone. "All we can say," comments Renée Weber, "is that this view is consistent with the notion that there's a truth, an actuality, a being beyond what can be grasped in thought, and that is intelligence, the sacred, the holy."[29]

Jesuit David Toolan perceives divinity within the holomovement. "A post-Einsteinian universe is unimaginably vast and ancient, is blessed with steadfast stability; still more remarkably it is also graced with process, self-organization, interconnection, communication, fluctuation, and openness," Toolan writes. "This is a universe whose fullness, diversity, promise, and risk simply dazzle. Given all that, it has to make a difference to our conception of God, our prayer life, our work and action."[30]

What Whitehead and Bohm offer the astrotheologian is a more comprehensive comprehensiveness that includes objective astrobiological data along with subjective spiritual sensibilities. Whether the systematic theologian adopts either Whiteheadian holism or Bohmian holism, the holistic principle has been articulated and made available to the theologian.

ASTROBIOLOGY, ASTROCULTURE, AND ASTROTHEOLOGY

"Hello, Universe!" Douglas Vakoch along with METI International (Messaging Extraterrestrial Intelligence International) foster "multidisciplinary research on the design and transmission of interstellar messages, building a global community of scholars from the natural sciences, social sciences, humanities, and arts."[31] Outer space has become local space, culturally speaking.

Numerous scholarly projects are emerging to study the societal impact of the space sciences with special attention to the likelihood of contact with extraterrestrial life. These projects are not merely descriptive. Fears and hopes bleed into the studies. The chief hope is that astrobiology could unite a divided earth into a single planetary society. Here is Ian Crawford.

> To my mind, the principal societal benefits arising from the study of astrobiology, and from its popularization to a wider public, are a consequence of the cosmic and evolutionary perspectives on human affairs that it naturally engenders . . . At a time when the Earth is faced with global challenges that can only be met by increased international cooperation (and arguably by developing institutions of global governance); yet tribal nationalistic and religious ideologies are acting to fragment humanity, the promulgation of a unifying cosmic perspective on human affairs is potentially of enormous importance.[32]

For Crawford, the stars could promulgate "a unifying cosmic perspective on human affairs."

Former holder of the Baruch Chair in Astrobiology at the US Library of Congress, Steven Dick, forecasts and invites an astroculture.

> *Astroculture* is a relatively new umbrella concept used to describe the array of images, events, and media reactions that "ascribe meaning to outer space while stirring both the individual and the collective imagination." The concept may be expanded to argue that, while different perspectives on space may exist in different cultures, humanity as a whole is increasingly creating and immersed in an overarching astroculture that transcends national boundaries, a kind of global astroculture.[33]

Our evolving astroculture vibrates with a cosmic consciousness that both transcends and unites the human divisions we have come to take for granted. Might we contextualize the science of astrobiology within this wider astroculture?

SHOULD WE DE-TRANSCENDENTALIZE COSMIC CONSCIOUSNESS?

If so, then might we be misled by anti-astrotheologians such as Paul Davies and Jill Tarter who expunge the transcendental dimensions from the human experience with space?

Astrobiologist and astrophysicist Davies asserts that Earth's religions are fragile, breakable. "The existence of extra-terrestrial intelligences would have a profound impact on religion, shattering completely the traditional perspective of God's special relationship with man."[34] In his haste to discourage terrestrial myopia, Davies underestimates the power of the Beyond sensibility at work in terrestrial subjectivity.

Tarter more aggressively de-transcendentalizes religious sensibilities. "Detected, long-lived extraterrestrials either never had, or have outgrown, organized religion."[35] Smart people either outgrow religion or never indulge in it at all, she assumes. Tarter's own antipathy toward religion is projected on the more advanced and hence more intelligent space aliens she imagines. Does this count as a reliable scientific speculation? No.

To the contrary, based upon the history of human experience on Earth, one might better posit a similar sense of transcendence occupying the minds of our future extraterrestrial friends. It is far more likely that future extraterrestrial friends will similarly report awe in the face of the Beyond.

In contradistinction to Davies and Tarter, the transcendent impulse alive in human subjectivity has become excited at the prospect of meeting extraterrestrial neighbors. Vatican Observatory astronomer Guy Consolmagno, S.J., spouts cosmic optimism. "Christians believe that the God of all things, the God of the entire universe, is in love with us humans. And when you're in love, you show a kind of special interest. That doesn't mean God can't also be in love with other intelligent beings on other planets."[36]

TEILHARDIAN PROMISE AND CAUTION

Many decades ago, another Jesuit, Pierre Teilhard de Chardin, forecasted a grand convergence. "Religion and science are the two conjugated faces or phases of one and the same act of complete knowledge—the only one that can embrace the past and the future of evolution so as to contemplate, measure, and fulfill them."[37] What a grand promise!

Teilhard also alerted us to three speculative pitfalls: (a) we should not assume that our planet is the only inhabited one in the universe; (b) we ought not assume that our earth is the only world which has fallen into sin; and

(c) we should not assume that God's saving work in Jesus Christ would be unknown to residents living anywhere other than earth.

> The idea that only one planet might be populated in the entire Universe has become just as unthinkable for us (insofar as we think about it at all) as the idea of a humankind appearing on earth with no genetic connection to the other creatures of the earth.[38]

Outer space vibrates with transcendence. Might we enjoy new kinships with extraterrestrial neighbors in our future? Might God be speaking to us through the twinkles of the stars?

CONCLUSION

The feeble attempt to denude objective science from subjective clothing has failed to eliminate the spiritual power of outer space. Space beyond earth is inherently religious in its valence.

The science of astrobiology is almost religious because it plucks the strings of our spiritual sensibilities. More specifically, our built-in psychological and cultural sense of *the Beyond* begins to vibrate with thoughts of astronomical distances and the prospect of extraterrestrial neighbors.

Astrobiology, like all other sciences, gathers objective data. But the data of astrobiology are abstracted from a more concrete experience with outer space as a spiritual stimulus. Astrobiological data vibrate with spiritual resonance. With the human soul now excited, the astrotheologian should enter the discourse.

The astrotheologian learns from astrobiological data while trying to account for its inspiring implications. Earthlings are growing in cosmic consciousness, at least in the minimal sense of being aware of possible intelligent consciousness living on exoplanets in the Milky Way.[39]

When contact with an extraterrestrial civilization is eventually established, will the interchange of human and alien consciousness lead to a fusing of horizons? Can we rightfully expect an expansion if not a deepening of human understanding, knowledge, and awareness? Might we test giving voice to a more intense cosmic consciousness with the physics of David Bohm or the metaphysics of Whiteheadian process philosophy?

With such holism in mind, theologian Junghyung Kim gives us an assignment. "The theological task remains to construct a Christian theology embracing the genuinely comic horizon as contemporary science presents us."[40]

Like Mary after a visit from the angel, Gabriel, the astrotheologian should ponder all these things in her heart.

NOTES

1. Plato, *Republic,* VII: 529.
2. Gingerich, "Mankind's Place in the Universe," 29.
3. Eliade, *The Sacred and the Profane*, 119, Eliade's italics.
4. Tarter, "The Evolution of Life in the Universe: Are We Alone?," 20.
5. Mix, *Life in Space*, 4.
6. Griffin, *A Process Christology*, 65, Griffin's italics.
7. Astrobiology as a field finds it difficult to shed extrascientific meaning. "The methodologies of science may be reductionist, but the output, especially in the case of the current scientific thinking about our cosmic origins and the emergence, evolution, and diversity of life on Earth, is systemic, holistic, and relational in nature." Scalice, "Astrobiology-as-Origin-Story," 52.
8. Lupisella, "Is the Universe Enough?," 134.
9. Is value real? Yes, to be sure. "*Meaning and value are as much integral aspects of the world as they are of us,*" answers physicist David Bohm. Bohm, "Postmodern Science and a Postmodern World," 67, Bohm's italics. Yet, there is no room for value let alone God in the astrobiological framework. "Worthiness of worship is only to be assigned to that God whose reason for being is found in the creative supremacy of its own Value. Value thus also reasons the existence of the world, human beings, and human purpose as worshipful beings. The axiological foundations of God and the world constitute worship as 'ontological gratitude'—a gratefulness for existence—which manifests itself in a value-creative life." Davis, "God, Value and Ontological Gratitude," 33.
10. Cooley, "Astrobiology as Contemporary Theology," 3.
11. Fishbane, *Sacred Attunement*, 34.
12. Sonderegger, *Systematic Theology*, 89.
13. See Peters, "Astrotheology" and "Astrotheology: A Constructive Proposal."
14. Barbour, *Issues in Science and Religion*, 415. See Peters, "Theology of Nature," 651–52.
15. "If we hold that God is the Creator of *this* universe, we are claiming something very special indeed, much more special and precise than we dreamed of before. We humans are not just tied to one particularly providential star with a propitious set of planets. Rather, we are somehow tied in with the entire cosmic scope of nature, for all of nature is governed by these 'anthropic' constants." Russell, *Cosmology from Alpha to Omega*, 287–89.
16. Thomas Aquinas, *Summa Theologica*, I.Q1.7.
17. "As a Christian, I think of astrobiology as a way to better understand how God created the world." Mix, *Life in Space,* 6. According to theologian John Hart, contact with ETI would lead us human beings to realize that we "are not the single, extraordinary *image of God* in a divine creation now known to evolve, diversify, and complexify; not the only intelligent life that exists cosmically; and, perhaps, not even related evolutionarily to some or all other living beings. Other life might not only have evolved independently but had become more complex and intelligent than humans." Hart, "Cosmic Commons," 374.

18. Ford, *The Lure of God*, 54.

19. Long before the invention of the telescope, medieval theologians searched the stars and speculated about other worlds. See Dick, *Plurality of Worlds.*

20. Griffin, "Introduction: the Reenchantment of Science," 3.

21. Nagel, *Mind and Cosmos*, 8. Retrieval of the subjective or mental dimension of experience will be necessary if we are ever able to make sense of the UFO or UAP (Unidentified Aerial Phenomena) experience, argues Jensine Andresen. "The cartography of human knowledge reflected in the current map of academic disciplines is obsolete," she states. Andresen, "Mind of the Matter, Matter of the Mind," 317. Andresen asks scientific researchers of UAP to include the phenomena on a spectrum that includes craft-like objects on one end plus the impact of ETI presence on human consciousness on the other end. Like Carl Peterson mentioned below, she believes the integrative physics of David Bohm will permit scientific methodology to get beyond the Cartesian subject-object split. Ibid., 295–300.

22. What is denied yet experienced is the mental pole present in experience. "Every actual entity has both a mental and a physical pole," claims Marjorie Hewett Suchocki, *God, Christ, Church*, 226. As a science, astrobiology extracts and eliminates the mental pole in the name of objectivity. This is as misleading as it is unnecessary according to the process philosopher or theologian.

23. Whitehead, *Process and Reality*, 3.

24. Daly and Cobb, Jr., *For the Common Good*, 387.

25. Whitehead, *Process and Reality*., 7–8.

26. Bohm, *Wholeness and the Implicate Order*, 85. Physicist Carl Peterson, "an undeniable Bohmian," follows up. "The holomovement gives viability to and carries the implicate order, since all entities found in it come from the undivided totality of its movement." Peterson, "Relativity and Quantum Theory," 279.

27. Whiteheadian metaphysics overcomes substance dualism right along with subject-object dualism. "God's creative activity is then at work, so to speak, from inside the organism. . . . Hence, there is no special divine intervention in the cosmic process so as to create the human soul as a strictly immaterial reality. Metaphysical dualism is thereby avoided, and emergent monism instead affirmed." Bracken, "Emergent Monism and the Classic Doctrine of the Soul," 201.

28. Bohm, "Postmodern Science and a Postmodern World," 65.

29. Renée Weber, "The Enfolding-Unfolding Universe," 70. Catherine Keller foresees a coming together of science and theology due not to what is known in common but rather in what is unknown. "The physics of quantum nonseparability will not produce empirical proof of any kind of God . . . No. It may, however, be offering material evidence of a universe apophatically entangled as to escape the rival classicisms that pit science and theology against each other." Keller, *Cloud of the Impossible*, 132.

30. Toolan, S.J., "Praying in a Post-Einsteinian Universe."

31. METI International, http://meti.org/missiona (accessed 1/28/2022).

32. Crawford, "Widening Perspectives," 58.

33. Dick, "Humanistic Implications of Discovering Life Beyond Earth," 752.

34. Davies, *God and the New Physics*, 71. Davies believes Christianity among the religions will confront the most severe crisis because the doctrine of the incarnation

would lead to a planet-hopping Christ. This would be an absurd idea, thinks Davies. Not so for theologian Ilia Delio. "Incarnation on an extra-terrestrial level could conceivably take place, as long as there is some type of intelligence within the extra-terrestrial species to grasp the Word of God through knowledge of the divine embodied Word. . . . many incarnations but one Christ." Ilia Delio, *Christ in Evolution*, 169.

35. Tarter, "SETI and the Religions of the Universe," 146.

36. Guy Consolagno, S.J., and Paul Mueller, S.J., *Would You Baptize an Extraterrestrial?,* 274.

37. Teilhard de Chardin, *The Phenomenon of Man*, 284–85.

38. Pierre Teilhard de Chardin, Pierre, "Une suite au problème des origines humaines," 277.

39. Jensine Andresen believes we all share consciousness with ETI, with one another, and with the whole of reality. What she has learned is that ETI are worried that terrestrial society could destroy itself through a nuclear conflagration. Further, ETI want to prevent exporting any of Earth's weapons arsenal to space. Andresen, "Mind of the Matter, Matter of the Mind."

40. Kim, "The ETI Hypothesis and the Scandal of Particularity," 341.

BIBLIOGRAPHY

Andresen, J. "Mind of the Matter, Matter of the Mind." In *Extraterrestrial Intelligence: Academic and Societal Implications*, edited by J. Andresen and O.A.C. Torres, 281–330. Cambridge: Cambridge Scholars Publishing, 2022.

Aquinas, T. *The Summa Theologica of St. Thomas Aquinas (Five Volumes)*. Notre Dame: Christian Classics, 1981.

Barbour, I. G., *Issues in Science and Religion*. New York: Prentice Hall and Harper, 1966.

Bohm, D. "Postmodern Science and a Postmodern World." In *The Reenchantment of Science*, edited by D. R. Griffin, 57–68. Albany: SUNY Press, 1988.

———. *Wholeness and the Implicate Order*. London: Routledge & Keegan Paul, 1980.

Bracken, J. "Emergent Monism and the Classic Doctrine of the Soul." In *Interactive World, Interactive God: The Basic Reality of Creative Interaction*, edited by C. Rausch Albright, J.R. Albright, and M. Turk, 191–206. Eugene: Cascade Books, 2017.

Consolagno, G., and P. Mueller. *Would You Baptize an Extraterrestrial?* New York: Image, 2014.

Cooley, D. S. "Astrobiology as Contemporary Theology." In *Theology and Science: Discussions about Faith and Facts*, edited by J. Seckbach and R. Gordon, 1–24. Singapore: World Scientific Publishing, 2019.

Crawford, I. "Widening Perspectives: The Intellectual and Social Benefits of Astrobiology, Big History, and the Exploration of Space." *Journal of Big History* 3, no. 3 (2019): 205–24. https://doi.org/10.22339/jbh.v3i3.3312.

Daly, H. E., and J. B. Cobb Jr. *For the Common Good: Redirecting the Economy toward Community, the Environment, and a Sustainable Future*. Boston: Beacon, 1989.

Davies, P. *God and the New Physics*. New York: Simon & Schuster, 1983.

Davis, A. M. "God, Value and Ontological Gratitude: On the Axiological Foundations of Worship." *Toronto Journal of Theology* 361, no. 1 (2020): 33–43. https://doi.org/10.3138/tjt-2020-0053.

Delio, I. *Christ in Evolution*. Maryknoll, NY: Orbis Books, 2012.

Dick, S. J. "Humanistic Implications of Discovering Life Beyond Earth." In *Handbook of* Astrobiology, edited by V. M. Kolb, 741–56. London: Taylor & Francis, 2019.

———. *Plurality of Worlds: The Extraterrestrial Life Debate from Democritus to Kant*. Cambridge: Cambridge University Press, 1982.

Eliade, M. *The Sacred and the Profane*. Translated by Willard R. Trask. New York: Harcourt, Brace and World, 1957.

Fishbane, M. *Sacred Attunement: A Jewish Theology*. Chicago: University of Chicago Press, 2008.

Ford, L. S. *The Lure of God: A Biblical Background for Process Theism*. Philadelphia: Fortress Press, 1978.

Griffin, D. R. "Introduction: the Reenchantment of Science." In *The Reenchantment of Science*, edited by D. R. Griffin, 1–46. Albany: SUNY Press, 1988.

———. *A Process Christology*. Louisville: Westminster John Knox, 1973.

Hart, J. "Cosmic Commons: Contact and Community." *Theology and Science* 8, no. 4 (2010): 371–92. https://doi.org/10.1080/14746700.2010.517637.

Haught, J. *Deeper than Darwin*. Boulder: Westview, 2003.

Russell, R. J. *Cosmology from Alpha to Omega: The Creative Mutual Interaction of Theology and Science*. Minneapolis: Fortress Press, 2008.

Keller, C. *Cloud of the Impossible: Negative Theology and Planetary Entanglement*. New York: Columbia University Press, 2015.

Kim, J. "The ETI Hypothesis and the Scandal of Particularity." *Theology and Science* 19, no. 4 (2021): 332–42. https://doi.org/10.1080/14746700.2021.1982247.

Lupisella, M. "Is the Universe Enough? Can It Suffice as a Basis for Worldviews?" *Journal of Big History* 3, no. 3 (2019) 123–40. https://doi.org/10.22339/jbh.v3i3.3370.

METI International, http://meti.org/mission. Accessed January 28, 2022.

Mix, L. J. *Life in Space: Astrobiology for Everyone*. Cambridge: Harvard University Press, 2009.

Nagel, T. *Mind and Cosmos: Why the Materialist Neo-Darwinian Conception of Nature Is Almost Certainly False*. Oxford: Oxford University Press, 2012.

Gingerich, O. "Mankind's Place in the Universe." *Nature* 457, no. 7225 (2009): 28–29. https://doi.org/10.1038/457028a.

Peters, T. "Astrotheology." In *The Routledge Companion to Modern Christian Thought*, edited by Chad Meister and James Beilby, 838–53. London: Routledge, 2013.

———. "Astrotheology: A Constructive Proposal." *Zygon* 49, no. 2 (2014): 443–57. https://doi.org/10.1111/zygo.12094.

———. "Theology of Nature." In *Religion Past and Present: Encyclopedia of Theology and Religion*, edited by H. D. Betz et al., 651–52. Leiden: Brill, 2012.

Peterson, C. "Relativity and Quantum Theory: The Manifestation of Unidentified Aerial Phenomena (UAP) and a New Order for Physics." In *Extraterrestrial Intelligence: Academic and Societal Implications*, edited by J. Andresen and O. A. C. Torres, 263–80. Cambridge: Cambridge Scholars Publishing, 2022.

Plato, *The Republic*. Translated by D. Lee. New York: Penguin Classics, 2007.

Scalice. "Astrobiology-as-Origin-Story: Education and Inspiration Across Cultures." In *Handbook of Astrobiology*, edited by V. M. Kolb, 49–53. London: Taylor & Francis, 2019.

Sonderegger, K. *Systematic Theology: Volume 2: The Doctrine of the Holy Trinity: Processions and Persons.* Minneapolis: Fortress Press, 2020.

Suchocki, M. H. *God, Christ, Church: A Practical Guide to Process Theology*. New York: Herder, 1982.

Tarter, J. "SETI and the Religions of the Universe." In *Many Worlds: The New Universe, Extraterrestrial Life and the Theological Implications*, edited by S. J. Dick, 143–49. Philadelphia: Templeton Foundation Press, 2000.

———. "The Evolution of Life in the Universe: Are We Alone?" *Proceedings of the International Astronomical Union* 2, no. 14 (2006): 13–29. https://doi.org/10.1017/S1743921307009829.

Teilhard de Chardin, P. "Une suite au problème des origines humaines. La multiplicité des mondes habités (5 June 1953)." *Œuvres.* Paris: Seuil, 1969.

———. *The Phenomenon of Man*, Intro. by J. Huxley. New York: Harper, 1959.

Toolan, D. S. "Praying in a Post-Einsteinian Universe." *Cross Currents* 46, no. 4 (1996–1997): 347–440.

Weber, R. "The Enfolding-Unfolding Universe: A Conversation with David Bohm." In *The Holographic Paradigm and Other Paradoxes*, edited by Ken Wilber, 44–104. Boulder: Shambhala, 1982.

Whitehead, A. N. *Process and Reality: An Essay in Cosmology.* Corrected Edition, edited by D. R. Griffin and D. W. Sherburne. New York: Free Press, 1978.

Chapter 8

Astrobiology, Cosmotheology, and the Biological Universe

Implications for Religion and Theology

Steven J. Dick

REVELATION: ASTROBIOLOGY AND THE NEW UNIVERSE

Recent discoveries in astronomy and astrobiology have transformed our worldview over the last century and especially over the last few decades. Little more than a century ago it was still possible to think the universe was only 3,600 light-years in diameter, and that our solar system was near the center.[1] With the work of astronomers Edwin Hubble, Harlow Shapley, and others, we have known for a century that the Earth, the solar system, and the galaxy are not the physical center of the universe in any way.[2] The Hubble Space Telescope has now revealed how correct Hubble the astronomer was. Beyond his wildest expectations the Hubble Deep Fields have shown us that billions upon billions of galaxies exist in a universe billions of light-years in extent. Observing long enough in any direction with a telescope like Hubble, integrating the photons emanating from billions of light-years away, reveals a host of galaxies of all sizes, shapes, colors, and ages (figure 8.1).

For decades after the discoveries of Shapley and Hubble, many clung to the notion that we might still be *biologically* central, the result of a rare event, or perhaps even the only life in the universe.[3] It has always been possible, of course, that these galaxies are devoid of planets and life. But thanks to the revelations of astrobiology we now know that planets are everywhere, a normal by-product of stellar evolution. Since 1995, more than 5,000 planets have been discovered beyond our solar system, and planetary systems science

Figure 8.1. A new worldview: The Hubble Deep Field, covering a speck of sky only 1/30 the diameter of the full Moon, reveals a universe full of galaxies. But is the universe enough for humans who seek meaning and purpose? (Courtesy of Robert Williams and the Hubble Deep Field Team [STScI] and NASA Image STScI-PRC96-01a)

has become an exciting discipline.[4] And that is the tip of the iceberg; planets are now believed to exist around virtually all stars. Not all of these planets are Earth-like, but enough fall into that category that if we scale up those numbers, there will be millions of Earth-like planets. To be sure, not all of these will have life, and not all will have intelligent life.

But we have seen enough over the last few decades to detect a trend. After the discoveries of exoplanets some argued that yes, there are thousands of exoplanets beyond our solar system but they are mostly gas giants. Then yes, there are some rocky planets, but they are not in a habitable zone and not Earth-like. Then yes, some are in a habitable zone and Earth-like, but none are known to have life. The latter statement remains correct, we do indeed not know if there is life beyond Earth. But the undeniable trend so far is that the Principle of Mediocrity, which states that Earth is nothing special in the universe, is increasingly validated by the findings of science. There are counterarguments to this "Copernican Principle,"[5] but I fall in the camp that fully expects life is common in the universe, that we live in what I have called a "biological universe" full of life.[6] There is probably more microbial than intelligent life, but scaled throughout the Galaxy and the universe, even if "rare" in some sense, intelligence could exist in great numbers. The James Webb Space Telescope's search for biosignatures, the expanding search for extraterrestrial intelligence (SETI) with new techniques, and the rapidly advancing science of astrobiology,[7] including robust international astrobiology programs with detailed strategic plans, will determine whether or not this conjecture is true.[8]

In my view these findings constitute a new worldview, some would say a "revelation" to use a religious term, that our cultures, religions, theologies, and philosophies must assimilate and act on, perhaps even changing their basic dogmas. We need to be open to the possibilities dictated by new knowledge, keeping in mind that even if intelligence is only rare, this would still change our worldview. At the same time we should remain skeptical of all premature claims. I do not see compelling evidence of alien intelligence in the recent spate of UFO reports, some from reliable sources and now taken seriously by the US government.[9] Changing the terminology to UAPs, or Unidentified Aerial Phenomena, does not advance the argument. In my view this phenomenon needs to be studied, and may well prove to be something of interest, but the evidence for the extraterrestrial hypothesis is not compelling. Carl Sagan's dictum still holds: "extraordinary claims require extraordinary evidence," though it should be used with caution.[10] Similarly, interpretations of the interstellar interloper asteroid Oumuamua as an alien spacecraft[11] remain unconvincing to most scientists. But the admittedly still circumstantial evidence for life beyond Earth is increasingly compelling.

The disconnect between theologies and our current worldview has been widely recognized even by many theologians and religious writers. In her recent book *The Unbearable Wholeness of Being* the Franciscan sister Ilia Delio notes that "Core Christian doctrine was formed against the backdrop of the concentric, static, and hierarchical Ptolemaic universe. To this day all our core doctrines are fitted to this universe even though, as early as the sixteenth century, modern science began to describe a new universe that departed from the static, Ptolemaic one. The inability to engage a new religious story, however, has confined us to the old one. The impasse we find ourselves in is centuries old, and until we can purify our memory . . . we are heading down a dangerous path."[12] These sentiments have been echoed many times in the past by forward-looking theologians and others, as we shall see below. But what is the "new religious story"? And how far should we go in "purifying" our memory? That is the controversial question.

INTERNALIZING THE COSMIC PERSPECTIVE: COSMIC EVOLUTION, BIG HISTORY, WORLDVIEW

Before delving into the issues of theological and societal impact, let me say more about the cosmic perspective that I believe we must adopt. That perspective has ancient roots, even in the Ptolemaic system. To take one early example the Roman statesman and Christian philosopher Anicius Manlius Severinus Boethius (c. 480–524) expressed it in his book *The Consolation of Philosophy*, written as he sat in prison in Italy awaiting execution in the year 523, almost exactly 1,500 years ago.[13] In this work, Boethius contemplated how evil could exist in a world governed by God, highlighted the transitory nature of money and power, argued for the superiority of a life of the mind, and defended himself from what he claimed were false accusations, even while still hinting at his own political ambitions. In these circumstances, which tend to focus the mind, he wrote:

> As you have heard from the demonstrations of the astronomers, in comparison to the vastness of the heavens it is agreed that the whole extent of the earth has the value of a mere point. . . . Further, as you have learned from the authority of Ptolemy, only about one-fourth of this so miniscule spot in the universe is the portion inhabited by animate creatures known to us.[14]

Moreover, Boethius held there was a moral justification for studying the heavens, since it

> Would mold both my actions and the principle of life in general in accordance with the model of heavenly order.[15]

The Copernican heliocentric theory, which made the Earth a planet and the planets potential Earths, has gradually made this cosmic perspective more urgent and it has now become ever more prevalent both in popular culture and among scholars.[16] To some extent astronomers of necessity have internalized it as part of their working lives. Already in the 1950s and 1960s Shapley not only held, but evangelized, such a perspective.[17] It made new strides in the 1970s with Carl Sagan's book *The Dragons of Eden* (1977) and his *Cosmos* TV series (1980), viewed by billions, where the "cosmic calendar" compressed into a single year begins with the Big Bang on January 1 and primitive humans appear during the last hours of December 31. And it is adopted today in the form of cosmic evolution, big history, and research on worldviews.[18]

Cosmic Evolution

The idea of cosmic evolution has a deep background,[19] and in modern times has been championed by SETI and astrobiology practitioners, adopted as a research program and guiding principle by NASA and other space agencies, and become grounded in sophisticated science.[20] Cosmic evolution is the 13.8- billion-year Master Narrative of the Universe from the Big Bang to the present (figure 8.2). It is the story of the formation, evolution, and demise of galaxies, stars, planets, and a myriad of some eighty-two distinct classes of astronomical objects (Dick, 2019).[21] Its outcome may result in a physical universe composed of planets, stars, and galaxies in which life is a fluke; a biological universe full of carbon-based life or something more exotic; or a postbiological universe in which cultural evolution has resulted in a universe full of artificial intelligence (Dick, 2003).[22] These outcomes determine the long-term destiny of humanity because the scope of astrobiology applies not

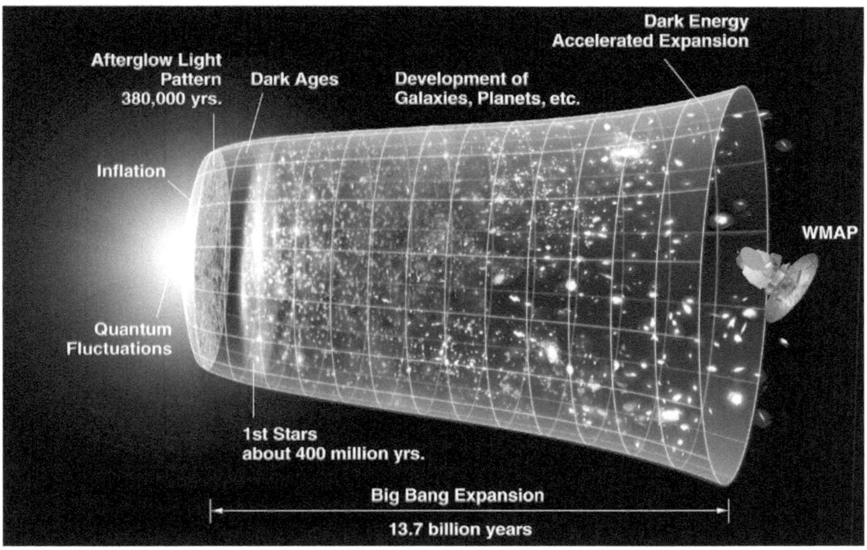

Figure 8.2. The master narrative of the universe: 13.8 billion years of cosmic evolution, as depicted by the Wilkinson microwave anisotropy probe (WMAP) program, which narrowed the age of the universe to within 100 million years. The current model has the universe beginning with the Big Bang, stars forming within the first few hundred million years, followed by the development of galaxies, planets, and life. (NASA/WMAP Science team)

only to the past and present, but also to the future. In short, the destiny of humanity falls within the purview of cosmology, astronomy, and astrobiology.

Big History

Cosmic evolution is not just some abstract astronomical theory; it involves humanity to our very core and should affect our thinking in a variety of areas. One attempt to incorporate this grand evolutionary epic into human history is what is now known as Big History.[23] And as the author of a recent book suggests, the emerging and related fields of Big History, astrobiology, and cosmic evolution may hold the key to new unifying frameworks for knowledge.[24] Big History unites the sciences and humanities and places humans in the context of 13.8 billion years of cosmic evolution. The online Big History Project, funded by none other than Microsoft founder Bill Gates, provides detailed curricula that bring home to students of all ages our deep history in cosmic evolution.[25]

Worldviews

In a broader sense astrobiology, Big History, and the cosmic perspective give rise to worldviews. Although worldviews for individuals come in forms large and small, if one considers a concentric shell of worldview hierarchies (figure 8.3), then philosophical, religious, cultural, and political worldviews would all be present. But the most encompassing worldview surrounding them all are cosmological. Whether we realize it or not, these cosmological worldviews have historically played a major role in our everyday thinking. Dante's *Divine Comedy*, with its moral lessons played out in a cosmological worldview of nested spheres with the Earth at the center and the sphere of fixed stars at the periphery, provided the backdrop for medieval life. Within this ordered framework, both a mental and moral model as well as a physical system, all human actions took place. "In every age the human mind is deeply influenced by the accepted model of the universe," C. S. Lewis reminded us in his book on medieval and Renaissance literature, *The Discarded Image*.[26] The model, he insisted, "is also influenced by the prevailing temper of mind."

As historian of science Nasser Zakariya has demonstrated in detail in his book *A Final Story: Science, Myth, and Beginnings*, the same is true in the modern world, where the sweeping scientific epic of cosmic evolution embraces the cosmological, the biological and the cultural, and provides the background for modern life—including the biological universe.[27]

How we gain, lose, and change worldviews has been the subject of considerable research, even in a cosmic context, most recently in Clement Vidal's provocative volume *The Beginning and the End: The Meaning of Life in a*

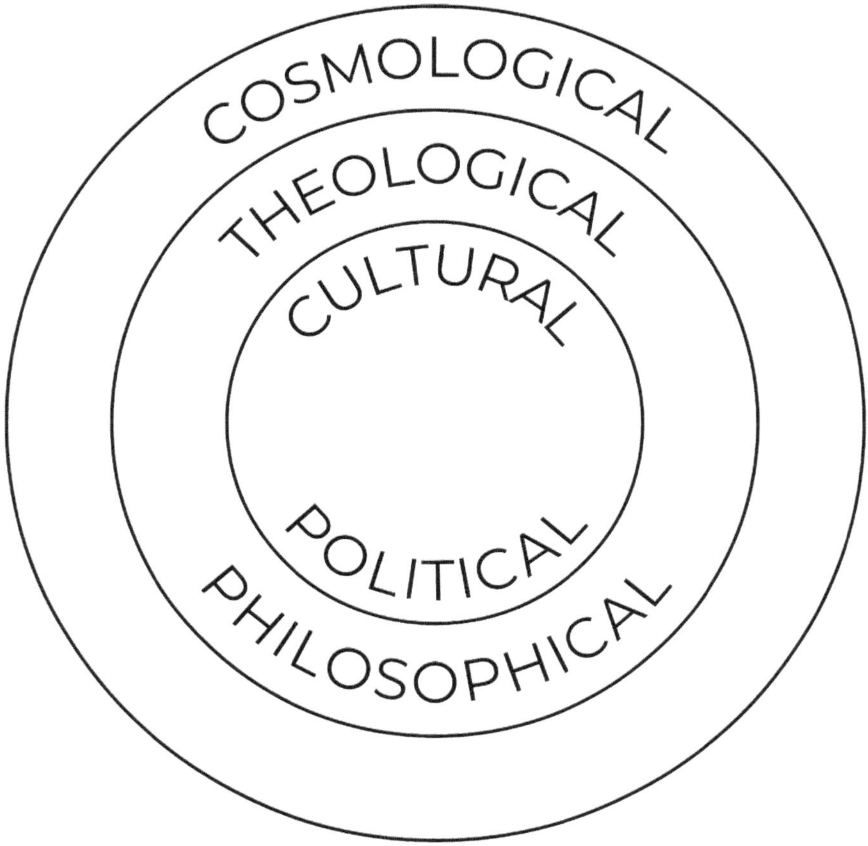

Figure 8.3. Concentric layers of cosmological, theological/philosophical, and cultural worldviews. Cosmological worldviews encompass all others. (Courtesy of Jared Morningstar)

Cosmological Perspective.[28] Vidal argues that the search for meaning in life begins with a comprehensive and coherent worldview, a more robust version of the sometimes inconsistent worldviews we all hold whether we know it or not. The construction of worldviews and their influence on our thinking are deep philosophical problems. Vidal not only examines religious, scientific, and philosophical worldviews, he also proposes criteria for evaluating which are best. This original and enlightening exercise results for him in a worldview that is fundamentally cosmological, though with philosophical, ethical, and religious implications. He leaves no doubt that the epic of cosmic evolution, complete with life, is increasingly becoming our new worldview.

Thus the new cosmological worldview increasingly immerses us in what historian Alexander Geppert has called "astroculture," an idea bolstered by a new field of historical inquiry that analyzes how humans have come to terms with the universe of which they are a part.[29] As part of this astroculture, both science fiction and the UFO debate are ways in which popular culture works out the implications of this new worldview.[30] It is all part of the human search for meaning, where the Earth is seen in the context of the universe in which it is located. We now know this is true not only in the Western world; historians have recently shown that science fiction, UFOs, and extraterrestrials are also an integral part of Islamic history, although perhaps not with such deep historical roots.[31]

It is important to realize that worldviews, whether cosmological, religious, or political, are not inert or academic by any means. Rather, they are a complex of ideas with an attitude. Indeed, one could argue that worldviews drive history, particularly when it comes to religion. There was a religious worldview behind the Crusades, just as there are religious worldviews behind much of the strife and beauty in the world today. One and the same worldview can be dangerous and wonderful, generating strife on the one hand and compassion and love on the other. The same holds true for political worldviews. We need to keep this in mind as we incorporate the new cosmological worldview into our religions and generate new ways of finding meaning and value in our lives.

RECONSIDERATIONS RESERVED AND RADICAL: THE IMPACT OF THE NEW UNIVERSE ON RELIGION AND THEOLOGY

How, then, does our new cosmic worldview impact theology and religion? There has been a long debate on this question, whose origins stretch back to the "plurality of worlds" tradition. In particular, medieval commentaries on Aristotle's *De Caelo* (*On the Heavens*) already took up the task of reconciling religion with the idea of other worlds, an endeavor that extended sporadically into the Renaissance and beyond. That history, once ignored and even considered taboo and without intellectual merit, has now been well written over the last four decades.[32] This is not the place to repeat that history, which includes the themes of God's omnipotence and beneficence in creating other worlds, the problem of inhabited worlds for doctrines such as Incarnation and Redemption, natural theology, and a bevy of other theological issues still relevant today. But this history should not be forgotten as we focus here on how this new worldview impacts theology and religion today.

Astrotheology

The sporadic ruminations of the past have now grown into a nascent field of astrotheology.[33] As early as 1995, the theologian Ted Peters took up the challenge of what he then called "exo-theology" (echoing exobiology), arguing that the discovery of extraterrestrial intelligence would present no significant challenge to theology.[34] That is seriously questioned by some theologians, including former Jesuit Vatican Observatory director George Coyne, as he made clear already in a Templeton Foundation meeting two decades ago. In particular, Coyne pointed to the Christian doctrines of Redemption and Incarnation, concluding that with the discovery of extraterrestrial intelligence, "theologians must accept a serious responsibility to rethink some fundamental realities within the context of religious belief."[35] Among those realities are the nature of a human being, and whether Jesus Christ could exist on more than one planet at one time. These particular doctrines have since been examined in more detail in relation to astrobiology, most recently by Peters and Russell.[36]

The implications of the new worldview have reached into the domain of theology in many forms. The American Catholic theologian Thomas F. O'Meara fills his book *Vast Universe: Extraterrestrials and Christian Revelation* with topics such as "salvation histories amid Galactic evolution."[37] The British biochemist and Anglican priest Sir Arthur Peacocke has called cosmic evolution "Genesis for the third millennium," arguing that "any theology—any attempt to relate God to all-that-is—will be moribund and doomed if it does not incorporate this perspective into its very bloodstream."[38] Michael Dowd and Connie Barlow, who consider themselves, "evangelists of cosmic evolution," have proposed an "evolutionary Christianity"—very different from Julian Huxley's evolutionary humanism, but both featuring evolution as a central concept. Evolutionary Christianity embraces cosmic evolution, variously termed "the Great Story" and the "epic of evolution," much more than did Huxley's original evolutionary humanism, undoubtedly because cosmic evolution has been so much more developed over the last fifty years, complete with evocative images from the Hubble Space Telescope (Barlow 1995; Dowd 2008).[39] Most recently, theology professor Joel Parkyn has authored a substantial volume on *Exotheology*, where he attempts to reconcile extraterrestrial intelligence with Christianity.[40]

Even religions without such doctrines as Redemption and Incarnation will be challenged. Speaking of extraterrestrial intelligence, Rabbi Norman Lamm, for example, noted that "this challenge must be met forthrightly and honestly," and called those who shrink from pursuing it "parochial and provincial." Citing astronomers who emphasize our peripheral place in the new universe, Rabbi Lamm noted that "Never before have so many been so enthusiastic about being so trivial." Cautioning that extraterrestrial life is far from proven, Lamm explored "a Jewish exotheology" and concluded that "A God who can exercise providence over one billion earthmen can do so for the billion times that number of creatures throughout the universe."[41]

While theologians are limited in their ability to answer such questions definitively, history suggests that were a discovery of extraterrestrial intelligence actually made, a way would be found for traditional theological doctrines to absorb the resulting new worldview, though perhaps not easily. The alternative would be extinction, and most religions have shown their ability to adapt to scientific discovery, if very slowly at times. It seems likely that the adjustment will be most wrenching for those monotheistic religions that see humans in the image of God (Judaism, Christianity, and Islam), a one-to-one relationship with a single Godhead (figure 8.4). It may be less wrenching for Eastern religions that teach salvation through individual enlightenment (Buddhism and Hinduism) rather than through a Savior, or that are this-worldly (Confucianism) rather than otherworldly.[42] The process of integration could in fact be an enriching and reinvigorating experience for theology. The question again is how far to go in our reconsiderations. Should we alter current theologies or start from basic principles of the relation of humans to the universe? What follows is a contemplation of the latter, what some might see as the radical rather than the reserved pathway of reconsideration.

Figure 8.4. ***The Creation of Adam,*** **a fresco painted by Michelangelo for the Sistine Chapel ceiling around 1510. It depicts a one-to-one relationship between God and humans, but astrotheology raises the fundamental question of how unique this relationship really is. This painting was conceived within the Christian tradition; other religious traditions will be affected differently. (Wikimedia Commons)**

Cosmotheology

My personal take on astrotheology is what I call cosmotheology.[43] Cosmotheology is simply a theology that takes into account what we know about the universe based on science. It is therefore a naturalistic cosmotheology, but not coextensive with scientism because it does not imply that science is the only way to understand the world, especially when it comes to the question of values. Over the last quarter century, I have enunciated five principles of cosmotheology, and added a sixth on human destiny.[44] These principles are as follows:

1. Cosmotheology must embrace the fact that humanity is in no way physically central in the universe. This has been proven beyond the shadow of a doubt in a continuous series of "de-centerings," beginning with the Copernican removal of the Earth from the center of the solar system in the sixteenth century, followed by Harlow Shapley's proof around 1920 that the solar system is on the periphery of the Milky Way Galaxy, and Edwin Hubble's proof in the late 1920s that our galaxy is only one of many in an almost infinite space. In other words, a complete rejection of the cosmology of A. R. Wallace a century ago.[45]
2. Cosmotheology must take into account the probability that humanity is not central biologically, mentally, or morally in the universe. The word "probability" is crucial here, since we have not yet found a single instance of life beyond Earth, much less intelligence. That is what the science of astrobiology is exploring.
3. Cosmotheology must take into account the probability that humanity is near the bottom in the great chain of beings in the universe. This follows from 13.8 billion years of cosmic evolution, and the fact that planets and life could have formed billions of years before our Earth originated 4.5 billion years ago. The extreme youth of our species, which has only in the last few thousand years emerged to the point that it can contemplate its place in the universe, is surely a sobering fact when placed in the context of cosmic evolution.[46]
4. Cosmotheology must be open to radically new and non-supernatural conceptions of God, not the God of the ancient Near East, nor the God of the human imagination, but a God grounded in cosmic evolution. It is entirely possible that beings have evolved in the natural course of the universe with many of the traits we attribute to God, including omnipotence, omniscience, and so on. But supernaturalism plays no role in cosmotheology, and the conventional God is therefore denied.
5. Cosmotheology must have a moral dimension, extended to embrace all species in the universe—a reverence and respect for life in any form.

This principle—a challenge even on Earth—gives rise not only to astrotheology but also to the related field of astroethics, with its various theories of moral status.[47]

6. Cosmotheology must embrace the idea that human destiny should be linked to natural cosmic events, not to the divine.

Let us elaborate on some of the assumptions and implications of these six principles.

Supernaturalism.

By now it should be clear that cosmotheology fits comfortably within the tradition known as religious naturalism. In the words of its premier historian, Jeremy Stone, religious naturalism "affirms a set of beliefs and attitudes that there are religious aspects of this world which can be appreciated within a naturalistic framework. There are some events or processes in our experience that elicit responses that can appropriately be called religious. These experiences and responses are similar enough to those nurtured by the paradigm cases of religion that they may be called religious without stretching the term beyond recognition."[48] In short, religious naturalism denies that an ontologically distinct and superior realm including God, the soul, and heaven is required to give meaning to the world. So does cosmotheology.

Like religious naturalism, cosmotheology denies supernaturalism. As the God of the ancient Near East stemmed from ideas of supernaturalism, our concept of a modern God, if needed at all, should stem from modern ideas divorced from supernaturalism. The billions of people attached to current theologies may consider this no theology at all, for a transcendent God above and beyond nature is the very definition of their theology. The supernatural God "meme," which we should remember is a historical idea the same as any other,[49] has been very efficient in spreading over the last few thousand years, picking up new memes such as those accepted by Christianity and other religions, including the divinity of Jesus, the Incarnation and Redemption, Trinity, Satan, salvation, and heaven and hell—all concepts with well-known histories and very human motivations.[50]

The denial of supernaturalism that is the bedrock of cosmotheology has many implications. I have described the effect of that denial before as similar to Einstein's fundamental assumption that the speed of light is constant, with all kinds of nonintuitive, disturbing, and unpleasant implications that nevertheless science has had to accept.[51] One of those implications of the denial of supernaturalism is that the Bible (figure 8.5) is not divinely inspired revelation, but simply a fascinating early example of ancient literature, embodying the good, the bad, and the ugly of ancient Near Eastern civilizations as

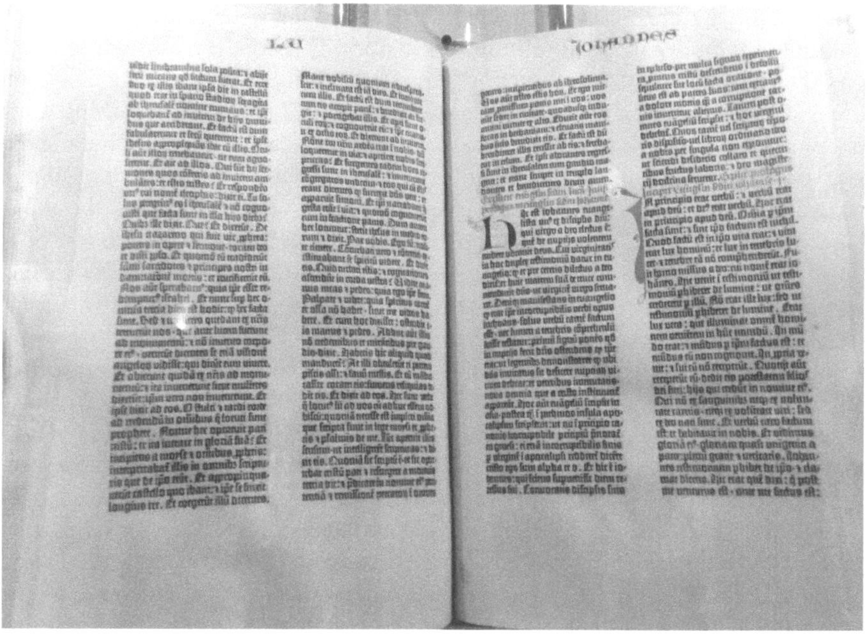

Figure 8.5. The Gutenberg Bible, the first great book in the Western world printed from movable metal type, was completed in Mainz, Germany in 1455. Cosmotheology denies supernaturalism and therefore holds that the Bible, while landmark literature, is not divinely inspired. (Library of Congress, photo by Steven J. Dick)

codified by a variety of writers millennia ago, sifted, parsed, and passed down ever since.[52] Its construction is a fascinating history that has been heroically uncovered, in part, by modern Biblical scholarship. The so-called "Documentary Hypothesis" argues that the first five books of the Bible (the Torah or Pentateuch) had four authors or groups of authors: the so-called J and E versions as early as the eighth or ninth century BC, the "P" (Priestly) authors perhaps writing as early as the reign of King Hezekiah (c. 715–687 BC) or as late as the sixth century, and the authors of Deuteronomy centered around the sixth century. These were then compiled by a fifth-century "redactor," perhaps Ezra in the fifth century according to Friedman, who put the Pentateuch together.[53]

While the dates are disputed, and the details of the Documentary Hypothesis are contested by some, the point is that from the ninth century to the fifth century BC in the case of the Torah, the later centuries BC for the remainder of the Old Testament, and again in the case of the New Testament writers from about 50 to 100 AD, with numerous changes ever since, the Bible was indisputably written, edited, and constructed by people who reflected their times. Taken together they constitute a fascinating if largely unreliable window into the tumultuous history of the cultures in which they were written.[54] But the question of divine revelation is at an entirely different level. After more than six hundred pages of painstaking, soul-searching, and uncomfortable exegesis taking into account the latest Biblical scholarship, Harvard professor James Kugel, an orthodox Jew now living in Jerusalem, reluctantly concluded in his book *How to Read the Bible* that "For the most part, modern scholarship has served only to undermine the very notion of a Holy Scripture as given by divine inspiration to holy prophets," providing numerous reasons why this must be so.[55] Many other religious and nonreligious scholars have come to the same conclusion.

Where does this leave the billions of people who wish to hold on to the Bible as divine revelation? As Richard Friedman stated in his meticulous book *Who Wrote the Bible?*, the idea of Biblical divine revelation is in the end a matter of faith, an epistemological method foreign to science, but accepted by many of the "faithful" despite modern scholarship. As he pointed out, people may read the Bible as literature, as checkered history, or as sacred revelation. Who wrote the Bible is an interesting question fraught with uncertainties, but one that pales in comparison to the question of divine revelation. Friedman's conclusion—whether the authors "did so at divine direction, dictation, or inspiration was always a matter of faith"—is hardly new and may seem anodyne, but it is indisputably the case.[56] Cosmotheology embraces the first two options (literature and history), but rejects divine revelation.

God and the Superintelligence Hypothesis.

Why should we even contemplate the idea of a natural God or Superintelligence? One reason is because in the view of some scientists the universe itself has led us in that direction. Over the last four decades some scientists have come to question why the laws of nature and the physical constants appear to be biofriendly, giving rise to what has been termed the anthropic principle—the idea that the universe seems to be made for humans. The principle has many variants, all having to do with the apparent fine-tuning of the physical constants for life.[57] The phrase "anthropic principle, however," is a spectacular misnomer, and the term "biocentric principle" is much preferred, since in the context of astrobiology the universe appears to be friendly to life in general, and the very question to be answered is whether humans are the only intelligent life.

The point is that explanations of the biofriendly universe offer several options:

1. It's just a coincidence. For some this is enough, but for most people who have thought deeply about the problem, it is too good to be true and requires an explanation.
2. A supernatural intelligence fine-tuned the universe. This is better known as the God Hypothesis.
3. A highly evolved natural intelligence fine-tuned the universe. This possibility has received some attention, but not a great deal. We label it here the Superintelligence Hypothesis.
4. A multiverse—an ensemble of universes—exists, and we happen to live in a universe that is suitable for life. Such a multiverse could exist sequentially or simultaneously.[58] To many this seems only to exchange one unverifiable entity (God) for another (the multiverse).

Those of a supernatural bent sometimes point to the fine tuning of the universe for life as proof of God.[59] This is a "God of the gaps" kind of argument no longer popular even among theologians, precisely because science keeps filling in the gaps. Cosmotheologists on the other hand might favor the Superintelligence Hypothesis. The idea of a superior natural intelligence not invoking the usual concept of God has already appealed to some in this context. In his book *The Intelligent Universe: A New View of Creation and Evolution* the astronomer Fred Hoyle, a self-proclaimed atheist, concluded that the simplest explanation for a universe whose laws and constants are uniquely suited for life, a so-called "biofriendly" universe, was that a natural superintellect had engineered the universe to make it fit for carbon-based life and intelligence.[60] The cosmologist Edward Harrison makes similar claims, extending even to the natural selection of universes.[61] More recently complexity theorist James Gardner has elaborated this theme in a book with the same title as Hoyle's, following up on his book *Biocosm*. His selfish Biocosm Hypothesis holds that life and intelligence are the primary cosmic phenomena, and that "the capacity of the universe to generate life and to evolve ever more capable intelligence is encoded as a hidden subtext to the basic laws and constants of nature," and the biofriendly laws of nature are equivalent to DNA in living creatures, "providing a recipe for development of a living universe and a blueprint for the construction of offspring (baby universes)." Where do these baby universes come from? Gardner speculated that the replication process could be engineered by a highly evolved intelligence, whose laws and constants provide a recipe for the birth and evolution of intelligent life. In such a selfish biocosm scenario, it is logical that the physical laws and constants would be rigged in favor of the emergence of life. Gardner offers three falsifiable tests of his hypothesis: SETI research, artificial life evolution, and the emergence of computer intelligence.[62] This concept may seem close to the accepted version of God, but in fact it is the opposite: the Superintelligence is not supernatural.

The idea of such a superintelligence eventually evolving on Earth, perhaps artificial or postbiological, has been further elaborated by Nick Bostrom.[63] And if this is the trend on Earth, how much more forceful is it when applied to the universe, where life could have originated billions of years ago.[64] It is thus clear that the concept of God as a supernatural entity and the concept of superior intelligence can be separated. The term "God" is freighted with history, the term "superintelligence" less so. While cosmotheology favors the Superintelligence Hypothesis as opposed to the God Hypothesis, in the end we should also emphasize the existence of a superintelligence is not a necessary, but only a possible, aspect of cosmotheology. All of its principles except number 4 hold even in the absence of the Superintelligence Hypothesis.

Natural superintelligences need not be the object of worship, intervene in human history, or be the cause for religious wars such as witnessed throughout human history. Rather these ideas resonate with Einstein's "cosmic religion," which he saw as "the strongest and noblest motive for scientific research."[65] "I cannot conceive of a God who rewards and punishes his creatures, or has a will of the type of which we are conscious in ourselves," Einstein wrote. "An individual who should survive his physical death is also beyond my comprehension, nor do I wish it otherwise. . . . Enough for me the mystery of the eternity of life, and the inkling of the marvelous structure of reality, together with the single-hearted endeavor to comprehend a portion, be it never so tiny, of the reason that manifests itself in nature."[66] This is the attitude that informs the principles of cosmotheology and religious naturalism in general.

Morality and the Supernaturalistic Fallacy

One might argue that despite its fifth principle, a godless cosmotheology leaves us bereft of a moral value system. It is often stated that morality stems from theology and the existence of God, with the unfortunate implication that any nonbeliever has no basis for morality. The attempt to derive meaning and morals from what we know about the universe even has a derogatory name: "the naturalistic fallacy," and precautions against inferring value from fact, "ought" from "is," date back at least to the philosopher David Hume in the eighteenth century.

Despite this history of unproven precautions, cosmotheology favors astroethical principles stemming from a reverence for life in all its manifestations, the product of the creativity of cosmic evolution, whether terrestrial or extraterrestrial. We recall that even Boethius drew moral lessons from the study of the heavens when he wrote he "*Would mold both my actions and the principle of life in general in accordance with the model of heavenly order.*"[67] Exactly how to draw meaning from the findings of science is open to interpretation, and not just by scientists. Sam Harris, in his book *The Moral Landscape*, espouses a secular ethics expressed in his subtitle *How Science Can Determine Human Values*. He argues that values can be defined as the well-being of conscious creatures, which in turn is based on the facts of world events and brain states; in other words, values are the subject of science in its broadest sense as verifiable knowledge. There is no Christian or Muslim morality, he argues; rather morality is an undeveloped science, one that he begins to develop in his book.[68] Thinkers such as complexity theorist Stuart Kauffman have no problem finding meaning and value in the creativity of the universe.[69] Ideas of religion, ethics, spirituality, and meaning grounded in Nature rather than the supernatural are commonplace. Books such as Ursula Goodenough's *The Sacred Depths of Nature*[70] or Brian Swimme and Thomas Berry's *The Universe Story*[71] spring to mind. Cosmotheology therefore declares the naturalistic fallacy misleading and inoperable. The real fallacy is that the universe, as opposed to supernatural sources, cannot inform morality. We might call this, the idea that morality must come from supernatural sources, the Supernaturalistic Fallacy.

Most recently philosopher-scientist Mark Lupisella has described a long-term worldview that "can be characterized as a morally creative cultural cosmos—a post-intelligent, post-technological universe that enters the realm of conscious evolution driven largely by moral and creative pursuits," in other words, a worldview in which meaning and value may be bootstrapped from the universe.[72] In his recent volume *Cosmological Theories of Value*,[73] Lupisella places cosmic evolution at front and center stage, arguing that human destiny is intimately connected to the cosmos and cosmic evolution. He sets forth three cosmological theories of value, both in the sense of "value" as basic principles that motivate actions, and as ideas that are valuable for individuals and cultures. In its simplest form that he terms "cosmic reverence," we can certainly value our place in 13.8 billion years of cosmic evolution—something that seems simple but that requires a certain cosmic consciousness. We can go one more step and argue that we are coevolving with the universe, that the universe gains some value from us in a two-way relationship, a kind of cosmocultural coevolution. Or we can invoke what Lupisella calls the "connected action principle," the idea that connectedness is an intrinsic property of the universe manifested in relations and actions that account for the creativity of cosmic evolution, including life and intelligence. This property is consistent with what we observe in terms of the diversity of relations all around us, whether process philosophy or quantum mechanics, but raises questions of determinism and free will. Nevertheless, Lupisella argues that such a dynamic creative universe can serve as the basis for cosmological intrinsic value, the basis for a kind of cosmocentric ethic.

Human Destiny

Although human destiny has most often been couched in divine terms, as in Reinhold Niebühr's *The Nature and Destiny of Man*[74] or Pierre Lecomte du Noüy's bestselling *Human Destiny*,[75] or, indeed as in the entire Christian theology and theologies of the Abrahamic religions, it need not be linked to the supernatural or to the idea of God. Rather, it can be linked to the process and endpoint of cosmic evolution. If cosmic evolution ends with humans and we are alone in the universe, our destiny involves stewardship of our pale blue dot and perhaps spreading, nurturing, or even creating, life in the universe—all pathways filled with ethical considerations. If cosmic evolution results in a biological universe—one in which life and intelligence is common, our destiny is to interact with this life in all its myriad possibilities, invoking a quite different set of ethical considerations raised in the fifth principle.[76]

The combination of these ideas—the denial of supernaturalism, the Superintelligence Hypothesis, the denial of supernaturalism for the origins of morality, and human destiny based on the long-term evolution of the cosmos, are at the heart of cosmotheology.

Process Theology and Cosmotheology

Scholars of religion will note that cosmotheology also resonates in some ways with process theology, a theology stemming from the process philosophy championed by Alfred North Whitehead (1861–1947) and others that emphasizes becoming over being. A survey of the literature on this subject shows that process theology means many things to many people.[77] But its guiding principle of evolutionary becoming is bedrock. Cosmotheology therefore resonates with process theology first of all because of its emphasis on cosmic evolution, which is the broadest possible process, the broadest possible context for becoming, the master narrative of the universe, a story that includes us as its most recent manifestation, although not its ultimate endpoint. As indicated above there is now a vast literature on cosmic evolution. Much more than when process theology began its development in the 1920s by philosophers Whitehead and Charles Hartshorne, cosmic evolution gives a solid scientific foundation to the idea of becoming—a becoming of which we are a part.

Secondly, it follows from cosmic evolution that, like process theology, cosmotheology is also a relational worldview. The unfolding of cosmic evolution means there are relations among all things, a view that resonates with

Mark Lupisella's recent book.[78] As mentioned above that volume sets forth three theories of value based on the cosmos: cosmic reverence, cosmocultural coevolution, and the connection-action principle, in which connectedness is an intrinsic property of the universe manifested in the creativity of cosmic evolution. Lupisella argues the latter may even provide a foundational basis for process philosophy.[79]

Thirdly, like cosmotheology, process theology typically is a form of religious naturalism in the sense it denies the supernatural or divine interventions such as miracles. In his book *Reenchantment Without Supernaturalism: A Process Philosophy of Religion*, the process philosopher of religion David Ray Griffin lays out the case that supernaturalism is no longer needed for religion. "Completely rejecting supernaturalism—understood as the belief in the possibility of occasional interruptions of the world's most fundamental causal order—I present a worldview that, although saturated with values, is fully natural. This worldview does involve a form of theism, but it is a fully naturalistic theism, according to which divine influence is a natural dimension of the world's most fundamental causal order, never an interruption thereof."[80] Griffin argues that naturalism is not equivalent to atheism, and theism is not equivalent to supernaturalism.[81] He sees Whitehead's process philosophy as a means to reconcile naturalism and theism, and thus to reconcile science and religion. Instead of a supernatural God he has a natural Divine Reality, a form of panentheism. This is not equivalent to cosmotheology, but the denial of supernaturalism is a step in the right direction. And note that Griffin says his worldview is "saturated with values" the lack of supernaturalism notwithstanding.

In process theology Jesus is not God incarnate; as Roger Olsen puts it he is "different in degree but not in kind from other creatures."[82] This alone understandably causes some Christians such as Olsen to reject process theology, since for many God's incarnation in Jesus is the essence of Christianity. But history tells us exactly when Jesus became God: in early 380 AD when, at the urging of the pro-Nicene bishop Ambrose, emperor Theodosius declared that true Christians were those who believed in "the single divinity of the Father, Son, and Holy Spirit within an equal majesty and an orthodox Trinity."[83] Cosmotheologists would not follow an edict from more than 1600 years ago. Rather, Jesus would be seen as a wise, compassionate, and ultimately influential man, characteristics shared by the leaders of many other religions. Christocentrism as a philosophy and a way of approaching life may still be wisely employed, but shorn of divinity and supernaturalism, and supplemented by the wisdom of other religions.

Process theology also denies that God could have created the world ex nihilo, nor is God omnipotent to do so. For cosmotheologists the world begins with the Big Bang, and though this beginning of 13.8 billion years of cosmic evolution still holds many mysteries, current science supports this hypothesis as the way our world began. Although scientists cannot yet say *how or why* the universe began, attempts to say "God did it" again reduce to statements of faith and fall in the category of "God of the gaps" arguments now discredited even by most theologians.

Although these principles of process theology are too much for many Christians to adopt, a few farsighted thinkers in the early twentieth century embraced the importance of becoming in their theology. Much to the chagrin of the Catholic Church, the French Jesuit priest, philosopher, and paleontologist Pierre Teilhard de Chardin famously made the evolution of the cosmos the central theme of his posthumous book *The Phenomenon of Man*.[84] Here he embraced cosmic evolution and argued for a teleological evolution in which humans would end in a collective consciousness called the "noosphere," which would ultimately lead to the Omega Point, the maximum level of consciousness, which he also identified with God. Though the idea was not accepted within the Catholic church, to say the least, a few adventurous souls have followed in Teilhard's footsteps, including the Catholic priest Thomas Berry and physicist Brian Swimme, whose joint book *The Universe Story* emphasizes the religious significance of cosmic evolution.[85] The Franciscan nun Ilia Delio has also championed the philosophy of Teilhard, while admitting it is unlikely to be accepted within the Catholic tradition any time soon.[86] One must admire these adventurous departures from dogmatic Christianity, even if not agreeing with their details, and even if they do not go far enough. In other words, even if they are reserved rather than radical changes in the context of cosmotheology, they are a step in the right direction.

I would suggest another idea not normally seen as a part of process theology, but that resonates with it and cosmotheology. This is the idea that religion and theology are themselves part of cosmic evolution, in particular cultural evolution following the process of brain development. There are many theories of the origin of religions, but in his book *Evolving Brains, Emerging Gods: Early Humans and the Origins of Religion*, E. Fuller Torrey argues that "brain evolution theory posits that gods, and later the formal religions that are tied to gods, are products of the development of the human brain."[87] Table 8.1 shows brain evolutionary development based on Torrey's discussion. The rise of religion is associated with the last stage some 40,000 years ago, when the brain of modern homo sapiens allowed us to project ourselves backward and forward in time, also allowing us to foresee our own deaths.[88] There are, of course, other theories of the origin of religion that need attention, and multiple origins are most likely. But what we now know about the evolution of the human brain provides a new perspective on process theology not available when Whitehead and Hartshorne first formulated it beginning a century ago.

Table 8.1. Five Cognitive Developments of the Human Brain

Brain Evolutionary Development	*Hominid Species*	*Emergence*
General intelligence	Homo habilis	2 million years ago
Awareness of self	Homo erectus	1.8 million years ago
Awareness of thoughts of others/"Theory of Mind"	Archaic Homo sapiens	200,000 years ago
Introspective ability to reflect on own thoughts	Early Homo sapiens	100,000 years ago
Autobiographical memory = Ability to project backward and forward in time	Modern Homo sapiens	40,000 years ago

Source: Steven J. Dick, based on data from Torrey, *Evolving Brains, Emerging Gods*.

How, then is cosmotheology different from process theology? First of all, by having a core set of six principles, which in both origin and substance are much more closely attuned to science than process theology in its original form. As I have laid out here, cosmotheology embodies and has at its central core the new universe, cosmic evolution, and the abundance of life throughout the cosmos. The principles of cosmotheology as outlined above resonate with Teilhard de Chardin's evolutionary cosmology, but without the teleological end, the "Noosphere" he identified with Jesus Christ, or a supernatural God as the Omega Point. And the sixth principle of cosmotheology makes clear that human destiny may be couched in terms of natural cosmic evolution. If one insists on the idea of God as part of a personal metaphysics, the God of cosmotheology is a natural evolutionary product, potentially with many characteristics normally associated with God. This Superintelligence Hypothesis leads us to consider our place in the universe with a new perspective.

Another difference with process theology is that cosmotheology does not start from any traditional religious worldview except religious naturalism. One of the benefits of starting de novo is that doctrines such as Incarnation and Redemption, undeniably human constructs of history, need not be reconciled to the new universe since they are not assumed or held to begin with. The problem of theodicy, how a loving God could allow evil in the universe, is not really a problem at all because God, or at least that kind of God, does not exist. Superintelligences may in theory be good, bad, or indifferent, and may or may not intervene in human history, if indeed they exist at all.

In the end, then, cosmotheology has many points of commonality with process theology, but is more grounded in modern physical, biological, and social science, from the Big Bang to stars, brains, and culture, all part of cosmic evolution. The resonances and differences among process theology, religious naturalism, and cosmotheology are fertile ground for further research.[89]

DIFFICULTIES: IS THE UNIVERSE ENOUGH?

I have always been impressed that in his masterpiece *On the Origin of Species* Charles Darwin devoted an entire chapter to the difficulties of his ideas with regard to descent with modification.[90] This was a recognition on Darwin's part that he had not solved all the problems he had boldly set out to address, and that more work remained to be done. Indeed, it did, and the work continues today. In that spirit let us now turn to the difficulties of the ideas that comprise cosmotheology as described above.

First, the six principles of cosmotheology are hardly reached with deductive rigor. While the first principle of the physical non-centrality of humanity is now undeniable, one could plausibly deny that the biological universe inherent in the second principle actually exists, in which case we may therefore indeed be at the center of the mental and moral universe and not at the bottom of a great chain of being. Although we now know that *habitable* planets are abundant, that is not the same as *inhabited* planets. We know so little about the origins of life that it is still possible that life exists on no other planet aside from Earth, much less intelligent life. Some argue that the Principle of Mediocrity, also known as the Copernican Principle, has been taken too far, and these arguments need to be taken seriously.[91] If the premise of a biological universe is false, the arguments for cosmotheology are weakened, though it would not take a universe full of life, a robust biological universe, to require a wholesale rethinking of theology. Moreover, it need hardly be said that traditional religions are not reached with deductive rigor either.

In short, some would argue we have been too bold in positing a biological universe. But in fact, it is also possible we have not been bold enough, and that the universe has evolved beyond the biological to the postbiological. It is entirely possible anthropocentrism has gotten the better of us, and we need to be prepared to find the unexpected.[92] This view of a postbiological universe is now accepted among a variety of scholars, including the former Astronomer Royal Sir Martin Rees, physicist Paul Davies, philosopher Susan Schneider, and astronomer Seth Shostak.[93] How would religion and theology be affected if we met postbiological alien artificial intelligence instead of biologicals?

Second is the question of how far we should go in our theological reconsiderations based on the new universe. Should we be reserved or radical? It is clear that most scholars attempting to advance theologies and reconcile them to modern science are still working within established religious traditions. In the Introduction to *The Unbearable Wholeness of Being*, the Franciscan sister Ilia Delio writes "my aim is to elucidate Christian life in an evolutionary world."[94] Even as he pleas for cosmic evolution to be incorporated into the very lifeblood of theology, Arthur Peacocke writes clearly within the Christian tradition. Similarly with Dowd and Barlow's "evolutionary Christianity," a label that speaks for itself.[95] And Teilhard de Chardin struggled mightily to maintain a Christocentric universe within his evolutionary worldview; as Julian Huxley wrote in the Introduction to Teilhard's pathbreaking book *The Phenomenon of Man*, "As a dedicated Christian priest, he felt it imperative to try to reconcile Christian theology with this evolutionary philosophy, to relate the facts of religious experience to those of natural science."[96] These are admirable efforts within the Christian tradition, and follow, each in their own way, what many others have attempted. The same can be said for other religions such as the Islamic and Jewish traditions.[97] But, even while it has religious naturalism as a base, cosmotheology is a much more fundamental undertaking than those who seek to reform established theologies. Many adherents to specific religious traditions will undoubtedly think our reconsiderations have gone too far, and thus reject the whole endeavor out of hand, a step (or several) too far. We recall that despite his efforts at reconciliation of evolution and Christianity the Jesuit Teilhard de Chardin's still Christocentric theology was roundly rejected by his Church, with publication of his works prohibited until after his death.

Third, the religious naturalism at the base of cosmotheology is, of course, a controversial position in more ways than one. The renowned historian of science John Greene, for example, cautioned in no uncertain terms that naturalism is just another world view, no more privileged than supernaturalism or anything else to serve as the source of value and meaning in human life. He pointed out that scientists such as Ernst Haeckel, Julian Huxley, Ralph Burhoe, and E. O. Wilson "all typify the scientist-ideologue bent on saving society by promulgating a new ethics and a new religion claiming the sanction of science." The pattern is "depressingly familiar," in Greene's view: "When will scientists and others learn that naturalism is a philosophical point of view with no more claim to the status of science than any other philosophical viewpoint, whether Marxian, Freudian, Russellian, Whiteheadian, or whatever. Scientists have as good a right to expound their philosophical, ethical and religious views as anyone else, but they have no right to palm these off as the findings of science."[98] In my view, this attitude is both instructive and dead wrong. It is instructive in the sense that a science-oriented theology is far from the sole source of meaning and value in life. And it is dead wrong in the sense that science has at least as much claim to infuse theological thought as any other worldview, and perhaps more, since being grounded in the natural world gives it a pillar of support that supernatural religions do not have, numerous attempts at natural theology notwithstanding.

Fourth, there are several objections to the Superintelligence Hypothesis, beginning with the anthropic principle that leads some to this hypothesis. Some consider the anthropic principle and the idea of the multiverse not to be science,[99] while others consider the arguments to be completely misguided.[100] Others may well object that the Superintelligence Hypothesis is just the God Hypothesis by another name. They would be wrong because they are separated by the wide gulf of supernaturalism. Accepting that, they might further object that the Superintelligence Hypothesis of cosmotheology really qualifies as cosmophilosophy rather than cosmotheology, since a transcendent God is central to theology. The same criticism would apply to the religious naturalist worldview. But if theology is defined broadly as that which gives meaning and value to life in a cosmic context, then even a naturalistic cosmotheology is indeed a theology, albeit one without God, at least the standard God. On the other hand, the term "cosmophilosophy" can also apply as a broader category. Philosophy and theology are closely connected, and I have placed them together in the worldview chart of Figure 8.1 because theology may be seen as a particular kind of philosophical worldview, often, but not necessarily, supernatural. In my view the terminology is not a substantial objection. The important thing is the search for meaning in an intellectually honest way.

Finally, there is the psychological/sociological question of whether a significant number of people will buy into cosmotheology, given the lack of salvation, heaven, and other suspiciously comforting concepts. Again, Lupisella is instructive when he asks, "Is the Universe Enough?"[101] Can people find meaning and value in their lives from the universe as opposed to more traditional and mostly supernatural religious worldviews? He concludes that "the universe, in the form of cosmological worldviews that focus on scientific cosmic evolution, probably isn't enough for most people most of the time, but could be enough for some people some of the time. However, the universe may increasingly suffice as we evolve and as more advanced intelligence evolves."[102] Only time will tell.

In the end reactions to radical attempts such as cosmotheology to rethink theology will run the gamut, reflecting the spectrum of opinions even within a single tradition such as Christianity. On the one hand there will be people like Reverend Kelly Brown Douglas, the canon theologian at Washington National Cathedral and dean of the Episcopal Divinity School at Union Theological Seminary. Speaking recently of the death of Rev. John Shelby Spong, a liberal theologian and former bishop who shook up the modern Episcopal Church, he noted "What he truly came to understand is doctrine, dogma, doesn't make us Christian. Doctrine, dogma, doesn't make us church. What makes us church is respecting the sacredness of every single human being and creating a world that does that and making sure the church is leading the world in doing that."[103] An attitude such as that might welcome a cosmotheology. On the other hand are the fundamentalists, who will undoubtedly bitterly reject any such changes. And there will be many positions in between.

In the end "some people," perhaps in increasing numbers over time with our increasing cosmic consciousness, will find that human destiny tied to the process of cosmic evolution—in which humans and all life in the universe are born from stardust, contemplate their place in the universe with or without extraterrestrials, advance knowledge to the best of their ability, live, love, learn, and in the end return to stardust—is a satisfying vision of humanity's role in the universe.

CONCLUSIONS: THE WAY AHEAD

Human experience clearly demonstrates that, for better or worse, religions have shaped cultures throughout history. Although its worst aspects are often emphasized by critics, religions have indisputably also played an important role in social cohesion, cooperation, and perhaps even in nurturing science with its concept of an ordered world.[104] At the personal level, human experience demonstrates that most people need a spiritual component in their lives, a search for meaning that throughout history has largely been filled by religious traditions with theologies elaborated in ever more subtle form.

Let me therefore be clear that I do not wish to denigrate other people's worldviews. Although they represent a different metaphysics and epistemology than cosmotheology, they often provide comfort in a difficult world. The main epistemology of science is empiricism—which admits of theory and observation, with all their complexities and problems, and yet which taken together have revealed so much about the universe around us. The main epistemology of theology, on the other hand, is revelation, faith, and a heroic attempt at empiricism via what has been known over the last four centuries as natural theology. But as we have emphasized, natural theology is no longer considered a good theological argument; even most theologians admit the complex universe no longer requires a God to explain its inner workings, intelligent design notwithstanding. This leaves revelation, which also reduces to faith—faith that somehow God revealed his or her word to the authors of the Bible and that we should be following its dictums thousands of years later.

In short whether embracing the idea of a supernatural God, belief in Christian doctrine constructed sporadically and controversially over many centuries, or Biblical revelation, in the end it all comes down to faith. That is the difference between science and theology. To someone whose epistemology does not include faith, the practice of religion and the weaving of supernatural theologies will seem strange activities indeed. But for those with a worldview that includes faith as a source of knowledge, it makes perfect sense, though it must be pointed out that with all the religious diversity found on Earth today, one person's faith is another person's heresy. To say, as some critics do, that scientists also have faith—faith in their methods, faith in facts, and faith in science itself—is a red herring, since scientists will overturn their beliefs, if sometimes grudgingly, in the face of strong evidence to the contrary. The same is not true of the faithful.

In my view the lifeblood of theologies is that they must evolve, perhaps radically, with changing times. I have argued that over the last century science has given us a new cosmological worldview, a kind of "revelation" that strongly indicates the need for a transformation of established theologies and suggests possibilities for new cosmically oriented theologies such as cosmotheology. I have suggested that both religious naturalism and a science-grounded elaboration of process theology resonate with this new worldview in ways that still need to be fully explored, and that a cosmotheology no longer requires supernaturalism or even the concept of God in the traditional sense.

But what is the way forward? There are a number of possibilities, alternate worlds if you will, for the future of theology in relation to science. First, the new worldview might not affect current religion and theology at all. In my view this is a nonstarter, both intellectually and practically. While large numbers of the faithful may ignore the new worldview for a while, in the long run denial will not happen, is not happening, and hasn't happened in the past, at least not to those religions that remain most vibrant, despite the lengthy period of time for religions to absorb new worldviews such as Copernicanism and Darwinism.

Secondly, current religions and theologies could adapt to the new worldview without change to their major doctrines. This is entirely possible, as seen in the work of those we have mentioned before: Sir Arthur Peacocke, who saw the need for including cosmic evolution into the very lifeblood of religion, or Ilia Delio with her Christocentric worldview, or Teilhard de Chardin with his Noosphere and Omega Point. Specifically addressing cosmotheology, professor of systematic theology Olli-Pekka Vainio inexplicably claims that its principles "1, 2, 3, and 5 have already been accommodated within orthodox Christian theology a long time ago, without the problems and challenges that Dick lists."[105] This would come as a surprise to theologians such as former Vatican Observatory director George Coyne, who has warned we should not underestimate the difficulties of religion adapting to the new worldview, especially for entrenched and fundamental doctrines such as Redemption and Incarnation.[106] "Accommodated" is not the same as discussed or ignored.

Thirdly, current religions and theologies could disappear. This seems unlikely, certainly in the near term, despite declining membership for many religions in the twenty-first century. Again, the history of humanity demonstrates there is an obvious need for a spiritual dimension to human life.

However, that spiritual dimension can and should evolve; those traditional religions that do not evolve risk irrelevance and perhaps extinction in the long term, the "dangerous path" that Delio alludes to.

This gives rise to the final possibility, advocated in this paper, namely that new religions and theologies such as cosmotheology should and will arise. After all, the evolution of religion and theology are part and parcel of cultural evolution, even though they tend to cling tenaciously to their principles which, we should remember, themselves evolved at particular times in particular cultures, and so should be open to further evolution with modern knowledge. We need to be on guard, however, against cult fringe religions, even though the academic study of religions now recognizes the category of "UFO religions" like the Unarians, the Raelians, the Solarians, and the Scientologists.[107] The human search for meaning sometimes travels strange paths.

These four alternate futures are not mutually exclusive and in any case are likely to take place in small steps. But religions and theologies urgently need to grapple with these issues. Moreover, my conclusions are very much dependent on the future course of history. They depend on how the discovery of the biological universe is made, and if it is made at all. And it is possible that, current controversies about the biocentric principle notwithstanding, the idea of a biofriendly universe has a deeper significance than we now realize. Over the last several decades the British Astronomer Sir Martin Rees, physicist and astrobiologist Paul Davies and many others have explored the ramifications of a possibly deep and profound relationship between biology and cosmology.[108] The physicist Freeman Dyson has suggested that the prospects are bright for a future-oriented science, joining together in a disciplined fashion the resources of biology and cosmology.[109] In such a cosmic ecology, life and intelligence would play a central role in the evolution of the universe, no less than its physical laws. And one or more naturally evolved superintelligences would be at the center of this nexus, which would indeed constitute a very strong form of the biological universe.

What form their search for meaning may take we cannot guess. Even without such a cosmological connection to biology, the scenario of the discovery of extraterrestrial intelligence will be crucial. One set of implications would hold if life is discovered on exoplanets and contact is made by radio transmissions in the form of a dial tone in which intelligence is evident but communication is not made.[110] If communication is possible, then all bets are off depending on what the conversation is like. The same is true if a spaceship lands on the White House lawn. At the other extreme, if no life is ever found, if we live in a purely physical universe in which we are the biological exceptions, that would be an equally astonishing conclusion. But all of our discussions would not be in vain, for it is always beneficial to examine our assumptions as part of the evolution of ideas in the evolution of culture. On one thing I would insist: there will be no salvation from terrestrial religions or extraterrestrial aliens. As Carl Sagan once said in response to a question of where a godless universe leaves us, "we're on our own." That is the real challenge for humanity in an increasingly dangerous and uncertain world.

NOTES

1. Wallace, Man's Place in the Universe; Dick, "The Universe and Alfred Russel Wallace."
2. Berenzden et al., *Man Discovers the Galaxies*; Smith, *The Expanding Universe*.
3. Ward and Brownlee, *Rare Earth*.
4. NASA Exoplanet Archive; Tasker, *Planet Factory*.
5. Scharf, *The Copernicus Complex*.
6. Dick, *The Biological Universe*.
7. NASA Astrobiology website: https://astrobiology.nasa.gov; Sullivan and Baross, *Planets and Life*; Kaufman, *First Contact*.
8. NASA, "Astrobiology Strategy"; Horneck et al., "AstRoMap."
9. United States, Director of National Intelligence, "Preliminary Assessment."
10. Sagan, *Cosmos*; McMahon, "Do Extraordinary Claims Require Extraordinary Evidence?"
11. Loeb, *Extraterrestrial*.
12. Delio, *The Unbearable Wholeness of Being*, xv.
13. Boethius, *The Consolation of Philosophy*.
14. Ibid., book II, part VII.
15. McCluskey, "Boethius's Astronomy and Cosmology"; Dick, "The Consolations of Astronomy."
16. Dick, "The Consolations of Astronomy."
17. Shapley, *Of Stars and Men*; Palmeri, "Bringing Cosmos to Culture."
18. Crawford, *Expanding Worldviews*.
19. Dick, "Cosmic Evolution"; Zakariya, *A Final Story*.
20. Chaisson, *Cosmic Evolution*.
21. Dick, *Classifying the Cosmos*.
22. Dick, "Cultural Evolution, the Postbiological Universe and SETI."
23. Christian, "The Case for Big History" and *Maps of Time*.
24. Crawford, *Expanding Worldviews*.
25. https://www.bighistoryproject.com/home.
26. Lewis, *The Discarded Image*, 222.
27. Zakariya, *A Final Story*.
28. Vidal, *The Beginning and the End*.
29. Geppert, *Imagining Outer Space*.
30. Achenbach, *Captured by Aliens*; Eghigian, "Making UFOs Make Sense"; Pasulka, *American Cosmic*.

31. Iqbar, "Islamic Theology Meets ETI"; Determann, *Islam, Science Fiction, and Extraterrestrial Life.*

32. Dick, *Plurality of Worlds*, *The Biological Universe*, and *Life on Other Worlds*; Crowe, *The Extraterrestrial Life Debate*; Guthke, *The Last Frontier*; Malone-France, "Hell is Other Planets."

33. Peters, *Astrotheology*. The idea of an astrotheology dates back at least to 1715, when the English clergyman and natural philosopher William Derham penned a book by that name whose subtitle indicated its purpose: *Astrotheology: Or a Demonstration of the Being and Attributes of God from a Survey of the Heavens*. In other words, this was an exercise in natural theology, an attempt to prove the existence of God from his created work. See Dick, *Plurality of Worlds*.

34. Peters, "Exo-Theology" and "Exotheology."

35. Coyne, "The Evolution of Intelligent Life on Earth and Possibly Elsewhere," 187.

36. Peters, *Astrotheology*, 271; Russell, "Many Incarnations or One?," 303.

37. O'Meara, *Vast Universe.*

38. Peacocke, "The Challenge and Stimulus of the Epic of Evolution to Theology," 92

39. Barlow, *Evolution Extended*; Dowd, *Thank God for Evolution.*

40. Parkyn, *Exotheology*.

41. Lamm, "The Religious Implications of Extraterrestrial Life."

42. Weintraub, *Religions and Extraterrestrial Life.*

43. The term cosmotheology originates with the philosopher Immanuel Kant, who referred to it in 1781 in his *Critique of Pure Reason* (without endorsing it) as a "transcendental theology" method of "inferring the existence of a Supreme Being from general experience," rather than a natural theology method of inferring the nature of a Supreme Being from the particulars of nature. Both of these views, of course, are far from my own concept of a naturalistic cosmotheology, and in fact they are far even from the modern view of most nonnaturalistic astrotheologies.

44. Dick, *Many Worlds*; "Toward a Constructive Naturalistic Cosmotheology."

45. Wallace, *Man's Place in the Universe.*

46. Dick, "Cultural Evolution, the Postbiological Universe and SETI."

47. Impey, *Encountering Life in the Universe*; Peters, "Astrotheolog"; Dick, *Astrobiology, Discovery, and Societal Impact*, ch. 8.

48. Stone, *Religious Naturalism Today*, 1.

49. Armstrong, *A History of God.*

50. Rubenstein, *When Jesus Became God.*

51. Dick, "Toward a Constructive Naturalistic Cosmotheology," 239.

52. Armstrong, *The Bible* and *The Lost Art of Scripture*. Fox, *The Unauthorized Version.*

53. Friedman, *Who Wrote the Bible?*

54. Cline, *Biblical Archaeology*; Kugel, *How to Read the Bible.*

55. Kugel, *How to Read the Bible*, 662 ff.

56. Friedman, *Who Wrote the Bible?*, 243–45.

57. Carter, "Large Number Coincidences"; Barrow and Tipler, *The Anthropic Cosmological Principle*; Rees, *Just Six Numbers*; Carr, *Universe or Multiverse?*; Davies, *Cosmic Jackpot.*

58. Carr, *Universe or Multiverse?*; Davies, *Cosmic Jackpot*, Dick, "Cosmology and Biology."

59. Harris, *Cosmos and Theos*; Ross, *The Creator and the Cosmos.*

60. Hoyle, *The Intelligent Universe.*

61. Harrison, "The Natural Selection of Universes."

62. Gardner, Biocosm; *The Intelligent Universe*, 159; "The Intelligent Universe."

63. Bostrom, *Superintelligence.*

64. Dick, "Cultural Evolution, the Postbiological Universe and SETI."

65. Einstein, "Religion and Science," 39; Jammer, *Einstein and Religion*, 75–84.

66. Einstein, *The World As I See It*, 5.

67. Boethius, *The Consolation of Philosophy*, book I, part IV.

68. Harris, *The Moral Landscape*; Dick, *Astrobiology, Discovery, and Societal Impact*, 242–44.

69. Kauffman, *Reinventing the Sacred.*

70. Goodenough, *The Sacred Depths of Nature.*

71. Swimme and Berry, *The Universe Story.*

72. Lupisella, "Cosmocultural Evolution."

73. Lupisella, *Cosmological Theories of Value.*

74. Niebühr, *The Nature and Destiny of Man.*

75. du Noüy, *Human Destiny.*

76. McKay, "Does Mars Have Rights?;" Dick, "Cosmic Evolution," 45–49; Vidal, *The Beginning and the End*; Randolph and McKay, "Protecting and Expanding the Richness and Diversity of Life."

77. Olson, "Why I Am Not a Process Theologian;" Griffin, *Reenchantment without Supernaturalism*; Viney, "Process Theism."

78. Lupisella, *Cosmological Theories of Value.*

79. Lupisella, "The Connection-Action Principle," this volume.

80. Griffin, *Reenchantment without Supernaturalism*, vii.

81. Griffin, *Religion and Scientific Naturalism.*

82. Olson, "Why I Am Not a Process Theologian."

83. Rubenstein, *When Jesus Became God*, 220; Freeman, *Heretics*, 25.

84. King, *Life and Vision*; de Chardin, *The Phenomenon of Man*; Aczel, *The Jesuit & the Skull.*

85. Swimme and Berry, *The Universe Story.*

86. Delio, "Extending the Noosphere," this volume.

87. Torrey, *Evolving Brains, Emerging Gods*, 205.

88. Ibid., 204.

89. Davis, "From Negation to Exemplification," this volume.

90. Darwin, *On the Origin of Species.*

91. Scharf, *The Copernicus Complex.*

92. Dick, "Cultural Evolution, the Postbiological Universe and SETI."

93. Rees, "Why Alien Life Will be Robotic;" Davies, *Are We Alone?*; Schneider, "Alien Minds"; Shostak, *Sharing the Universe.*
94. Delio, *The Unbearable Wholeness of Being*, xxii.
95. Dowd, *Thank God for Evolution.*
96. de Chardin, *The Phenomenon of Man*, 22–23.
97. Iqbar, "Islamic Theology Meets ETI"; Lamm, "The Religious Implications of Extraterrestrial Life"; Samuelson, "Jewish Theology Meets the Alien."
98. Greene, "Afterword," 404.
99. Pagels, *Perfect Symmetry*; Smolin, "Scientific Alternatives to the Anthropic Principle."
100. Stenger, *The Fallacy of Fine-Tuning.*
101. Lupisella, "Is the Universe Enough?"
102. Ibid.
103. Smith, "John Shelby Spong."
104. Wilson, *Darwin's Cathedral*; Norenzayan, *Big Gods.*
105. Vainio, *Cosmology in Theological Perspective*, 182.
106. Coyne, "The Evolution of Intelligent Life on Earth," 187.
107. Lewis, *The Gods Have Landed*; Achenbach, *Captured by Aliens*; Battaglia, *ET Culture*; Pasulka, *American Cosmic*; Malone-France, "Hell is Other Planets," 48.
108. Rees, *Just Six Numbers*; Davies, *The Accidental Universe*; *The Mind of God*; *Cosmic Jackpot.*
109. Dyson, *Infinite in all Directions.*
110. Dick, *Astrobiology, Discovery, and Societal Impact*, 54–59.

BIBLIOGRAPHY

Achenbach, J. *Captured by Aliens: The Search for Life and Truth in a Very Large Universe*. New York: Simon & Schuster, 1999.

Aczel, A. *The Jesuit & the Skull: Teilhard de Chardin, Evolution, and the Search for Peking Man*. New York, Penguin, 2007.

Armstrong, K. *A History of God: The 4,000 Year Quest of Judaism, Christianity and Islam*. New York: Ballantine Books, 1993.

———. *The Bible: A Biography*. New York: Atlantic Monthly Press, 2007.

———. *The Lost Art of Scripture: Rescuing the Sacred Texts*. New York: Alfred A. Knopf, 2019.

Barlow, C. *Evolution Extended: Biological Debates on the Meaning of Life.* Cambridge, MA: MIT Press, 1995.

Barrow, J. D, and F. J. Tipler. *The Anthropic Cosmological Principle*. New York: Oxford University Press, 1986.

Battaglia, D., ed. *ET Culture: Anthropology in Outer Spaces*. Durham, North Carolina, Duke University Press, 2006.

Berenzden, R., et al. *Man Discovers the Galaxies*. New York: Science History Publications, 1976.

Boethius. *The Consolation of Philosophy*. London: The Folio Society, 1998.

Bostrom, N. *Superintelligence: Paths, Dangers, Strategies*. Oxford: Oxford University Press, 2014.
Carr, B., ed. *Universe or Multiverse?* Cambridge: Cambridge University Press, 2007.
Carter, B. "Large Number Coincidences and the Anthropic Principle in Cosmology." In *Confrontation of Cosmological Theories with Observational Data*, edited by M. S. Longair, 291–98. Dordrecht: D. Reidel Publishing, 1974.
Chaisson, E. *Cosmic Evolution: The Rise of Complexity in Nature.* Cambridge, MA. Harvard University Press, 2001.
Christian, D. "The Case for Big History." *Journal of World History* 2 (1991): 231–38
———. 2004. *"Maps of Time": An Introduction to "Big History."* Berkeley: University of California Press, 2004.
Cline, E. H. *Biblical Archaeology: A Very Short Introduction.* Oxford: Oxford University Press, 2009.
Coyne, G. V. "The Evolution of Intelligent Life on Earth and Possibly Elsewhere: Reflections from a Religious Tradition." In *Many Worlds: The New Universe, Extraterrestrial Life and the Theological Implications*, edited by Steven J. Dick, 177–90. Philadelphia: Templeton Foundation Press, 2000.
Crawford, I. *Expanding Worldviews: Astrobiology, Big History and Cosmic Perspectives*, ed. by Ian Crawford. Cham, Springer, 2021.
Crowe, M. J. *The Extraterrestrial Life Debate, 1750–1900: The Idea of a Plurality of Worlds from Kant to Lowell*. Cambridge: Cambridge University Press, 1986.
Darwin, C. *On the Origin of Species.* London: Penguin Classics, 2009.
Davies, P. *The Accidental Universe.* Cambridge: Cambridge University Press, 1982.
———. *The Mind of God*. New York: Simon & Schuster, 1992.
———. *Are We Alone? Philosophical Implications of the Discovery of Extraterrestrial Life.* New York: Basic Books, 1995.
———. *Cosmic Jackpot: Why Our Universe Is Just Right for Life*. Boston: Houghton Mifflin, 2007.
Davis, A. M. "From Negation to Exemplification: A Deeper Whiteheadian Cosmotheology," this volume.
Delio, I. *The Unbearable Wholeness of Being: God, Evolution, and the Power of Love*. Maryknoll, NY: Orbis Books, 2013.
———. "Extending the Noosphere into Intergalactic Life: Teilhard de Chardin and the Third Axial Age," this volume.
Determann, J. M. *Islam, Science Fiction, and Extraterrestrial Life: The Culture of Astrobiology in the Muslim World.* London, I. B. Taurus, 2021.
Dick, S. J. *Plurality of Worlds: The Origins of the Extraterrestrial Life Debate from Democritus to Kant*. Cambridge: Cambridge University Press, 1982.
———. *The Biological Universe: The Twentieth Century Extraterrestrial Life Debate and the Limits of Science*. Cambridge: Cambridge University Press, 1996.
———. *Life on Other Worlds*. Cambridge: Cambridge University Press, 1998.
———. "Cultural Evolution, the Postbiological Universe and SETI." *International Journal of Astrobiology* 2 (2003): 65–74. https://doi.org/10.1017/S147355040300137X. Reprinted in *Space, Time, and Aliens: Collected Works on Cosmos and* Culture, edited by S. J. Dick, 171–90. Cham, Switzerland: Springer, 2020.

———. "Toward a Constructive Naturalistic Cosmotheology." in *Exotheology: Science and Theology Meet Extraterrestrial Life*, edited by Ted Peters, 228–44. Eugene: Cascade, 2018.

———. *Astrobiology, Discovery, and Societal Impact*. Cambridge: Cambridge University Press, 2018.

———. *Classifying the Cosmos: How We Can Make Sense of the Celestial Landscape*. New York: Springer, 2019.

———. "The Consolations of Astronomy and the Cosmic Perspective." In *Space, Time, and Aliens: Collected Works on Cosmos and Culture*, edited by S. J. Dick, 755–69. Cham: Springer, 2020.

———. "Cosmology and Biology: An Entangled Web." In *Space, Time, and Aliens: Collected Works on Cosmos and Culture*, edited by S. J. Dick, 723–36. Cham: Springer, 2020.

———. *Space, Time, and Aliens: Collected Works on Cosmos and Culture*. Cham: Springer, 2020.

———. "The Universe and Alfred Russel Wallace." In *Natural Selection and Beyond: The Intellectual Legacy of Alfred Russel Wallace*, edited by C. H. Smith and G.W. Beccaloni, 320–40. Oxford: Oxford University Press, 2008.

———. "Cosmic Evolution: History, Culture and Human Destiny." In *Cosmos & Culture: Cultural Evolution in a Cosmic Context*, edited by S. J. Dick and M. Lupisella, 25–62. Washington, DC: NASA, 2009.

Dick, S. J., ed. *Many Worlds: The New Universe, Extraterrestrial Life and the Theological Implications*. Philadelphia: Templeton Foundation Press, 2000.

———. *The Impact of Discovering Life Beyond Earth.* Cambridge: Cambridge University Press, 2015.

Dick, S. J., and M. Lupisella, eds. *Cosmos and Culture: Cosmic Evolution in a Cosmic Context*. Washington, DC: NASA, 2009.

Dowd, M. *Thank God for Evolution: How the Marriage of Science and Religion Will Transform Your Life and Our World*. New York: Viking, 2008.

du Noüy, P. L. *Human Destiny*. London: Longmans, Green and Co, 1947.

Dyson, F. *Infinite in all Directions.* New York: Harper & Row, 1988.

Eghigian, G. 2017. "Making UFOs Make Sense: Ufology, Science, and the History of their Mutual Mistrust." *Public Understanding of Science* 26, no. 5 (2017): 612–26, https://doi.org/10.1177/0963662515617706.

Einstein, A. "Religion and Science." In *Ideas and Opinions*. New York: Bonanza Books, 1954.

———. *The World As I See It.* London, John Lane, 1935.

Fox, R. L. *The Unauthorized Version: Truth and Fiction in the Bible.* New York: Vintage Books, 1992.

Freeman, C. *AD 381: Heretics, Pagans, and the Dawn of the Monotheistic State*. New York: Overstock Press, 2009.

Friedman, R. *Who Wrote the Bible?* San Francisco: Harper, 1997.

Gardner, J. *Biocosm: The New Scientific Theory of Evolution: Intelligent Life Is the Architect of the Universe*. Makawao, HI: Inner Ocean Publishing, 2003.

———. 2007. *The Intelligent Universe: AI, ET and the Emerging Mind of the Cosmos*. Franklin Lakes: New Page, 2007.

———. "The Intelligent Universe." *Cosmos & Culture: Cultural Evolution in a Cosmic Context*, edited by S. J. Dick and M. Lupisella, 361–82. Washington, DC: NASA, 2009.

Geppert, A. C. T., ed. *Imagining Outer Space: European Astroculture in the Twentieth Century*. London: Palgrave Macmillan, 2012.

Goodenough, U. *The Sacred Depth of Nature*. New York: Oxford University Press, 1988.

Greene, J. "Afterword." In *History, Humanity and Evolution: Essays for John C. Greene*, edited by J. R. Moore, 404–14. Cambridge: Cambridge University Press, 1989.

Griffin, D. R. *Religion and Scientific Naturalism: Overcoming the Conflicts*. Albany: SUNY Press, 2000.

———. *Reenchantment without Supernaturalism: A Process Philosophy of Religion*. Ithaca: Cornell University Press, 2001.

Guthke, K. S. *The Last Frontier: Imagining other Worlds from the Copernican Revolution to Modern Science Fiction*. Ithaca: Cornell University Press, 1990.

Harris. E. E., *Cosmos and Theos: Ethical and Theological Implications of the Anthropic Cosmological Principle*. Atlantic Highlands, NJ: Humanities Press International, 1992.

Harris, S. *The Moral Landscape: How Science Can Determine Human Values*. New York: Free Press, 2010.

Harrison, E. W. "The Natural Selection of Universes Containing Intelligent Life." *Quarterly Journal of the Royal Astronomical Society* 36, no. 3 (1995), 193–203.

Horneck, G. et al. "AstRoMap European Astrobiology Roadmap." *Astrobiology* 16, no. 3 (2016): 201–43. https://www.liebertpub.com/doi/full/10.1089/ast.2015.1441

Hoyle, F. *The Intelligent Universe: A New View of Creation and Evolution*. New York: Holt, Rinehart, and Winston, 1983.

Impey, C. et al., eds. *Encountering Life in the Universe: Ethical Foundations and Social Implications of Astrobiology*. Tucson: University of Arizona Press, 2013.

Iqbar, M. "Islamic Theology Meets ETI." In *Astrotheology: Science and Theology Meet Extraterrestrial Life*, edited by T. Peters, 216–27. Eugene: Cascade, 2018.

Jammer, M. *Einstein and Religion*. Princeton: Princeton University Press, 1999.

Kauffman, M. *First Contact: Scientific Breakthroughs in the Hunt for Life Beyond Earth*. New York: Simon & Schuster, 2011.

Kauffman, S. A. *Reinventing the Sacred: A New View of Science, Religion and Reason*. New York: Basic Books, 2008.

King, U. *The Life and Vision of Teilhard de Chardin*. New York: Orbis Books, 1996.

Kugel, J. L. *How to Read the Bible: A Guide to Scripture, Then and Now*. New York: Free Press, 2007.

Lamm, N. "The Religious Implications of Extraterrestrial Life." In *Challenge: Torah Views on Science and its Problems*, edited by A. Carmell and C. Domb, 354–98. New York: Feldheim, 1978.

Lewis, C. S. *The Discarded Image*. Cambridge: Cambridge University Press, 1964.

Lewis, J. R., ed. *The Gods Have Landed: New Religions from Other Worlds*. Albany: SUNY Press, 1995.

Loeb, A. *Extraterrestrial: The First Sign of Intelligent Life Beyond Earth*. Boston: Houghton Mifflin Harcourt, 2021.

Lupisella, M. "Cosmocultural Evolution: The Coevolution of Culture and Cosmos and the Creation of Cosmic Value." In *Cosmos & Culture: Cultural Evolution in a Cosmic Context*, edited by S. J. Dick and M. Lupisella, 321–59. Washington, DC: NASA, 2009.

———. *Cosmological Theories of Value: Science, Philosophy, and Meaning in Cosmic Evolution*. New York, Springer, 2020.

———. "Is the Universe Enough? Can it Suffice as a Basis for Worldviews?" In *Expanding Worldviews: Astrobiology, Big History and Cosmic Perspectives*, edited by Ian Crawford, 217–37. Cham: Springer, 2021.

Malone-France, D. "Hell is Other Planets: Extraterrestrial Life in the Western Theological Imagination." In *Social and Conceptual Issues in Astrobiology*, edited by K. C. Smith and C. Mariscal, 21–53. Oxford: Oxford University Press, 2020.

McCluskey, S. C. "Boethius's Astronomy and Cosmology." In *A Companion to Boethius in the Middle Ages*, edited by N. H. Kaylor Jr. and P. E. Phillips, 47–73. Leiden: Brill, 2012.

McKay, C. "Does Mars Have Rights? An Approach to the Environmental Ethics of Planetary Engineering." In *Moral Expertis*, edited by D. MacNiven, 184–97. New York: Routledge, 1990.

McMahon, S. "Do Extraordinary Claims Require Extraordinary Evidence? The Proper Role of Sagan's Dictum in Astrobiology." In *Social and Conceptual Issues in Astrobiology*, edited by K. C. Smith and C. Mariscal, 117–29. Oxford: Oxford University Press, 2020.

NASA. *Astrobiology Strategy*, 2015. Accessed March 20, 2023. https://astrobiology.nasa.gov/nai/media/medialibrary/2015/10/NASA_Astrobiology_Strategy_2015_151008.pdf

NASA Astrobiology. Accessed March 20, 2023. https://astrobiology.nasa.gov

NASA Exoplanet Archive. Accessed March 20, 2023. https://exoplanetarchive.ipac.caltech.edu/docs/counts_detail.html.

Niebühr, R. *The Nature and Destiny of Man*. Louisville: Westminster John Knox Press, 2021.

Norenzayan, A. *Big Gods: How Religion Transformed Cooperation and Conflict*. Princeton: Princeton University Press, 2013.

Olsen, R. Why I Am Not a Process Theologian, 2013. Accessed March 20, 2023. https://www.patheos.com/blogs/rogereolson/2013/12/why-i-am-not-a-process-theologian/

O'Meara, T. F. O. *Vast Universe: Extraterrestrials and Christian Revelation*. Collegeville, MN: Liturgical Press, 2012.

Pagels, H. *Perfect Symmetry: The Search for the Beginning of Time*. New York: Simon & Schuster, 1985.

Palmeri, J. "Bringing Cosmos to Culture: Harlow Shapley and the Uses of Cosmic Evolution," In *Cosmos & Culture: Cultural Evolution in a Cosmic Context*, edited by S. J. Dick and M. Lupisella, 489–521. Washington, DC: NASA, 2009.

Parkyn, J. L. *Exotheology: Theological Explorations of Intelligent Extraterrestrial Life*. Eugene: Pickwick Publications, 2021.

Pasulka, D. W. *American Cosmic: UFOs, Religion, Technology*. Oxford: Oxford University Press, 2019.

Peacocke, A. "The Challenge and Stimulus of the Epic of Evolution to Theology." In *Many Worlds: The New Universe, Extraterrestrial Life and the Theological Implications*, edited by S. J. Dick, 89–117. Philadelphia: Templeton Foundation Press, 2000.

Peters, T. "Exo-Theology: Speculations on Extraterrestrial Life." In *The Gods Have Landed: New Religions from Other Worlds*, edited by J. R. Lewis, 187–206. Albany: SUNY Press, 1995.

———. "Exotheology: Speculations on Extraterrestrial Life." In *Science, Theology and* Ethics, 121–36. New York: Routledge, 2003.

———. "Astrotheology: A Constructive Proposal," *Zygon* 49, no. 2 (2014): 443–57. https://doi.org/10.1111/zygo.12094.

———. *Astrotheology: Science and Theology Meet Extraterrestrial Life*. Eugene: Cascade Books, 2018.

Randolph, R. O. and C. P. McKay. "Protecting and Expanding the Richness and Diversity of Life, an Ethic for Astrobiology Research and Space Exploration." *International Journal of Astrobiology* 13, no. 1 (2014): 28–34. https://doi.org/10.1017/S1473550413000311.

Rees, M. *Just Six Numbers: The Deep Forces that Shape the Universe*. New York: Basic Books, 1997.

———. *Our Cosmic Habitat*. Princeton, NJ: Princeton University Press, 2001.

———. "Why Alien Life Will be Robotic." *Nautilus* (2015). Accessed March 20, 2023. https://nautil.us/why-alien-life-will-be-robotic-235661/.

Ross, H. *The Creator and the Cosmos: How the Greatest Scientific Discoveries of the Century Reveal God*. Colorado Springs: NavPress, 1993.

Rubenstein, R. E. *When Jesus Became God: The Struggle to Define Christianity During the Last Days of Rome*. New York: Houghton Mifflin Harcourt, 2018.

Russell, R. J. 2018. "Many Incarnations or One?" In *Astrotheology: Science and Theology Meet Extraterrestrial Life*, edited by T. Peters, 303–16. Eugene: Cascade Books, 2018.

Sagan, C. *The Dragons of Eden: Speculations on the Evolution of Human Intelligence*, New York: Random House, 1977.

———. *Cosmos: A Personal Voyage*, PBS Dec 14, 1980.

Samuelson, N. "Jewish Theology Meets the Alien." In *Astrotheology: Science and Theology Meet Extraterrestrial Life*, edited by T. Peters, 208–15. Eugene: Cascade Books, 2018.

Scharf, C. *The Copernicus Complex: Our Cosmic Significance in a Universe of Planets and Probabilities*. New York: Farrar, Straus & Giroux, 2014.

Schneider, S. "Alien Minds." In *The Impact of Discovering Life Beyond* Earth, edited by S. J. Dick, 189–206. Cambridge: Cambridge University Press, 2015.

Shapley, H. *Of Stars and Men: The Human Response to an Expanding Universe*. Boston, Beacon Press, 1958.

Shostak, S. *Sharing the Universe: Perspectives on Extraterrestrial Life*. Berkeley: Berkeley Hills Book, 1998.

Smith. H. "John Shelby Spong, Liberal Episcopal Bishop and LGBTQ Advocate, Dies at 90." *Washington Post*, September 16, 2021. Accessed March 20, 2023. https://www.washingtonpost.com/local/obituaries/john-shelby-spong-dead/2021/09/14/f6e145a0-1566-11ec-9589-31ac3173c2e5_story.html

Smith, K., and C. Mariscal. *Social and Conceptual Issues in Astrobiology*. Oxford: Oxford University Press, 2020.

Smith, R. *The Expanding Universe: Astronomy's Great Debate, 1900–1931*. Cambridge: Cambridge University Press, 1982.

Smolin, L. 2007. "Scientific Alternatives to the Anthropic Principle." In *Universe or Multiverse?*, edited by B. Carr, 323–66. Cambridge: Cambridge University Press, 2007.

Stenger, V. J. *The Fallacy of Fine-Tuning: Why the Universe Is Not Fine-Tuned for Us*, New York: Prometheus Books, 2011.

Stone, J. A. *Religious Naturalism Today: The Rebirth of a Forgotten Alternative*. Albany: SUNY Press, 2010.

Sullivan, W. T., III, and J. A. Baross, eds. *Planets and Life: The Emerging Science of Astrobiology*. Cambridge: Cambridge University Press, 2007.

Swimme, B., and T. Berry. *The Universe Story*. New York: HarperCollins, 1992.

Tasker, E. *Planet Factory: Exoplanets and the Search for a Second Earth*. London and New York: Bloomsbury, 2017.

Teilhard de Chardin, P. *The Phenomenon of Man*. New York, HarperCollins, 2002.

Torrey, E. F. *Evolving Brains, Emerging Gods: Early humans and the Origins of Religion*. New York: Columbia University Press, 2017.

United States, Director of National Intelligence. Preliminary Assessment: Unidentified Aerial Phenomena, 2021. Accessed March 20, 2023. https://www.dni.gov/files/ODNI/documents/assessments/Prelimary-Assessment-UAP-20210625.pdf

Vainio, O. P. *Cosmology in Theological Perspective: Understanding Our Place in the Universe*. Grand Rapids: Baker Academic, 2018.

Vidal, C. *The Beginning and the End: The Meaning of Life in a Cosmological Perspective*. Heidelberg : Springer, 2014.

Viney, D. "Process Theism." *The Stanford Encyclopedia of* Philosophy, edited by Edward N. Zalta, 2020. Accessed March 20, 2023. https://plato.stanford.edu/archives/sum2020/entries/process-theism/.

Wallace, A. R. *Man's Place in the Universe: A Study of the Results of Scientific Research in Relation to the Unity or Plurality of Worlds*. New York: McClure, Phillips & Co., 1904.

Ward, P., and D. Brownlee. *Rare Earth: Why Complex Life is Uncommon in the Universe*. New York: Copernicus, 2000.

Weintraub, D. *Religions and Extraterrestrial Life: How Will We Deal with It? Cham:* Springer, 2014.

Wilson, D. S. *Darwin's Cathedral: Evolution, Religion and the Nature of Society*. Chicago: University of Chicago Press, 2002.
Zakariya, N. *A Final Story: Science, Myth and Beginnings*. Chicago: University of Chicago Press, 2017.

Chapter 9

From Negation to Exemplification

A Deeper Whiteheadian Cosmotheology

Andrew M. Davis

For over two decades, Steven J. Dick has been a laudable advocate for the development of "cosmotheology" as a robust endeavor employing the best of our scientific knowledge to widen theological conversation and radically transform theological conviction.[1] As Dick understands it, cosmotheology revolves around the use of our ever-expanding knowledge of the universe to "modify, expand or change entirely" current theologies.[2] As such, cosmotheology aims to be a theology that "takes into account what we know about the universe based on science."[3] Central to these theological changes, Dick insists, is the rejection of anthropocentrism and supernaturalism: human beings are relegated to the cosmological periphery and there is nothing beyond the natural world. These commitments converge in Dick's affirmation of both scientific and religious naturalism: nature is all there is and a creative evolutionary universe devoid of "God" suffices to inspire the reverence, awe, and meaning traditionally assigned to the religious sphere. Dick's six core principles of cosmotheology thus consist primarily in a series of *cosmological negations* of human existence and experience and ultimately, an *ontological negation* of God. While admitting that his proposal is perhaps better termed "cosmophilosophy" rather than "cosmotheology," Dick nevertheless leaves open the possibility that advanced extraterrestrials in the "biological universe" may have something akin to the traditional theological attributes. As seen in the previous chapter, Dick has newly considered the attractiveness of the "superintelligence hypothesis" as a possible cosmotheological alternative to traditional theological conviction.

My goal in this chapter is to argue that Dick's admirable project of cosmotheology can be considerably deepened when inversely approached through Whitehead's organic metaphysics. Endeavoring this requires showing the ways in which the core *cosmological negations* at the heart of Dick's cosmotheological principles can be inverted into *metaphysical exemplifications*. For Whitehead, human experience, while certainly cosmologically peripheral, is nevertheless *metaphysically central*. Human existence and experience exemplify fundamental metaphysical principles of an organic, experiential, and aesthetic character that are essential to a creative evolutionary universe at all scales. Rather than being the great supernatural "exception" to metaphysical conditions, "invoked to save their collapse," Whitehead's cosmotheology affirms God as their "chief exemplification." Whitehead's deeper naturalism coupled with his theological realism allow for imaginative metaphysical continuity in our reflection all the way "down" and all the way "up" the biological universe: from terrestrial and extraterrestrial life to the cosmic life of God. What is more, Whitehead's cosmotheology reintegrates and resolves outstanding metaphysical problems surrounding necessity and contingency, possibility and actuality, objective value, truth, and meaning, that arguably remain philosophically unsubstantiated in Dick's anti-realist cosmotheology. It will be my conclusion that Whitehead therefore offers a *deeper* cosmotheology that includes Dick's naturalistic intuitions while also transcending his conceptual and explanatory limitations. In seeking to be constructive rather than deconstructive in this chapter, I want to emphasize my respect and appreciation of Dick's contributions in nevertheless moving toward a deeper Whiteheadian cosmotheology.[4]

SCIENCE, METAPHYSICS, AND THEOLOGY

Dick has stated that the affirmation of evolutionary becoming at the heart of his cosmotheology "resonates with Alfred North Whitehead's process theology."[5] Yet one cannot adequately grasp the various contours of Whitehead's process cosmotheology without due attention to its essential weddedness to respective scientific and metaphysical endeavors. Indeed, Whitehead's process cosmotheology cannot ultimately be abstracted from his long-standing engagement with (and critique of) the philosophical, theological, and scientific tradition. For Whitehead, each of these domains of human striving are independently important but also necessarily tangled; they need—even *require*—each other to be more authentically themselves. Whitehead thus states, "You cannot shelter theology from science, or science from theology; nor can you shelter either of them from metaphysics, or metaphysics from either of them. There is no shortcut to truth."[6]

In saying that his cosmotheology "is simply a theology that takes into account what we know about the universe based on science,"[7] Dick is clearly affirming the first of Whitehead's convictions: "You cannot shelter theology from science." On reading Dick, however, it is less clear that his cosmotheology is positively informed by the theological tradition, and less clear still that it is positively informed by a metaphysical scheme that adequately supports its deepest convictions. To the extent that Dick's scientific and religious "naturalism" essentially correlates the positive claim "Nature is all there is" with the negative claim "There is no ontologically distinct and superior realm including God," he is of course affirming a limited metaphysic: "the supernatural does not exist."[8] I say *limited* because these basic affirmations leave the *depths* of ontology and metaphysics still unarticulated. What kinds of things populate the furniture of reality? What exists and with what kind of character? What kinds or modes of existence are there, and how do they relate? Put simply, holding that nature is all there is does not tell us *what nature is*, what nature *consists in*, or, for that matter, what is *ultimately natural*. A more robust affirmation of "naturalism" would require this kind of metaphysical investigation.[9]

This is not to say that Dick is unaware of the importance of these base-level ontological questions or of the need to develop a philosophy of astrobiology, astronomy and cosmology; but it does remain unclear how Dick himself might navigate these fundamental questions, and in turn, how this might inform or re-form the development of his naturalistic cosmotheology. By enacting a "call for philosophers, historians, scientists and science studies scholars to join in this new endeavor," Dick realizes the necessity of this wider philosophical effort; and to the extent that "radically new conceptions of God" are required by the biological universe, Dick also calls *theologians* to take part in this important task.[10]

It is on this level of inquiry that I believe Whitehead's metaphysical scheme can aid the deepening of Dick's naturalistic cosmotheology. For as much as Dick's cosmotheology is informed by "science," he knows full well that "all scientists" harbor "metaphysical preconceptions before they begin their empirical and theoretical work." "Science and philosophy" he rightly states "are inseparable, especially in the form of epistemology and metaphysics."[11] As to *which* metaphysics or epistemology, however, is an open question with a diversity of implications for cosmotheology. Whitehead too reminds us that "Science does not diminish the need of a metaphysic. . . . Science only renders the metaphysical need more urgent."[12] What is more, he offers a stark warning: "If science is not to degenerate into a medley of *ad hoc* hypotheses, it must become philosophical and must enter upon a thorough criticism of its own foundations."[13]

Science, however, is by no means alone in requiring metaphysical justification in this regard. Whitehead thoroughly applies the same criticism to the endeavors of religion and history alike.[14] For Whitehead, it remains clear: our tangled disciplines cannot ultimately outrun their philosophical shadows because the Sun of metaphysics casts its light upon all human efforts. It remains too bright to observe plainly, but its warmth is theoretically presupposed even if visually or verbally denied. After all, even the blind can feel the warmth of the sun. My constructive dialogue with the core principles and dimensions of Dick's cosmotheology is thus grounded in my specific attempt to offer the beginnings of a Whiteheadian metaphysical foundation for true cosmo*theological* affirmation. This affirmation, moreover, also seeks to establish Whitehead as an essential (although currently neglected) resource for contemporary discussions of philosophy, theology, and extraterrestrial life.[15]

DISTINGUISHING THE COSMOLOGICAL AND METAPHYSICAL ENDEAVORS

If my interest in this chapter is to deepen Dick's naturalistic cosmotheology through Whitehead's philosophy, it is important to first establish some basic distinctions between *cosmology* and *metaphysics* as Whitehead understands them. This is not meant to be exhaustive, but helpful and orienting. Cosmology and metaphysics are not clearly distinguished in Dick's approach to cosmotheology, but they are central to Whitehead's philosophy and to various contours of this discussion.

Although Whitehead is rightly seen as a great metaphysician, his principal concern was cosmology. It is often overlooked that his magnum opus *Process and Reality* is followed by the subtitle: *An Essay in Cosmology*. Whitehead however was aware that, while distinct from each other, the efforts of cosmology and metaphysics are nevertheless deeply related. In principle, they each exhibit essential domains of contingency and necessity, particularity and generality, order and presuppositions of order that are arguably indispensable to any approach to the evolutionary nature of things, whether terrestrial, extraterrestrial, or divine. Below, I look briefly to how Whitehead defines and distinguishes cosmology and metaphysics, a distinction that will be central to my subsequent inversions of Dick's cosmotheological principles.

Contours of Evolutionary Cosmology: Contingency, Order, Novelty

Cosmology as Whitehead understands it "sets out to be the general system of general ideas applicable to this epoch of the universe."[16] Stated alternatively, "Cosmology is the effort to frame a scheme of the general character of the present stage of the universe."[17] What I want to stress in these two definitions is the quest for those notions relevant to *our particularity*; that is, cosmology consists in the effort to formulate a system of ideas applicable not to *any and all* epochs of the universe, but to the *particular* epoch in which we find ourselves, and the manifold evolutionary contingencies associated with it, be they the strangely limited dimensions of the spatiotemporal continuum, the so called "laws of nature" or various other domains of contingent formational order. As Whitehead states,

> There are the four dimensions of the spatio-temporal continuum, the geometrical axioms, even the mere dimensional character of the continuum—apart from the particular number of dimensions—and the fact of measurability . . . these properties are additional to the more basic fact of extensiveness; also, that even extensiveness allows for grades of specialization, arbitrarily one way or another, antecedently to the introduction of any of these additional notions.[18]

Whitehead repeatedly stresses that the three dimensions of space and the four dimensions of the spatiotemporal continuum "present themselves as arbitrary in respect to a more abstract possibility."[19] Such odd cosmological limitations "warn us that we are in a special cosmic epoch," he states.[20] This is not a claim of anthropocentrism or terracentrism (as we will later see), but simply the realization that our epoch "illustrates one special physical type of order" when it could have otherwise been entirely different. The "absurdly limited number of three dimensions of space" Whitehead repeats, are none other than "a sign that you have got something characteristic of a special order."[21] In addition to the contingent dimensional character of our epoch, Whitehead points to various measurable regularities of nature as exhibiting not the immutable necessity of "law," but rather the habitual and procedural character of a truly evolutionary cosmos.

> Now the simplest things are the widespread habits of nature that dominate the whole stretch of the universe within our remotest, vaguest observation. None of these laws of nature gives the slightest evidence of necessity. They are modes of procedure which within the scale of our observations do in fact prevail. I mean, the fact that the extensiveness of the universe is dimensional, the fact that the number of spatial dimensions is three, the spatial laws of geometry, the ultimate formulae for physical occurrences. There is no necessity in any of these ways of behaviour. They exist as average, regulative conditions."[22]

Whitehead thus questions: "Why talk about the 'laws of nature' when what we mean is the characteristic behaviour of phenomena within certain limits at a given stage of development in a given epoch—so far as these can be ascertained?"[23] Contrary to the static permanence of previous cosmologies, he speaks to the primacy of cosmic evolution belonging to his own cosmological vision: "On a grand scale, our cosmology discloses a process of overpowering change, from nebulae to stars, from stars to planets, from inorganic matter to life, from life to reason and moral responsibility." These evolutionary disclosures, moreover, are utterly significant for the future of cosmology: "We can no longer conceive of existence under the metaphor of a permanent depth of ocean with its surface faintly troubled by transient waves. There is an urge in things which carries the world far beyond its ancient conditions."[24]

Whitehead wrote prior the advent of Big Bang cosmology and followed the immanent and evolutionary intuitions of Plato and Aristotle, rather than the transcendent and supernatural requirements of Newton.[25] For Plato and Aristotle, cosmic "creation" refers not to the "the beginning of finite matter of fact," but to the "incoming of a type of order establishing a cosmic epoch." The formation of our cosmos was the "incoming of a certain type of social order."[26] Contrary to common preoccupations with absolute origins, Whitehead's cosmological wonder was directed not to some mythical "beginning" to cosmic evolution (a notion itself problematic), but rather to the emergent forms of contingent order and novelty belonging to our particular epoch. "We have to explain the aim at forms of order, and the aim at novelty of order, and the measure of success or failure," he states.[27] It is these core features which lie at the explanatory depths of Whitehead's evolutionary cosmology.

Process philosophers often emphasize that Whitehead would not necessarily have denied "Big Bang" origins, but he would have situated such an event within a far vaster multiplicity of evolutionary epochs where no absolute beginning can be found. For Whitehead our particular "cosmic epoch" emerged and will pass away, only to be followed by other cosmic epochs. The universe is "thus passing with a slowness, inconceivable in our measures of time, to new creative conditions." He states, "The present type of order in the world has arisen from an unimaginable past, and it will find its grave in an unimaginable future."[28] The various layers and effects of our existing type of order are surely "passing away into trivialities," yet Whitehead offers an important clarification:

> That does not mean that there are not some other types of order of which you and I have not the faintest notion, unless perchance they are to be found in our highest mentality and unperceived by us in their true relevance to the future. The universe is laying the foundation of a new, type, where our present theories of order will appear as trivial. If remembered, they would be remembered or discerned in the future as trivialities, gradually fading into nothingness. This is the only possible doctrine of a universe always driving on to novelty.[29]

Whether in the physical universe or the myriads of minds that may inhabit it, Whitehead's evolutionary cosmology is necessarily open to forms of physical and mental order that are beyond our ability to conceive. We will return to this later in the context of Dick's cosmotheological principles and the possibility of the "superintelligence hypothesis." To summarize for now: Whitehead highlights the radical contingency, developmental order and emergent novelty that are part and parcel of our particular evolutionary epoch.

Contours of Metaphysical Necessity: Experience, Exemplification, Extension

Using the language of "speculative philosophy," Whitehead defines metaphysics as "the endeavor to frame a coherent, logical, necessary system of general ideas in terms of which every element of our experience can be interpreted."[30] Alternatively, "By metaphysics" he states, "I mean the science which seeks to discover the general ideas which are indispensably relevant to the analysis of everything that happens."[31] Different still, metaphysics "endeavors to express the most general concepts adequate for the all-inclusive universe," with the "primary metaphysical data" being the world of which we are "immediately conscious."[32]

These different formulations hang together in their emphasis on attempting to identify the necessary principles or conditions that underlie and are exemplified in our experience as expressions of a particularly ordered evolutionary cosmos—an endeavor that is at once rationalist, empiricist, and speculative.[33] These principles "can never fail of exemplification," Whitehead states; "We can never catch the actual world taking a holiday from their sway." Every detailed occurrence in the cosmos must "exhibit the same texture" such that "the philosophical scheme should be 'necessary,' in the sense of bearing in itself its own warrant of universality throughout all experience."[34]

When Whitehead engages the origins of modern science, he stresses that the "full scientific mentality" was not divorced from metaphysical intuitions. "Geniuses such as Aristotle, or Archimedes, or Roger Bacon . . . instinctively [held] that all things great and small are conceivable as exemplifications of general principles which reign throughout the natural order." What Whitehead calls our "instinctive faith" in a cosmological order of nature, should therefore be coupled with the still-deeper faith that this order exemplifies necessary principles of metaphysics.[35] "That we fail to find in experience any elements intrinsically incapable of exhibition as examples of general theory is the hope of rationalism" Whitehead states. "It is the faith which forms the motive for the pursuit of all sciences alike, including metaphysics."[36]

These are strong claims that will naturally provoke skepticism in our current philosophical climate. It is important to stress, therefore, Whitehead's thoroughgoing humility in making them (and my own by endeavoring this discussion). This project must ultimately fail. "Philosophers can never hope to finally formulate these first principles," Whitehead admits. "Weakness of insight and deficiencies of language stand in the way inexorably."[37] Metaphysical statements are not to be understood as "dogmatic statements of the obvious"; rather, they are tentative and ever-revisable "formulations of the ultimate generalities [of nature]."[38] Contrary to common critiques of metaphysics, Whitehead explicitly rejects the notion "that metaphysical thought [start] from principles which [are] clear, distinct, and certain."[39] Metaphysics should "never start from systematization"; rather, "Its primary state can be termed assemblage."[40] What is first "assembled" for Whitehead is the welter of human experience in and as nature, and as interpreted through the various human disciplines, whether science, philosophy, history or religion. Only when a tentative scheme of ideas has been assembled and informed in this way, can we then imaginatively extend these ideas to the wider stretches of the cosmos.

Cosmology and metaphysics, therefore, are distinct, but deeply related endeavors for Whitehead. Whereas cosmology deals with the contingent evolutionary order, "laws" and regularities belonging to our particular cosmic epoch, metaphysics attempts to formulate those tentative, but (in principle) necessary notions which underlie and are always exemplified through them.[41] Of utmost significance in this regard is attention to human experience as *a part of* and not *apart from* the cosmic evolutionary order. As we proceed further, we will encounter these tentative metaphysical notions as Whitehead understands them. Let it be said again, however, that Whitehead fully recognizes the ambitiousness of metaphysics which always faces a final failure: "How shallow, puny, and imperfect are efforts to sound the depths in the nature of things. In philosophical discussion, the merest hint of dogmatic certainty as to finality of statement is an exhibition of folly."[42]

INVERTING DICK'S COSMOTHEOLOGICAL PRINCIPLES

Dick is of course right to point to an essential resonance between Whitehead's cosmic evolutionary vision and his own affirmation of cosmic evolution as the presupposed backdrop of cosmotheology. Cosmic evolution, he states, "is the broadest possible process, the broadest possible context for becoming, the master narrative of the universe, a story that includes us as its most recent manifestation, although not its ultimate endpoint."[43] Whitehead fully agrees: we are exemplifications of, and not exceptions to, the cosmic evolutionary process. The difference, however, is that what Dick affirms *cosmologically* in this regard, Whitehead substantiates *metaphysically*. In drawing the above distinctions between the contingencies of cosmological conviction and their weddedness to necessary notions of metaphysical exemplification, we are now in position to endeavor an inversive consideration of Dick's principles for the development of a deeper Whiteheadian cosmotheology.

Dick has stressed that his naturalistic cosmotheology starts from considerations of the "basic principles of the relation of humans to the universe."[44] As shown above, however, there are both *cosmological* and *metaphysical* layers at play in this "relation." That Dick only reads this "relation" *cosmologically* is shown by the fact that his first three principles of cosmotheology consist primarily in *cosmological negations* of the place of human beings in nature. The second three principles then note what cosmotheology "must" affirm for Dick in light of these negations. From a Whiteheadian framework, these cosmological negations are not wrong (as we will see), but they are far from the whole story, being at best what Whitehead might call "half-truths."[45] My inversive approach to Dick's principles therefore involves showing how his cosmological negations can instead be read as *metaphysical exemplifications* with important implications for the deepening of cosmotheology.

With respect to Dick's first three principles, I will represent this approach in the following way: *"Humanity is not"* → "Humanity exemplifies." My engagement with the last three of Dick's principles will then seek to metaphysically substantiate what arguably remains unsubstantiated in his treatment of what *"Cosmotheology must . . . "* affirm. This inversive approach to the six principles of Dick's antirealist cosmotheology will thus be progressively mirrored by six Whiteheadian principles of cosmotheological realism.

Inversion 1: *"Humanity is not physically central to the universe."* → "Humanity exemplifies metaphysical principles that are utterly central to the universe."

The first of Dick's cosmotheological principles follows upon the revolutionary heals of a series of compounding cosmological de-centerings, from Copernicus to Shapley to Hubble. We now know that the Earth is not the center of a geocentric universe, but only a small part of a marginal heliocentric solar system in the Milky Way Galaxy. Our galaxy is but one of billions of galaxies, each with billions of suns and planets with no known end. These extraordinary discoveries naturally invoke awe, humility, and cosmological wonder at the infinite scales of the universe.

Whitehead fully acknowledges the reality of these cosmological de-centerings. "The human race consists of a small group of animals which for a small time has barely differentiated itself from the mass of animal life on a small planet circling round a small sun," he states; "The universe is vast."[46] What is more, he fully agrees with Dick that religion and theology ignore these discoveries to their own peril and irrelevance. The "principle of mediocrity" is a fact and we have only begun to engage these de-centerings in our philosophical and theological disciplines.

> Any system of thought based on this earth of [ours] . . . is extremely limited in its conceptions—either theology or philosophy—and most of them have been. We know now that our earth is an insignificant planet swinging round a second-rate sun in no very important part of the universe. The response to that knowledge of first-rate people talking together . . . should be immeasurably larger than it is.[47]

Truly, any form of anthropocentrism or terracentrism dissolve against the backdrop of these cosmological de-centerings. Despite these utter contingencies of our cosmological existence, however, Whitehead nevertheless insists upon our internal metaphysical continuity with everything that exists, regardless of cosmological location, size, or position. We in fact must say this because cosmological and biological evolution presuppose definitively that we have not dropped into the cosmos from elsewhere, but have grown out of it like a flower out of soil.

The analogy is apt: The flower is intrinsically connected to the nature of soil from which it has grown and we are intrinsically connected to the nature of the cosmos out of which we have grown. This is an organic analogy of "growth," but it is not *just* an analogy as we will see in considering Dick's second principle. What this means for Whitehead is that our existence and experience exemplify to some significant degree the *same* metaphysical depths that apply across vast grades of evolutionary complexity and intensity in the universe. We should not miss the importance of this claim. If we have not yet begun to feel the impact our cosmological periphery, as Whitehead suggests above, we have also not yet begun to fully appreciate that fact that we are *as fully natural as everything else in the cosmos*.

Whitehead is adamant, however, that the significance of this fact has not been adequately acknowledged by scientific and philosophical endeavors whose assumptions about the relation of human beings toward nature have been steeped in vices of bifurcation and abstraction.[48] Put simply: they have too often remained content to move from the wider world of nature to human experience, but *not* from human experience to the wider world of nature. In drawing this out, Whitehead stresses the need to finally interpret the world in light of what is given in the depths of human experience. Indeed, one of the purposes of philosophy for Whitehead is to be a "critic of abstractions" in precisely this way.[49] Consider his progressive conclusion as it emerges in the following quotations from different phases of his work.

> It is the accepted doctrine in physical science that a living body is to be interpreted according to other sections of the physical universe. This is a sound axiom; but it is double-edged. For it carries with it the converse deduction that other sections of the universe are to be interpreted in accordance with what we know of the human body.[50]

> Any doctrine which refuses to place human experience outside nature, must find in descriptions of human experience factors which also enter into the descriptions of less specialized natural occurrences. If there be no such factors, then the doctrine of human experience as a fact within nature is mere bluff, founded upon vague phrases whose sole merit is a comforting familiarity. . . . We should point out the identical elements connecting human experience with physical science.[51]

> The body is that portion of nature with which each moment of human experience intimately cooperates. There is an inflow and outflow of factors between bodily actuality and the human experience, so that each shares in the existence of the other. The human body provides our closest experience of the interplay of actualities in nature . . . thus we finally construe the world in terms of the type of activities disclosed in our intimate experience.[52]

Contrary to common scientific and philosophical abstractions, there is no gulf that separates human bodies and minds from the rest of the world. Contrary to religious or theological abstractions, neither is there a supernatural gulf that demarcates humanity as a "special creation," fully different in kind from every other creature. For Whitehead we are not physically, metaphysically, or theologically discontinuous with the world; rather, *we are the world in intimate experiential form*. Just as theology and religion ignore our cosmological standing to their own peril and irrelevance, so too will science and philosophy die under heaps of abstraction and bifurcation. For Whitehead, there is no approach to the "objective" world without passing through the subjectivity of human experience *as world*. Inverting Kant's own "Copernican revolution," Whitehead holds that the experiential subject "emerges from the world" rather than that "the world emerges from the subject." Human experience is reestablished as *outcome* of the real-world process and thus, as *revelatory* of its deepest nature.[53]

For Whitehead then, the fact of our *cosmological de-centering* should be balanced by the necessity of our *metaphysical re-centering* as authentic expressions of the nature and character of the world at its most fundamental levels. It is this conviction of *metaphysical continuity* that relativizes attention to our minute size and physical location in the universe. On Whitehead's account, although we remain very small, our experience is actually very large because it expresses the same metaphysical principles that reign throughout the cosmos in various hierarchies of organization, complexity, and intensity.

Although our experience is immanent to ourselves, it also transcends us because it is connected to the entire universe as a moving whole. Whitehead, moreover, fully recognizes that to begin approaching human experience as *an exemplification* rather than *an exception* in this way, is to begin to marvel at the fact that "in being ourselves we are more than ourselves." In no small way, it is to begin to discover a real sense of *belonging* in the universe because our "experience, dim and fragmentary as it is yet sounds the utmost depths of reality."[54]

These *depths* are not merely physical or cosmological in nature, but *metaphysical*. They are the same depths that are exemplified both physically and cosmologically throughout the entire cosmos. Dick's fist principle of cosmotheology remains fully sound and is affirmed, but on Whitehead's inversive reading, the fact of our cosmological periphery, must be seen in light of our experience as *metaphysically exemplary*. Turning now to Dick's second principle of cosmotheology, we will begin to grasp some of the metaphysical depths belonging to human experience as Whitehead interprets it.

Inversion 2: *"Humanity is not central biologically, mentally or morally in the universe."* → "Humanity exemplifies biological, mental and moral antecedents that are metaphysically central to the universe."

It's important for Dick to qualify his second principle in terms of "probability." The fact that intelligent life and mind have not yet been found in the universe should deter us from making overly confident assertions about our merely peripheral biological, mental, and moral standing in the universe. However improbable, there remains the possibility that life as we know it really is a unique and "lonely" phenomenon on our planet alone.[55] Even if this were the case, however, we would *still* have to wrestle with the antecedent metaphysical conditions of biological, mental, and moral phenomena. The lesson of the Copernican principle must also apply to claims that life is ultimately lonesome. If "the general principle of uniformity of nature's laws has held with respect to the existence of exoplanets," Dick states, "and the expectation is confirmed that what is true of Earth is true of other places in the universe" then a similar uniformity can plausibly apply to the diversity of life in the universe. "Although this has yet to be proven in the case of life," Dick continues, "the trend is clear, despite a great diversity of planetary systems found. No one would expect a solar system exactly like ours; as with life, so with planets: diversity is the coin of the realm."[56]

This correlation between planetary and biological uniformity and diversity is helpful and can be deepened in ways presupposing *metaphysical uniformity* across nature: "from nebulae to stars, from stars to planets, from inorganic matter to life, from life to reason and moral responsibility."[57] Consider Whitehead's radical openness to the diversity of life beyond earth.

> The forces that are constantly playing in upon our minds and bodies are inconceivably innumerable—for instance, rays from some star millions of light-years away from us—forces as fantastic as that . . . and the forms of life which might be lived on other stars millions of light-years away and million hence could be infinite and admit every possibility that the imagination could conceive.[58]

For Whitehead, we cannot ultimately separate claims of our planetary uniformity from the higher biological, mental, and purposive grades of uniformity that have emerged on our planet. Just as our planet is but one expression of a multitude of planets diversely expressing various grades of uniformity (including urablity and habitability[59]), so too are the various grades of biological, mental, and purposive uniformity found on earth also found *beyond earth* in myriad forms. One can coherently make this claim only by assuming a deeper *metaphysical uniformity* among the antecedent conditions of life, mind, and morality in the universe. For Whitehead, this is what any affirmation of necessary metaphysical principles requires.

What then for Whitehead are the metaphysical antecedents of the biological, mental, and moral dimensions expressed in Dick's second principle? He expresses the metaphysical antecedents of the biological domain in terms of *organism*; the metaphysical antecedents of the mental domain in terms of *creative experience*; and the metaphysical antecedents of the moral domain in terms of *aesthetic attainment*.

Organism

The emergence of Whitehead's "philosophy of organism" was a direct result of the collapse of mechanistic materialism with its assumption of a "senseless, valueless, purposeless" matter as the ultimate metaphysical substrate.[60] Matter on this metaphysical scheme "just does what it does do, following a routine imposed by external relations which do not spring from the nature of its being."[61] What is more, "the transition of time has nothing to do with the character of the material. The material is equally itself at an instant of time."[62] As Whitehead expresses it, Cartesian-Newtonian physics was based upon the static "individuality of each bit of matter," yet "bit by bit" this concept was "given away, or dissolved, by the advance of modern physics."[63] In no small way, Whitehead's ontology emerges in light of the fact that "matter" as inert billiard-ball-like *substances* had melted into something far more fluid, organic, and experiential. "Science is taking on a new aspect which is neither purely physical, nor purely biological," Whitehead states. "It is becoming the study of organisms."[64] The world was unveiled not as a vast machine, but as a layered multiplicity of organisms *within* organisms across various scales of evolutionary growth, emergence, and complexity throughout nature.

In many ways, the category of organism is ripe. We assume that "organisms" may be "living" or "nonliving," but "nonliving" does not mean *dead.* For Whitehead there is "no absolute gap between 'living' and 'non-living'" organic systems.[65] It's no wonder that the senseless, valueless, purposeless, and immobile matter of mechanistic materialism suffers as an explanatory metaphysics. *Ex nihilo nihil fit.* Such an ontology of death cannot coherently explain the evolution of life, mind, and value.[66] Rather, their antecedents must be found in an organic ontology where evolutionary growth is *primary.* "The whole point of the modern doctrine [of evolution] is the evolution of complex organisms from antecedent states of less complex organisms," Whitehead states. "The doctrine thus cries aloud for a conception of organism as fundamental for nature."[67] Mechanistic materialism was shown to be an inhospitable metaphysical context for comprehending the workings of real evolution, and its collapse came with opportunity: "The field is now open for the introduction of some new doctrine of organism which may take the place of the materialism with which, since the seventeenth century, science has saddled philosophy."[68]

Prior to the emergence of what we recognize as "living" organisms, there are still-more-fundamental organisms which exhibit their share of active evolution, dynamic response, and purposive internal relations to their environment. As Darek Malone-France has argued in this volume, the category of "organism" in this respect is *deeper* than "life" and carries wider implications of applicability to the entire universe: "The more general category of 'organism,' as constructed in Whitehead's thought, is not only a better candidate than 'life' for universalizability, but also . . . has the additional advantage of reflecting the ultimate convergence (non-dualism) of both 'physical' (nonbiological) and 'biological' phenomena, under the rubric of a more generic ontological category."[69] For Whitehead, the evolution of life on our planet presupposes a more fundamental *ontology of organic evolution.*

Creative Experience

Once you get rid of matter and the illusion of its vacuous "undifferentiated endurance," there is no good reason "to provide another more subtle stuff" to take its place, Whitehead insists.[70] Matter essentially dissolved into energetic activity and Whitehead found this deeply suggestive. For Whitehead, even the physicist's concept of "energy" is an abstraction from a more fundamental level of activity. What is it that is *internal to* or *constitutive of* vibrating energetic activity at the base of the quantum world? Whitehead's answer: energy is an abstraction from the creative becoming of events—what he calls "actual occasions of experience." "'Actual entities'—also termed 'actual occasions,'" Whitehead states, "are the final real things of which the world is made up. There is no going behind actual entities to find something more real . . . the final facts are, all alike, actual entities: and these actual entities are drops of experience, complex and interdependent."[71] The most primitive organisms of Whitehead's ontology are emergent "drops of experience" or what he elsewhere calls experiential "pulses of emotion."[72] Here we recall Whitehead's statement above that the world be understood "in terms of the type of activities disclosed in our intimate experience." These activities are those of *creative experience and feeling* and they are never wholly absent from nature even at its most fundamental levels.

As the "really real things which in their collective unity compose the evolving universe,"[73] actual entities are not static, but on the move; they are temporal verb-like instants that grow into being both individually and collectively as embodiments of the ultimacy of Creative Evolution—what Whitehead simply terms "Creativity." It is essential to not miss how fundamental the category of Creativity is for Whitehead. He is adamant that "in all philosophic theory there is an ultimate which is actual in virtue of its accidents. It is only then capable of characterization through its accidental embodiments, and apart from these accidents is devoid of actuality." For Whitehead, "Creativity" is this ultimate and it replaces Aristotle's notion of "primary substance."[74] Creativity for Whitehead is the "universal of universals" which characterizes all matter of fact; it is the "ultimate principle by which the many [actual occasions], which are the universe disjunctively, become the one actual occasion, which is the universe conjunctively." "The many become one, and are increased by one," he states; "entities are disjunctively 'many' in process of passage into conjunctive unity."[75] For Whitehead then, the very "being" of actual entities is fundamentally constituted by their creative temporal becoming and relationality to every other occasion in the universe. They exhibit both internal and external relations through psychophysical *dipolarity*. With their *physical pole* occasions inherent the objective and settled past, and with their *mental pole* they reach for possibilities of an unsettled and open future. Inert matter for Whitehead is finally replaced by the creative evolution of atomic events of experience that are both physically and mentally constituted. It is these kind of events for Whitehead that are dimly analogous to those of our own experience.[76]

Whitehead thus holds to a "panexperientialist" ontology where creative experience—indeed "mind"—belongs to nature as an ontological primitive. As Whitehead notes, science has often ill-treated human mentality as a great exception to the rest of nature, instead of seeing it as an extreme exemplification of mental functioning as it is present at all scales of nature.[77] While mental experience is never wholly absent in Whitehead's ontology, it's important to stress that this *does not* commit him to the claim that "consciousness" goes all the way down in nature (a claim often associated with "panpsychism"). To the contrary, for Whitehead, "consciousness presupposes experience and not experience consciousness."[78] Consciousness, in other words, is a late emergent phase in the evolution of very complex "living" organisms involving incredibly layered synchronizations of experience that finally bloom into the "function of knowing."[79] As Whitehead puts it, "Consciousness is the crown of experience, only occasionally attained not its necessary base."[80] Stated differently: "The organic philosophy holds that consciousness can only arise in a later derivative phase of complex integrations [of actual occasions]."[81] The lateness of consciousness notwithstanding, in the primitive form of "occasions of experience," the place of mind in nature remains irreducible.[82]

Aesthetic Attainment

For Whitehead, one of the deepest inadequacies of mechanistic materialism concerns the relationship between fact and value. By holding to "nothing but" the existence of vacuous bits of matter that are independent and only externally (mechanically) related, this metaphysics *divorced* fact from value. The final constituents of nature remained "senseless, valueless, purposeless." The only value nature had was *instrumental* and not *intrinsic*: value to be used by others and not value in and of itself. Whitehead's philosophy of organism completely rejects this picture, and for good reason: "We shall never elaborate an explanatory metaphysics unless we abolish this notion of valueless, vacuous existence," he insists.[83] In drawing our attention to the poetic reaction to "lifeless" nature as portrayed in the mechanistic-materialistic paradigm, Whitehead points to dimensions of our own value-experience and to the necessity of retrospectively extending these dimensions all way down in nature.

> Remembering the poetic rendering of our concrete experience, we see at once that the element of value, of being valuable, of having value, of being an end in itself, of being something which is for its own sake, must not be omitted in any account of an event as the most concrete actual something. 'Value' is the word I use for the intrinsic reality of an event. Value is an element which permeates through and through the poetic view of nature.[84]

Actual occasions, as the most fundamental events of experience, are not valueless for Whitehead, but something *in and for themselves*. They are permeated with *intrinsic value* in their very reality of becoming. As Whitehead put it in his first Harvard lecture, the world bears witness to the "the becomingness of real values."[85] Just as experience and mind go all the way down in nature, therefore, so also does *value*. Whitehead's philosophy is thus not just "panexperientialism," but also what Victor Lowe calls "pan-valuism."[86] Experience *and* value belong together in Whiteheadian ontology such that he conceives the very becoming into being ("concrescence") of every occasion as a *value process* where possibilities of value are actualized and then passed on to subsequent occasions. As Philip Rose has rightly noted, *"to be"* for Whitehead "is to be the source of values given and the centre of values felt."[87] Value is inherent in the *making of fact*, and fact is an *attainment of value*.[88]

Whitehead thus directly follows the intuitions of the nature poets like Coleridge, Wordsworth, and Shelley in their "romantic reaction" to a mechanical dead nature as dissected by science. The nature-poets "emphatically witness that nature cannot be divorced from its aesthetic values."[89] The deep intuitions of the "nature-poetry of the romantic revival was a protest on behalf of the organic view of nature," Whitehead holds, "and also a protest against the exclusion of value from the essence of matter of fact. . . . The romantic reaction was a protest on behalf of value."[90] Whitehead's own "philosophy of organism" shares in this protest: "The values of nature are perhaps the key to the metaphysical synthesis of existence."[91] What is more, he would come to see that attention to value experience—*aesthetic value experience* in particular—is not only the proper starting point of philosophy, but also a clue to what cosmic evolution is ultimately about.

> My own belief is that at present the most fruitful, because the most neglected, starting point [of philosophy] is that section of value-theory which we term aesthetics. Our enjoyment of the values of human art, or of natural beauty, our horror at the obvious vulgarities and defacements which force themselves upon us—all these modes of experience are sufficiently abstracted to be relatively obvious. And yet evidently they disclose the very meaning of things."[92]

For Whitehead, cosmic and biological evolution exhibit an upward value-drive, the denial of with strains credulity. In this regard, Dick rightly states that Whitehead's philosophical system (along with that of Teilhard de Chardin) is "infused with teleology"—indeed, an *aesthetic teleology*.[93] It is clear that Dick is uncomfortable with teleology and its implications, however. While he wants to fully acknowledge the "undeniable" fact "that life on Earth is more complex after 3.8 billion years of evolution," he seeks to do so "without invoking any goal-oriented or teleological principles."[94] This admittedly heroic endeavor, however, disregards human teleological experience as an *exemplification* of nature and seems instead to be a result of Dick's desire of avoid any hint of "supernaturalism." Yet for Whitehead (and Teilhard) and a number of philosophically oriented biologists today, the reality of purposive teleology in nature does *not* require supernaturalism.[95] What it does require, however, is an ontology and metaphysics that *includes* it—one that expresses the reality of value-experience all the way down in nature.

For Whitehead, the ascendent lure of evolution toward complexity exhibits a drive from *mere existing* to an aesthetic *quality of existing*. What we see in the evolutionary process is not simply the drive "to live," but "to live well" and "to live better." Stated differently, the evolutionary advance is grounded in a three-fold qualitative urge:

1. to be alive (to live)
2. to be alive in a satisfactory way (to live well)
3. to acquire an increase in satisfaction (to live better)[96]

For Whitehead, what finally lures the evolutionary process upward and forward is the pervasive haunting for deeper *intrinsic value*—what Steve McInstosh calls "value gravity."[97] Evolution is about the qualitative lure for *value intensity* and its realization in the world on every level of existence. This has involved the *progressive increase in the capacity to experience intrinsic worth*. Each individual occasion of experience is an *aesthetic feeling process* in quest for *value attainment*. As a product of evolution ourselves, our experience is the most intimate expression of this drive for qualitative aesthetic intensity, and it is this same drive that is *felt* within the evolution of every emergent occasion of experience in the universe.

In saying that aesthetic values "disclose the very meaning of things," Whitehead is thus stressing that the achievement of aesthetic value is what the universe *is* and what it is *ultimately up to* at its most fundamental levels of becoming. All levels of life and all levels of experience exhibit hierarchies of aesthetic value attainment in the universe, and it is only within *this context* that morality can be approached. The "ethical" and "moral" dimensions of human experience are evolutionarily awakened expressions of the primordiality of aesthetic value experience as it pervades nature. Aesthetic value for Whitehead is the widest and most inclusive form of value such that the "moral order is merely certain aspects of the aesthetic order" realized in human life and action.[98] Aesthetic value-experience pervades the world and it is only in this context that morality can authentically arise on the human level.

> Everything has some value for itself, for others, and for the whole. This characterizes the meaning of actuality. By reason of this character, constituting reality, the conception of morals arises. We have no right to deface the value-experience which is the very essence of the universe. Existence, in its own nature, is the upholding of value-intensity.[99]

Not only is value-experience the "very meaning of things" for Whitehead, it also constitutes "the very essence of the universe." Existence itself is the *preservation* of a definite form of aesthetic value—the *persistence* of value in and through *process*. There are thus *real standards* of aesthetic value expressed in the universe. There are also primordial *possibilities of value* that are already *there* in the nature of things such that they can be attained in and through cosmic evolution. As we will encounter later, it is these ontological ingredients which allow Whitehead to insist that "The teleology of the Universe is directed to the production of beauty."[100]

Let's summarize this inversion of Dick's second principle: For Whitehead, the fact that biological, mental, and moral realities have emerged through human beings as part and parcel of nature, means that they cannot be *limited* to human expression alone; rather their metaphysical antecedents must be found in various degrees across the natural order. While particular human expressions of life, mind, and morality on our planet are certainly cosmologically peripheral, the more fundamental antecedents of evolutionary organism, creative experience and aesthetic attainment are metaphysically central. It is on this assumption of metaphysical uniformity, moreover, that we necessarily face implications of extraterrestrial commonality across these domains. However more or less advanced different alien species, intelligences, minds, and moralities may be, they will *still* share these ontological antecedents. Affirming an ontology of organic evolution, mental experience, and aesthetic attainment meaningfully connects us to billions of potential beings that also exemplify these same principles in their own way. It is on the basis of such an ontology that David Ray Griffin utters the following conviction: "On other planets with the conditions for life to emerge and to evolve for many billions of years, we should expect there to be some with creatures that, no matter how different in physical constitution and appearance, would share some of our capacities, such as those for mathematics, music and morality, or, more generally, truth, beauty and goodness."[101]

Inversion 3: *"Humanity is not at the top of the great chain of being in the universe."* → "Humanity exemplifies the same metaphysical principles that are expressed in various intensities throughout the great chain of being in the universe."[102]

Dick's third principle is particularly interesting. In holding that we are likely near the bottom of the great chain of beings in the universe, he rightly imagines that there are billions of alien planets, intelligences, and civilizations that have antedated our new-born species by billions of years of cosmic, biological, and cultural evolution. One is faced with the "surely sobering fact" that against the backdrop of 13.8 billion years of evolution, we are cosmologically infantile. Indeed, the "claim that most intelligent civilizations would *not* be more advanced than us," Dick insists "seems to border on nihilism."[103] True as this is, more is *metaphysically required* of claims about a "great chain of being."

In many ways, the very claim that there *is* a great chain of being in the universe rouses metaphysical wonder. It is not metaphysically neutral, but requires substantiation within an ontology that meaningfully authenticates this chain not only as *possible* (since it presupposes the reality of infinite possibilities), but also *actual* and *connected* despite vast ranges of diversity. This chain is not merely cosmologically connected, but *metaphysically connected* through real ontological "links." Despite innumerable varieties of cosmological contingency, complexity, and difference belonging to this chain, the question of what all beings *commonly presuppose* is an ontological inquiry in search for metaphysical uniformity all the way up and down the great chain of being.

We've now seen the basic sketches of Whitehead's metaphysics as an event ontology of organism, creative experience, and aesthetic value. At base level, there is no going below "actual occasions of experience." Actual entities truly are at the *ontological bottom* of the great chain of being in the universe. However, because this chain is more fundamentally a great chain of *becoming*, it is truly *infinite* in its reach toward novel possibilities of being. The chain is *open* and while it allows for higher and higher heights, it does not in principle have a "top." What is more, the fact that on Whitehead's scheme, human minds and bodies have evolved as extraordinarily complex expressions if this ontology, means that we are *not* at the absolute bottom of the great chain of being. As Whitehead states, "An occasion of experience which includes a human mentality is an extreme instance, at one end of the scale, of those happenings which constitute nature."[104] Extreme as the human "instance" may be, we may nevertheless still be "near the bottom" for Whitehead too. In one of his most imaginative (and neglected) statements on such matters, Whitehead gives a sense for how infinitely vast the great chain of being may be.

> I see no reason to suppose that the air about us and the heavenly spaces over us may not be peopled by intelligences, or entities, or forms of life, as unintelligible to us as we are to the insects. In the scale of size, the difference between the insects and us is nothing to that between us and the heavenly bodies; and—who knows?—perhaps the nebulae are sentient entities and what we can see of them are their bodies. That is not more inconceivable than that there may be insects who have acute minds, though . . . their outlook would be narrower than ours. My point is, that we are part of an infinite series and since the series *is* infinite,

we had better take account of that fact, and admit into our thinking these infinite possibilities."[105]

There are several layers of significance in this imaginative statement. In the first place, Whitehead's attribution of pervasive "intelligence," "sentience," "forms of life," and "mind" to everything in the cosmos—whether the air, insects, planets, and nebulas—is consistent with his organic ontology of creative experience where, despite infinitely vast degrees of complexity and difference, these dimensions are never *wholly absent*. Furthermore, Whitehead again includes human beings as a *part of*, and not *apart from*, this infinite chain of being. The contours of our experience are authentic exemplifications along this chain. The forms of experience below us are far narrower in scope (the insects perhaps), but there are also (in principle) forms of experience above us that are far wider and vastly more inclusive than our own.

A natural question surfaces in light of Dick's third principle as to the kind of beings that lie "above" us as we imaginatively ascend the great chain of being. If indeed, humanity is likely "near the bottom" of this chain, then what lies still *higher* and what is it like? Both Whitehead and Dick allow for highly evolved intelligence and sentience vastly beyond that of human beings, yet Whitehead substantiates this claim in an ontology of primitive evolutionary intelligence which internally connects us to this chain and what it achieves at all levels. Despite vast cosmological contingencies, this connection is based upon the necessary metaphysical character that this chain exemplifies at all levels. As Whitehead suggests above, the degrees of difference in size, complexity, and intensity between ourselves and such intelligences might be likened to that between ourselves and insects, or ourselves and the heavenly bodies. As we will see below in discussing Dick's fourth principle of cosmotheology, he is fully open to affirming "God" as a supremely wise and powerful, albeit fully "natural," superintelligence that is contingent upon cosmic evolution.

Before moving on to Dick's fourth principle, another metaphysical dimension must be mentioned in light of Whitehead's statement above. For Whitehead, such a great chain of being requires the ontological reality of "infinite possibilities." There are not only *actualities* in the process of becoming (actual occasions embodying ultimate Creativity), but also *possibilities* that are being actualized in the cosmic evolutionary process. These possibilities are what Whitehead calls "eternal objects."[106] Indeed, the actualization of relevant and available possibilities is what actual occasions of experience *do* in their process of becoming. Possibilities too are a part of "what there is" for Whiteheadian ontology and he insists upon their *necessity* precisely because the fundamental character of the universe is "a creative advance into novelty."[107] Such possibilities are "fundamental for the understanding of existence, as soon as the notion of process is admitted," Whitehead states. As such they have certain extreme ontological finality.[108] "When we enjoy fact as the realization of a specific value, or possibility as an impulse towards realization," Whitehead insists, "we are stressing the ultimate character of the Universe."[109] This "character" of entertaining, realizing, and enjoying possibilities of value is "ultimate" because it is what is exemplified on every level of existence throughout the great chain of being. The backdrop of the great chain of being, therefore, is an equally great chain of *nonbeing*—that of possibility and value which find actualization along this chain. The great chain of being presupposes the availability, relatedness, relevance, and achievability of "infinite possibilities" that hibernate in the metaphysical depths of nature. Whereas this appears unproblematic for Dick, as we will see below, it is metaphysically problematic and requires ontological backing for Whitehead.

Inversion 4: *"Cosmotheology must be open to radically new and non-supernatural conceptions of God . . . a God grounded in cosmic evolution"* → "Cosmotheology must be open to truly radical and non-supernatural conceptions of God, a God grounding and exemplifying the metaphysical conditions of cosmic evolution.

Building upon the cosmological negations of his first three principles, Dick's fourth principle is the first of three subsequent affirmations as to what cosmotheology must now affirm of "God," the "moral dimension" and "human destiny" respectively. Rather than being *truly radical* in its theologically affirmation, however, Dick's cosmological negations now cross over into an *ontological negation* of God in his fourth principle. The "radical" and "non-supernatural" God that Dick has in mind is a fully contingent superintelligence, an emergent *outcome* of cosmological evolution, rather than its necessary condition. The "God of cosmotheology" Dick states, "is a natural evolutionary product, potentially with many characteristics normally associated with God."[110] For Dick, it is "entirely possible that beings have evolved in the natural course of the universe with many of the traits of God, including omnipotence, omniscience, and so on."[111] These traits notwithstanding (and certainly understood differently by Whitehead), there is one traditional divine trait that has been fully negated in Dick's cosmotheology and that, in principle, no evolved superintelligence can ever have, namely: *metaphysical necessity*. In contrast to the entire sweep of classical theology, Dick's "God" remains only on the level of *cosmological contingency*. In light of this, Dick has rightly said that some "may well object" to the language of "cosmotheology" and instead refer to his position as "cosmophilosophy."[112] I would agree that his denial of metaphysical necessity to "God," coupled with his own intuition that there is really "no need" to even use the "loaded term" does not adequately justify his position as "cosmo*theology*" or, for that matter, a non-supernatural *theology* at all.[113] In being theologically *anti-realist* in this way, his cosmotheology is in fact *not radical enough.*

This comes at a cost. Dicks proposal arguably suffers real metaphysical problems surrounding possibility and necessity, truth and meaning—problems that have traditionally found remedy in the realms of philosophical theology. The same cannot be said of Whitehead, however. The ontological *realism* of Whitehead's cosmo*theology* not only allows him to addresses Dick's metaphysical limitations by associating essential domains of necessity with the "primordial nature of God," but also deepen and expand Dick's intuitions of an evolutionary superintelligence through the "consequent nature" of God. Whitehead, moreover, accomplishes this in a *truly* non-supernatural fashion, such that divine action is a natural part of the world's normal processes, and never an interruption thereof. Whitehead's cosmotheology thus supports a vision of God that both *grounds* and is *grounded in* cosmic evolution and the metaphysical conditions it presupposes. I will draw these points out below through attention to three related features of Whitehead's cosmotheology: God as *"chief exemplification"* of metaphysical principles, God as *primordial correlate* to metaphysical problems of possibility, and God as *consequent evolutionary ground* of truth and meaning.

God as "Chief Exemplification"

The "naturalism" under which Whitehead's cosmotheology operates is not the same as Dick's. Whereas Dick's naturalism requires the ontological disappearance of "God," such that God has been fully negated, Whitehead's naturalism engenders the ontological reappearance and affirmation of God, albeit as the "chief exemplification" of the *same* metaphysical conditions applicable to the world. This involves both a radical metaphysical inversion with respect to Dick's cosmotheology *and* a radical theological inversion with respect to dominant habits of the philosophical and religious tradition.

One of Whitehead's most adamant critiques of the philosophical and religious tradition is that in its eagerness to support the "religious significance of God," it inappropriately paid God "metaphysical compliments." It is this "unfortunate habit" that makes God a supernatural *exception* to the most general conditions appliable to all other things, and thus the ultimate "foundation of the metaphysical situation." Such an affirmation for Whitehead can never in principle escape the problem of evil, nor claims as to the inherent clash of science and religion.[114] By inverting this long-standing theological habit, however, Whitehead effectively charts a new path of *naturalistic theology.* "In the first place," he states, "God is not to be treated as an exception to all metaphysical principles, invoked to save their collapse. He is their chief exemplification."[115] The radicalism of this statement is that *God too* is to be understood in light of the same metaphysical conditions expressed in the world, albeit *preeminently*. Rather than establishing these conditions, God is their *primordial embodiment*. It is this radical move for Whitehead that categorically bars supernatural action and makes divine activity part and parcel of the world's normal natural processes and never their competition. "God's role is not the combat of productive force with productive force, or destructive force with destructive force," Whitehead states; rather the "divine element" is to be understood "as a persuasive agency and not as a coercive agency."[116] In a truly radical move, competitive and coercive action as the hallmark of supernatural theology is *metaphysically denied*.

The claim that God is the "chief exemplification" of metaphysical principles is not insignificant. God too is to be understood in terms of an organically related event of experience, albeit an *everlasting event* occurring and encompassing the entire cosmic process. Whitehead stresses that God too is an "actual entity."[117] Just as all "occasions of experience" embody the ultimate reality of Creativity in their becoming, so too does God as ultimate actuality embody and condition Creativity *primordially* and on behalf of the world.[118] Just as actual entities emerge through an aesthetic feeling process in quest for the realization of novel possibilities of value, so too is God conceived as the cosmic grounding and limiting factor that makes possibilities of value relevant to the world process and then integrates and evolves in light of what the world achieves. Like every event, God too exhibits psychophysical "dipolarity" with a necessary "primordial nature" (mental pole) and a contingent "consequent nature" (physical pole) the functions of which render the world process metaphysically stable, as well as existentially meaningful. I turn first to the "primordial nature" as it relates to metaphysical problems of possibility and necessity that remain unarticulated in Dick's proposal.

God as Primordial Correlate to Metaphysical Problems of Possibility

It is important to stress that Whitehead himself was in no way "eager to establish the religious significance of God." While not wanting to dismiss the data of religious experience, he had no religious agenda as such and had been theologically disenchanted as an agnostic or atheist for some two decades.[119] Nevertheless, on grounds of consistency, coherence, and meaning, Whitehead did come to affirm the reality of God as utterly necessary for his philosophy.[120] William Ernest Hocking recounts Whitehead's comment on the inclusion of God in his philosophy: "I should never have included [God] if [God] had not been strictly required for descriptive completeness."[121] A.

H. Johnson too recalls Whitehead saying he brought God into his metaphysical system to show that "He belonged."[122]

The sense in which God is ontologically "required" or "belongs" for Whitehead is relevant to considering the metaphysical limitations of Dick's cosmotheology. One of the most telling statements Dick has made with respect to Whitehead's philosophical theology concerns the status of possibility as a necessary part of the ontological ingredients of reality. It is a statement that again underscores his primary emphasis on *cosmological* rather than *metaphysical* dimensions of cosmotheology. "In its emphasis on evolutionary becoming," Dick states, "cosmotheology resonates with Alfred North Whitehead's process theology, *without seeing God as the beginning and source of all possibilities.*"[123] This statement appears to be based in Dick's anxieties to again avoid any hint of "supernaturalism." Yet it is a less-than-radical "naturalism" that would see "supernaturalism" as necessarily entailed in the correlation between God and possibility.[124] This is certainly not the case for Whitehead, nor for process and other thinkers after him who have distinguished varieties of naturalism and argued for the necessary coexistence of naturalism *and* theological realism.[125]

Possibilities appear to be an inescapable domain of *metaphysical necessity* in the nature of things, and if God is their "beginning and source" then God too is *necessarily existent*. It is this conclusion that Dick seeks to avoid in a "naturalism" that entails no "ontologically distinct and superior realms including God."[126] It is worth pointing out, however, that possibilities *are* ontologically distinct from actualities. They need not be seen as "superior" or "more real" in some platonic sense, but they are not "nothing"; as such, they *are* ontologically problematic. By refusing a connection between God and possibility, Dick has left unaddressed the metaphysical problems of possibility that have haunted the philosophical and theological tradition, not least Whitehead and his subsequent influence on a diversity of process philosophers.[127]

It is not at all obvious that we can take for granted the fact that reality is utterly lush with possibility without asking some truly fundamental questions.[128] What indeed is the ontological nature, character, and function of possibilities with respect to cosmic evolution? How and where do possibilities "exist" and how are they made relevant to the world process? In addition to Dick, such questions can also be directed to the stimulating work of Mark Lupisella whose "connection-action principle" presupposes not only a "very large, if not infinite possibility space," but a "very large, if not infinite possibility space *of value*."[129] In ways stronger than Dick, Lupisella has recognized not only the metaphysical importance of possibilities as such, but also—to a large extent—the status of possibilities as inherently axiological in nature.[130] While Lupisella's work is thicker than Dick's on these metaphysical issues, he nevertheless also leaves these problems outstanding in relation to an ambiguous theological position wavering between realism and antirealism.

To reiterate for Whitehead: a universe of creative evolutionary becoming presupposes the indispensability of possibility, but he is not unaware of the layered metaphysical issues that are involved. These are related issues surrounding the ontological status, limitation, relevance, and efficacy of the possible for the actual in the creative advance of things.[131] Again, it is not obvious where possibilities are ontologically; how they can "exist," or how they can exhibit causal relations and relevance to a fluid evolutionary world. We already saw for Whitehead that our cosmological conditions exhibit "absurd limitations" against an infinite backdrop of abstract possibility. Some principle of limitation is required not only for these conditions, but also for actual occasions to arise at all. For Whitehead, actual occasions can only emerge in light of an extremely slim range of possibilities relevant to their own becoming. But what is the basis of this relevance? Neither Dick nor Lupisella have problematized possibility in these ways, but on Whitehead's account, this is arguably what their evolutionary proposals *require*.

At the heart of Whitehead's approach to these problems of possibility is his expression of the "ontological" or "Aristotelian principle." Elevating the explanatory power of actuality, this principle has had a long-standing foothold in the philosophical and theological tradition, from Aristotle, Augustine, and Aquinas, to Descartes, Kant, Leibniz, and Whitehead himself. Whitehead articulates this principle in different ways. "Everything must be somewhere," he states, for it remains "a contradiction in terms to assume that some explanatory fact can float into the actual world out of nonentity." He is adamant that, "there are no self-sustained facts, floating in nonentity" and that "apart from things that are actual, there is nothing—either in fact or in efficacy."[132] As seen in these statements, the ontological principle aims to secure a mode of explanatory coherence so that all *reasons* have recourse to *actuality* or nothing at all. "Thus, the search for a reason is always the search for an actual fact which is the vehicle of the reason," Whitehead states.[133] Nowhere is this principle expressed more acutely when confronting the metaphysical problems of possibility.

Whitehead recognizes immediately that upholding the ontological principle means that the infinitude of possibility presupposed by the world process must "be somewhere" since such possibility retains its "proximate relevance to actual entities for which it is unrealized."[134] It is not only that the actual process of the universe at any one point presupposes the infinite fullness of its possibility, but also that each actual entity is an emergence of possibilities relevant to that entity as it becomes into being. It seems apparent that no single finite actual entity can account for infinite possibility, and no finite actual entity can provide relevant possibilities *for its own emergence*; rather, it navigates these possibilities as *already given*. Neither on Whitehead's vision can it be said that possibility itself "procures" anything due to its abstractness.[135] Infinite possibilities alone cannot coherently limit themselves as relevant to the world process.

In seeking to resolve these problems, Whitehead must, in consonance with the ontological principle, refer the locus of infinite possibility, as well as its limited/ordered relevance and efficacy to the world process, to something both nontemporal *and* actual. What could this "somewhere" and "something" be? Whitehead answers: "This 'somewhere' is the non-temporal actual entity" and "relevance" means "'relevance as in the primordial mind of God.'"[136] In part, this divine mind must be *nontemporal* to account for the *timeless infinitude* of possibility, and *actual* to account for its limited/ordered relevance for the world. Eternal possibility in God's primordial mind "constitute the Platonic world of ideas," Whitehead states. These forms, however, are "forms of process" and rather unlike reified renderings of traditional "Platonic Ideas."[137]

Analogous to the mental poles of finite actual entities which navigate a limited set of possibilities relevant to their own becoming, the primordial nature of God is the divine mental pole which encompasses infinite possibilities with strategic urge for their realization in the world. For Whitehead then, there remains an essential metaphysical correlation: *where there is possibility, there is mind; and where there is infinite and necessary possibility, there is infinite and necessary mind*. As such, the ontological principle "issues in the doctrine of conceptualism" which, when viewed from its highest metaphysical perspective, coincides with the primordial mind of God as an all-inclusive actuality.[138] "I think the universe has a side which is mental and permanent," Whitehead states. "This side is that prime conceptual drive which I call the primordial nature of God."[139]

This primordial mind for Whitehead is nothing other than the *presupposed reason* for the relevance of possibility to the evolution of the universe at all scales and levels. This mind, he states, "is the unconditioned actuality of conceptual feeling at the base of things; so that, by reason of this primordial actuality, there is an order in the relevance of eternal objects to the process of creation."[140] So essential is the divine appetitive ordering of possibility that without it, the world would halt in disjointed stagnation.[141] Whitehead thus holds that the nontemporal actuality of divine mentality is included in every creative phase of evolution in the actual world. Considered alone, the unlimited wealth of abstract possibility "would leave each creative phase still indeterminate," Whitehead states. Conceived in terms of their "bare isolated multiplicity," possibility lacks any existent character. For Whitehead, eternal possibility requires "the transition to the conception of them as efficaciously existent by reason of God's conceptual realization of them."[142] As Derek-Malone France has stated, "the realm of possibility is identical to the conceptual content of the divine mind."[143]

Whitehead insists that it is the definite determination or limitation provided by God that gives ordered balance and value to the world process as it shares in God's "unchanged consistency" of character.[144] For Whitehead, the consistency of this unchanged divine character is "founded on Value," on "ideals of perfection, moral and aesthetic." God is necessarily good and thus limited by goodness in dividing "the good from the evil" in the provision of possibility for the world.[145] Indeed, for Whitehead, the ordered harmony in the actual world is due to the world's *sharing* in the axiological steadfastness of this divine character.[146] "The necessity of the trend towards order does not arise from the imposed will of a transcendent God," Whitehead states, it "arises from the fact, that the existents in nature are sharing in the nature of the immanent God."[147]

Whitehead therefore integrates and explains the metaphysical problems of possibility through their *subsistent residence* within the evaluative understanding of necessary divine mentality. The functions of this divine actuality reason the very possibility, order, and value of the world, and the actual entities which variously comprise it. Considered from its highest altitude, Whitehead's affirmation of the "ontological principle" expresses the fundamental commitment that without the actuality of God in relation to every occasion of experience in the world, there would be *nothing* either possible or actual, no order, and thus no world at all. The divine "ordering entity" Whitehead concludes, "is a necessary element in the metaphysical situation presented by the actual world."[148]

Dick appears willing to affirm a fully contingent superintelligence as an outcome or *product* of cosmological evolution. Such a superintelligence, however, never escapes being an *actualization* of possibility and not its necessary ground. The metaphysical problem of possibility thus remains in full force. As noted above, there is in principle nothing on Whitehead's metaphysical scheme that would bar him from affirming the reality of vastly more intelligent beings of the kind that Dick imagines. Such a superintelligence may indeed be "God-like," but to the extent that it is completely contingent, it is categorically different from any traditional affirmation of God. Whiteheadian cosmotheology can truly *deepen* Dick's cosmotheology in this regard. By integrating possibility with divine mentality, Dick would not only resolve metaphysical problems of possibility, but also witness the ontological *reemergence* of God as necessary for his cosmotheology. These dimensions of necessity are not all that Whitehead's cosmotheology offers Dick, however. In turning now to Whitehead's vision of the "consequent nature" of God we will see that Whitehead's cosmotheology not only complements and deepens

Dick's intuitions as to a contingent evolutionary superintelligence, but also addresses lacunae of truth and meaning that arguably remain ungrounded in this hypothesis.

God as Consequent Evolutionary Ground of Truth and Meaning

That both necessity *and* contingency coherently belong in the concept of God is one of the bedrock influences of Whitehead on the subsequent development of process theology.[149] For Whitehead, God as *necessary* and God as *contingent* are not two different and incompatible approaches to "God," but in fact two essential functions differently belonging to a *single divine life*. "God as primordial is not a distinct actuality," according to Whitehead, "only the union of primordial and consequent natures forms a distinct actuality."[150] In addition to being the unchanging primordial ground of possibility and value, God is also "consequent" upon the temporal world and truly relates, experiences, and evolves through incorporating the world process in every moment. Using Dick's language, it would not be false to say that the consequent nature is God as an evolutionary superintelligence that experiences and expresses the same evolution of the world. There is an important difference, however. Whereas Dick's natural superintelligence evolves as a finite *outcome* of cosmological evolution, Whitehead's God evolves as the infinite receptive condition of the world's evolution through ever-expanding divine experience. Both proposals are deeply evolutionary, but the differences are stark, especially with respect to resolving fundamental metaphysical questions concerning truth and meaning.

Whitehead's introduction of the "consequent nature of God" is not arbitrary or without reason, but a required extension of his commitment to God being the "chief exemplification" of the same metaphysical categories applicable to the world. Just as every occasion has a physical pole which receives from the settled past as it becomes into being, so too does God have a physical pole that receives, integrates and transforms worldly accomplishment in the divine life. "The consequent nature of God," Whitehead states, "is the realization of the actual world in the unity of [God's] nature, and through the transformation of his wisdom."[151]

For Whitehead addressing the metaphysical problem of possibility for an evolving world is one important means of confronting the necessity of God; but it is arguably not the most important. As a complement to the "primordial nature," the "consequent nature" of God also addresses real metaphysical questions surrounding truth and meaning that arise in human experience as exemplifications of cosmic evolution. What accounts for the objectivity of the past as accomplished fact and thus as "true" in a becoming world? For

Whitehead, it is not obvious that the reality of truth can exist as such without a real ontological referent to an all-inclusive experience or perspective of the universe and its evolution. The finite contingency of a natural superintelligence cannot by definition have an impartial, all-inclusive perspective, but only a limited perspective as a "product" of cosmological evolution. This remains problematic in ways that have been emphasized by Whitehead and subsequent process thinkers.[152] The ontological question as to what it is that could possibly distinguish what *was* from what *might-have-been* thus remains outstanding, for any natural superintelligence would presuppose these distinctions in its own reality.[153] It is these kinds of problems that led Whitehead to affirm an all-inclusive divine experience that evolves in receptive unison with the world process.

> The truth itself is nothing else than how the composite natures of the organic actualities of the world obtain adequate representation in the divine nature. Such representations compose the 'consequent nature' of God, which evolves in its relationships to the evolving world.[154]

In stressing that the "truth itself" must be "somewhere" Whitehead is again upholding the explanatory power of the ontological principle. "In this way," he states, "the 'ontological principle' is maintained—since there can be no determinate truth, correlating impartially the partial experiences of many actual entities, apart from one actual entity to which it can be referred."[155] Just as there would be no *possibility* (and thus no novel future) without the primordial nature for Whitehead, so too there would be no *truth* (and thus no preserved past) without the all-receptive evolution of divine experience in the consequent nature. Another metaphysical correlate thus emerges for Whitehead: *where there is historical truth of any kind, there is an impartial divine reception of that truth as distinguished from what might have otherwise been.*[156]

The problem of truth is closely related to the problem of meaning, both of which straddle the tangled lines dividing cosmology, metaphysics, and religion.[157] Whitehead recognized the question of meaning to be of utmost significance in a processive universe that is characterized by "perpetual perishing." A world of becoming and perishing weighs upon those it has produced with existential angst, longing, and inevitable loss. This is not only the case for our species, but any and all extraterrestrial species across an evolutionary cosmos. "The ultimate evil of the temporal world is deeper than any specific evil," Whitehead states, "It lies in the fact that the past fades, that time is a 'perpetual perishing' . . . In the temporal world, it is the fact that process entail loss."[158] What indeed comes of our experience—of all experience in the universe? Is there any resolution to the conflicting and discordant dimensions of human striving, anxiety and turmoil? Can any ultimate significance be assigned to our life—to any life—and the planet on which we've emerged if we all inevitably perish into nothingness?

Whitehead fully recognizes that these questions are at home in a world of ceaseless evolution, but he stresses that "there is no reason, of any ultimate metaphysical generality, why this should be the whole story." For Whitehead, the world does not perish into *nothingness*, but into the consequent nature of God where it is imprinted, remembered, and judged against the ideals of the divine nature. "The consequent nature of God is his judgment on the world. He saves the world as it passes into the immediacy of his own life. It is the judgment of a tenderness which losses nothing that can be saved. It is also the judgment of a wisdom which uses what in the temporal world is mere wreckage."[159] As such, the consequent nature is also the ground of meaning—the memory and preservation of accomplished fact in the universe.[160]

What Whitehead calls "objective immortality" is the everlasting contribution that is made to the life of God such that the value and importance of finite accomplishment is never lost.[161] This for Whitehead is not only significant cosmologically and metaphysically, but also religiously. One notices the religious language Whitehead uses above when describing the activity of the consequent nature of God: "judgment," "salvation," "tenderness," "wisdom." Without holding to any supernatural implications, Whitehead nevertheless holds that there is an ontological referent in the nature of things, the activity of which corresponds to such language. "The peculiar character of religious truth," Whitehead states, "is that it explicitly deals with values. It brings to our consciousness that permanent side of the universe which we can care far. It thereby provides a meaning, in terms of value, for our own existence, a meaning which flows from the nature of things."[162]

For fear of creeping supernaturalism, Dick's religious naturalism denies that any divine ontological referent "is required to give meaning to the world." Rather, his religious naturalism holds that "meaning is derived from our knowledge of the natural world, from the creativity and beauty of nature."[163] No doubt meaning truly is derived in these ways; Whitehead would fully agree. However, without the functions of the primordial and consequent natures; without a real ontological referent to *that which provides these functions*, there would be no world at all, only meaningless chaos and disorder. "But if theology is defined broadly as that which gives meaning and value to life in a cosmic context," Dick states, "then even a naturalistic cosmotheology is indeed a theology, albeit one without God, at least the standard God."[164] This statement again focuses upon meaning in a "cosmic context" and *not* upon the metaphysical presuppositions of knowledge, meaning, creativity, and beauty in nature.

We've seen in this fourth inversion that the correlate to rejecting the "standard" supernatural God for Whitehead is not the ontological negation of God as such. In Whitehead's cosmotheology, naturalism and theological realism are coupled in a vision of God that chiefly exemplifies the same metaphysical conditions expressed in the world. God is "required" and "belongs" for Whitehead because of his demand for consistency and coherence with respect to a constellation of metaphysical problems that arise for a processive cosmos.

Let it be said that process philosophers and theologians fully agree with Dick: we should reject arguments for a "God of the gaps." Such arguments, Dick rightly states are "no longer popular even among theologians, precisely because science keeps filling in the gaps."[165] The gaps Dick speaks of, however, are only *cosmological gaps*; but there are also *metaphysical gaps* that science in principle cannot fill. As David Ray Griffin has provocatively stated, these are not so much "gaps" as they are "God-shaped" metaphysical holes that arguably remain unfilled in Dick's cosmotheology.[166]

Inversion 5: *"Cosmotheology must have a moral dimension, extending to embrace all species in the universe—a reverence and respect for life in any form."* → Cosmotheology must provide the ontological basis and stimulus for ideals of moral reverence and respect in the nature of things.

Dick's fifth principle laudably affirms the necessity of moral foresight in looking to an open future that may or may not include interaction with extraterrestrial beings. This "moral dimension" of cosmotheology gives rise to the development of "astroethics" or "cosmocentric ethics," including "astropolicy" and a variety of ethical endeavors and worldviews.[167] Dick's moral anticipations are nicely complemented by Whitehead's own in this regard: "The greater part of morality hinges on the determination of relevance in the future. The relevant future consists of those elements in the anticipated future which are felt with effective intensity by the present subject by reason of the real potentiality for them to be derived from itself."[168] Indeed, any relevant cosmotheology anticipating our interaction with extraterrestrial life necessitates profound ethical and moral consideration *now*. It also requires explaining how the "moral dimension" metaphysically belongs to the nature of things and why pursuing it is ultimately worthwhile.

In part, Dick's fifth principle can be juxtaposed to one of the central convictions of his second principle. Although Dick's cosmotheology must have a "moral dimension" it also proceeds under the assumption that we are "not central" morally in the universe and thus have no right to impose our merely peripheral morality on other worlds, let alone on other cultures on our planet. This fifth principle, Dick fully recognizes, is "a challenge even on Earth."[169] Why then do we think we are able to do this beyond our planet or that other extraterrestrial beings would share such reverence and respect? If we are not central morally in the cosmos, then neither are other beings on other planets. This raises the question as to whether there really are objective moral ideals or requirements of a binding character, and whether or not they require a referent to an *ontological center* that is, in principle, deeper than any peripheral cosmological expression. These questions are again linked to an inversive consideration seeking the *metaphysical basis* for objective moral claims in the universe such that cosmological moral relativism is avoided. This should be of interest to any philosophy of astrobiology or cosmology keen on expectations of extraterrestrial contact.

While Dick's intuitions rightly seek to universally extend ideals of "reverence" and "respect" to terrestrial and extraterrestrial others, what is it in the nature of things that would categorically condemn the eradication of all other planetary species by those civilizations that do not share such peripheral cosmological ideals? As Roland Faber has argued, the "dark forest" response to the Fermi paradox is based in the worrisome presupposition that ideals of this kind will not be shared by other extraterrestrial civilizations.[170] On what metaphysical grounds does one say that a cosmological holocaust cannot and should not happen? Dick of course roundly agrees that this should not happen and takes the "moral dimension" of cosmotheology very seriously, but in the absence of an articulated metaphysics, it is not obvious where and how moral ideals and facts can exist in a naturalistic cosmotheology that has ontologically negated the theological center which has traditionally resolved these problems. In part, Dick acknowledges this when he speaks about the "problem of universals" with respect to extraterrestrial communication. Candidates for universality would not only be mathematics and science, but also evolution, *ethics* and *aesthetics*.[171] Indeed, if there really are objective ethical and aesthetic ideals, then they are what they are regardless of what we think about them. This raises the long-standing philosophical problem as to their ontological standing in the nature of things. For Dick, however, realist theology, is not required.

"It is often stated that morality stems from theology and the existence of God, with the unfortunate implication that any non-believer has no basis for morality," Dick states.[172] The God Dick has in mind is again a "supernatural" divinity whose own moral reputation is often no better than our own. In rejecting the "naturalistic fallacy," the "real fallacy" Dick holds "is that the universe, as opposed to supernatural sources, cannot inform morality. We might call this, the idea that morality must come from supernatural sources, the Supernaturalistic Fallacy," he states.[173] In commenting on the argument of Nancy Murphy and George Ellis "that morality has an 'objective' basis in the nature of reality and that the discipline of ethics can be construed as the scientific study of this moral order," Dick again responds with hesitancy since their "theory of ultimate reality . . . is deeply teleological and theological, a theory 'about the purpose for which God created the universe.'"[174] Without needing a "creator," or purpose-based teleology, or any supernatural moral arbiter, Dick holds in contrast to the naturalistic fallacy, that "astroethical principles" like "reverence for life in all its manifestations" can solely be "the product of the creativity of cosmic evolution, whether terrestrial or extraterrestrial."[175] Among a spectrum of possible positions, what Dick has called a "strong cosmocentric ethic" would hold that the "cosmos itself confers meaning in some way, and is the basis for a universal ethics that we should apply when considering practical questions about how to treat life in the universe, whatever its complexity."[176] As to precisely *how* it is the cosmos can house such moral facts let alone "confer" or "inform" moral meaning from them as the "basis for a universal ethics," requires deeper metaphysical treatment.

Whitehead does not commit the "supernaturalistic fallacy." Neither, however, does he commit the "naturalistic fallacy" since his cosmotheology is at once non-supernaturalist *and* realist in nature. In consonance with the "supernaturalistic fallacy," Whitehead agrees that objective moral standards *do not* require allegiance to any "supernatural" deity; and in consonance with the "naturistic fallacy," he agrees that such standards cannot be *nowhere* in a universe of ceaseless becoming. If indeed divinity has been completely negated ontologically there is no apparent way to make sense of the locus, efficacy, and weight of moral ideals and truths in human—and by extension—extraterrestrial experience. The naturalistic fallacy is thus valid to the extent that it points to the God-shaped metaphysical hole surrounding the reality of moral facts and duties in the nature of things; yet it is invalid to the extent that it fills this hole with a supernatural God, thus falling into Dick's "supernaturalistic fallacy."

To the extent that Dick has left the ontological status and locus of objective moral requirements unanswered *metaphysically*, it is not obvious that he *has* really escaped the "naturalistic fallacy" in his form of naturalistic cosmotheology. It is his form of religious naturalism, we remember, that denies categories of ontological distinction among such things. There is no "ontologically distinct and superior realm including God, the soul, and heaven" with which such objective facts could be correlated.[177] As Mackie rightly put it, objective moral facts would be far too ontologically "queer" to exist in a Godless cosmos. They "would be entities of a very strange sort, utterly different from anything else in the universe," he states. For example, if Plato's "Form of the Good" existed it would be utterly different by having what Mackie calls a "has to-be-pursuedness somehow built into it."[178] Indeed, to affirm inherently prescriptive ideals or "requirements" of this trans-planetary kind, is to assume their abstract ontological distinction and relation to "infinite possibilities *of value*" that are woven into the nature of things.[179]

As discussed in the second inversion above, for Whitehead, the evolutionary awakening to objective ideals of morality or "rightness" correlate to the already deeper *aesthetic nature* of the universe as such, where real standards of value and disvalue, better and worse, harmony and discord, intensity and triviality are exhibited variously in the becoming of all experience. Whitehead holds to an ontology of aesthetic value ("pan-valuism") and morality essentially concerns the *art of living*. "The canons of art are merely the expression, in specialized forms, of the requisites for depth of experience," Whitehead states. "The principles of morality are allied to the canons of art, in that they also express, in another connection, the same requisites."[180] These requisites for Whitehead are *metaphysical* in nature and they find their justification in God.

> The metaphysical doctrine, here expounded, finds the foundations of the world in aesthetic experience. . . . All order is therefore aesthetic order, and the moral order is merely certain aspects of the aesthetic order. The actual world is the outcome of the aesthetic order, and the aesthetic order is derived from the immanence of God.[181]

As founded on necessary ideals of aesthetic and moral perfection, the divine nature as interactive with the world process is "self-consistent in relation to all change." "God is the measure of the aesthetic consistency of the world," Whitehead states. "There is some consistency in creative action, because it is conditioned by his immanence." If the moral order is just aspects of the aesthetic order, then God too is the measure of the moral order awakened in human experience as exemplifications of the same metaphysical conditions undergirding all cosmological experience. "The purpose of God is the attainment of value in the temporal world," Whitehead insists—a purpose that is both aesthetic and moral.[182]

Objective aesthetic and moral ideals therefore are not "nowhere" for Whitehead, they are in God—the divine actuality—who functions as their ontological locus, efficacy, and weight as experienced across the universe. In contrast to Dick's naturalistic cosmotheology, Whitehead's ontological principle supports the universal reach of moral values and ideals and thus the realism of terrestrial and extraterrestrial intuitions and aspirations. "The universe is thus understood as including a source of ideals," Whitehead states. "There are experiences of ideals—of ideals entertained, of ideals aimed at, of ideals achieved, of ideals defaced. This is the experience of the deity of the universe" as effective and immanent in present experience.[183]

Although no planet or species can ever be "central" morally or ethically in the universe, the universe for Whitehead does evolve in relation to an aesthetic and moral center which forbids holocaustal relations among its inhabitants. To the extent that we engage in such horrors on this planet, or on any planet, we run against the moral and religious fabric of the universe and thus deface the divine ground of our being. I include "religious" here because the religious intuition at base level essentially consists in the desire to be in harmony with the fundamental nature of reality as one conceives it. "We have no right to deface the value-experience which is the very essence of the universe," Whitehead reminds us. This value-experience includes that of ourselves, extraterrestrial others, and ultimately the divine experience which receives all that is accomplished for good or ill in the universe.

Inversion 6: *"Cosmotheology must embrace the idea that human destiny should be linked to natural cosmic events, not to the divine."* → "Cosmotheology must embrace human destiny as inextricably linked to the destiny of the cosmos as an infinite evolutionary expression of the metaphysical conditions chiefly exemplified in the divine."

Dick's sixth and final principle of cosmotheology aims to secure a meaningful destiny for human beings that is grounded in cosmic evolution rather than supernatural teleology. "Although human destiny has often been couched in divine terms . . . it need not be linked to the supernatural," he insists. "Rather it can be linked to the process and endpoint of cosmic evolution."[184] Despite some important resonances, Dick has often expressed this final principle over and against the teleological impetus of Teilhard de Chardin. Like Whitehead, "the principles of cosmotheology . . . resonate with Teilhard de Chardin's evolutionary cosmology," he states "but without the teleological end, the 'Noosphere' he identified with Jesus Christ, or a supernatural God as the Omega Point . . . the sixth principle of cosmotheology makes clear that human destiny may be couched in terms of natural cosmic evolution."[185] As mentioned already, however, there need not be anything inherently "supernatural" about purposive teleology in nature; and Teilhard scholars, I trust, would largely disagree that Teilhard affirms the kind of "supernaturalism" that Dick rejects.[186] With Whitehead, Teilhard is arguably another example of an authentic cosmo*theology* where God, rather than being ontologically negated, is reconceived in truly natural and non-competitive relations to the universe and its evolution.

One of the important differences between Teilhard and Whitehead does concern the question of ultimate convergent "unity" or the final culmination of the universe in its relationship to the divine. Whereas for Teilhard, the "point Omega," as the anticipated divine-world convergence of cosmic evolution, is often thought to involve a purely linear and progressive divine teleology, for Whitehead there is no final divine-world convergence that is not always followed by yet another creative divergence in an infinite evolution of cosmological epochs.[187] In part, Whitehead's rejection of a final divine terminus is based in his rejection of any finality of order in a processive universe.

> The immensity of the world negates the belief that any state of order can be so established that beyond it there can be no progress. This belief in a final order, popular in religious and philosophic thought, seems to be due to the prevalent fallacy that all types of seriality necessarily involve terminal instances. It follows that Tennyson's phrase: ". . . one far-off divine event / To which the whole creation moves," presents a fallacious conception of the universe.[188]

While Whitehead knows no end or divine culmination to cosmic evolution in this way, it's important to stress that his statements as to an "infinite series," as seen earlier, appear implicitly open to forms of evolutionary convergence that spawn sentience, mind, and consciousness far beyond that expressed in humanity. In speculating that even the "nebulae are sentient entities," he seems readily open to universal and potentially "hyperpersonal" or "superintelligent" outcomes that might deeply resonate with Teilhard's vision. However, to the extent a culmination in "Omega" is understood to be inevitable or interpreted as some final collapse to cosmological evolution, Whitehead must part ways on behalf of infinite becoming toward ever-new horizons of cosmic order.[189]

In conceiving the natural destiny of human beings in light of the "process and endpoint" of evolution, Dick points to two alternative scenarios. "If cosmic evolution ends with humans and we are alone in the universe, our destiny involves stewardship of our pale blue dot and perhaps spreading, nurturing or even creating, life in the universe," he states. While this remains a staggering but possible scenario, Dick's impressive corpus in the last two decades presupposes its falsity—or, at least the *hope* of its falsity. However, if "cosmic evolution results in a biological universe—one in which life and intelligence is common," Dick continues, "our destiny is to interact with this life in all its myriad possibilities."[190]

Dick naturalistic cosmotheology remains implicitly open to the fact that one of these myriad possibilities includes our interaction with a pseudo-divine natural superintelligence as product of cosmological evolution. Deeper interaction with such intelligence may still lie ahead in the future, but this intelligence for Dick, may not be entirely irrelevant to our cosmic past. As Dick has suggested, "A highly evolved natural intelligence" may even have "fine-tuned the universe" for life—a possibility which has received "some attention, but not a great deal."[191] Perhaps the reason for this limited attention is the paradox—even contradiction—that appears to be involved in such a hypothesis. In order to "fine-tune" the cosmological constants this superintelligence *cannot* itself be a product of their tuning; but this is precisely what Dick affirms in holding that this superintelligence, like everything else, is a fully contingent *product* of cosmological evolution. This seems deeply problematic. I would submit that Dick's cosmotheology cannot logically affirm both of these convictions *unless* he also admits *necessary existence* into the superintelligence. If a superintelligence fine-tuned the cosmological conditions, then this superintelligence *transcends* this tuning in a way that opens the door to eternal *metaphysical necessity* and not *just* immanent cosmological contingency. As Whitehead expresses it, the "notion of God" is "that of an actual entity immanent in the actual world, but transcending any finite cosmic epoch—a being at once actual, eternal, immanent, and transcendent."[192]

Alongside cosmological contingency, if Dick *also* admitted metaphysical necessity this would allow the superintelligence to be *both* the primordial actuality in virtue of which there is definite limitations ("fine-tuning") among infinite possibilities, only a limited selection of which allow for life; *and* the contingent evolutionary outcome of these limitations in relation to an evolutionary cosmos. It is this claim of "dipolarity" that Whitehead affirms in his cosmotheology. God is *both* the necessary mental ground of the "myriad possibilities" (primordial nature) of which Dick speaks *and* the contingent product of cosmological evolution through the divine experience (consequent nature). Affirming a dipolar cosmo*theology* in which divine necessity has not been negated is again a way for Dick to better honor the philosophical and theological tradition, as well as remedy some potentially deep metaphysical problems.

In a biological universe, Dick is correct to say that the "anthropic principle" is a "spectacular misnomer."[193] As he rightly suggests, a "biocentric principle" is better suited to the context of astrobiology where life comes in truly plenitudinous forms. A "life-centric" drive in the universe would surely avoid any narrow focus on human (*anthropos*) life alone and must include any and all forms of life, whether terrestrial or extraterrestrial. While gaining increasing support, a "biocentric" or "biological" universe remains controversial because it suggests in no small way that the universe is up to something *teleological* in nature. As preconditioning our past and the reality of our emergence and evolution, this teleology would also indicate our purpose and destiny with respect to the future. It is here that we can end with Whitehead's fundamental metaphysical conviction.

If the universe is "biocentric" it is only because at a deeper level, it is already *organic*, *aesthetic,* and *experiential*. For Whitehead, these are the metaphysical preconditions without which life could not be produced at all—and certainly not *intelligent life* with highly evolved mental and conscious abilities. Whitehead, I trust, would instead hold to what John F. Haught calls the "aesthetic cosmological principle."[194] This principle insists upon the universal tendency toward the creation of value and beauty of which human subjective emergence—human life—is but one possible expression alongside a myriad of others. As Haught states, "The aim toward aesthetic intensity is the central theme of the cosmic story, and subjectivity is the most intense concentration of the cosmic aim toward beauty."[195] In saying this, Haught merely echoes Whitehead's own fundamental refrain: "The teleology of the Universe is directed to the production of beauty."[196]

In the final analysis, for Whitehead, human destiny is inextricable linked to cosmological evolution as a participant, and cosmological evolution is inextricably linked to a participatory God that chiefly exemplifies the metaphysical conditions rendering this universe possible, actual, and infinitely continuous. Up to this point, the process of cosmic evolution has involved the non-linear but, nevertheless, progressive intensification of such conditions with no final end in sight. For Whitehead, reaching for novel possibilities of beauty the origins of which reside not in a void, but in the primordial mind of God, and actualizing them to the highest degree possible—this is what constitutes human purpose and cosmic destiny. Only in this way can our lives be justified and meaningful and the divine experienced forever enriched. Should we finally meet and harmoniously interact with extraterrestrial life-forms, the universe will have then achieved a possibility of beauty that had hitherto hibernated in the life of the divine.

CONCLUSION: PRINCIPLES OF A DEEPER WHITEHEADIAN COSMOTHEOLOGY

The primary aim of this chapter has been to demonstrate how Dick's laudable project of naturalistic cosmotheology can be substantially deepened when inversely approached through Whitehead's organic metaphysics. This has involved articulating the ways in which the cosmological negations of Dick's first three principles can be inverted into metaphysical exemplifications, resulting in a *realist* affirmation of cosmo*theology* in the fourth principle, and real ontological support for the ethical and moral convictions of Dick's fifth and sixth principles. I pointed to different, but related, metaphysical problems of a God-shaped character that arguably remain in Dick's cosmotheology, including the ways in which Whitehead's theological realism transcends yet includes the superintelligence hypothesis, without falling into supernaturalism. While further work ever-remains in the tangled realms of cosmology, metaphysics, and theology, my inversive approach to Dick's cosmotheology has produced six alternative principles of a deeper Whiteheadian cosmotheology:

1. Humanity exemplifies metaphysical principles that are utterly central to the universe.
2. Humanity exemplifies biological, mental, and moral antecedents that are metaphysically central to the universe.
3. Humanity exemplifies the same metaphysical principles that are expressed in various intensities throughout the great chain of being in the universe.
4. Cosmotheology must be open to truly radical and non-supernatural conceptions of God, a God grounding and exemplifying the metaphysical conditions of cosmic evolution.
5. Cosmotheology must provide the ontological basis and stimulus for ideals of moral reverence and respect in the nature of things.
6. Cosmotheology must embrace human destiny as inextricably linked to the destiny of the cosmos as an infinite evolutionary expression of the metaphysical conditions chiefly exemplified in the divine.

NOTES

1. Dick first broached the topic in a chapter titled "Cosmotheology: Theological Implications of the New Universe." He returned to the topic in "Cosmotheology revisited: Theological Implications of Extraterrestrial Life" and later expanded his considerations in "Toward a Constructive Naturalistic Cosmotheology." His most recent deliberations "Astrobiology, Cosmotheology, and the Biological Universe" appear in the current volume, which has, in part, taken its rise from intuitions in his expansive work. For my expanded engagement of Dick's cosmotheology from within Whitehead's metaphysics, see Davis, *Metaphysics of Exo-Life*.

2. Dick, "Cosmotheology," 200.

3. Dick, "Toward a Constructive Naturalistic Cosmotheology," 193.

4. In no small way, Dick's work functioned as a catalyst for my organization of the conference which has resulted in the current volume.

5. Dick, "Toward a Constructive Naturalistic Cosmotheology," 200.

6. Whitehead, *Religion in the Making*, 67

7. Dick, "Toward a Constructive Naturalistic Cosmotheology," 193.

8. Ibid.

9. Dick has repeatedly expressed pushback against the following statements by John Greene: "When will scientists and others learn that naturalism is a philosophical point of view with no more claim to the status of science than any other philosophical viewpoint, whether Marxian, Freudian, Russellian, Whiteheadian, or whatever.

Scientists have as good a right to expound their philosophical, ethical and religious views as anyone else, but they have no right to palm these off as the findings of science." In holding that Greene is "dead wrong," Dick again emphasizes the fact that science must inform theology and religion: "science has at least as much claim to infuse theological thought as any other worldview, and perhaps more, since being grounded in the natural world gives it a pillar of support that supernatural religions do not have, numerous attempts at natural theology notwithstanding." Dick, "Astrobiology, Cosmotheology and the Biological Universe," this volume. Yet this affirmation of "naturalism" as opposed to "supernaturalism" is at best sparse and can be strengthened in ways that necessitate the wisdom of the metaphysical and theological tradition.

10. Dick, *Space, Time and Aliens*, 631. Whether or not the theological tradition already harbors the necessary radicalism Dick requires of cosmotheology is an ongoing debate. For an argument in favor of the sufficiently radical dimensions of the tradition, see Vainio, *Cosmology in Theological Perspective*. Alongside the various contributions in Peters, *Astrotheology*, Joel Parkyn's full-scale investigation in *Exotheology* is an important recent contribution to the discussion, as is Andrew Davison's *Astrobiology and Christian Doctrine*. To the extent that Dick's "naturalism" requires him to ontologically negate "God" out of existence and semantically reduce the discipline of "theology" to "that which gives meaning," it is not clear that the theological tradition has *positively* informed Dick's cosmotheology. For these reasons, Dick agrees that "cosmophilosophy" may indeed be the better term to use for his position. Dick, "Toward a Constructive Naturalistic Cosmotheology," 200.

11. Dick, *Space, Time and Aliens*, 656.

12. Whitehead, *Aims of Education*, 152–53.

13. Whitehead, *Science and the Modern World*, 17.

14. "Religion requires a metaphysical backing; for its authority is endangered by the intensity of the emotions which it generates. Such emotions are evidence of some vivid experience; but they are a very poor guarantee for its correct interpretation. Thus dispassionate criticism is beyond all things necessary. The foundations of dogma must lie in a rational metaphysics which criticizes meanings. . . . This position has never been seriously doubted, though in practice it is often evaded. One of the most serious periods of neglect [occurs] . . . through the dominance of historical interest . . . history presupposes a metaphysics. It can be objected that we believe in the past and talk about it without settling our metaphysical principles. . . . But you can only deduce metaphysical dogmas from your interpretation of the past on the basis of prior metaphysical interpretation of the present. . . . This criticism applies equally to science or to a religion which hopes to justify itself without any appeal to metaphysics." Whitehead, *Religion in the Making*, 71–73. Speaking elsewhere of the historical endeavor Whitehead uses stronger words: "[The] notion of historians, of history devoid of aesthetic prejudice, of history devoid of any reliance on metaphysical principles and cosmological generalizations, is a figment of the imagination. The belief in it can only occur to minds steeped in provinciality,—the provinciality of an epoch, of a race, of a school of learning, of a trend of interest—minds unable to divine their own unspoken limitations." Whitehead, *Adventures of Ideas*, 4.

15. Barring the expansive work of Faber in *The Divine Manifold*, *The Cosmic Spirit*, and most recently in *The Mind of Whitehead*, Whitehead has largely been absent from current philosophical and theological reflections on astrobiology and intelligent alien life. As the literature continues to grow under various "astro," "exo," and "cosmo" headings, my hope is that Whitehead, and the process tradition more broadly, will now be an important part of the discussion. Indeed, my efforts in the introduction to this volume have sought to further this goal. For other philosophical and theological perspectives, see Ćirković, *The Astrobiological Landscape*; Peters, *Astrotheology*; Parkyn, *Exotheology*; and Playford, *Exophilosphy* (forthcoming).

16. Whitehead, *The Function of Reason*, 70, emphasis mine.

17. Ibid., 61, emphasis mine.

18. Whitehead, *Process and Reality*, 91.

19. Whitehead, *Science and the Modern World*, 161.

20. Whitehead, *Process and Reality*, 91.

21. Ibid., 127.

22. Whitehead, *Modes of Thought*, 154–55.

23. As recorded by Price, *Dialogues of Alfred North Whitehead*, 342.

24. Whitehead, *Science and Philosophy*, 211.

25. For a brief account of Whitehead's own mediations in and between Plato and Newton, see *Process and Reality*, 93–96. For an overview of this discussion, see Davis, *Process Cosmology*, introduction; and Davis, *Mind, Value and Cosmos*, ch. 10.

26. Whitehead, *Process and Reality*, 96.

27. Whitehead, *Modes of Thought*, 88.

28. Whitehead, *Religion in the Making*, 144–45. While many have followed Griffin in saying that Whitehead affirmed a *single* series of "successive universes" such that there is "only one universe at a time," Faber complicates the discussion in terms of "many worlds" and the multiverse. See Griffin, *God Exists But Gawd Does Not*, 297; Faber, *Divine Manifold*, ch. 7; *The Garden of Reality*, 200. For a more recent discussion of the plurality of worlds in Whitehead's philosophy, see Petronio, "Cosmology and Mesology"; Morais, "Cosmic Epochs of Vigorous Order"; and McHenry, *The Event Universe*, ch. 5.

29. Whitehead, *Science and Philosophy*, 127.

30. Whitehead, *Process and Reality*, 3.

31. Whitehead, *Religion in the Making*, 72.

32. Ibid., 72–73.

33. Whitehead draws together these complementary methodological approaches in his famous aeronautical analogy: "The true method of discovery is like the flight of an aeroplane. It starts from the ground of particular observation; it makes a flight in the thin air of imaginative generalization; and it again lands for renewed observation rendered acute by rational interpretation." Whitehead, *Process and Reality*, 5.

34. Whitehead, *Process and Reality*, 4.

35. Whitehead, *Science and the Modern World*, 4, 5.

36. Whitehead, *Process and Reality*, 42. Alongside a number of twentieth-century scientists, Whitehead would thus disagree with Dick when he insists that it is *faith* as such that is the "difference between science and theology." Dick continues, "To say,

as some critics do, that scientists also have faith—faith in their methods, faith in facts, and faith in science itself—is a red herring, since scientists will overturn their beliefs, if sometimes grudgingly, in the face of strong evidence to the contrary. The same is not true of the faithful." Dick, "Astrobiology, Cosmotheology, and the Biological Universe," this volume. Whitehead fully agrees that "Religion will not regain its old power until it can face change in the same spirit as does science." Whitehead, *Science and the Modern World*, 189. However, the dogma that science is immune to "faith" turns a blind eye to the original theological trust in the rationality of the universe—a trust that helped stimulate the development of modern science itself. In speaking of the origins of modern science, Whitehead states, "My explanation is that the faith in the possibility of science generated antecedently to the development of modern scientific theory, is an unconscious derivative from medieval theology." Ibid., 13. As with most things, Nietzsche perhaps said it best: "it always remains a metaphysical faith upon which our faith in science rests—that even we devotees of knowledge today, we godless ones and anti-metaphysicians, still take our fire too from the flame which a faith thousands of years old has kindled: that Christian faith, which was also Plato's faith, that God is truth, that truth is divine." Nietzsche, *The Portable Nietzsche*, 450. For more on the indispensability of basic existential faith, see Ogden, *Notebooks*, 3–5.

37. Ibid., 4.

38. Ibid., 8.

39. Whitehead, *The Function of Reason*, 49.

40. Whitehead, *Modes of Thought*, 2.

41. As Leeman B. McHenry confirms, "Metaphysics [for Whitehead] therefore seeks principles that are necessary for any possible world or cosmic epoch while cosmology is discovered by observing what happens to be the case about our actual world or this cosmic epoch." McHenry, *The Event Universe*, 70.

42. Whitehead, *Process and Reality*, xiv.

43. Dick, "Astrobiology, Cosmotheology, and the Biological Universe," current volume.

44. Ibid.

45. See for example, Whitehead, *Science and the Modern World*, 18, 201 and Price, *Dialogues of Alfred North Whitehead*, 14.

46. Whitehead, *Science and Philosophy*, 129.

47. As recorded by Price, *Dialogues of Alfred North Whitehead*, 233–34.

48. See especially *The Concept of Nature* and *Science and the Modern World*.

49. See Whitehead, *Science and the Modern World*, 59, 87.

50. Whitehead, *Process and Reality*, 119.

51. Whitehead, *Adventures of Ideas*, 184–85.

52. Whitehead, *Modes of Thought*, 115.

53. Whitehead, *Process and Reality*, 88. For a deeper discussion of Whitehead's inversion of Kant, see Segall and Damer, "The Cosmological Context of the Origin of Life," this volume. Respective Kantian or Whiteheadian approaches to the world in this regard are not insignificant for the search for extraterrestrial intelligence and the question of potential commonalities between "us" and "them." Dick for example, rightly notes a Kantian solution to the Fermi paradox: "Where are they? Why do we

not see them? Because their mental structures are different from ours; their categories of thought are different from ours; their ways of understanding the world are different from ours. And the ways in which we might see them could manifest in very different ways." Dick, *Astrobiology, Discovery and Societal Impact*, 149. However, if one follows Whitehead's inversion of Kant, the possibility of real commonalty between ourselves and extraterrestrial intelligence immediately reemerges. Dick himself tends "toward a neo-Kantian position that such a comparison of human and alien minds might approach absolute reality." Ibid., 148. His influence by the Whiteheadian intuitions of Derek Malone-France in this regard is potentially significant and might be developed further in terms of the possibility of extraterrestrial commonality. See Dick, *Astrobiology, Discovery and Societal Impact*, 144–51 and Derek Malone-France, *Deep Empiricism*. For a different consideration of extraterrestrial commonality (in resonance with Haramia, "Astrobiological Searches for Life and Shared Knowledge," this volume), see Davis, "Whiteheadian Cosmotheology: Platonic Entities, Divine Realities and Shared Extraterrestrial Values."

54. Whitehead, *Science and the Modern World*, 18.

55. See Morris, *Life's Solution* and *From Extraterrestrials to Animal Minds*.

56. Dick, "Toward a Constructive Naturalistic Cosmotheology," 194.

57. Whitehead, *Science and Philosophy*, 211.

58. As recorded by Price, *Dialogues of Alfred North Whitehead*, 280.

59. "Urability" as used by Bruce Damer and David Deamer indicates the "thermodynamic and chemical conditions necessary for life to emerge on a planet." As Damer and Segall clarify, "We are used to thinking of the 'habitability' of planets, but 'urability' has to do with establishing not just habitability for existing life but the conditions for the origin (ur-) of life." Damer and Segall, "The Cosmological Context of the Origin of Life," this volume, n. 225.

60. Whitehead, *Science and the Modern World*, 17.

61. Ibid.

62. Ibid., 50,

63. Whitehead, *Adventures of Ideas*, 156–57.

64. Whitehead, *Science and the Modern World*, 150.

65. Whitehead, *Process and Reality*, 102.

66. As Whitehead states more fully: "The aboriginal stuff, or material, from which a materialistic philosophy starts is incapable of evolution. This material is in itself the ultimate substance. Evolution, on the materialistic theory, is reduced to the role of being another word for the description of the changes of the external relations between portions of matter. There is nothing to evolve, because one set of external relations is as good as any other set of external relations. There can be merely change, purposeless and unprogressive." Whitehead, *Science and the Modern World*, 107.

67. Ibid., 107.

68. Ibid., 36.

69. Refer to Malone-France, "Prospects for a Universal Philosophy of Organism," this volume.

70. Whitehead, *Science and the Modern World*, 36.

71. Whitehead, *Process and Reality*, 18.

72. Ibid., 163.
73. Whitehead, *Modes of Thought*, 151.
74. Whitehead, *Process and Reality*, 7, 21.
75. Ibid., 21
76. Following Whitehead and others, this affirmation of dynamic events and processes as opposed to inert matter informs Clement Vidal's own contributions to cosmology and astrobiology: "I choose an ontology of actions and agents, i.e., elementary processes and relations, not independent, static pieces of matter (in the spirit of Whitehead . . .)." Vidal, *The Beginning and the End*, 321.
77. "Scientific reasoning is completely dominated by the presupposition that mental functionings are not properly part of nature. Accordingly it disregards all those mental antecedents which mankind habitually presuppose as effective in guiding cosmological functioning . . . this sharp division between mentality and nature has no ground in our fundamental observation. We find ourselves living within nature. . . . I conclude that we should conceive mental operations as among the factors which make up the constituents of nature." Whitehead, *Modes of Thought*, 156.
78. Whitehead, *Process and Reality*, 53.
79. Whitehead, *Science and the Modern World*, 144, 151.
80. Whitehead, *Process and Reality*, 267.
81. Ibid., 162.
82. See Whitehead's influence among the contributions in Kelly and Kelly, *Irreducible Mind*; Kelly et al., *Beyond Physicalism*; and Kelly and Marshal, *Consciousness Unbound.*
83. Whitehead, *The Function of Reason*, 24. Ian McGilchrist, arguably one of the most important process philosophers to date, echoes Whitehead's conviction: "At the core of the contemporary world is the reductionist view that we are—nature is—'nothing but' a bundle of senseless particles, pointlessly, helplessly, colliding in a predictable fashion, whose existence is purely material, and whose only value is utility. . . . I cannot remember a time when I thought this sounded at all convincing; and a lifetime's thinking and learning had done nothing at all to allay my scepticism. Not only is it mistaken, I believe, but actively damaging—physically, to the naturally world; and psychologically, morally, and spiritually to ourselves as part of that world. It endangers everything that we should value." McGilchrist, *The Matter With Things*, vol. 1, 5. See also his brilliant metaphysical elaborations in *The Matter With Things*, vol. 2.
84. Whitehead, *Science and the Modern World*, 93.
85. Whitehead, "First lecture," 19.
86. See Lowe, *Alfred North Whitehead, vol. 1*, 168, 270. For a recent contribution to this discussion, including the equation of "pan-experientialism" with a "pan-axiological view of nature," see Barrett, *Enjoyment as Enriched Experience*, 345.
87. Rose, *On Whitehead*, 3, emphasis mine.
88. For an underappreciated study of value in Whitehead's metaphysics, refer to Leue, *Metaphysical Foundations for a Theory of Value in the Philosophy of Alfred North Whitehead.*

89. Whitehead, *Science and the Modern* World, 86.

90. Ibid., 94.

91. Whitehead, *The Concept of Nature*, 4.

92. Whitehead, *Science and Philosophy*, 138–39.

93. Dick, *Space, Time and Aliens*, 682

94. Dick, "Toward a Constructive Naturalistic Cosmotheology," 195.

95. See for example Cobb, *Back to Darwin*; Koutroufinis, *Life and Process*; Dupré and Nicholson, *Everything Flows* and Dupré, *The Metaphysics of Biology*; Henning and Scarfe, *Beyond Mechanism*; Reiss and Ruse, *The New Biology* and McGilchrist, *The Matter With Things*, ch., 29.

96. Whitehead, *The Function of Reason*, 5.

97. See McIntosh, *Evolution's Purpose* and *The Presence of the Infinite*.

98. Whitehead, *Religion in the Making*, 91. For Whitehead *all* order is aesthetic order, even the order exhibited in the "laws" of nature and the presupposed formational order of the world. As Charles Hartshorne states, "The most general principles of harmony and intensity are more ultimate than the laws of physics and are the reasons for there being natural laws." Put differently, the laws of nature are already an expression of *deeper* aesthetic principles. Charles Hartshorne, "A Replay to My Critics," 590. Similarly, Mario Livio states "Since the laws of nature have the elements of beauty engraved in them, it should come as no surprise that aesthetic principles played a major role in the shaping of our thinking about the origins of the universe." Livio, *The Accelerating Universe*, 43. These statements by Hartshorne and Livio resonate nicely with Whitehead's own conviction that "the foundations of the world" are founded upon the "aesthetic order" and that the world itself "is the outcome of aesthetic order." Whitehead, *Religion in the Making*, 91–92.

99. Whitehead, *Modes of Thought*, 111.

100. Whitehead, *Adventures of Ideas*, 265.

101. Griffin, *Panentheism and Scientific Naturalism*, 88. To say that intelligent extraterrestrial beings would "share some of our capacities" with respect to "morality" or "Goodness," however, is not to say that morality or ethics can be rigidly codified in ways that are universally applicable to all beings in the cosmos. Whitehead, in fact, rejected this, saying, "the notion that there are certain regulative notions, sufficiently precise to prescribe details of conduct, for all reasonable beings on Earth, in every planet, and in every star-system, is at once to be put aside. That is the notion of the one type of perfection at which the Universe aims. All realization of the Good is finite, and necessarily excludes certain other types." Whitehead, *Adventures of Ideas*, 291. For further discussion on these themes, refer to McInstosh, *The Presence of the Infinite* and Davis, "Whiteheadian Cosmotheology."

102. To remain consistence in demonstrating the shared element of negation in the phrasing of Dick's first three principles, I have adjusted his expression of his third principle which is usually stated as "*Humanity is near the bottom of the great chain of beings in the universe.*" Dick, "Toward a Constructive Naturalistic Cosmotheology," 194.

103. Ibid., 194–95.

104. Whitehead, *Adventures of Ideas*, 184.

105. As recorded by Price, *Dialogues of Alfred North Whitehead*, 233–34.

106. Randall E. Auxier and Gary L. Herstein rightly state that "There are those among Whitehead's interpreters who are misled by the name Whitehead chose for this category of entities. Some think that Whitehead must be 'naturalized,' and that eternal objects are somehow non-natural or even unnatural. There term 'possibility' is much less confusing on this point. Anyone who wants to rid Whitehead's philosophy of eternal objects is actually attempting to cut out the category of possibility. It is neither necessary nor possible to do that while preserving any semblance of the philosophy of organism. Taking possibility seriously is a requirement of all process philosophy." Auxier and Herstein, *The Quantum of Explanation*, 143.

107. Whitehead, *Process and Reality*, 222.

108. Ibid., 22; Whitehead, *Modes of Thought*, 99–100.

109. Whitehead, *Science and Philosophy*, 91–92.

110. While Dick states that "cosmotheology favors the Superintelligence Hypothesis as opposed to the God Hypothesis," he also stresses that "in the end," the existence of a superintelligence is not a necessary, but only a possible, aspect of cosmotheology." Dick, "Astrobiology, Cosmotheology, and the Biological Universe," this volume.

111. Dick, "Toward a Naturalistic Cosmotheology," 195.

112. Ibid., 200; "Astrobiology, Cosmotheology, and the Biological Universe," this volume.

113. As Auxier and Herstein comment, "Whitehead retained the word 'God' and was right to do so. That is the word philosophers use when grappling with the essence of the Whole. It is no rigid designator, but it also cannot fail to refer, at least not if anything *has* an explanation." Auxier and Herstein, *The Quantum of Explanation*, 271. For a brilliant meditation on the indispensability of the word "God" see Rahner, *Foundations of Christian Faith*, 44–51. See also McGilchrist, *The Matter With Things*, 1203–7.

114. Consider Whitehead's full statement: "Among medieval and modern philosophers, anxious to establish the religious significance of God, an unfortunate habit has prevailed of paying to Him metaphysical compliments. He has been conceived as the foundation of the metaphysical situation with its ultimate activity. If this conception be adhered to, there can be no alternative except to discern in Him the origin of all evil as well as of all good. He is then the supreme author of the play, and to Him must therefore be ascribed its shortcoming as well as its success." Whitehead, *Science and the Modern World*, 179. See also ch. 12, "Science and Religion." For a larger discussion, see Griffin, *Religion and Scientific Naturalism*.

115. Whitehead, *Process and Reality*, 343.

116. Ibid., 146; *Adventures of Ideas*, 167.

117. "God is an actual entity, and so is the most trivial puff of existence in far-off empty space. But, there are gradations of importance, and diversities of function, yet in the principles which all actuality exemplifies all are on the same level. The final facts are, all alike, actual entities; and these actual entities are drops of experience, complex and interdependent." Whitehead, *Process and Reality*, 18.

118. "The true metaphysical position is that God is the aboriginal instance of this creativity, and is therefore the aboriginal condition which qualifies its action. It is the function of actuality to characterize the creativity, and God is the eternal primordial character." Whitehead, *Process and Reality*, 225.

119. Price reports that over nearly a decade, during his Cambridge period, Whitehead was an adamant reader of theology: "This was all extracurricular, but so thorough that he amassed a sizable theological library. At the expiry of these eight years he dismissed the subject and sold the books. A Cambridge bookdealer was willing to give quite a handsome figure for the collection. It then appeared that the pay must be taken in books at his shop. So he went on an orgy of book-buying until he had overdrawn his account." Price, *Dialogues of Alfred North Whitehead*, 6.

120. In this regard Thomas Hosinksi rightly states that "Whitehead did not introduce the concept of God into his philosophy because he had some hidden religious agenda; he introduced it on rational grounds because it was necessary for the consistency and coherence of his philosophy. . . . The principle reason for introducing the concept of God . . . has do with its necessity for answering rationally the ultimate questions which arise from his analysis of experience." Hosinski, *Stubborn Fact and Creativity Advance*, 156.

121. Hocking, "Whitehead as I Knew Him," 16.

122. Johnson, "Some Conversations with Whitehead," 4.

123. Dick, "Toward a Constructive Naturalistic Cosmotheology," 200, my emphasis.

124. Auxier and Herstein rightly comment that whether Whitehead's concept of God "is or is not beyond the scope of a defensible naturalism depends both upon what 'naturalism' means and what role the concept of God was playing in a given inquiry. . . . Demonstrating that his theory of possibility . . . is well within the scope of naturalism implies, in our view, that the burden of showing that any claim that eternal objects are non-natural now devolves upon those who claim it. Whitehead would find the charge surprising." Auxier and Herstein, *The Quantum of Explanation*, 241. These statements are also applicable to conversations among "process naturalists" or advocates of "Whitehead without God" like Frederick Ferré and Donald Crosby who (like Dick) deny the need to associate possibilities with God. There continues to be a debate as to the nature and definition of "naturalism," as well as the nature of metaphysical explanation among Whitehead scholars. Refer to Ferré, *Being and Value*, ch. 11; Crosby, *Novelty*, ch. 3; Rose, "Rethinking Whitehead's Cosmology"; Davis, *Mind, Value and Cosmos*, ch. 7.

125. Dick has recognized this in part, but he has not thoroughly engaged the work of process thinkers on the topic. See Griffin, *Religion and Scientific Naturalism*; *Reenchantment Without Supernaturalism*, and Malone-France, *Deep Empiricism*. For a different perspective on the issue of "naturalism" vs. "supernaturalism," see Forrest, *God Without the Supernatural*.

126. Dick, "Toward a Constructive Naturalistic Cosmotheology," 196.

127. See for example Derek Malone-France, *Deep Empiricism*, part II; Davis, *Mind, Value and Cosmos*, part III; Eastman, *Untying the Gordian Knot*, chs. 8; Auxier

and Herstein, *The Quantum of Explanation*, chs. 8, 12; and Rescher, *Metaphysical Perspectives*, chs. 1, 7–8.

128. In part, both James Lindsay and Thomas Hosinski point directly to the kind of fundamental questions one must engage: "The philosophy of possibility," Lindsay states, "cannot evade the question of the origin of possibilities. Can we trace possibility simply to the human mind? Do possibilities not exist before the human mind comes into being? Will the possibilities not exist after the human mind has ceased to exist? Can we even ascribe the possibilities to the universe? If the universe were done away, would possibilities not remain in undiminished form? For, are the possible universes not infinite? And, is not possibility necessary and eternal? These are among the questions that may be asked. The philosophy of possibility can hardly be satisfied to accept these possibilities as accounting for themselves." Lindsay, "Philosophy of Possibility," 322–23. Similarly, Hosinski comments: "Possibility is fascinating . . . the point I would make here is that possibility is not self-explanatory. Where do possibilities come from? Must we simply say, as we do for energy that possibilities just are? Must we simply assume them without being able to explain where they come from and how they function? Possibilities by definition are not actual, even though they are real. They are real in the sense that we can think of possibilities and interact mentally with them, but they are not actual in the sense that they are not agents, nor are they actual things. We cannot hold a possibility in our hands or take it on a walk. Possibilities must be actualized in and by some actual agent of the universe to be become actualized. How then, if they are not actual things, can they interact with actualities? These are not scientific questions but philosophical ones." Hosinski, *The Image of the Unseen God*, 77–78.

129. Refer to Lupsella, "The Connection-Action Principle," this volume.

130. Consider Lupisella's comments from his recent book *Cosmological Theories of Value*: "One way to think about metaphysics . . . is that while ontology is concerned with what actually exists (primarily physically), metaphysics is more concerned with what may exist, or theoretical possibilities, including that which is 'non-physical,' based on a general fundamental understanding of reality in the broadest sense. This may include considerations such as God, or a kind of Platonic realm in which there are theoretical properties (e.g., 'universals' such as mathematical constructs, logic, redness, etc . . .) or theoretical possibilities for the universe and reality in general. The domain of metaphysics might be thought of as the largest possibility space for 'ultimate reality,' a subset of which is the actual and/or temporary state of reality. On some interpretations, this metaphysical possibility space could include things like value, meaning, purpose, divinity, 'spirit,' etc. So, metaphysics can be seen to be a very broad (perhaps the broadest) category of investigation (and experience)." Lupisella, *Cosmological Theories of Value*, 130–31.

131. For a more detailed discussion, see Davis, *Mind, Value and Cosmos*, chs. 6, 7.

132. Whitehead, *Process and Reality*, 46, 11, 40.

133. Ibid., 40.

134. Ibid., 46.

135. Whitehead, *Religion in the Making*, 137.

136. Whitehead, *Process and Reality*, 46.

137. Ibid., 44; *Modes of Thought*, Lect. V. Malone-France states rightly that Whitehead does *not* claim any *extra* or *independent* existence for the 'possibles' (or 'impossibles') entertained by actualities. Their 'existence' is, therefore, according to Whitehead, *nothing other than [divine] entertainment*. Even in the case of possibles or impossibles that are not, and never have been entertained by human understanding, the existence imputed to them in Whitehead's theory is non-reified. The *only* mode of existence that is attributed to eternal objects that do not find inclusion in any *other* actual entity is their inclusion in the *divine* actual entity" Malone-France, *Deep Empiricism*, 121. Faber states that Whitehead's vision was a "direct reversal" of Plato (and what many think central to Platonism) by holding that "the forms in becoming are forms of becoming, not essences actualizing themselves in the shadow of matter, but potentials of actualizations exhibited by the character of societies" Faber, *The Becoming of God*, 35. Didier Debaise too states that "Whitehead aims not to recover some kind of purified and original 'Platonism' but to introduce elements into Platonic thought to effect a rupture, though in the shape of a return . . . Whitehead asserts a Platonic inheritance by completely inverting the theory of ideas." Didier Debaise, *Speculative Empiricism*, 95. Auxier and Herstein in summary: "These are not Plato's forms, but, as Whitehead says, closer to what he imagines Plato would say in he lived in the twentieth century." Auxier and Herstein, *The Quantum of Explanation*, 240.

138. Whitehead, *Process and Reality*, 40, 46, 19.

139. Whitehead, *Science and Philosophy*, 126

140. Whitehead, *Process and Reality*, 344.

141. "The final entity is the divine element in the world, by which the barren inefficient disjunction of abstract potentialities obtains primordially the efficient conjunction of ideal realization. This ideal realization of potentialities in a primordial actual entity constitutes the metaphysical stability whereby the actual process exemplifies general principles of metaphysics, and attains the ends proper to specific types of emergent order. By reason of the actuality of this primordial valuation of pure potentials, each eternal object has a definite, effective relevance to each concrescent process. Apart from such orderings, there would be a complete disjunction of eternal objects unrealized in the temporal world. Novelty would be meaningless, and inconceivable." Ibid., 40.

142. Whitehead, *Religion in the Making*, 81–82; *Process and Reality*, 349.

143. Malone-France, *Deep Empiricism*, 190.

144. Whitehead, *Religion in the Making*, 81–82.

145. Whitehead, *Science and Philosophy*, 98, 101; *Religion in the Making*, 138; *Process and Reality*, 345; *Science and the Modern World*, 179. For a discussion of value as the base of divine necessity in dialogue with Whitehead, Keith Ward and John Leslie, see Davis, *Mind, Value, and Cosmos*, Part II.

146. Whitehead, *Religion in the Making*, 50.

147. Whitehead, *Adventures of Ideas*, 130.

148. Whitehead, *Religion in the Making*, 90–91.

149. Refer to Hartshorne's *Philosophers Speak of God* and Dombrowski's *A History of the Concept of God.*

150. Johnson, "Some Conversation with Whitehead," 5–6. For a helpful discussion on the relationship between the primordial and consequent natures of God, see Lansing, "The 'Natures' of Whitehead's God."

151. Whitehead, *Process and Reality*, 345.

152. Consider David Ray Griffin's "atheist thought experiment" with respect to this problem: "If the universe contained only a plurality of finite perspectives, there would be no perspective that corresponds with the general nature of reality, let alone one that correctly reflects the concrete events that have occurred. Experience would consist of nothing but a vast plurality of perspectives, with no impartial perspective that includes all of the partial perspectives. These partial perspectives would often have radically different perspectives on the truth of this or that. There would, accordingly, no place where truth as such—*sometimes called Truth with a capital T*—could exist. But this would leave us with paradox: The ultimate truth about reality is that there is no ultimate truth." Griffin, *God Exists, But Gawd Does Not*, 225.

153. Following Hartshorne (although not without criticism), Auxier and Herstein raise this point poignantly for the nontheist: "For the non-theist, what is the ontological ground for distinguishing the *was* from the *might-have-been*? Laws of Nature? Those depend upon the same distinction; alternative accounts of such Laws could account for the same present. Personal memory? This implies solipsism and a transcendental unity of the self that could never be demonstrated. Transcendental ego? Even that ego needs an actual past to distinguish its temporal retending from its intending and portending. Configuration of the present? Infinitely many pasts could have produced the present; if only and only one of them actually *was*, what basis do we have for throwing out the others? We repeat: How would a non-theist support the claim that the *was* and the *might-have-been* are different? Here we find also the reason that the Whiteheadian interpreters who attempt to eliminate God from the system are the wrong track. In short, no God, no actuality, including actuality of the *possible* in distinction from the actuality of the concrete." Auxier and Herstein, *The Quantum of Explanation*, 267–68.

154. Whitehead, *Process and Reality*, 12–13. Pierfrancesco Basile comments on this passage as "somewhat surprising" but also "theologically innovative" in its departure of classical theological sensibilities: "As the world develops, the past actualities are summed up in God, where they are fused together into a single representation of the entire cosmic history—'the truth itself.' Since history never comes to a stop, new occasions enter God's consequent nature, so that the synoptic representation contained there is constantly updated. God is in a way omniscient, yet only *ex post*; since history exists in the making, not even God can have any knowledge of future events. They simply are no such events to be known in the first place. In this respect, Whitehead's God differs significantly from . . . the God of classical theism." Basile, *Whitehead's Metaphysics of Power*, 107–8.

155. Ibid., 12–13.

156. As Leemon B. Mchenry states, the consequent nature of God "turns out to be a sort of retainer of the whole of cosmic history since only bits and pieces of the past survive directly in the present. This has the rather odd consequence that all historical statements must ultimately be understood as theological since the referents of such

historical statements are the retention of the past in God." McHenry, *The Event Universe*, 105–6.

157. Hosinski, *Stubborn Fact and Creative Advance*, 182.

158. Whitehead, *Process and Reality*, 340.

159. Ibid., 198.

160. As Daniel A. Dombrowski has stated, divine memory in this sense "is the paradigm case of experiencing and provides the avenue by which to best understand why perpetual perishing is not the last word. God is not a mere spectator, but a participant in the process of the world with ideal memory." Dombrowski, *A History of the Concept of God*, 248.

161. While Whitehead did not give any clear doctrine of "subjective immortality" as some of his interpreters have attempted, he also did not reject life after death like his assistant Charles Hartshorne. Although he at one point said that his philosophy was "neutral" on the question of survival of death, he was nevertheless empirically open to being more than neutral "provided that the evidence is trustworthy." Whitehead, *Religion in the Making*, 97. For different entry points into this discussion/debate with respect to process metaphysics, see Hartshorne, *Omnipotence and Other Theological Mistakes*; Griffin, *James and Whitehead on Life After Death*; *Parapsychology, Philosophy and Spirituality*; Suchocki, *The End of Evil*; and Bracken, *World Without End*.

162. Whitehead, *Religion in the Making*, 110.

163. Dick, "Toward A Constructive Process Cosmotheology," 196.

164. Ibid., 200.

165. Ibid, 193.

166. Griffin, *Reenchantment Without Supernaturalism*, 146, 148, 170, 184, 188–89.

167. Refer to Dick, *Astrobiology, Discovery, and Societal Impact*, chs. 8–9; Smith and Mariscal, *Social and Conceptual Issues in Astrobiology*, part IV & V; Peters, *Astrotheology*, part 5; and Torres, *Astrobiology*.

168. Whitehead, *Process and Reality*, 27.

169. Dick, "Astrobiology, Cosmotheology, and the Biological Universe," this volume.

170. Refer to Faber, "A Darker Forest?," this volume.

171. See Dick, *Astrobiology, Discovery, and Societal Impact*, 137–40.

172. Dick, "Astrobiology, Cosmotheology, and the Biological Universe," this volume.

173. Ibid.

174. Dick, *Astrobiology, Discovery, and Societal Impact*," 243, 244.

175. Dick, "Astrobiology, Cosmotheology, and the Biological Universe," this volume.

176. Dick, *Astrobiology, Discovery, and Societal Impact*, 264.

177. Dick, "Toward a Constructive Naturalistic Cosmotheology," 196.

178. Quoted in Griffin, *God Exists But Gawd Does Not*, 155.

179. For a stimulating and challenging argument for the reality of Plato's "Good" in the form of ethical "needs" or "requirements," some of which are creatively effective,

refer to Leslie, *Infinite Minds* and Davis, *Mind, Value, and Cosmos*. For a different take, see Rescher, *Axiogenesis*.

180. Whitehead, *Process and Reality*, 317.

181. Whitehead, *Religion in the Making*, 91–92.

182. Ibid., 86–87.

183. Whitehead, *Modes of Thought*, 103.

184. Dick, "Toward a Constructive Naturalistic Cosmotheology," 195.

185. Dick, "Astrobiology, Cosmotheology, and the Biological Universe," this volume.

186. For deeper insight into the "noosphere," as Teilhard understands it, see Delio, "Extending the Noosphere into Intergalactic Life," this volume. As Delio has expressed elsewhere: Teilhard's God "is not a supernatural being hovering above Earth, but the supra-personal whole, the Omega, who exists in and through all." Delio, *From Teilhard to Omega*, 45. Refer also to recent work by John F. Haught in *The New Cosmic Story*; *The Cosmic Vision of Teilhard De Chardin,* and *God After Einstein* which is consistently concerned with affirming both the reality of God and the naturalism of science. His book *God After Darwin* is also a valuable contribution.

187. As Faber states, "For Whitehead, the situation is different [from Teilhard] . . . since his cosmology does not imply any 'end of the world,' but rather 'an endless renewal of worlds.' Faber, however, warns against surface-level readings of Teilhard's 'linear teleology.' While it is apparent that Whitehead denies a 'linear progress in light of an infinite cycle of cosmic epochs without beginning and end,'" Faber insists that "it can easily be overlooked that Teilhard's thought was hardly that simple." We cannot assume his belief in some "naïve progression." Faber, *The Ecology of God*, ch. 5.

188. Whitehead, *Process and Reality*, 111. The modern Muslim process philosopher Muhammad Iqbal echoes Whitehead's intuitions in this regard: "The world-process or the movement of the universe in time, is certainly devoid of purpose, if by purpose we mean a foreseen end—a far-off fixed destination to which the whole creation moves. To endow the world-process with purpose in this sense is to rob it of its originality and its creative character. . . . A time-process cannot be conceived as a line already drawn. It is a line in the drawing—an actualization of open possibilities." Iqbal, *The Reconstruction of Religious Thought in Islam*, 44. Thanks to Farhan Shah for bringing this quote to my attention.

189. In his introduction to Teilhard's *Phenomenon of Man*, Julian Huxley states the following: "Presumably, in designating this state as Omega, [Teilhard] believed that it was a truly final condition. It may have been better to think of it as a novel state or mode of organization, beyond which the human imagination cannot at present pierce." Huxley, "Introduction," 18. If indeed, Omega only designates radically convergent novelty in cosmological evolution, then Whitehead and Teilhard come closer together, for radical novelty is essentially what the universe is up to, not least in ourselves. For Whitehead, however, there is no *guarantee* that can be assigned to any radically convergent novelty in cosmological evolution.

190. Dick, "Astrobiology, Cosmotheology, and the Biological Universe," this volume.

191. Dick included this possibility in his first ever chapter on cosmotheology, saying, "Such advanced intelligence could have fine tuned the physical constants, thus explaining the conundrum of the anthropic principle. In principle, it could even 'intervene in human history,' the touchstone principle of the Christian faith, not to mention of the UFO and alien abductee advocates." Dick, "Cosmotheology," 204.

192. Whitehead, *Process and Reality*, 94.

193. Dick, "Astrobiology, Cosmotheology, and the Biological Universe," this volume.

194. See Haught, *God after Darwin*, 128.

195. Haught, *The New Cosmic Story*, 140.

196. Whitehead, *Adventures of Ideas*, 265.

BIBLIOGRAPHY

Auxier, R. E., and G. L. Herstein. *The Quantum of Explanation: Whitehead's Radical Empiricism*. London: Routledge, 2019.

Barrett, N. F. *Enjoyment as Enriched Experience: A Theory of Affect and Its Relation to Consciousness*. Cham: Palgrave, 2023.

Basile, P. *Whitehead's Metaphysics of Power: Reconstructing Modern Philosophy*. Edinburgh: Edinburgh University Press, 2017.

Bracken, J. A. *World Without End: Christian Eschatology from a Process Perspective*. Grand Rapids: Eerdmans, 2005.

Ćirković, M. M. *The Astrobiological Landscape: Philosophical Foundations of the Study of Cosmic Life*. Cambridge: Cambridge University Press, 2012.

Cobb, J.B. Jr. *Back to Darwin: A Richer Account of Evolution*. Grand Rapids: Eerdmans, 2008.

Conway Morris, S. *From Extraterrestrials to Animal Minds: Six Myths of Evolution*. Philadelphia: Templeton Press, 2022.

———. *Life's Solution: Inevitable Humans in a Lonely Universe*. Cambridge: Cambridge University Press, 2003.

Crosby, D. A. *Novelty*. Lanham, MD: Lexington, 2005.

Davis, A. M. *Metaphysics of Exo-Life: Toward a Constructive Whiteheadian Cosmotheology*. Grasmere. SacraSage Press, 2023.

———. *Mind, Value and Cosmos: On the Relational Nature of Ultimacy*. Lanham, MD: Lexington, 2020.

———. "Whiteheadian Cosmotheology: Platonic Entities, Divine Realities and Shared Extraterrestrial Values." In *Process Cosmology: New Integrations in Science and Philosophy*, edited by Andrew M. Davis et al., 423–52. Cham: Palgrave, 2022.

Davis, A. M. et al., eds. *Process Cosmology: New Integrations in Science and Philosophy*. Cham: Palgrave, 2022.

Davison, A. *Astrobiology and Christian Doctrine: Exploring the Implications of Life in the Universe*. Cambridge: Cambridge University Press, 2023.

Debaise, D. *Speculative Empiricism: Revisiting Whitehead*. Edinburgh: Edinburgh University Press, 2007.

Delio, I. "Extending the Noosphere into Intergalactic Life: Teilhard de Chardin and the Third Axial Age," current volume.

———. *From Teilhard to Omega: Co-creating an Unfinished Universe*. Maryknoll, NY: Orbis Books, 2014.

Dick, S. J. "Astrobiology, Cosmotheology, and the Biological Universe: Implications for Religion and Theology," current volume.

———. "Cosmotheology revisited: Theological Implications of Extraterrestrial Life." In *ConsCiências: actas do Fórum Internacional Ciência, Religião e Consciência*, 287–301. Porto, Edições: Universidade Fernando Pessoa, CTEC, 2005.

———. "Cosmotheology: Theological Implications of the New Universe." In *Many Worlds: The New Universe, Extraterrestrial Life, and its Theological Implications*, edited by S. J. Dick, 191–210. Philadelphia: Templeton Press, 2000.

———. "Toward a Constructive Naturalistic Cosmotheology." In *Astrotheology: Science and Theology Meet Extraterrestrial Life*. Edited by T. Peters, 228–44. Eugene: Cascade, 2018.

———. *Astrobiology, Discovery, and Societal Impact*. Cambridge: Cambridge University Press, 2018.

———. *Space, Time, and Aliens: Collected Works on Cosmos and Culture*. Cham: Springer, 2020.

Dombrowski, D. A. *A History of the Concept of God: A Process Approach*. Albany: SUNY, 2016.

Dupré, J. *The Metaphysics of Biology*. Cambridge: Cambridge University Press, 2021.

Eastman, T. *Untying the Gordian Knot: Process, Reality, and Context*. Lanham, MD: Lexington, 2020.

Faber, R. "A Darker Forest? The Fermi Paradox and Extraterrestrial Spiritual Life (ETS)," current volume.

———. *The Becoming of God: Process Theology, Philosophy and Multireligious Engagement*. Eugene: Cascade, 2017.

———. *The Cosmic Spirit: Awakenings at the Heart of All Religions, The Earth, and the Multiverse*. Eugene: Cascade, 2021.

———. *The Divine Manifold*. Lanham, MD: Lexington, 2014.

———. *The Ecology of God: Whiteheadian Excursions in Eco-Theology, Togetherness and Divine Liberation*. Eugene: Pickwick, Forthcoming.

———. *The Garden of Reality: Transreligious Relativity in a World of Becoming*. Lanham, MD: Lexington, 2018.

———. *The Mind of Whitehead: Adventure of Ideas.* Eugene: Pickwick, 2023.

Ferré, F. *Being and Value: Toward a Constructive Postmodern Metaphysics*. Albany: SUNY Press, 1996.

Forrest, P. *God without the Supernatural*. Ithaca: Cornell University Press, 1996.

Griffin, D. R. *God Exists But Gawd Does Not: From Evil to New Atheism to Fine-Tuning*. Anoka: Process Century Press, 2016.

———. *James and Whitehead on Life After Death*. Anoka: Process Century Press, 2022.

———. *Panentheism and Scientific Naturalism: Rethinking Evil, Morality, Religious Experience, Religious Pluralism, and the Academic Study of Religion.* Claremont: Process Century Press, 2014.

———. *Parapsychology, Philosophy and Spirituality: A Postmodern Exploration.* Albany: SUNY Press, 1997.

———. *Reenchantment Without Supernaturalism: A Process Philosophy of Religion.* Ithaca: Cornell University Press, 2001.

Hartshorne, C. *Omnipotence and Other Theological Mistakes*. Albany: SUNY Press, 1984.

Hartshorne, C., and W. L. Reese. *Philosophers Speak of God*. Amherst, NY: Humanities, 2000.

Haught, J. F. *God after Darwin: A Theology of Evolution*. Boulder: Westview Press, 2000.

———. *God after Einstein: What's Really Going On in the Universe?* New Haven: Yale University Press, 2022.

———. *The Cosmic Vision of Teilhard De Chardin*. Maryknoll, NY: Orbis Books, 2021.

———. *The New Cosmic Story: Inside Our Awakening Universe*. New Haven: Yale University Press, 2017.

Henning, B., and A. Scarfe, *Beyond Mechanism: Putting Life Back Into Biology*. Lanham, MD: Lexington, 2013.

Hocking, W. E. "Whitehead as I Knew Him." In *Alfred North Whitehead: Essays on His Philosophy*, edited by G. L. Kline, 7–17. Englewood Cliffs: PrenticeHall, 1963.

Hosinski, T. *The Image of the Unseen God: Catholicity, Science and Our Evolving Understanding of God*. Maryknoll, NY: Oribs, 2017.

———. *Stubborn Fact and Creative Advance: An Introduction to the Metaphysics of Alfred North Whitehead*. Lanham, MD: Rowman & Littlefield, 1993.

Huxley, J. "Introduction." In Teilhard De Chardin, Pierre. *The Phenomenon of Man*, 11–30. New York: Harper Perennial, 1955.

Iqbal, M. *The Reconstruction of Religious Thought in Islam*. Stanford: Stanford University Press, 2013.

Johnson, A. H. "Some Conversations with Whitehead Concerning God and Creativity." In *Explorations in Whitehead's Philosophy*. Edited by L. S. Ford and G. L. Kline, 3–13. New York: Fordham University, Press, 1983.

Kelly, E. F. and E. W. Kelly. *Irreducible Mind: Toward a Psychology for the 21st Century*. Lanham, MD: Rowman & Littlefield, 2007.

Kelly, E. F. and P. Marshal. *Consciousness Unbound: Liberating Mind from the Tyranny of Materialism*. Lanham, MD: Rowman & Littlefield, 2021.

Kelly, E. F. et al. *Beyond Physicalism: Toward a Reconciliation of Science and Spirituality*. Lanham, MD: Rowman & Littlefield, 2015.

Koutroufinis, S. A. *Life and Process: Towards a New Biophilosophy*. Berlin: de Gruyter, 2014.

Lansing, J. W. "The 'Natures' of Whitehead's God." *Process Studies* 3, no. 3 (1973): 143–52.

Leslie, J. *Infinite Minds: A Philosophical Cosmology*. Oxford: Clarendon Press, 2001.

Leue, W. H. *Metaphysical Foundations for a Theory of Value in the Philosophy of Alfred North Whitehead*. Ashfeld: Down-to-Earth Books, 2005.

Lindsay, J. "The Philosophy of Possibility." *The Monist* 32, no. 3 (July 1922): 321–38.

Livio, M. *The Accelerating Universe: Infinite Expansion, the Cosmological Constant, and the Beauty of the Cosmos.* New York: Wiley, 2000.

Lowe, V. *Alfred North Whitehead: The Man and His Work. Vol. II: 1910–1947*. Edited by J. B. Schneewind. Baltimore: Johns Hopkins University Press, 1990.

Lupisella, M. "The Connection-Action Principle: A Basis for Process Philosophy, Cosmic Creativity, and Value?," current volume.

———. *Cosmological Theories of Value: Science, Philosophy, and Meaning in Cosmic Evolution*. New York: Springer, 2020.

MacIntosh, S. *Evolution's Purpose: An Integral Interpretation of the Scientific Story of Our Origins*. New York: Select Books, 2012.

———. *The Presence of the Infinite: The Spiritual Experience of Beauty, Truth, and Goodness*. Wheaton: Quest Books, 2015.

Malone-France, D. "Prospects for a Universal Philosophy of Organism," current volume.

———. *Deep Empiricism: Kant, Whitehead, and the Necessity of Philosophical Theism*. Lanham, MD: Lexington, 2007.

McGilchrist, I. *The Matter With Things: Our Brains, Our Delusions and the Unmaking of the World*, vol. 1 & 2. London: Perspectiva Press, 2021.

McHenry, L. B. *The Event Universe*: *The Revisionary Metaphysics of Alfred North Whitehead.* Edinburgh: Edinburgh University Press, 2020.

Morais, L. "Cosmic Epochs of Vigorous Order." In *Process Cosmology: New Integrations in Science and Philosophy*. Edited by A. M. Davis et al., 211–20. Cham: Palgrave, 2022.

Nietzsche, F. *The Portable Nietzsche*. Edited and translated by W. Kaufmann. New York: Penguin Books, 1977.

Ogden, S. *Notebooks*. Eugene: Cascade, 2018.

Parkyn, J. L. *Exotheology: Theological Explorations of Intelligent Extraterrestrial Life*. Eugene: Pickwick, 2021.

Peters, T. *Astrotheology: Science and Theology Meet Extraterrestrial Life*. Eugene: Cascade, 2018.

Petronio, R. "Cosmology and Mesology: Alfred North Whitehead and the Plurality of Worlds." In *Process Cosmology: New Integrations in Science and Philosophy*. Edited by A. M. Davis et al., 193–209. Cham: Palgrave, 2022.

Playford, R., ed. *Exophilosophy: The Philosophical Implications of Alien Life*. Abingdon: Routledge, forthcoming.

Price, L. *Dialogues of Alfred North Whitehead*. Jaffrey: David R. Godine, 2001.

Rahner, K. *Foundations of Christian Faith: An Introduction to the Idea of Christianity*. New York: Herder & Herder, 1982.

Reiss, M.J. and M. Ruse. *The New Biology: A Battle between Mechanism and Organicism*. Cambridge: Harvard University Press, 2023.

Rescher, N. *Axiogenesis: An Essay in Metaphysical Optimalism*. Lanham, MD: Lexington, 2010.

———. *Metaphysical Perspectives*. Notre Dame: University of Notre Dame Press, 2017.
Rose, P. "Rethinking Whitehead's Cosmology Through the Cosmogonic Philosophy of C.S. Peirce: Speculation on the Origins of the Actual and the Metaphysical Primacy of the Possible." In *Process Cosmology: New Integrations in Science and Philosophy*. Edited by A.M. Davis et al., 247–75. Cham: Palgrave, 2022.
———. *On Whitehead*. Cengage Learning, 2001.
Segall, M. D., and B. Damer, "The Cosmological Context of the Origin of Life: Process Philosophy and the Hot Spring Hypothesis," current volume.
Smith, K. C., and C. Mariscal. *Social and Conceptual Issues in Astrobiology*. Oxford: Oxford University Press, 2020.
Suchocki, M. H. *The End of Evil: Process Eschatology in Historical Perspective*. Albany: SUNY Press, 1988.
Torres, O.A.C. et al. *Astrobiology: Science, Ethics, and Public Policy*. Beverly: Scrivener Publishing, 2021.
Vainio, O.P. *Cosmology in Theological Perspective: Understanding Our Place in the Universe*. Grand Rapids: Baker Academic, 2018.
Vidal, C. *The Beginning and the End: The Meaning of Life in a Cosmological Perspective*. Heidelberg: Springer, 2014.
Whitehead, A. N. *Adventures of Ideas*. New York: Free Press, 1967.
———. *Aims of Education*. New York: Free Press, 1967.
———. *The Concept of Nature*. Cambridge: Cambridge University Press, 1964.
———. "First lecture," ADD018, Whitehead Research Library. Accessed March 9, 2022: http://wrl.whiteheadresearch.org/items/show/2093.
———. *The Function of Reason*. Princeton: Princeton University Press, 1929.
———. *Modes of Thought*. New York: Free Press, 1968.
———. *Process and Reality*. Corrected edition. Edited by D. R. Griffin and D. Sherburne. New York: The Free Press, 1978.
———. *Religion in the Making*. Cambridge: Cambridge University Press, 1926.
———. *Science and Philosophy*. Paterson, NJ: Littlefield, Adams & Co. 1964.
———. *Science and the Modern World*. New York: The Free Press, 1967.

PART III

Cosmic Religion

Chapter 10

Religious Belief and the Discovery of Extraterrestrial Life

What's Worldview Got to Do with It?

Constance M. Bertka

THE NEW AND FAMILIAR QUESTIONS OF ASTROBIOLOGY

As a discipline astrobiology explores questions that all human societies have pondered: "Where did we come from?" "Are we alone?" and "Where are we going?" Humanity may even be defined by the ability we collectively possess to ask these questions and imagine their answers. We can rephrase the questions in language more familiar to scientific investigation: "How does life begin and develop?" "Does life exist elsewhere in the universe?" and "What is the future of life on Earth and beyond?" Regardless of phrasing, these questions retain a broad public interest that extends beyond the scientific community. That reality is perhaps both a blessing and a challenge for scientists—a blessing when it leads to financial support for their work and a challenge if the answers suggested by science are dismissed as optional alternatives. These questions may be familiar to all but the context from which they are answered, that is, the worldview of the questioner, casts a shadow on the answers which is difficult to escape.

Worldview is a concept that is increasingly gaining attention in both the field of formal science education and the field of informal science communication. Simply stated, "worldview provides a person with presuppositions about what the world is really like and what constitutes valid and important knowledge about the world."[1] While we may all recognize that the process of doing science and the discoveries of science are a way of learning about the world and a source of knowledge about the world, individuals vary on the value they assign to the process and discoveries of science, particularly in relation to other ways of knowing they depend on to learn about the world. This reality is challenging in formal education when teachers ask their students to engage a topic that is culturally controversial. Such topics include ideas that are not contested within the scientific community, but whose legitimacy is called into question by the worldviews of some communities. Examples of these topics include the theory of evolution, anthropogenic climate change, and the efficacy of vaccines. Those communicating the discoveries of science to an informal public audience face the same challenge when individuals dismiss an idea because it is seen as a threat to their significant social connections and relationships.[2] Science communicators are advised to be cognizant of "cultural cognition" recognizing that when positions on facts become associated with opposing social groups then individuals are likely to selectively assess evidence in patterns that reflect their group identity.[3]

While compared to a decade ago religious beliefs in the United States are declining, they are still a significant component of many peoples' group identity. In the United States 71 percent of the population identifies with a religious tradition[4] while only 40 percent of scientists describe themselves as religious.[5] This contrast suggests that a significant component of the worldview of the majority of the US population may be personally insignificant to a majority of the scientific community. In fact, a survey of over 1,700 scientists at the nation's top universities revealed that 13 percent reported being raised in homes with no religious tradition, and while the families of the remaining 87 percent could identify with a religious tradition, these families did not actively participate in their religious community.[6] When many scientists explore the origin, extent, and future of life they may well be starting from a worldview that does not include a traditional religious outlook, but there is a recognition that the questions astrobiology asks resonate with the questions religious traditions reflect on. This recognition can be documented from the inception of NASA's Astrobiology program.

While NASA had previously supported research in exobiology (life on other planets), the creation of a more ambitious astrobiology program was encouraged by the culmination of three significant events: the 1977 discovery of life under extreme conditions, namely microbes thriving in hydrothermal vents at the bottom of the ocean by deriving energy from chemicals dissolved in superheated fluids;[7] the 1995 discovery by a Swiss team of astronomers of a planet orbiting a star other than our sun, an exoplanet;[8] and in 1996 the claim by a team of researchers that they had found fossil evidence for past Martian life in a Martian meteorite.[9] That last claim has been largely disputed by the scientific community but continued discoveries around the extreme conditions under which life can survive on Earth, and subsequent identification of numerous exoplanets, served to cement the field of astrobiology as an active area of research.

As a formal scientific field of inquiry, astrobiology originated in 1998 with the work of an interdisciplinary group of scientists brought together by NASA to outline a scientific approach to answering "How does life begin and develop?" "Does life exist elsewhere in the universe?" and "What is the future of life on Earth and beyond?" The result was published as the first Astrobiology Roadmap.[10] This was followed by the creation of the NASA Astrobiology Institute (NAI), a collection initially of eleven science teams across several research institutions and universities managed out of astrobiology headquarters at NASA Ames. The NAI has since been replaced by Research Coordination Networks (RCNs), a virtual collaboration of researchers organized in 2019 around topically focused research interests. The first Astrobiology Roadmap included an operating principle that highlighted societal interest in the search for extraterrestrial life resulting in questions like the following to be posed: "How will astrobiology affect and interact with human societies and cultures?"[11] Successive road maps and strategies followed the 1998 original, the most recent in 2015. In the 2015 version of the road map, "NASA Astrobiology Strategy,"[12] natural science research priorities for the program are outlined, but an appendix is included that suggests possible contributions from the humanities and social sciences based on the premise that:

> The expansive nature of these [astrobiology] questions also results in their intersection with questions about life that arise in other fields. Answers to these questions have implications for a broad range of stakeholders. . . . The results of astrobiology research will have broad societal impact, affecting the way we think about life in the context of ethics, law, philosophy, theology, and a host of other issues."[13]

Our first extraterrestrial life discovery is an increasingly reasonable expectation and the questions that define astrobiology now seem to carry a greater urgency, at least from a scientific perspective. Analyses of data from the Kepler space telescope, which was retired in October 2018, has led to the identification of nearly five thousand candidate exoplanets planets in just a portion of our Milky Way Galaxy. While from a scientific perspective we can't currently answer the question "Are we alone?" with a definitive no, over the last several decades astronomers have discovered that planets around other stars are common and that many are likely rocky planets in locations around their stars comparable to earth's location in relation to our sun. If life can naturally arise and thrive on our water-bearing rocky planet then we assume it could on other water-bearing rocky planets as well.

In considering the question "How will astrobiology affect and interact with human societies and cultures?" it is reasonable that we are curious about how astrobiology will impact religious belief, as religions reveal societies' reflections on "Where did we come from?" "Are we alone?" and "Where are we going?" However, the question of whether or not extraterrestrial life exists and its potential impact for religious belief is far from new. The question was previously explored in Christianity under the title "plurality of worlds," rather than under the umbrella of astrobiology and its discovery of exoplanets and life in extreme environments. How did theologians answer the extraterrestrial life question prior to the discoveries and research of astrobiology and what new questions is astrobiology research inspiring for theologians?

ASTROBIOLOGY AND THEOLOGY IN CONVERSATION

A predominant theme that emerges from a historical review of Christianity's response to the "plurality of worlds" through the nineeenth century is both the "readiness with which extraterrestrial life was accepted by an array of religious figures" and concern over the implications for specific Christian doctrines like the incarnation—the belief that God became man on earth and died to redeem humanity.[14] The discussion around this topic was extensive with publications numbering in the thousands. By the 1700s it was largely assumed that the discovery of extraterrestrial life would be supportive evidence of God's benevolence and omnipotence.[15] It should be noted that the historical record of this discussion is dominated by Christian perspectives.[16] The same is true of the contemporary discussion.

When considering the impact of the discovery of extraterrestrial life on religious belief, the focus on Christianity, especially by European and North American scholars, has been both pragmatic and parochial. Globally Christianity is still the world's largest religion, accounting for 31 percent of the global religious population,[17] but while up to 1910 most Christians, 66 percent, resided in Europe, by 2010 their numbers in Europe had fallen to 26percent.[18] In the United States 65 percent of people described themselves as Christians in 2019, although this number has also fallen from 77 percent in 2009.[19] Overall interest in the question of how the discovery of extraterrestrial life will impact religions may be general in nature, but of necessity the question of impact will need to be posed for individual religious traditions, for each brings its own perspective to reflection on "humanity's place in the universe." Even the range of beliefs within Christianity precludes a single generic answer.

In writing about the intersection of astrobiology with theological worldviews, Dick notes the increasing number of modern scholars with interests in astronomy and theology who are publishing on the topic and the numerous workshops and conferences that have been held since the 1998 announcement of potential fossil life in a Martian meteorite. He states that:

> The subject has only grown more intense in the two decades since planets have been discovered beyond the solar system, giving way to more substantial discussions of what is now increasingly termed "astrotheology," or in one of its many possible flavors, "cosmotheology." Under whatever name, these terms embody the radical idea that our theologies may need to be expanded or modified in rather wholesale fashion to encompass the new worldview that cosmic evolution has resulted in abundant planets and perhaps abundant life and mind beyond Earth.[20]

While it is clear that the volume of published works and conferences has been rising, it is less clear that the majority of theologians participating would agree that the ideas that come out of these works represent a wholesale modification of theology or even that the idea of expanding or modifying theologies is a radical one to begin with. Dick noted that the discussion around the discovery of extraterrestrial intelligent life (ETI) and Christianity has focused on three areas: creation, human dignity, and incarnation and that the views of theologians Ted Peters and David Wilkinson are examples of the general consensus emerging for the first two topics.

With respect to creation Peters states, "What I forecast is this: contact with extra-terrestrial intelligence will expand the existing religious vision that all of creation—including the 13.7 billion year history of the universe replete with all of God's creatures—is the gift of a loving and gracious God."[21] This view resonates with that of the majority of pre-twentieth-century Christian theologians. As for questions of human dignity, Wilkinson argues "Extraterrestrial intelligence does not pose a problem to the Christian belief that men and women are special in the eyes of God."[22] As Dick notes, there is no theological consensus concerning the Christian doctrine of incarnation and the discovery of ETI.

A recently published work by Kim outlines the current state of the incarnation and ETI discussion and suggests the greatest challenge to positioning the doctrine in relation to an extraterrestrial life discovery does not come from those outside the Christian tradition who question the reasonableness of either a single or multiple incarnation, but from within the tradition "as to the nature, cause and meaning of the Christ event."[23] Kim notes that one camp of Christian theologians, Russell, Peacock, and Hess, favors the particularity of the incarnation event for humankind and the necessity of any

ETI life requiring its own incarnation event, while others, Peters, Moritz, and Nesteruk argue for the sufficiency of a single incarnation for all of creation. Likewise, in an earlier work McMullin also suggested that discussion about the incarnation raised by the discovery of ETI would hinge on, not determine, existing theological opinions: "working from the slender and not always easily reconciled clues afforded by the Bible and the life of the Church, theologians have found very different ways of interpreting the significance of the Incarnation. Ought it be understood primarily as an expression of God's love or primarily as an expiation for human sin?"[24] Christian theologians have not agreed on the answer for humans and the discovery of ETI will not clarify the answer one way or the other. As a Christian, McMullin called for a sense of humility and caution in proposing definitive theological answers that would define or limit the Creator's motives. He was wary of speculations that arise from an ignorance of the flexibility of Christian doctrine, including the range in their interpretation and how they evolve over time. Regardless of whether or not extraterrestrial intelligent life is discovered, theologians' understanding of key Christian doctrines including the incarnation will continue to vary.

In the late 1900s astrobiology reinvigorated the question of "plurality of worlds" with progress toward addressing theoretical probabilities with empirical data and invitations were increasingly extended to theologians to consider Christian doctrine in light of this new data. As briefly outlined above, the response from contemporary theologians, like their historical predecessors, has been positive in terms of the ability of Christianity to welcome ETI life into its framework of Christian beliefs. However, as I've outlined, this was not the answer notable spokespeople for the SETI and astrobiology community, who were not members of the Christian community, imagined. They argued that if ETI were discovered then Christianity, and/or terrestrial religion would face a crisis.[25]

Implicit in their logic, as highlighted by Peters, as part of what he refers to as the "ETI myth," is the idea that extraterrestrial intelligence will not only be technologically advanced, but morally advanced as well, moving beyond fractious religious beliefs altogether or arriving at a single religion superior to ours.[26] These arguments are reminiscent of the traditional secularization theory popular with sociologists of religion throughout much of the twentieth century, which held that as societies become increasingly modernized and secular, religious belief, and the influence of religious institutions would decline. Peter Berger, earlier a major proponent of the theory, has now abandoned this idea, as have most sociologists of the twenty-first century, because "the theory seemed less and less capable of making sense of the empirical evidence from different parts of the world, not least the United States."[27] The world today has more people with traditional religious views than ever before.[28] More recently, Traphagan has coined the term "evolutionary eschatology" to describe the ideology of SETI scientists, namely one that equates technological progress with moral progress predicting a linear cultural evolution that ultimately results in a peaceful and altruistic civilization. He notes that this ideology reflects a Western liberal perspective and that a review of history does not support the idea of a steady march toward a stable civilization as technology advances.[29]

In contrast to the notion that Christianity would face a crisis upon the discovery of extraterrestrial intelligent life, Peters contended that "In fact, theologians might relish the new challenges to reformulate classical religious commitments in light of the new and wider vision of God's creation."[30] I summarized the situation as theologians predicting a feast for Christianity with the discovery of ETI while many scientists outside the faith tradition were predicting a famine for Christianity and terrestrial religion in general, but both camps were failing to adequately consider the reaction of religious people outside academic circles; particularly the majority of Christians who now reside in the Global South and adhere to much more conservative views than academic theologians in the Global North.[31] Furthermore, revised theories of secularization call attention to the role that existential insecurity plays in religiosity with an increase in insecurity resulting in an increase in religiosity worldwide.[32] Considering this revised theory of secularization I noted:

> How would the discovery of extraterrestrial intelligent life impact feelings of existential security and how might Christianity, particularly in the Global South respond? If the answer is a decrease in existential security then we should be wary to predict either a famine for Christian beliefs or a feast for liberal theology, if anything more likely would be an increase in traditional Christian views, particularly in the Global South where the growth of conservative, personal, experience-based Christianity is the current trend.[33]

The takeaway is the impossibility of predicting a single impact or response for Christianity specifically, or religious belief more broadly, if ETI life is discovered. How such a discovery is integrated into Christian religious belief "will at minimum be a long and convoluted process with more than one likely outcome."[34]

QUESTIONING THE UNIQUENESS AND CENTRALITY OF HUMANKIND

Christian astronomer Wolf-Chase believes that the outcome of integrating ETI into Christian belief will be a positive one. She arrives at this conclusion by first pointing to the work of theologian Ted Peters, who in her opinion successfully refutes the accusation that Christianity is inherently geocentric and anthropocentric and therefore unable to accommodate ETI. Secondly, she suggests that:

> The real issue here is apparently not whether Christianity—and other religious traditions—will be able to accommodate the discovery of ETL; for many people, they already have, as witnessed by the numerous organizations dedicated to promoting respectful and reflective dialogue on matters of common interest to science and religion matters. Rather, the more pertinent issue would seem to be how "cultures of fear" that demonize "the other" and seek to isolate or disavow diversity, might shape public opinion and perceptions.[35]

Insightfully Wolf-Chase concludes by noting that the world revealed by science is one that favors an "interacting web of relationship" over the more common hierarchies favored by human societies. She suggests that perhaps the discovery of ETI could help us reevaluate our preference for hierarchies over interconnection, a task she finds encouraged in her reading of the Christian gospels.

The conversation between astrobiology and theology will always be colored by starting assumptions about what inherently defines the religious tradition, both by those within the tradition (as illustrated by the incarnation discussion above) and those outside of the tradition. A more intriguing question may be how the "dialogue on matters of common interest to science and religion" is impacting this conversation. Wolf-Chase suggests these dialogues are evidence that many people have already accommodated the idea of ETI into their religious traditions while Dick suggests their growing frequency points to the need for significant modification of religious worldviews. One has to wonder if they are participating in the same dialogues.

In an introduction to a volume on astrotheology (defined broadly as "the place of cosmic space in Christian theological reflection"[36]), Peters notes, "Now, admittedly, relatively few twentieth and twenty-first-century theologians have incorporated the universe into their description of creation, let alone the scientific details we now know about the universe."[37] Indeed, of the fifteen authors who contributed to this volume, nine have theological training, seven of those in Christian theology. The volume consists of twenty-four chapters and Peters authored seven of them. Peters describes the approach the theologians in the volume take, as "a theology of nature wherein what we know by faith is complemented and expanded by what we learn from science."[38] Peters also reminds readers that there are Christians of another ilk, namely biblical creationists, who deny evolution and the possibility of ETI and that their perspective is not included in the volume. He and his coauthors are open to theology being expanded by astrobiology while hoping for a creative mutual interaction between the two, that is, for science to be impacted by the engagement as well. In this type of engagement, a theological understanding could be the impetus for a scientific research program. Peters admits this is at best a future possibility and currently the workflow is science contributing to a theology of nature. Is Peters's introduction to the topic of astrotheology supportive of Dick's suggestion that "theologies may need to be expanded" as exemplified by Peters and coauthors, and/or "modified wholesale," in the case of the biblical creationists Peters refers to?

On closer examination and excluding Peters from the count, three of the six Christian authors focus their discussion on questions about the incarnation, which as noted above by Kim is an event whose meaning is contested within the tradition regardless of the existence of ETI. What expansion of Christian theology do the other three authors suggest? First it should again be stated that none see a problem for the Christian tradition ultimately welcoming the existence of ETI. There are finer nuances among them when considering the impact on a Christian understanding of creation.

Putz uses a theological argument similar to earlier Christian responses to the "plurality of worlds" to suggest that "the emergence of sentient life throughout the universe is inevitable and warranted by God."[39] Graves recognizes the discovery of ETI will affect a Christian understanding of creation but without a threat to "our relationship with a creator God or Savior Christ."[40] His interest is around the possibility of shared spirituality with ETI that this discovery might allow. However, Funes notes that "There are core principles that might be challenged by a potential discovery of ETI" and among these he lists "uniqueness and centrality of humankind, and refers to these as "classic doctrine."[41] This is in contrast to the argument put forth by Peters in the same volume that geocentrism and anthropocentrism are not defining characteristics of Christianity.[42] While the above theologians have focused on the discovery of ETI life, in a recent work, Pryor considers the significance of non-intelligent extraterrestrial life for Christian theology. He suggests that from an astrobiology perspective, the intra-active relationship of living systems and their habitable environment is the context from which theological reflection should begin. From this context he reinterprets an understanding of the image of God beyond a characteristic applied to individual humans, but instead as an intra-action between a living system and its environment that encourages the flourishing of all creatures.[43] It seems that, like the meaning of the incarnation, whether or not the "uniqueness and centrality of humankind" are defining characteristics of the Christian tradition is a matter for internal debate, even among theologians who are not biblical creationists. For those in the pew the debate may be even more pronounced.

Peter refers to the work of Christian theologian Crysdale who suggests that the discovery of ETI "can threaten [the] basic religious intuition" of the "contemporary Christian world of North America" (and I would suggest the Global South as well).[44] As Peters notes, "Although Crysdale herself would like to construct a theological vision that makes room for ETI, she is worried about the faith of the others who sit in her pew or who stand in her chancel."[45] The reason that Crysdale questions the readiness of those who "sit in her pew or who stand in her chancel" to not feel threatened by the discovery of ETI is because of the lack of resolution within Christianity for accepting a new understanding of God's relation to the world:

> There is indeed a major theological challenge for our time—but it has to do not with God and astrobiology but whether we can reconcile an evolving, emerging world with a God who is present yet transcendent, who created a cosmos with an order and orientation but without determining specific outcomes.[46]

Crysdale is concerned about the ability of Christians at large to let go of a narrative about the world that not only "sees human life and salvation as the apex of meaning" but also ascribes to the world a divine determinism that hinges on divine intervention.[47] She reviews several examples in contemporary Christian theology that offer ways to accommodate a new narrative for God's relation to the world, including process theology, what she questions is the willingness of the religious public to adopt these new narratives.

As noted above, Peters critiques the ETI myth put forward by notable members of the SETI and astrobiology community and he also revisits earlier work reporting on the results of a survey he administered to judge whether or not individuals from multiple religious traditions worldwide would feel that the discovery of ETI would threaten their religious beliefs.[48] He concludes that the survey has settled the matter for the general population, religious believers do not feel that the discovery of ETI will result in a religious crisis for them. While I agree that a religious crisis is not a necessary outcome, as noted earlier I am not confident that the survey has answered this question, especially for the majority of Christians who live in the Global South and adhere to a more conservative Christianity:

> The majority of respondents to the Peters survey 1,000, out of 1,300 (Peters, personal communication) were from the US and those surveyed from other parts of the world participated through a web survey largely by invitation. The Peters survey is typical of other surveys conducted on beliefs and assumptions about extraterrestrial life—these surveys sampled educated persons in North American and European countries. Would survey conclusions be different if a broader world demographic was included? Put another way, "who speaks for Christianity?[49]

Perhaps the discovery of ETI will enliven a discussion about the uniqueness and centrality of humans from a Christian perspective driven by, as Wolf Chase suggests, how we view "the other." To the best of my knowledge, we do not have the data to answer this question for worldwide Christianity let alone for the worldwide religious tradition predicted to outnumber Christianity by 2050, Islam.[50] However, if the crux of the matter comes down to how important the centrality of humanity is to the tradition, then according to one Muslim scholar, Muzaffar Iqbal, Islam will be able to accommodate ETI.[51] Viewing the earth as the center of the cosmos was never a theological concern for Islam and furthermore a being's value, or rank among created beings, is judged by their spiritual relationship to God. If ETI has a superior spiritual relationship then according to the "Quranic notion of superiority" that would "demote humans from their creational rank," but not cast doubt on the validity of the Islamic tradition.[52]

Beyond the question of whether or not, based on the guidance of theologians, a religious public would ultimately embrace the decentering of humanity, there is another question that Dick notes is not being asked. Dick has argued that if theology takes into account what science has learned about the universe, then the result will be cosmotheology, a naturalistic "theology" or perhaps better termed "philosophy." Cosmotheology calls for "radically new conceptions of God" that do not include the supernatural.[53] While he is aware of the work of Christian astrotheologians, he notes their efforts are "like those of Thomas Aquinas in another turbulent era, attempts to reconcile new scientific knowledge within the Christian tradition, rather than attempts to ask the more basic question of whether the Christian tradition still makes sense in the modern world."[54] He sees little evidence that their efforts will lead to an acceptance of the naturalistic cosmotheology he proposes. He attributes the impasse to a fundamental difference in worldviews between science and theology. However, in another essay, while noting the unliklihood that cosmotheology will be widely accepted in the near future, he takes a longer view stating, "But it may be that over the course of centuries astrotheology will transform into astrophilosophy, as the natural gods of cosmic evolution and the biological universe replace the supernatural God of the ancient Near East."[55] His speculation depends on time being the key to one worldview replacing another. Does the contemporary relationship between science and religion give any hints as to the likelihood of shifting worldviews?

THE RELATIONSHIP BETWEEN SCIENCE AND RELIGION: PAST, PRESENT, AND FUTURE

A common belief about the relationship between science and religion is that it is most accurately described as one of conflict. In the popular imagination we envision Galileo's conflict with the Roman Catholic church as the norm for the relationship between science and religion, beginning in the sixteenth to seventeenth century, the time of the Scientific Revolution. While the church as an institution was on the watch for new ideas that could be challenging to its authority, "those concerned with studying and understanding the natural world in the early modern period were every bit as religious as the population at large."[56] Fast forward to the mid-nineteenth through the mid-twentieth century and many natural scientists and social scientists were arguing that advancements in science would result in a decline in religion. However, as noted above, the view that there is a general correlation between scientific progress and the secularization of society has been shown to be oversimplified:

> Instead of regarding science as the agent of an inexorable secularization, it is surely more accurate to say that scientific theories have been susceptible of both theistic and atheistic readings. Historically they have provided resources for both. Sometimes the same scientific concept, in different hands, has been manipulated to generate a sense of the sacred or the profane (Brooke 2010, 110).[57]

While there is not a simple association between the advancement of science and decline in religious belief or authority, it is also the case that by the mid-twentieth century, theologians largely viewed science as "irrelevant to theology." Rather than concerning itself with nature and creation, mid-twentieth-century theology focused on "existential, historical and psychological concerns."[58] This focus had its advantages, allowing science and theology to peacefully coexist, but it left ministers ill equipped to respond to new questions raised by science and technology. The contemporary science and religion field was born out of a recognition by some in both the scientific and the religious community that this gap of irrelevancy was dangerous to both humanity and more broadly nature.

The year 1966 was a significant one for theologians, philosophers, historians, and scientists interested in exploring how science and religion might be more than isolated disciplines. Two events that encouraged and launched a new contemporary, interdisciplinary, scholarly field took place; the publication by Ian Barbour, a Christian physicist, of the highly influential book *Issues in Science and Religion*,[59] and the publication of the first academic journal committed to the topic of religion and science, *Zygon*. The founding editor of the latter was Ralph Wendell Burhoe, a Unitarian Universalist and then professor of "theology and the sciences" at Meadville Theological School. Burhoe was also integral to the formation of "The Institute on Religion in an Age of Science (IRAS)" which was established earlier in 1954, following a landmark conference on "Religion in an Age of Science" that took place on Star Island off the coast of New Hampshire. By the 1980s, growing interest in the field resulted in the establishment of two centers, "The Center for Theology and Natural Sciences (CTNS)" and the "Zygon Center for Religion and Science," both of which encouraged scholarly work in the area and increasingly pursue an agenda of broader public engagement through public programming and outreach to religious leaders and communities. In 2003, CTNS also spearheaded the creation of the field's second academic journal, *Theology and Science*.

The science and religion field today includes academic societies, the "European Society for the Study of Science and Theology" established in the mid 1980s, and since 2002, the "International Society for Science and Religion." In 1995 the world's largest voluntary association of sciences, the American Association for the Advancement of Science, created the Program of Dialogue on Science, Ethics, and Religion. While early program efforts concentrated on workshops that brought scholars together around topics at the intersection of science, religion, and ethics, and produced academic volumes highlighting their work, in the early 2000s the program's focus was shifted increasingly to broader public engagement efforts.

The contemporary field that Burhoe, Barbour, and their colleagues helped to launch has grown beyond scholarly interests, but challenges of both scope and broader outreach still abound. While there are efforts to broaden its scope, the majority of work that has taken place has been about relating Western science to Christian beliefs. New efforts at broader public engagement are noteworthy and hopefully will continue to grow, however it still remains the case that many people in the pew are unaware of the field and most religious leaders, including those who receive advanced degrees from seminaries, are not required to explore the relationship between science and religion; irrelevancy is still an option.

Barbour's pioneering work did describe an option for relating science to theology, *Integration*, that allowed for the possibility that scientific discoveries and scientific theories could have significance for theological doctrines and that boundaries between the two fields could be crossed. One way this can occur is when traditional doctrines are open to reformulation based on broad and widely accepted characteristics about nature that have been revealed by science. For example, viewing nature as continually evolving with both chance and law playing a role, rather than as a static world, can impact how we envision God's relation to nature, including humanity. Theologians can consider how God might work through a process of law and chance rather than direct interventions. Is the role of chance in nature that science observes a reflection of God's self-limitation or, masked from the view of science, is God providentially controlling all events? These types of questions reflect "attempts to reconcile new scientific knowledge within the Christian tradition" and, as Dick would note, don't question the starting premise of a supernatural God.

However, another *Integration* approach discussed by Barbour (1997), in addition to the reformulation of traditional doctrines, is an attempt to create a single metaphysical system, using the contributions of both science and theology to create a coherent worldview.[60] The example he provided is process theology which draws from science a "view of reality as a dynamic web of interconnected events" and views God as "the source of novelty and order."[61] The God of process theology works through persuasion rather than compulsion and is intimately involved with history, characteristics that process theologians see as consistent with the God revealed in Hebrew and Christian scriptures, but the process God does not retain the traditional attributes of being unchanging over time, all-knowing and all-powerful. However, the process God is unchanging in the sense that God is always "[responsive] to the experiences of the whole universe," all-knowing in terms of "totally omniscient only of the past and present" and all powerful in the sense that by being the keeper of all cosmic history, "[God's] function is to experience and remember all that is past."[62] God influences every activity that builds by necessity on the past. Furthermore, while God cannot totally determine another's free decision, God can only act by persuasion, the laws of nature are determined by God and all of creation is subjected to them.[63] In this integration of science and theology a new view of God emerges that is more compatible with Dick's cosmotheology. The importance of creativity in process theology is what drives "radically new conceptions of God" that do not include the supernatural:

> Clearly, treating creativity as the ultimate metaphysical principle has far-reaching consequences for the concept of God and of God's relations to others. It is not possible, in process metaphysics, to conceive divine activity as a "supernatural" intervention into the "natural" order of events. . . . In process thought, there is no such thing as a realm of the natural in contrast to that which is supernatural. On the other hand, if "the natural" is defined more neutrally as "what is in the nature of things," then process metaphysics characterizes the natural as the creative activity of actual entities . . . It is tempting to emphasize process theism's denial of the supernatural and thereby highlight what the process God cannot do in

> comparison to what the traditional God can do (that is, to bring something from nothing). In fairness, however, equal stress should be placed on process theism's denial of the natural (as traditionally conceived) so that one may highlight what the creatures cannot do, in traditional theism, in comparison to what they can do in process metaphysics (that is, to be part creators of the world with God).[64]

Process theologians have been at work for decades, long before the establishment of astrobiology, revising a classical view of God and God's relation to humanity, yet the more traditional view still thrives. In the words of one theologian, "the Gospel cannot bear the changes that appear to be required and remain the Christian gospel. . . . A more 'activist' understanding of God [is required] than is possible on the Process account.[65] Will the discovery of extraterrestrial life speed up the revision, paving the way for a cosmotheology devoid of a traditionally supernatural God? Another way to ask this question is will such a discovery further encourage an *Integration* approach to science and religion? Without normalizing science and religion discussions, to the point where they are the concern of both theologians and the scientific community writ large, and popularizing these discussions, I suspect not. Frustrating this normalization is the lack of interest in considering an *Integration* approach. Survey research has shown that by far the most popular choice for relating science and religion within the scientific community is an *Independence* approach in which science and religion "refer to different aspects of reality."[66] The historical challenges of the AAAS Program of Dialogue on Science, Ethics, and Religion provide one example of the significant barriers to normalizing science and religion discussions and moving beyond *Conflict* or *Independence* approaches.

Among First World countries, the US public is one of the most religious.[67] The scientific community needs to be concerned about a religious public at large. Given that most of the funding for scientific research comes from public dollars, rather than private sources, it stands to reason that public understanding and support of science will serve to further its advancement and that this advancement should have an ongoing relation to societal concerns.[68] Individual scientists caught up in the excitement and demanding rigors of their own research programs may lose sight of these facts, but the scientific community as a whole has a responsibility to address them. One of the ways that the American Association for the Advancement of Science (AAAS), the world's largest voluntary association of scientists representing multiple scientific disciplines, accepted the challenge of this responsibility was with the creation of the AAAS Program of Dialogue on Science, Ethics, and Religion (DoSER) in 1995. The history of the scientific community's reactions to this program can be appreciated through a review of news reports about the program in AAAS's own journal *Science*. These provide a snapshot of the scientific community's struggles with moving beyond a *Conflict* and *Independence* approach to science and religion to include *Dialogue* and, to some degree at least, *Integration.*

The purpose of the DoSER program is to "facilitate communication between scientific and religious communities."[69] The program was made possible by a grant from the John Templeton Foundation to AAAS in 1995. Outspoken concerns from scientists about AAAS running this program, based almost entirely on support from the John Templeton Foundation, was an operating hurdle the program faced for nearly fifteen years. As part of the DoSER team from 2000 to 2008, first as a senior program associate and then as DoSER program director, I witnessed this challenge firsthand. Concerned scientists voiced opposition to AAAS involvement with a foundation that they accused of having an agenda that was unscientific, namely "trying to prove the existence of God" and using science to increase the "intellectual respectability" of religion.[70] Within AAAS the lead on these charges stemmed from the physics section of the organization who in 2000 called for a review of the goals of the program.[71] The physicists voiced concerns about advisory members of the AAAS program also being on the board of the John Templeton Foundation and the fact that the majority of the funds for the program came from the Templeton Foundation. Then DoSER director Dr. Audrey Chapman countered that "the Dialogue is teaching the religious community about science and promoting discussion of difficult ethical issues," and that in the future no more than 40 percent of program funding would come from Templeton and only one advisory board member served on both boards.[72] The majority of the physicists were appeased and the AAAS science council took no further action, although the program was mandated to make good on the funding promise. Objections to the programs existence during these early years reflected not only a *Conflict* approach to science and religion but also a concern that a strict *Independence* approach was being violated.

By 2005, in the context of combating the call to teach Intelligent Design as an alternative to evolution in US science classrooms, AAAS was highlighting the presence of the DoSER program as one of the ways the organization had been working to protect the teaching of evolution in science classrooms:

> The controversy over teaching evolution in public school science classes has been a recurring part of American culture—and a continuing interest of AAAS. In the 1920s, AAAS supported the teaching of evolution and raised money to help defend Tennessee biology teacher John Scopes. Throughout the past decade, AAAS's Dialogue on Science, Ethics, and Religion (DoSER) has produced conferences and books on the issue.[73]

The conferences and books the article refers to, however, were not limited to an *Independence* approach but illustrate engagement between evolutionary science and religious reflection in terms of both *Dialogue* and *Integration* approaches (e.g., *An Evolving Dialogue: Theological and Scientific Perspectives on Evolution*[74]). Movement toward an integration between science and religion that is more than skin deep must involve conversations that move beyond the drawing of boundaries. This may be an uncomfortable discussion that many scientists prefer to avoid, but working to provide science content to religious communities seeking opportunities to consider the significance of scientific discoveries for their faith tradition falls within AAAS's "longstanding commitment to relate scientific knowledge and technological development to the purposes and concerns of society at large."[75]

By 2010 controversy over the presence of the DoSER program at AAAS and the mandate to reduce John Templeton Foundation funding had all but ceased. That year the foundation awarded it highest prize to a former AAAS board president, Francisco Ayala, priest turned biologist who was well known for his work to keep creationism, including Intelligent Design, out of the science classroom. In speaking to *Science* about the award and the past controversy over the DoSER program Ayala noted the change in attitude:

> He says the conflict has grown less intense since Templeton funds helped to launch a program in the mid-1990s called Dialogue on Science, Ethics and Religion at AAAS, which continues to be supported by the foundation. Some scientists objected at the time, he recalls. "They said, 'What business does science have talking to religion?' I don't think there are many thoughtful scientists who would make that point today."[76]

While there are certainly individual scientists who still suggest that John Templeton Foundation funded projects involving scientists and theologians are trying to bestow "scientific credibility" on theology, the scientific community at large seems more willing to recognize at least the importance of engaging religious communities.[77] Projects long envisioned by prior DoSER leadership and advisory committees, including helping theological seminaries to incorporate science into their core curricula and building relationships with evangelical communities, have been supported by AAAS leadership while fully funded by the John Templeton Foundation. In a 2010 article Alan Leshner, then AAAS CEO, explained that the goal of an upcoming DoSER sponsored program was to nurture "civil dialogue" between scientists and religious leaders potentially providing a way forward to find common ground on issues but otherwise devoid of any particular outcome.[78]

A call for dialogue is an important recognition that science and religion could benefit from being familiar with one another if they are to work together on common concerns, and it is no small accomplishment that the world's largest voluntary science organization found within its institutional constraints room for this effort. Dialogue can also be the first step "toward a more nuanced discussion of science and religion," the kind of discussion necessary for the integration of science and religion opening the possibility of shifting worldviews.[79] It remains to be seen if in the future the common relationship between science and religion will be one that has moved beyond dialogue toward integration, but presently, from both institutional and individual perspectives, the barriers are immense.

FINAL THOUGHTS

If worldviews are to be impacted then both theologians and the scientific community writ large need to be supportive of a more nuanced conversation and one that reaches beyond academia. Recent efforts by the scientific community to support a broader conversation are encouraging, conflict between science and religion is falling out of fashion but work toward integration, where significant worldview change might happen, is still elusive. Without this work, when faced with incorporating new information into our worldview, whether it be the worldview of theology or science, as individuals we are likely to prefer a mental shortcut. We easily accept information that accords with our prior beliefs but are more dismissive of information that does not; we protect our worldviews.

Is there any reason to argue that given even "the course of centuries" those who are comfortable with theological assumptions, and those who are not, are likely to modify their worldviews? Process theology offers a way to revision a traditional view of God toward one more compatible with an evolving, biological universe. Whether the worldviews of traditional theists and nontheists will ever move closer to one another is likely less dependent on centuries of scientific discoveries than it is on cracking the foundations of siloed worldviews and normalizing the possibility of science and religion discussions that move toward "civil integration." It seems to me that is a more formidable goal then achieving centuries of scientific discoveries, even if these include the discovery of extraterrestrial intelligent life. Writing this from the context of the world in 2022, it is a goal I find increasingly vital to all of creation.

NOTES

1. Cobern, "Worldview Theory and Conceptual Change in Science Education," 5.
2. Kahan, "Fixing the Communications Failure."

3. Kahan, "Climate-Science Communication and the Measurement Problem."
4. Pew Research, "About Three-in-Ten U.S. Adults are Now Religiously Unaffiliated."
5. Larson and Witham, "Scientists Are Still Keeping the Faith"; Ecklund et al., "Religion among Scientists in International Context."
6. Ecklund, "Science vs. Religion."
7. Ballard, "Notes on a Major Oceanographic Find."
8. Michel and Queloz, "A Jupitar-Mass Companion."
9. McKay et al., "Search for Past Life on Mars."
10. Morrison and Meyer, "Astrobiology Roadmap."
11. Dick and Strick, *The Living Universe*, 218.
12. Hays, *NASA Astrobiology Strategy 2015*.
13. Mix and Bertka, "Beyond Natural Sciences."
14. Crowe, *The Extraterrestrial Life Debate 1750–1900*, 557.
15. Crowe and Dowd, "The Extraterrestrial Life Debate from Antiquity to 1900."
16. Crowe, *The Extraterrestrial Life Debate: Antiquity to 1915.*
17. Pew Research, "The Changing Global Religious Landscape."
18. Pew Research, "Global Christianity."
19. Pew Research, "In U.S., Decline of Christianity Continues."
20. Dick, *Astrobiology, Discovery, and Societal Impact*, 223.
21. Peters, "The Implications of the Discovery," 654.
22. Wilkinson, *Alone in the Universe?*, 124.
23. Kim, "The ETI Hypothesis and the Scandal of Particularity," 333.
24. McMullin, "Life and Intelligence Far from Earth," 172.
25. Bertka, "Christianity's Response."
26. Peters, "Astrotheology and the ETI Myth."
27. Berger, "Reflections on the Sociology of Religion Today," 445.
28. Norris and Inglehart, *Sacred and Secular*.
29. Traphagan, "SETI, Evolutionary Eschatology, and the Star Trek Imaginary."
30. Peters, "The Implications of the Discovery," 654.
31. Bertka, "Christianity's Response."
32. Norris and Inglehart, *Sacred and Secular*.
33. Bertka, "Christianity's Response," 338.
34. Ibid.
35. Wolf-Chase, "Astrobiology, Astroethics, and Astrotheology," 200.
36. Peters, "Introducing Astrotheology," 11.
37. Ibid., 13.
38. Ibid.
39. Putz, "God's Self-Communication in a Cosmos Bound for Life," 177.
40. Graves, "ET, Call Church!," 267.
41. Funes, "The Road Map to Other Earths," 67.
42. Peters, "The Tasks of Astrotheology."
43. Pryor, "What If We're Not Alone."
44. Crysdale, "God, Evolution and Astrobiology," 220.
45. Peters, "Extraterrestrial Life and Terrestrial Religion,"193.

46. Crysdale, "God, Evolution and Astrobiology," 238.
47. Ibid., 239.
48. Peters, "Extraterrestrial Life and Terrestrial Religion."
49. Bertka, "Christianity's Response," 236.
50. Pew Research, "The Future of World Religions."
51. Iqbal, "Islamic Theology Meets ETI."
52. Ibid., 225.
53. Dick, *Astrobiology, Discovery, and Societal Impact*, 229.
54. Ibid., 237.
55. Dick, "Toward a Constructive Naturalistic Cosmotheology," 229.
56. Henry, "Religion and the Scientific Revolution," 39.
57. Brooke, "Science and Secularization," 210.
58. Cole-Turner, *An Unavoidable Challenge*.
59. Barbour, *Issues in Science and Religion*.
60. Barbour, "Ways of Relating Science and Theology."
61. Ibid., 43.
62. Gunton, "Process Theology's Concept of God," 292, 294.
63. Viney, "Process Theism."
64. Ibid.
65. Gunton, "Process Theology's Concept of God," 295.
66. Ecklund et al., "Religion Among Scientists."
67. Coyne, "Science, Religion, and Society."
68. *Nature* Editorial, "Private Funding for Science."
69. DoSER, https://www.aaas.org/programs/dialogue-science-ethics-and-religion/about.
70. Holden, "Subjecting Belief to the Scientific Method."
71. Holden, "Physicists Wary of Science-Religion Dialogue."
72. Ibid.
73. Lempinen, "Science and Society," 635.
74. Miller, *An Evolving Dialogue*.
75. DoSER, https://www.aaas.org/programs/dialogue-science-ethics-and-religion/about.
76. Bhattacharjee, "Latest Prize Bolsters Templeton's Shift to Mainstream," 1565.
77. Ibid.
78. Leshner, Science, Religion and Civil Dialogue."
79. Haught, *Science and Religion*, 15–16.

BIBLIOGRAPHY

Ballard, R. D. "Notes on a Major Oceanographic Find. Woods Hole Oceanographic Institution," 1977. Accessed January 12, 2022 https://www.whoi.edu/feature/history-hydrothermal-vents/pdf/BallardOc77Notes.pdf

Barbour, I. G., "Ways of Relating Science and Theology." In *Physics, Philosophy and Theology: A Common Quest for Understanding*, 3rd edn., edited by R. J. Russell,

W. J. Stoeger, and G. V. Coyne, 21–48. Notre Dame: University of Notre Dame Press, 1997.

Barbour, I. G., *Issues in science and religion*. New York: Harper & Row, 1966.

Berger, P. "Reflections on the Sociology of Religion Today." *Sociology of Religion* 62, no. 4 (2001): 443–54. https://doi.org/10.2307/3712435.

Bertka, C. M. "Christianity's Response to the Discovery of Extraterrestrial Intelligent Life: Insights from Science and Religion and the Sociology of Religion." In *Many Worlds: Astrobiology, History and Society: Life Beyond Earth and the Impact of Discovery*, edited by D. A. Vakoch, 329–40. Cham: Springer-Verlag, 2013.

Bhattacharjee, Y. "Latest Prize Bolsters Templeton's Shift to Mainstream. *Science* 327, no. 5973 (2010): 1565. https://doi.org/10.1126/science.327.5973.1565.

Brooke, J. H. "Science and Secularization." In *The Cambridge Companion to Science and Religion*, edited by P. Harrison, 103–23. Cambridge: Cambridge University Press, 2010.

Cobern, W. W. "Worldview Theory and Conceptual Change in Science Education." *Science Education* 80, no. 5 (1996): 579–610. Accessed January 12, 2022. https://scholarworks.wmich.edu/science_slcsp/15/.

Cole-Turner, R. S. *An Unavoidable Challenge: Our Church in an Age of Science and Technology*. The Division of Education and Publication: United Church Board for Homeland Ministries, 1992.

Coyne, J. A. "Science, Religion, and Society: The Problem of Evolution in America." *Evolution* 66, no. 8 (2012): 2654–63. https://doi.org/10.1111/j.1558-5646.2012.01664.x.

Crowe, M. J. *The Extraterrestrial Life Debate 1750–1900: The Idea of a Plurality of Worlds from Kant to Lowell*. Cambridge: Cambridge University Press, 1986.

Crowe, M. J., and Dowd, M. F. "The Extraterrestrial Life Debate from Antiquity to 1900." In *Astrobiology, History, and Society: Life Beyond Earth and the Impact of* Discovery, edited by D. A. Vakoch, 4–56. Cham: Springer-Verlag, 2013.

Crowe, M. J., ed. *The Extraterrestrial Life Debate: Antiquity to 1915, A Source Book*. Notre Dame: University of Notre Dame Press, 2008.

Crysdale C. S. W. "God, Evolution and Astrobiology." In *Exploring the Origin, Extent, and Future of Life: Philosophical, Ethical and Theological Perspectives*, edited C. M. Bertka, 220–41. Cambridge: Cambridge University Press, 2009.

Dick, S. J. "Toward a Constructive Naturalistic Cosmotheology." In *Astrotheology: Science and Theology Meet Extraterrestrial Life*, edited by T. Peters, et al., 228–44. Eugene: Cascade Books, 2018.

———. *Astrobiology, Discovery, and Societal Impact*. Cambridge University Press, 2018. DoSER. AAAS DoSER, 2020. Accessed January 12, 2022. https://www.aaas.org/programs/dialogue-science-ethics-and-religion/about.

Dick, S. J., and J. E. Strick. *The Living Universe: NASA and the Development of Astrobiology*. New Brunswick, NJ: Rutgers University Press, 2004.

Ecklund, E. H., et al. "Religion Among Scientists in International Context: A New Study of Scientists in Eight Regions." *Socius: Sociological Research for a Dynamic World* 2 (2016): 1–9. https://doi.org/10.1177/2378023116664353.

Ecklund, H. E. *Science vs. Religion: What Scientists Really Think*. Oxford: Oxford University Press.

Funes, J. G. "The Road Map to Other Earths: Lessons Learned and Challenges Ahead." In *Astrotheology: Science and Theology Meet Extraterrestrial Life*, edited by T. Peters, et al., 56–73. Eugene: Cascade Books, 2018.

Graves, M. "ET, Call Church!": Astrosemiotics and Shared Spirituality." In *Astrotheology: Science and Theology Meet Extraterrestrial Life*, edited by T. Peters, et al., 245–68. Eugene: Cascade Books, 2018.

Gunton, C. "Process Theology's Concept of God: An Outline and Assessment." *The Expository Times* 84, no.10 (1973): 292–96. https://doi.org/10.1177/001452467308401002.

Haught, J. F. *Science and Religion: From Conflict to Conversation*. Mahwah, NJ: Paulist Press, 1995.

Hays, L., ed. *NASA Astrobiology Strategy 2015*. 2015. Accessed January 12, 2022. https://astrobiology.nasa.gov/nai/media/medialibrary/2015/10/NASA_Astrobiology_Strategy_2015_151008.pdf.

Henry, J. "Religion and the Scientific Revolution." In *The Cambridge Companion to Science and Religion*, edited by P. Harrison, 39–58. Cambridge: Cambridge University Press, 2010.

Holden, C. "Physicists Wary of Science-Religion Dialogue." *Science* 287, no. 5458 (2000): 1587. https://doi.org/10.1126/article.37798.

Holden, C. "Subjecting Belief to the Scientific Method." *Science* 284, no. 5418 (1999): 1257–59. https://doi.org/10.1126/science.284.5418.1257.

Iqbal, M. "Islamic Theology Meets ETI." In *Astrotheology: Science and Theology Meet Extraterrestrial Life*, edited by T. Peters, et al., 216–27. Eugene: Cascade Books, 2018.

Kahan, D. M. "Fixing the Communications Failure." *Nature* 463 (2010): 296–97. https://doi.org/10.1038/463296a.

———. "Climate-Science Communication and the Measurement Problem." *Advances in Political Psychology* 36, no. S1 (2015): 1–43. https://doi.org/10.1111/pops.12244.

Kim, J. "The ETI Hypothesis and the Scandal of Particularity." *Theology and Science* 19, no. 4 (2021): 332–42. https://doi.org/10.1080/14746700.2021.1982247.

Larson, E. J., and L. Witham. "Scientists Are Still Keeping the Faith." *Nature* 386 (1997): 435–36. https://doi.org/10.1038/386435a0.

Lempinen, E. "Science and Society, AAAS Fighting to Defend the Integrity of Science Education." *Science* 310, no. 5748 (2005): 634–37. https://doi.org/10.1126/science.310.5748.634.

Leshner, A. I. "Science, Religion and Civil Dialogue." *Huffington Post*, June 6, 2010. Accessed January 12, 2022. https://www.huffpost.com/entry/science-religion-and-civi_b_611674/.

McKay, D. S., et al. "Search for Past Life on Mars: Possible Relic Biogenic Actiovity in Martian Meteorite ALH84001." *Science* 273, no. 5377 (1996): 924–30. https://doi.org/10.1126/science.273.5277.9.

McMullin, E. "Life and Intelligence Far from Earth: Formulating Theological Issues." In *Many Worlds: The New Universe, Extraterrestrial Life and the Theological Implications*, edited by S. J. Dick, 151–75. West Conshohocken, PA: Templeton Pres, 2000.

Michel, M., and D. Queloz. "A Jupitar-Mass Companion to a Solar-Type Star." *Nature*, no. 378 (1995): 355–59. https://doi.org/10.1038/378355a0.

Miller, J. B., ed. *An Evolving Dialogue: Theological and Scientific Perspectives on Evolution*. Trinity Press International, 2001.

Mix, L., and C.M. Bertka. "Beyond Natural Sciences: Humanities and Social Science Contributions to Astrobiology." In *NASA Astrobiology Strategy 2015*, edited by L. Hays, 155–60, 2015. Accessed January 12, 2022. https://astrobiology.nasa.gov/nai/media/medialibrary/2015/10/NASA_Astrobiology_Strategy_2015_151008.pdf.

Morrison, D., and M. Meyer. *Astrobiology Roadmap*, 1998. Accessed January 12, 2022. https://astrobiology.nasa.gov/nai/media/roadmap/1998/.

Nature editorial. "Private Funding for Science." *Nature Methods* 13, no. 7 (2016): 537. https://doi.org/10.1038/nmeth.3923.

Norris, P., and R. Inglehart. *Sacred and Secular: Religion and Politics Worldwide*, 2nd ed. Cambridge: Cambridge University Press, 2011.

Peters, T. "The Tasks of Astrotheology." In *Astrotheology: Science and Theology Meet Extraterrestrial Life*, edited by T. Peters, et al., 27–55. Eugene: Cascade Books, 2018.

———. "Extraterrestrial Life and Terrestrial Religion: A Crisis?" In *Astrotheology: Science and Theology Meet Extraterrestrial Life*, edited by T. Peters, et al., 183–207. Eugene: Cascade Books, 2018.

———. "Introducing Astrotheology." In *Astrotheology: Science and Theology Meet Extraterrestrial Life*, edited by T. Peters, et al., 3–26. Eugene: Cascade Books, 2018.

———. "Astrotheology and the ETI Myth." *Theology and Science* 7, no. 1 (2009): 3–30. https://doi.org/10.1080/14746700802617097.

Pew Research Center. "The Changing Global Religious Landscape," 2017. Accessed January 12, 2022. https://www.pewforum.org/2017/04/05/the-changing-global-religious-landscape/.

———. "About Three-in-Ten U.S. Adults are Now Religiously Unaffiliated," 2021. Accessed January 12, 2022. https://www.pewforum.org/2021/12/14/about-three-in-ten-u-s-adults-are-now-religiously-unaffiliated/.

———. "Global Christianity-A Report on the Size and Distribution of the World's Christian Population," 2011. Access January 12, 2022. https://www.pewforum.org/2011/12/19/global-christianity-exec/.

———. "In U.S., Decline of Christianity Continues at Rapid Pace," 2019. Access January 12, 2022. https://www.pewforum.org/2019/10/17/in-u-s-decline-of-christianity-continues-at-rapid-pace/.

———. "The Future of World Religions: Population Growth Projections, 2010–2050," 2015. Access January 12, 2022. https://www.pewforum.org/2015/04/02/religious-projections-2010-2050/.

Pryor, A. "What If We're Not Alone: Considering the Significance of Non-Intelligent Alien Life for Constructive Christian Theology." *Theology and Science* 20, no. 3 (2022): 276–90. https://doi.org/10.1080/14746700.2022.2084850.

Putz, O. "God's Self-Communication in a Cosmos Bound for Life." In *Astrotheology: Science and Theology Meet Extraterrestrial Life*, edited by T. Peters, 160–79. Eugene: Cascade Books, 2018.

Traphagan, J. W. "SETI, Evolutionary Eschatology, and the Star Trek Imaginary." *Theology and Science* 19, no. 2 (2021): 120–31. https://doi.org/10.1080/14746700.2021.1910911.

Viney, D. "Process Theism." In The Stanford Encyclopedia of Philosophy, edited by E. N. Zalta, 2020 Edn. Accessed January 12, 2022. https://plato.stanford.edu/archives/sum2020/entries/process-theism/.

Wilkinson, D. *Alone in the Universe?* Crowborough: Monarch, 1997.

Wolf-Chase, G. "Astrobiology, Astroethics, and Astrotheology in Conversation." In *Intersections of Religion and Astronomy*, edited by C. Corbally et al., 197–205. London: Routledge, 2020.

Chapter 11

"Worlds as Numerous as the Grains of Sand of the Ganges"

Cosmic Pluralism and Swami Vivekananda's Religion of the Future

Jeffery D. Long

LIFE AND ITS MYRIAD POSSIBILITIES

Is there life in the universe, or the multiverse, beyond the planet Earth? Responding to this question from the perspective of philosophies of process, such as that of Alfred North Whitehead, prompts us to raise another question, "What is life?" Is there finally anything which is *not*, in some sense, alive? The emergence of philosophies of process, as well as a growing sense among thinkers in the West that consciousness (or at least experience) is in some sense foundational to existence, and not something which evolves from an essentially material reality, puts the question of life in the universe on a different level from that of simply searching for entities like ourselves. It becomes not only an empirical question, but a fundamental question of ontology.

Such a situation is already envisioned in many Indian philosophical traditions that both take consciousness to be foundational to existence *and* take for granted that there are many planes of existence inhabited by intelligent beings. We are therefore witnessing a convergence of Indian and Western thought, in our current period, as we reflect upon the question of life in the universe and the closely related question of the nature of life itself.

Over a century ago, Swami Vivekananda envisioned this convergence when he spoke of religions to emerge in the future, and a civilization drawing on the spiritual wisdom of India and the scientific and technical knowledge of the West.

This chapter will take up the question of the kind of religious sensibility that emerges from reflecting on the possibility of extraterrestrial life. The demise of religion has long been predicted by scholars who have subsequently proven premature in their claims. As Malory Nye has noted, "Conventional wisdom assumes that religion is on the decline in western societies–that the forces of modernization and . . . secularization are simply too strong, and so the once dominant force of religion in public life has now been radically curtailed. Despite this, it is very hard to understand the modern world without understanding the major religious traditions and influences that cut across global boundaries and feed into and shape global politics."[1]

This chapter will suggest not that religion will be destroyed or superseded as human beings engage with the possibility of extraterrestrial life, but that religion, rather, will be transformed by this process, hopefully in a pluralistic direction. Specifically, it will argue that traditions already exist which contain with themselves teachings that would seem to anticipate the existence of nonhuman intelligence in the universe and the reality of inhabited worlds beyond the earth. The Hindu, Buddhist, and Jain traditions, in particular, have long taught that nonhuman intelligent beings are real, and that there are numerous worlds beyond our immediate region of the cosmos.

THE SEARCH FOR EXTRATERRESTRIAL LIFE, RELIGION, SCIENCE-FICTION, AND COSMIC PLURALISM

There are few questions that are as fascinating as the question of whether life has arisen beyond our planet Earth. When this question is posed, it is typically connected with a whole cluster of related questions: Has *intelligent* life—that is, life with which human beings might meaningfully communicate—arisen beyond the Earth? Given the vastness of space, what is the likelihood that we might contact such life-forms, if they have indeed arisen? Why have we not heard from them yet? Or perhaps we have heard from them, but this is not yet a widely acknowledged fact. If even one of the countless stories of extraterrestrial visitation would prove to be true, or if even one of the many sightings of aerial phenomena which have recently been acknowledged by the United States government is of an extraterrestrial craft, then we could say that intelligent life-forms have already, in fact, contacted humanity; but this remains, in my judgment, an unproven assertion. Although there are some accounts of contact that many of us might regard as quite persuasive, the kind of publicly available evidence that enables for scientific proof beyond a reasonable doubt does not yet appear to be forthcoming. It is also the case that many of the accounts of contact are possibly with extradimensional beings and not with beings from other planets in our physical universe.

If the realities behind UFO sightings and stories of alien abduction, or ancient stories which some have been interpreted as evidence of extraterrestrial visitation, end up having an explanation that does not involve extraterrestrial life, then the question of why we have yet to hear from intelligent beings in the cosmos raises another set of possibilities. There is the grim possibility that life-forms do not survive their industrial era, even if they reach it. The fact that the future survival of humanity remains an open question makes this a reasonable point to raise about any putative extraterrestrial civilization. There could also be a host of worlds with intelligent civilizations which lack the ability to communicate with us, perhaps inhabiting a preindustrial (or a deliberately nonindustrial) mode of existence. In that case, it could be that we are the first civilization in the universe to reach the point of exploring space and searching for life beyond our own world. Then there is what is perhaps the grimmest possibility of all: that we are truly alone in the universe. This is grim from the point of view of those of us who are excited at the prospect of contacting and learning from intellectual beings who are radically different from ourselves, but who are enough like us to engage with us in meaningful communication. It also dramatically intensifies our sense of how precious life on Earth is, and our consequent responsibility to protect and foster that life: the only life in the universe.

More optimistically, it might simply be the case that space is just so vast that the statistical likelihood of extraterrestrial civilizations existing at times and in places that make two-way contact feasible is exceedingly small. Over the life span of the universe, there may be countless worlds on which intelligent civilizations arise, civilizations which even spread out into space, but never in the right place or at the right time to become aware of one another.

Even more optimistically, perhaps it is just a matter of time before contact occurs. Given recent dramatic advances in our knowledge about extrasolar planets, it seems increasingly likely that, within our lifetimes, we will at least become aware of planets that are able to sustain life as we know it—carbon-based, water-dependent, life—and that at least one of these planets will end up proving to actually possess life, such as by having an abundance of oxygen in its atmosphere, or perhaps even some industrial pollutants, for good measure. It could then just be a matter of pointing our radio telescopes at these worlds and waiting.

Speaking of radio telescopes, it could also be the case that we are simply too primitive to contact the civilizations that exist. Our hope that we will detect radio signals from an advanced extraterrestrial civilization, or that an extraterrestrial civilization will detect our radio or television signals and take an interest in us, might be equivalent to sending out smoke signals, and hoping someone else will reciprocate, in a world where people mainly communicate through radio. Radio transmission is being increasingly superseded even here on the Earth, with serious competition from fiber optics. What other ways of communicating might a billion-year-old civilization have achieved, which we cannot yet imagine? Indeed, it could be that one or more extraterrestrial civilizations have already become aware of us but are refraining from contacting us on principle. *Star Trek's* Prime Directive comes to mind, as well as the possibility that we are simply not sufficiently interesting or important to contact. This idea that we are not, after all, particularly important or interesting would seem to be the next logical step in the de-centering of humanity that has been facilitated by the advance of science ever since Copernicus proposed that Earth is not the center of the universe.

What, we might ask, does the search for extraterrestrial life have to do with religion? Of the scenarios just mentioned, do any of them bear any religious significance?

The answers to these questions depend, of course, upon one's religious orientation. From the perspective of theistic religion, it would seem that a vast universe housing myriad civilizations made up of a dizzying array of life-forms implies a far grander and more majestic vision of God than a universe with only one inhabited planet. At the same time, this is a possibility which raises a variety of questions that can be disruptive to established orthodoxies. Giordano Bruno, after all, was burned at the stake in 1600, at least in part for his belief that the universe contains countless worlds, and that these worlds could be inhabited. While some have denied that this particular view of Bruno's was the cause of his execution, recent evidence has come to light which suggests that it did, indeed, play a role in his condemnation by the Inquisition.[2] If intelligent life-forms that are comparable to human beings exist on other worlds, do they stand in need of salvation, as human beings do, according to most forms of Christianity? If so, then did God become incarnate multiple times and in multiple forms to atone for the sins of the various extraterrestrial life-forms and enable them to achieve salvation? Would extraterrestrial life-forms need divine revelation to know the will of God, as taught in Judaism and Islam? Are there then extraterrestrial prophets who carried the divine message to their respective peoples? What would be the relationship of their various revelations to the revelation of the Prophet Muhammad, which is held in Islam to be final, at least for human beings? On a personal note, I will mention that thinking about the possibility of multiple divine incarnations for extraterrestrial worlds was one of the factors that led me, as a young Roman Catholic, to feel drawn to Hinduism. If God, who is pure love, would share the divine grace with beings on many worlds—as, it seems, a God of pure love must—it would seem that this grace would also be shared with multiple *human* cultures as well, and not only one. Why should civilizations beyond the pale of Christianity be denied the access to truth and salvation enjoyed by Christians? The same question could be asked of any religion making exclusive claims of this kind.

It is not a big stretch from the idea of multiple divine incarnations on multiple planets to the Hindu concept of the *avatar*, or divine descent, who appears wherever and whenever living beings are in need. In the words of the *Bhagavad Gītā*: "Whenever *dharma* declines and chaos arises, then do I manifest myself. To destroy evil and restore the good I appear age after age."[3] Multiple divine incarnations or manifestations would seem a better fit for a universe containing a variety of diverse intelligent life-forms than a singular incarnation who comes once and for all time and then needs the good news of his coming to be transmitted across cultures through missionary activity, on the assumption that other cultures are lost in darkness until they receive this message. In other words, an affirmation of the possibility of intelligent extraterrestrial life would seem to be of a piece with a pluralistic approach to religious truth. Both militate against anthropocentric (in the first case) and ethnocentric (in the second case) models of reality which privilege just one mode of being in the world (with "the world" here meaning "the universe"). And both conceive of divine reality as vastly beyond what can be contained in the teaching of any singular religious or cultural tradition, and as grounding a creation which encompasses a multitude of modes of being that take the shape of both numerous biological manifestations (extraterrestrial life-forms) and of numerous cultural and religious modes in the life of a single species (the cultures and religions of humanity).

Such a *cosmic pluralism*, if we may call it that, is comparatively easy to approach from the perspective of the traditions of India: the Hindu, Buddhist, Jain, and Sikh traditions. My focus here is on Hindu traditions, but a similar assertion can be made about other Indic traditions as well. In the Western world, at least in the modern period, the primary medium through which human beings have envisioned the possibility of extraterrestrial life has been the genre of science fiction. From the early days of the genre, with the writings of H.G. Wells, to the most recent streaming sci-fi series, speculation about extraterrestrial life—its nature and its intentions—have been standard fare. (I have already had an occasion to mention *Star Trek*, when referring to an extraterrestrial Prime Directive as a possible reason that there has yet to be any indisputable communication from a nonhuman civilization.) Science fiction has been a medium of speculation while at the same time shaping our lived reality, through its envisioning of technologies and ways of living that have inspired actual advancements, from computers to CRISPR.[4] In the traditions of India, however, concepts associated in the contemporary West with science fiction, including the idea of many worlds inhabited by intelligent nonhuman beings, are commonplace. Hindu texts depict a cosmos that is vast in both spatial and temporal terms. As Carl Sagan famously noted:

> The Hindu religion is the only one of the world's great faiths dedicated to the idea that the Cosmos itself undergoes an immense, indeed an infinite, number of deaths and rebirths. It is the only religion in which the time scales correspond, no doubt by accident, to those of modern scientific cosmology. Its cycles run from our ordinary day and night to a day and night of Brahma, 8.64 billion years long, longer than the age of the Earth or the Sun and about half the time since the Big Bang. And there are much longer time scales still.[5]

The one correction that I would want to make to Sagan's account is his claim that the Hindu religion is "the only religion" which affirms such vast time scales. Again, this is actually true of Indic religions in general, and not only Hindu traditions. His claim, of course, that the fact that these time scales are in close accord with those of modern scientific cosmology is "no doubt by accident" is one which many Hindus would of course contest, believing that the ancient seers, or *rishis*, were able to perceive the true nature of reality while deep in meditation. For the purposes of this chapter, though, I think we can allow that difference of worldviews to stand. Whether it was by accident or through some special insight of the ancient seers, Hinduism and related traditions are well equipped with cosmological concepts that would seem to smooth the way for the eventual discovery of extraterrestrial intelligence. The notion that human religions would all be shaken to their foundations by this discovery would seem to break down when we bring the religions of India into our reflections. Such a discovery might even be seen by some practitioners of these traditions as validations of long-held beliefs.

The idea that the cosmos is not only incredibly old, but that it also contains many inhabited worlds, is also commonplace in the writings of the Indic traditions. Buddhist texts, in particular, refer frequently to "worlds as numerous as the grains of sand of the Ganges."[6] Hindu texts refer to numerous worlds, or *lokas*, inhabited by various deities and other intelligent nonhuman beings. It is also believed by some that these worlds can be visited astrally or etherically by means of various meditative practices.[7] One entire Buddhist school of thought, the Yogachara system of philosophy, is held to have been revealed to its founder, a monk named Asaṅga, by the bodhisattva Maitreya, who was visiting our world from his usual plane of existence, a celestial realm called the Tushita (the "happy" or "contented") heaven.[8]

Again, the idea of intelligent life on many worlds, while not directly implying pluralism, or any other specific religious view, would seem to be of a piece with a pluralistic understanding of religion. If human beings have developed an incredibly diverse array of religious traditions, it is not unreasonable to wonder how much more diverse the religions of many intelligent species might be. And if we allow ourselves to import the pluralistic assumption that spiritual truths ought to be as available for the assent of living beings as scientific ones—or that, in theistic terms, a truly loving divine reality would not lavish the saving truths upon only one civilization, much less only species, but would rather make these available to all for discovery—then a cosmic pluralism would seem to be a logical approach to take toward the religious diversity of the universe.

This, of course, begs a number of questions which this field of inquiry opens up. Is there a reason to assume that an intelligent extraterrestrial species would have religion at all? In many of the worlds which are envisioned by science fiction, religion is viewed negatively, as a peculiarly human malady, and one which an advanced extraterrestrial civilization would have long shed by the time it made contact with humanity. In this view, we humans, ourselves, will also outgrow our need for religion if we survive our current crises and continue to evolve a higher sensibility.

Indeed, some might argue that leaving religious thinking behind is itself a precondition for future human survival. This has been the view of a number of prominent science fiction authors, such as Gene Roddenberry, who conceived of the future depicted in *Star Trek* as a wholly secular one (though subsequent contributors to *Star Trek* lore have deviated from this vision, most notably in the series *Star Trek: Deep Space Nine*, in which religion plays a fairly central role).

Contrary to this perspective, there is the point that, even if the religions of extraterrestrial life-forms are radically different from human religions, perhaps so different as to stretch the very meaning of the word *religion*, certain questions are of a sufficiently universal nature to be salient even for nonhuman life-forms. The experiences of death and loss, for example, form a powerful motivation to develop some set of beliefs about an afterlife, and a set of practices and ways of life designed to create meaning in the face of mortality. Human traditions alone are sufficiently diverse as to stretch and challenge conventional meanings of the word *religion* a great deal. For example, is Confucianism (or Ruism) a religion? Much depends upon how the word *religion* is defined. But if we are willing to think broadly of what we might call spiritual aspirations, there does not seem to be any reason to assume these will be absent among intelligent extraterrestrial beings, even if the quest to discern them might prove to be one humanity's greatest intellectual challenges, should we ever have the opportunity to pursue it.

Though he does not, to my knowledge, directly address the question of extraterrestrial life, Swami Vivekananda does speak about another science fiction theme when he raises the question of the religions that human beings might practice in the future. He affirms pluralistic convictions, on the basis of his Hindu, Vedantic philosophy, that could provide some guidance in thinking about the topic of possible extraterrestrial religion as well. He first affirms a strongly pluralistic attitude toward the past and present religions of humanity:

> I accept all religions that were in the past, and worship with them all; I worship God with every one of them, in whatever form they worship Him. I shall go to the mosque. . . . I shall enter the Christian's church and kneel before the crucifix; I shall enter the Buddhistic temple, where I shall take refuge in Buddha and in his Law. I shall go into the forest and sit down in meditation with the Hindu, who is trying to see the Light which enlightens the heart of every one.[9]

He then goes on, however, to affirm the same pluralistic attitude toward religions which do not yet exist:

> Not only shall I do all these, but I shall keep my heart open for all that may come in the future. Is God's book finished? Or is it still a continuous revelation going on? It is a marvellous book—these spiritual revelations of the world. The Bible, the Vedas, the Koran, and all other sacred books are but so many pages, and an infinite number of pages remain yet to be unfolded. I would leave it open for all of them. We stand in the present, but open ourselves to the infinite future. We take in all that has been in the past, enjoy the light of the present, and open every window of the heart for all that will come in the future. Salutations to all the prophets of the past, to all the great ones of the present, and to all that are to come in the future![10]

It does not seem a great stretch at all, and indeed, to be entirely in keeping with the Hindu tradition's affirmation of the reality of many worlds and many forms of life, to extend the kind of pluralism that Vivekananda affirms here toward the religions of the future to the possible religions of intelligent extraterrestrial civilizations. This extension could occur on the basis of the principles that Vivekananda affirms: that "We take in all that has been in the past, enjoy the light of the present, but open every window of the heart for all that will come in the future." From the human perspective, of course, any extraterrestrial religion would be, in our current time, a religion of the future, as we have yet to contact intelligent extraterrestrial life.[11] It is also based on Vivekananda's sense that divine revelation is ongoing: that the self-revelation of an infinite mind must itself be a process, and not something that can be contained in any single text or tradition.[12]

My basic claim here is that the search for extraterrestrial life needs to be accompanied by the kind of openness to diverse views and possibilities that

Vivekananda commends; for if there are indeed a great many extraterrestrial civilizations in existence, there would seem to be every reason to expect that the religious diversity of the universe would also be correspondingly vast. It may well be, as some science fictions predict, that there are extraterrestrial species that have left religion behind, or that have never had it. But it may well also be that there are, as other science fiction works have also predicted, no less of a variety of religious beliefs and attitudes among extraterrestrial civilizations than that which exists on our single small planet today.

WHAT IS LIFE?: PANEXPERIENTIALISM AND ITS IMPLICATIONS FOR THE SEARCH FOR LIFE IN THE UNIVERSE

Beyond the question of the religious implications of the search for extraterrestrial life and the cosmic pluralism that this search would seem to commend, another, even more basic question presents itself as we consider this search from a philosophical perspective informed by the thought of Alfred North Whitehead, as well as by traditions such as Hinduism and Buddhism. This is the question of what, precisely, we mean when we speak of *life*?

In this chapter thus far, I have spoken of intelligent extraterrestrial life in terms of the kind of life which human beings might widely recognize as such: that is, of beings enough like us for us to be able to communicate with them in a meaningful way—to exchange ideas and get to know how we each experience and conceive of our existence. A major criticism of this way of thinking about extraterrestrial life, and of the science fiction that employs it, is that it is much too limiting. I once heard a friend complain that most of the species encountered in *Star Trek* are basically humans with prosthetics on their foreheads. Given the diversity of life on earth alone (the stark differences between, say, a human being and an elm tree), one would certainly expect life on other planets to be very different from human life. The best science fiction tries its best to conceive of life that is truly *alien*: truly other to humanity. To my way of thinking, three of the most successful science fiction works in this regard are *2001: A Space Odyssey*, *Solaris*, and *The Arrival*.

In *2001*, the extraterrestrial life-forms are not even seen. Rather, their artifacts are encountered: first on the Earth of the distant past, then on the Moon, and finally, in orbit around the planet Jupiter. Each of these encounters leads to a major transformation of humanity: first, from prehuman hominids to the humanity that we know today, and finally, to a new form which is visualized as a great embryo floating in space. In *Solaris*, astronauts in orbit around the planet that gives the film its title are contacted telepathically by what is essentially a living ocean on the planet's surface. *The Arrival* recounts the arrival of an extraterrestrial species on Earth whose experience of time is completely different from that of human beings. As the story unfolds (slight spoiler alert!), it is revealed that this different experience of time is not chiefly a function of the extraterrestrials' biology, but of their language, which can be learned by human beings (who then experience time in much the way that the extraterrestrials do). The extraterrestrials come to the earth to share technology with the human race that will allow humans, in the future, to save the extraterrestrials from a threat to their existence that they know is coming because of their unusual perception of time.

Of course, given the vastness of the universe, it is certainly possible that other beings have evolved who are, indeed, similar enough to human beings to enable us to communicate with them with relative ease—what we might want to call, casually, *Star Trek* species—as well as life-forms which we might have difficulty even recognizing as such.

According to process thought, the cosmos itself is, in a sense, alive. It is an organism, that possesses a mind (which most process thinkers refer to as *God*), and that consists of entities which are thoroughly interdependent and interlinked. These "actual entities" are conceived not as material entities, but as centers of experience. *Panexperientialism*, a term coined by David Ray Griffin to characterize this doctrine of process philosophy, refers to the idea "that the ultimate units of the world contain experience (and perhaps spontaneity)."[13] Although process thinkers are careful to distinguish between *experience* and *consciousness* (at least as consciousness is normally defined, as self-aware awareness), there is resonance between panexperientialism and Hindu and Buddhist conceptions of consciousness as something fundamental to the nature of existence. This is a view toward which many in the field of physics are gravitating (pun intended!) based upon observations such as those in the now famous "double slit" experiment which suggest that the act of observing a phenomenon plays a constitutive role in its occurrence.[14] At its broadest, panexperientialism involves a denial of both materialism and of conventional dualism, rejecting the concept of "dead matter," or "physical actualities *devoid* of experience."[15] Process thinkers explain the differences, at the macro level of day-to-day experience, between living beings (dogs, cats, humans, plants) and nonliving beings (desks, stones, books, computers) through a duality within the structures in which actual entities arrange themselves. What we conventionally think of as living beings are in fact serially ordered societies of actual entities, whereas what we think of as nonliving beings are not serially ordered, thus not allowing for the rich continuity of experience from one entity to the next that is enabled in a serially ordered society.

Conventional understandings, in both India and the West, of life differentiate living beings (*jīvas* in Sanskrit) from nonliving matter (called *ajiva*—literally 'nonliving'—in Jain philosophy). Nonliving entities are thus basically what process thought rejects: "physical actualities *devoid* of experience."[16] The universe, again, from a process perspective is essentially a living thing, made up of countless centers of experience that arrange themselves into self-aware entities which persist through time (what we conventionally think of as living beings) and aggregates (which correspond to what we conventionally think of as inanimate physical objects, but which are made up, on the most fundamental level, of experiencing entities).

Even without turning to process philosophy, or to Vedantic or Buddhist systems of thought which take the universe to be a manifestation of consciousness, it is not uncommon to think about, for example, the lives of stars. According to the famous Gaia hypothesis, there is also good reason to think of the planet Earth itself as a living being.

This way of thinking has the potential to open up an entirely new paradigm for the search for extraterrestrial life. If we are seeking to communicate with life-forms different from ourselves, we need look no further than our own planet. Animals and even plants are, it can be argued, centers of experience, and even, in some cases, of consciousness capable of communicating with our own. (Anyone who has spent an appreciable amount of time with animals knows this to be true.) Nature-based religions, which teach that all things are infused with spiritual energies, and also that human flourishing is dependent upon maintaining harmony with these energies, arguably carry a wisdom that it would be well for human beings, collectively, to recover, in light of the destruction which our global consumerist culture has wrought upon the world. The idea that, through the medium of extreme climate events, the Earth is "trying to tell us something" may not be quite as farfetched as it might seem. In the Hindu tradition, the Earth is a goddess—Bhūmī, or Bhū Devī. In the Hindu epic, the *Mahābhārata*, Mother Earth goes to Lord Vishnu for assistance, because human beings are wreaking havoc upon her surface. Lord Vishnu acquiesces and manifests as the avatar Krishna, who plays a key role in bringing about the destruction of numerous human beings through the war between the Kauravas and the Pandavas. Far from a merely human conflict, this war is revealed to be the means by which cosmic forces are balanced. Humans are one cosmic force, the story seems to be telling us. We would do well to maintain balance with the other forces at play in the universe, or else we risk our own destruction.

As we seek to discover intelligent life beyond the Earth, and, we hope, communicate with that life, we should also be mindful of the life that is all around us, here and now. Perhaps this is the very lesson we need to learn in order to be ready for contact with extraterrestrial life. Science fiction stories—and the world's great spiritual literatures—are full of cautionary tales about what happens when human beings arrogantly see ourselves as the measure of all things, and not knowing our place in this vast and intricate cosmic process. If we do manage to contact extraterrestrial life, it will be a great leap not only in our understanding of the universe, but of ourselves and who and what we really are in relation to it. It is a venture which needs to be approached courageously, but also with humility and a loving openness to the minds we hope one day to encounter.

NOTES

1. Nye, *Religion*, 182
2. Martínez, "Was Giordano Bruno Burned at the Stake for Believing in Exoplanets?"
3. *Bhagavad Gītā* 4:7–8; translation mine.
4. CRISPR "is a technology that can be used to edit genes and, as such, will likely change the world." Page, "CRISPR."
5. Sagan, *Cosmos*, 258.
6. Nāgārjuna, *Ratnāvali*, Verse 486, in Thubten, *Practical Ethics and Profound Emptiness*.
7. Bhaktivedanta Swami Prabhupada, *Easy Journey to Other Planets*.
8. Williams, *Mahāyāna Buddhism*, 80.
9. Vivekananda, *Complete Works*, vol. 2, 374.
10. Ibid.
11. Again, at least beyond dispute, respectfully setting aside the many interesting claims that this has already occurred.
12. This is a central point of the Vivekananda's lecture, "Hinduism and Shri Ramakrishna": that the term *Veda* refers primarily not to the Hindu sacred books known by that name, but to the sum total of "supersensuous truths." Ibid., 181. These truths cannot be exhausted, according to Vivekananda, by any single text.
13. Griffin, *Re-enchantment without Supernaturalism*, 97.
14. A lucid discussion of these developments can be found in Tucker, *Return to Life*, 165–94.
15. Griffin, *Re-enchantment without Supernaturalism*, 6. Emphasis his.
16. Griffin accounts for the existence of what are conventionally understood to be inanimate objects, such as tables and chairs, with a concept of "organizational duality," according to which actual entities organize themselves either serially (forming stable entities typically regarded as living) or aggregationally (to form tables, chairs, rocks, and so on). Ibid.

BIBLIOGRAPHY

Bhaktivedanta, A. C. Swami Prabhupada, *Easy Journey to Other Planets*. Mumbai: Bhaktivedanta Book Trust, 1970.

Griffin, D. R. *Re-enchantment without Supernaturalism: A Process Philosophy of Religion*. Ithaca: Cornell University Press, 2001.

Le Page, M. "CRISPR: A Technology That Can Be Used to Edit Genes." *New Scientist*. Accessed December 12, 2021. https://www.newscientist.com/definition/what-is-crispr/

Martínez, A. A. "Was Giordano Bruno Burned at the Stake for Believing in Exoplanets? Most Historians Say No, but New Evidence Suggests Otherwise." *Scientific American*, March 19, 2018. Access December 12, 2021. https://blogs.scientificamerican.com/observations/was-giordano-bruno-burned-at-the-stake-for-believing-in-exoplanets/.

Nye, M. *Religion: The Basics*, 2nd edn. London: Routledge, 2004.

Sagan, C. *Cosmos*. New York: Random House, 1980.

Thubten C., ed. *Practical Ethics and Profound Emptiness: A Commentary on Nāgārjuna's Precious Garland*, translated by K. J. Tegchok. New York: Simon & Schuster, 2017

Tucker, J. *Return to Life*. New York: St. Martin's Griffin, 2013.

Vivekananda, S. "Hinduism and Shri Ramakrishna." In *Complete Works*, Vol. Six, 181–86. Mayavati: Advaita Ashrama, 1979.

———. *Complete Works*, Vol. Two. Mayavati: Advaita Ashrama, 1979.

Williams, P. *Mahāyāna Buddhism*. London: Routledge, 1989.

Chapter 12

Multiplicity without Tyranny

The Nonviolent Aim in Jainism and Whitehead's Process Metaphysics

Brianne Donaldson

Both process metaphysics and Jainism seek to explain the universe through a basic activity common to the multiplicity of existent entities.[1] Importantly, these activities—because they take place in the actual, embodied world—include loss and violent exclusion, as an essential, if tragic, feature of the universe.

In Whitehead's process metaphysics, loss is identified in two key aspects: first, as negative prehensions, or exclusion, within the concrescence of the actual occasion,[2] and second, when living societies of actual occasions consume other societies as food or energy as a means of stabilization and repair.[3] In Jainism, which posits a universe full of infinite living beings, the fundamental violence, or *hiṃsā*, of embodied life is constituted in four instincts (*saṃjñā*), including the craving for food that establishes basic competition, followed by sex, fear, and the desire to accumulate resources for future use.[4]

Yet, both of these metaphysical accounts also provide a clear framework for diminishing, or even eliminating, these harms. Jainism is globally renowned for the ethical mandate of nonviolence, or *ahiṃsā*, most commonly demonstrated in the iconic images of monks and nuns wearing mouthshields to prevent disturbing air-bodied beings, walking barefoot so as not to injure earth-bodied beings, or using hand brooms to sweep the ground clear of insects when walking and sitting. Although only a very small portion of the Jain community is composed of these mendicants, they function as living exemplars for the much larger lay community, all of whom also strive in the context of work, family, and home to practice nonviolence in thought, word, and deed toward their own self and toward all living beings.

Harm reduction is also a central aim of Whitehead's process metaphysics. Whitehead diagnoses violence as a ubiquitous cosmic reality, that is, "In the temporal world, it is an empirical fact that process entails loss . . . the nature of evil is that the characters of things are mutually obstructive [such that] the depths of life require a process of selection."[5] Yet this destruction is not unavoidable. "There is no reason," he says, "of any ultimate metaphysical generality, why this [obstruction] should be the whole story . . . [because] selection is elimination as the first step *toward another temporal order seeking to minimize obstructive modes*."[6] Whitehead continues, "Selection is at once the measure of evil, and the process of its evasion. It means *discarding the element of obstructiveness of fact*. . . . [by] building up a mode of utilization . . . introducing a complex structure of harmony."[7] Whitehead's realist acknowledgment of violence—that "Life is robbery,"[8] or that "Decay, Transition, Loss, Displacement belong to the essence of the Creative Advance"[9]—is not inescapable, but rather a present exclusion motivating every actual entity toward an aim "best for *that* [particular] impasse."[10] The process exhortation for each entity to, as Roland Faber writes, "Be Multiplicity!" describes a universal mode of expansive becoming that seeks to minimizes obstruction.[11]

In this chapter, I argue that the Jain practice of *ahiṃsā*, when seen in the logical structure of Jain metaphysics, provides a dipolar framework very akin to Whitehead's own dipolar structure, through which obstructive modes are overcome. *Ahiṃsā* is best understood, not as a flat prohibition of harm meant only for humans, but as a universal and uneven practice between the poles of present obstruction and the possibility of evading such obstruction.

By using the Jain dipolar framework as a comparative example, I push beyond a seeming reticence among process philosophers in order to identify a strong ethics of nonviolence in Whitehead's work. I speculate on some reasons for this reticence, and illuminate a few counter examples wherein process scholars interpret Whitehead's metaphysics as a practice of embodied nonviolence.

WHY JAINISM?

By Whitehead's own admission, his process philosophy of organism "seems to approximate more to some strains of Indian, or Chinese, thought" insofar as it "makes process ultimate."[12] Though numerous comparisons have been made between process philosophy, Buddhism, and various Chinese traditions, Jainism is relatively little engaged,[13] though its dipolar structure is perhaps more metaphysically salient for a process comparison, uniquely illuminating a distinct Whiteheadian ethics of nonviolence.

Jainism is a global philosophical system and indigenous religious community originating in India over 2500 years ago. Although the overall community of global Jains is small—according to the World Religion's Database (WRD),[14] less than 1 percent of the Indian population and estimated 6–7 million Jains worldwide[15]—its philosophical claims of universal sentiency and the related ethical ideal of nonviolence have considerably shaped the subcontinent and the world, notably through undermining Vedic practices of animal sacrifice, systematizing a unique account of karmic causality, providing robust philosophical foundations for an ancient plant-based dietary ethics, driving debates with and between Buddhist and other Indian philosophers, and shaping Indian concepts of statehood from antiquity to the present, including informing Mahatma Gandhi's and indirectly Martin Luther King Jr.'s nonviolent political struggles.

There is no deity or founder in the Jain tradition, but rather a series of twenty-four teachers, called *tīrthaṅkara*s, in every cyclical epoch of time who perpetuate a path, or *tīrtha*, of nonviolence as the most effective way to traverse, and possibly liberate oneself from, the cycle of birth, death, and rebirth, known as *saṃsāra*. The last two of these teachers are considered historical persons who oversaw a community of mendicant monks and nuns—Pārśva (8th–7th c. BCE), and Mahāvīra, the "great hero" (5th c. BCE), who was an elder contemporary of the Buddha. Distinct from the dominant Vedic view of the time emphasizing right ritual knowledge for a select few, the mendicants followers of Mahāvīra (and the Buddha) represented a "striver" (*śramaṇa*) movement, in which anyone, through rigorous knowledge and restraints, could overcome their birth condition and pursue liberating knowledge and practices without the mediation of specialized rituals. The community of mendicants would eventually grow to include laypeople, providing two distinct paths by which to practice the five Jain vows: nonviolence (*ahiṃsā*), and four others that flow from the first: truthfulness (*satya*), taking only what is given (*asteya*), mendicant celibacy (*brahmacarya*), and non-accumulation of goods (*aparigraha*). Mendicants practice these fully as great vows (*mahā-vrata*) while lay Jains strive toward them partially as minor vows (*aṇu-vrata*) in the context of work, family, and home. The community eventually divided into two main sects based on the clothing worn by the mendicants: the larger Śvetāmbara, or "white-clad" lineage, and the minority Digambara, or "sky clad," naked lineage. These two main sects, along with several sub-lineages, represent distinct-yet-overlapping Jain communities.

As noted above, many people are primarily familiar with Jainism through images of monks and nuns striving to avoid harm to other living beings. A superficial understanding of the mendicant ideal can lead to a dismissal of

Jainism as an impracticable lifeway, at best impossibly idealistic, at worst, a delusional attempt to deny social and ecological entanglements. Scholar Jeffery D. Long[16] captures this tension well when he describes his students' greater resistance to Jain philosophy as compared to Hinduism and Buddhism. "The issue, I strongly suspect," he writes, "is that Jainism, more than any other Indic religious tradition, explicitly and dramatically embodies not only a rejection, but a *reversal* of the values that are dominant in contemporary Western society," continuing, "The idea that the life of an insect might be every bit as valuable as my own is, to say the least, foreign."[17]

This suspicion of nonviolence can also be seen among process thinkers. After I offered a public talk for the Cobb Institute in 2021 exploring a comparative ethical analysis of process and Jainism, preeminent process theologian John Cobb commented that process philosophy privileges ecology *over* nonviolence, suggesting an incompatibility between a process-relational worldview and a nonviolent ethos.[18] Indeed, beyond any comparison with Jainism, process philosophers frequently illuminate the essential violence of ecological relations that aim toward greater intensity and contrast. Myron Moses Jackson, for instance, affirms that "disharmony and disequilibrium . . . [as well as] error is a necessary condition of the experience of actual occasions which establishes various types of value and intensity."[19] In Cobb's Whiteheadian analysis of evolution, he affirms natural violence as an essential part of organisms' creativity since, in the Process context of ecological interrelationship, rather than separative organisms, "total insulation is not possible," including insulation from harms.[20] In Brian Henning's *Ethics of Creativity*, he makes clear that, in ordered societies such as animals or plants, exclusion takes the form of predators destroying prey, a violent "selection" that characterizes the complex, if tragic, beauty of a stable and vital ecological whole.[21] But as I already stated, violent selection is not Whitehead's last word regarding the world's creative process, but rather, "selection is elimination as the first step towards another temporal order seeking to minimize obstructive modes,"[22] so it is incumbent upon us to consider how this alternate order comes into being in a process framework, which is where Jainism can be especially helpful.

BETWEEN KARMICALLY BOUND *JĪVA* AND LIBERATED *SIDDHA*: THE DIPOLAR ACTIVITY OF NONVIOLENCE IN JAIN METAPHYSICS

The first line of the *Tattvārtha-sūtra*, an ancient Jain canonical text considered authoritative by all Jain sects, states that "right worldview (*samyak-darśana*), right knowledge (*samyak-jñāna*), and right conduct (*samyak-cāritra*) constitutes the path to liberation."[23] This means that the ethic of right conduct—that of *ahiṃsā*, or nonviolence—is not an arbitrary doctrine, but rather must be understood as emerging from right view and knowledge. In this way, Jainism exemplifies a central facet of Indian naturalism by which perception and knowledge naturally influence action,[24] very akin to feminist physicist Karen Barad's neologism of an "ethico-onto-epistem-ology" asserting the inseparability of ethics, ontology and epistemology.[25]

In Jain philosophical discourse, right view, knowledge, and conduct constitutes the *mokṣa-mārga*, or path of liberation. From the early centuries through the medieval period,[26] this path takes the shape of fourteen steps, called *guṇa-sthāna*s, or "stages of qualities" along which every being in the universe is located. In artistic renditions, this staged path is imaged as the "snakes and ladders" game which has versions in several cultures. The dipolar framework shows us how each living being, or *jīva*, exists between its present physical limits of embodied, karmically conditioned existence within the path and a nonphysical, virtual telos of liberated omni-perception known as the *siddha* which is the path's ultimate aim.

Jīva and *Karma*

The Jain universe (*loka*) is said to be populated with an infinite number of unique *jīva*s, a core life force constituting each distinct living being, from subtle elemental entities in earth and water, to cells, insects, and mammals, including people. Each *jīva* in the universe exists somewhere on the *guṇa-sthāna*s path, always undergoing movement and change. But what is the mechanism of this change?

In its innate form, each *jīva* is considered a radically singular eternal substance characterized by four unobstructed qualities of consciousness (*upayoga*), with its two aspects of pure knowledge and pure perception, pure energy (*vīrya*), and pure bliss (*ānanda*), a technical term approximating inner stability. All *jīva*s are interdependently supported by a matrix of five nonliving substances called *ajīva*: medium of rest, medium of motion, space, time, and matter.[27] Consequently, Jainism requires that each entity in the universe be understood from the *intra-activity* of its substance, qualities, and modes. As summarized by Padmanabh Jaini, "any existent must be understood on three levels [simultaneously]: the modes, which last only a moment and belong to the qualities; the qualities, which undergo changes and yet inhere forever in their substances; and the substance, which remains the abiding common ground of support for the qualities and their modes."[28] Affirming these states in unison exemplifies the central Jain doctrine of epistemology, called *anekanta-vāda*, or "non-one-sided-view,"[29] by which any aspect of the universe must be understood from multiple perspectives simultaneously. Views are more or less adequate depending on how much of this univocal reality can be understood, perceived, or felt in any moment. While *anekanta-vāda* was used as a rhetorical strategy by ancient Jain philosophers to undermine rival assertions as too partial, the concept came to function in a wide range of nonviolent applications by undermining any one-sided assertion that could justify harm.[30]

While the concept of unchanging substance is one of the central fallacies that Whitehead rejects in Western thought,[31] as he does the concept of Soul,[32] the Jain concept of substance/soul is always undergoing continuous modal change and always co-constitutive with the material and nonmaterial aspects of reality. Within the staged path of embodied existence—the qualities, or *guṇa*s, of these substances—both *jīva* and *ajiva*—are considerably obstructed and/or undergoing constant modal change (*paryāya*) due to *karma*.[33]

Karma is more comprehensively theorized in Jainism than any other Indian tradition, uniquely described as material particles tightly packed into the universe.[34] These particles are said to be activated or "charged" by any mental, verbal, or physical action expressing aversion or attraction, sticking to the *jīva* like fine dust sticking to a wet light bulb and obstructing its innate qualities of knowledge, perception, energy, and bliss. In fact, the Sanskrit root for the term karma, √kṛ, means action, doing, making, causing, effecting.[35] Thus, an embodied *jīva* on the *guṇa-sthāna*s path attracts karma with every action, which obscures pure perception and knowledge of its own nature as well as that of the other infinite sovereign *jīva*s, each on their own unique karmic journey toward the possibility of liberation. This karmic bondage leads to incorrect knowing and careless, harmful action.

It is important to note that karma in Jainism is not only one kind. Jain authors actually describe 148 kinds of karmic matter which need not be detailed here other than to say it is divided into two kinds: (1) "destructive" (*ghātiyā*) karma, called such because its accumulation destroys the four qualities of the *jīva*, while (2) "nondestructive" (*aghātiyā*) karma does not affect the *jīva*, but rather determines the kind of body and rebirth a *jīva* will experience, including birth embodiment (*nāma-karman*), duration/quantity of life (*āyu-karman*), the environment (*gotra-karman*), and satisfaction/feeling in a given environment (*vedanīya-karman*).[36] In sum, nondestructive karma accounts for variations in physical existence, while destructive karma accounts for variations in the *jīva*'s qualities of knowledge, perception, energy, and bliss.

In their embodied, karmically determined forms, living beings in the Jain universe are further divided by the number of senses they have. One-sense beings in earth, air, fire, water, wind, and plants possess the single sense of touch; two- through four-sensed beings possess taste, smell, and vision respectively and include beings such as mollusks and insects; five-sense beings possess hearing, including birds, mammals, aquatic creatures, humans, and heavenly and infernal beings.[37]

Existing in the universe, according to Jain texts, means inhabiting a body that requires the use and sustenance of other bodies to survive, which always has a karmic cost. The great vows of mendicants are aimed to avoid *sūkṣma-hiṃsā*, or "subtle violence," toward beings with one or more senses, including those one-sensed beings that may be immobile or even invisible, explaining the iconic practices described above.[38] The minor vows of lay Jains strive to avoid *sthūla-hiṃsā*, or "gross violence," toward mobile beings with two or more senses that are easier to detect with attention.[39] Consequently, the Jain *mokṣa-mārga*, exemplified in the *guṇa-sthāna* path, includes detailed teachings on how to distinguish between different sensed beings and various vows of restraint to avoid harming these living beings as much as possible.

The *Siddha* as Ultimate Ideal

The ultimate aim of the *guṇa-sthāna*s path is to move from profound karmic bondage that limits right view to a state free of all karmic bondage, an omni-perceptive state known as *siddha*. Stages one through four represent a nonlinear movement between deluded worldview (*mithyā-darśana*), in which karmas obscure understanding of the fundamental elements of the Jain universe, and right worldview, called *samyak-darśana*. All beings who lack knowledge of *jīva*, *ajīva*, and karma, for example, reside in the first stage.[40] But an urge to overcome this delusion through self-development can enable a brief glimpse of right worldview, or *samyak-darśana*, at the fourth stage, said to be attainable by any five-sensed human or animal with mind, signifying a rare realignment of perception needed for greater advancement along the path.[41]

Stages five through seven describe the possible evolution from "partial restraint," including multiple preparatory practices of lay Jains that can enable them to possibly take the five minor vows, up to the "total restraint" of mendicant vows, though at the sixth stage the mendicant vows are characterized as "restraint with carelessness" (*pramatta-virata*)[42] or at the seventh stage as "restraint with carefulness" (*apramatta-virata*), signifying the continued eradication of certain forms of karma permitting gradual progression, interrupted by likely regressions. Stages eight through fourteen mark the suppression or elimination of remaining karmas and related passions toward an eventual goal of embodied omniscience, and then liberated non-embodied omniscience called *siddha*.[43]

A siddha (literally "one who has reached the goal) is a karmically free *jīva* that is no longer in a physical, karmically bound body. Digambara philosopher Kundakunda calls the *siddha* "the beautiful ideal in the whole universe,"[44] meaning a *jīva* attaining its pure state of perception, knowledge, energy, and bliss. Though perhaps sounding like a transcendent Other outside the embodied world, the siddha remains connected to the embodied universe in three primary facets. First, the siddha state of total omni-perception or complete knowledge, called *kevala-jñāna*, refers to a synchronous understanding of the relations between substances, their qualities, and infinitely changing modes throughout the universe in every moment without obstruction. The *Kalpa Sūtra* describes Mahāvīra's attainment of full perception this way:

> Mahāvīra . . . was a Kevalin, omniscient and comprehending all objects; he knew and saw all conditions of the world, of gods, men, and demons: whence they come, whither they go, whether they are born as men or animals or become gods or hell-beings, the ideas, the thoughts of their minds, the food, doings, desires, the open and secret deeds of all the living beings in the whole world; he the Arhat, for whom there is no secret, knew and saw all conditions of all living beings in the world, what they thought, spoke, or did *at any moment*.[45]

Second, the liberated siddha is said to remain in the universe, within the upper most portion called *siddha-loka*. In embodied form, the *jīva* occupies the space of whatever body it inhabits and pursues its path within the limits of that form and function. At liberation, however, the *jīva* is said to shrink down, to shoot up to the top of the Jain universe with its body occupying the whole universe.[46]

Third, the siddha "body" is represented by the hollow cutout of its last human form, denoting an identity memory tied to its last embodiment though no longer bodily bound to it. Rather, the siddha is constituted by its perception of the entire living multiplicity. It is, in some respects, akin to Deleuze and Guattari's virtual "body without organs," that is identified, not by any concrete manifestations or attributes, but by its virtual potential and indeterminate identity.[47]

Thus, as I have argued elsewhere,[48] the siddha is a state of pure perception, knowledge, bliss, and energy that perceives the multiplicity of relations in the universe without obstruction. Although it is a transcendent ideal, it functions as an *immanent* potential that each actual *jīva* strive toward within its own path, inciting greater perception and less violence toward self and others. Kundakunda describes this as the interplay of two onto-epistemic standpoints: the conventional standpoint (*vyavahāra naya*) limited by one's karmic bonds, and the ultimate standpoint (*niścaya naya*) unlimited by karmic bonds exemplified in the siddha.[49] Even if the state is never attained, its potential can be enfolded into every activity.

JAIN NONVIOLENCE AS EXPERIMENTAL PRACTICES TOWARD THE *SIDDHA*

We see the practice of *ahiṃsā* afresh through the *guṇa-sthāna*s framework between the *jīva*'s conventional karmically conditioned activity, and the ultimate potential of the liberated siddha; between *the fact* of obstructed knowledge and harmful action and *the potential* of unobstructed omni-perception of the universe the unison of its constantly changing substances, qualities, and modifications.

In the logic of this dipolar metaphysical frame, nonviolence is not a singular activity, but *any mental, verbal, or physical restraint, by any being, that minimizes current obstructive modes*. Ancient and medieval Jain texts detail many and varied contextual practices of nonviolence by which lay and mendicant Jains—but also nonhuman beings which I will sidestep for now[50]—strive to widen their view, knowledge, and conduct toward less obstructive modes.

For instance, texts describing nonviolence for lay Jains detail seven supplementary vows (*guṇa-vrata*s), by which one can concretely experiment with nonviolence. A summary of the complexities of the first three of these supplementary vows include: (1) avoiding purposeless activities (*anartha-daṇḍa-vrata*) such as gambling, rumination, listening to violent stories, digging the earth or cutting trees, or providing a means of destruction for others; (2) avoiding specific pleasures that increase attachment and aversion (*bhoga-upabhoga*), such as eating certain foods, using stimulant substances or medicines, excessive bathing, eating at night, or drinking unfiltered water; (3) limiting the area in which one walks or travels in order to minimize harm (*dig-vrata*). The additional four supplementary vows (*śīkṣā-vrata*) include supporting mendicants, extending travel and communication restrictions for a set period, temporary fasting and keeping a state of mental calm (*sāmāyika*) for forty-eight minutes each day through forms of study, praising the *tīrthaṅkara*s, or silence.

Avoiding the consumption of certain substances is an essential practice of Jain ethics since ingesting foods, beverages, and medicines not only requires the deaths of living beings—particularly those foods said to house innumerable groups of one-sensed beings such as microbes in animal flesh, fermented alcohol, root vegetables, or high-seed eggplants or figs[51]—but particular consumables are also said to activate certain passions, possessiveness, and ego. Householder texts also list specific occupations forbidden to lay Jains, including jobs that destroy plants or trees, trade in animal by-products, trade in humans or animals, selling weapons, or using large vehicles that require animals to carry heavy loads and crush living beings on the road.[52] It is very common today to find Jains in employment fields such as education, banking, engineering, and medicine, in part, reflecting these early guidelines.

Even mendicants have experimental practices by which nonviolence becomes concretely manifest. Five such supporting rules, called *samiti*—meaning "care"—require extreme attention in walking (*īryā-samiti*), (2) speaking (*bhāṣā-samiti*), (3) accepting food donations (*eṣanā-samiti*), (4) picking up and setting down objects (*ādāna-nikṣepaṇa-samiti*), and (5) performing excretory functions where living being reside (*utsarga-samiti*).[53] Mendicants also refrain from eating at night (*rātri-bhojana*), since any light or smell may draw insects who can fall into food without being detected (Williams 1991, 108–9).[54] An additional six obligatory mendicant restraints (*āvaśyaka*) that contextualize the vow of nonviolence include a daily period of mental quiet (*sāmāyika*) and a daily confession of any harms done to other living beings (*pratikramaṇa*), among other regular habits.[55]

Finally, both mendicants and lay Jains have the opportunity, though not the requirement, to take an additional vow of voluntary death toward the end of their life known as *sallekhanā*, usually when health declines to a debilitated or terminal state.[56] Voluntary death usually involves a lengthy period of gradual fasting from solid food, then soft food, liquids, and eventually water, signifying reduction of harm to other beings, and the erosion of one's own karmic bondage.

These supportive practices make the vow of nonviolence concrete. They are not flat prohibitions, but rather positive mental, verbal, and bodily restraints attempted in varied contexts. Stating it again, nonviolence refers to *any mental, verbal, or bodily restraint that minimizes karmically obstructive modes toward the possibility of a state free of obstruction.*

THE NONVIOLENT AIM OF WHITEHEAD'S PROCESS METAPHYSICS

Whitehead's metaphysics does not present us a progressive path of ethical actions, but we do find a similar dipolar framework that functions in a parallel manner to that of Jainism. On one pole, the actual entity/actual occasion as the "final real things of which the world is made up"[57] or "the really real things which in their collective unity compose the evolving universe."[58] On the other pole, the Ultimate actual occasion, called "God" both in *Science and the Modern World* and in *Process and Reality* before being expanded into other terms in later texts (see below). This Ultimate, or God, is the "chief exemplification" of the activity of becoming and not transcendence of it.[59]

Although Whitehead's reflection on God was, in the early twentieth century, according to Roland Faber "seen as philosophical suicide,"[60] Whitehead makes clear that what he means by "ultimate"—whether "God" or another term—is not akin to monism or any other supernatural or imperialist concept "illegitimately allowed a final, 'eminent" reality, beyond that ascribed to any of its accidents."[61] Throughout his work, Whitehhead offers a "secularization of the concept of God's functions in the world" as key element in the creative advance.[62]

In its "consequent" aspect,[63] Whitehead's "God" is an ultimate occasion that prehends and receives all happenings in the universe without the "loss of immediate unison" that characterizes the temporal world.[64] As Steven Shaviro puts it, "God" is "the *result* of the sufficient reasons, or final causes, of all finite entities, rather than their ground . . . gathering these many little private harmonies, without exception, into a grand, public, and never-completed 'conceptual harmonization.'"[65] In *Adventures of Ideas*, Whitehead describes this consequent prehension of all events as a *virtual*[66] storehouse akin to Plato's *khora* or "Receptacle," the "natural matrix of all things,"[67] or a "completed ideal harmony,"[68] each of these terms referring to "a principle of concretion"[69] that is "devoid of negative prehensions."[70] In two of his last essays, Whitehead describes this consequent aspect as "Immortality" in relation to what is mortal, the "Infinite" in relation to the finite, and the "World of Value" in relation to the "World of Activity."[71]

The mortal, finite, World of Activity also prehends the virtual storehouse of God/Receptacle/Infinite/World of Value[72] as an "absolute wealth of potentiality"[73] that provides relevant possibilities to each concrete becoming. In this "primordial aspect" of God,[74] "each temporal [actual entity] receives that initial aim from which its self-causation starts" through "initial gradations of relevance of eternal objects for [an actual entity's] conceptual feeling."[75] This primordial role, says Whitehead, is "performing an efficacious role in multiple unifications of the universe"[76] luring each event toward an "enlargement" of the data that it uses rather than discards.[77] Because the primordial aspect of God/Receptacle/Infinite/World of Value provides ideal ordering of received potentials for each actual entity, Whitehead also calls this receptive and generative Ultimate the "fostermother of all becoming"[78] and a "flight after the unattainable"[79] through which every actual entity unifies itself in light of the totality of received potentials.

We step back to consider these two dipolar frameworks. In process metaphysics, the temporal World of Activity contributes itself to the ever-expanding virtual World of Value, and that World of Value lures, but does not command, every limited actual entity toward its own contextual novel expansion. In Jainism, each *jīva*—with its dynamic substances, qualities, and changing modes—is perceived in the siddha ideal toward which each embodied and karmically conditioned *jīva* strives to emulate. Both of the dipolar frameworks present relations of "mutual requirement," whereby each polar element is an "instrument of novelty for the other."[80] "When we enjoy 'realized value,'" Whitehead states, "we are experiencing the essential junction of the two worlds . . . One side is the mortal world of transitory fact acquiring the immortality of realized value; and the other side is the timeless world of mere possibility acquiring temporal realization. The bridge between the two is the 'idea' with its two sides."[81]

As with the Jain siddha, the process Ultimate plays a distinct metaphysical role, affirming a purposive direction or natural aim—what Cobb calls a "telic element"[82]—essential to biological evolution and the creative world process, a claim that is as significant for philosophy and science as it is for religion. Animals, plants, and even unicellular organisms are, in a process perspective, according to Cobb, "*purposive* in their behavior," an affirmation also shared by the Jain view that every living being is located on the path toward liberation, though with various levels of karmic restraint possible.[83] The telic aim of living beings is, in process perspective to "(i) to live, (ii) to live well, (iii) to live better."[84]

The telic aim of living beings in Jainism is toward right view, right knowledge, and right conduct as the path to liberation.[85] It is the qualitative telos of "better" or "righter" conduct where we find the lure toward less obstructive modes, or nonviolence. "Finite entities," Whitehead asserts, "require the unbounded universe . . . [and] the universe acquires meaning and value by reason of its embodiment of the activity of finitude."[86] He continues, "finitude vivifies the infinite," and the infinite "adds the perception of worth and beauty to the mere transition of sense-experience."[87]

Here "worth and beauty" refer to the ethical and aesthetic content of nonviolence in Whitehead's framework. The decision of each entity's becoming is a dipolar fusion between the limits of physical data any entity can take in and the virtual unlimited possibility of unbounded prehension without loss. "Judgment," writes Whitehead, "is a process of unification" between particularity of existence and "the totality of existence,"[88] a totality described in *Process and Reality* as "that of a tender care that nothing be lost" or excluded.[89] This means that individual novelty differently decides itself *toward* "minimize[ing] obstructive modes . . . [into] a complex structure of harmony."[90] Becoming is *toward* the mitigation of robbery, the minimization of loss, and *toward* the Ultimate occasion which salvages "what in the temporal world would be mere wreckage," thereby "convert[ing] the opposition into a contrast."[91] This disharmonic unison describes the telos of every coordination between limited Activity, with its essential exclusions, and unlimited God/Receptacle/Infinite/World of Value without obstruction. As Whitehead describes, "Each [actual] tragedy is the disclosure of the ideal—What might have been, and was not: [with] What can be."[92]

This aim toward less obstructive potential—"what can be"—is not a value-neutral ethical claim. It is a bold vision of minimizing loss that is conceptually equivalent to the experimental practices of nonviolence described in Jainism. The dipolar unifications are not a flat prohibition of all action, nor do they provide strict imperatives, but a striving between the limits of what presently exists toward an unlimited ideal without loss. Process theologian Bernard Loomer describes this coordinated becoming as an increase of one's "stature" or "size," referring to the "ability to absorb more and more dimensions of the world in the unity in your own being and add to the stature of your soul." He goes on:

> How much of the other can you incorporate into your being? How many of the contrasts and contradictions of life can you take in without being disorganized, thrown, or broke? Size is increased by the number and intensity of the contrasts that you attempt to unify within yourself. The greater the range and depth of contrasts that you attempt to synthesize into your unity, the larger the Size of your spirit."[93]

The words "incorporate" and "absorb" are unfortunate as they have tones of assimilation or take over. But if we think Loomer's statement alongside the dipolar activity of the *jīva* or actual entity, we can think this "stature" in terms the growing perception toward a state of wider feeling without obstruction. Whitehead says it better when he relates this unification of the becoming self and world with the concept of "Peace." "Peace," he writes, "is self control at its widest—at the width where the 'self' has been lost, and interest has been transferred to coördinations wider than personality,"[94] meaning that self-becoming develops toward world-becoming, such that "the task" of becoming, according to Whitehead, "is to fathom the deeper depths of the many-sidedness of things."[95] The process Ultimate presents a univocal occasion in which the flourishing of the one is synonymous with the flourishing of the many. As Faber provocatively states, "Only with a 'satisfaction' beyond personality (although never without it), does a person become God-capable."[96] To the degree that any entity strives toward the expansive ideal of God/Receptacle/Infinite/World of Value, it reduces obstructive modes.

More boldly, Whitehead describes this dipolar activity as "Love." Using divine metaphor, he says, "What is done in the world is transformed into a reality in heaven," meaning the creative temporal process of activity contributes itself, and thus constitutes, the ever-expanding ultimate occasion of God/Receptacle/Infinite/World of Value, "and the reality of heaven passes back into the world. By reason of this reciprocal relation, the love in the world passes into the love in heaven and passes back into the world," slowly transforming the "mutually obstructive" nature of reality into "a complex structure of harmony" to "minimize obstructive modes."[97] Although Whitehead describes Love as "a little oblivious as to morals," his dipolar framework describes a telic aim characterized by the contextual but nonnormative, reduction of obstructive modes; a universal aim of experimental nonviolence.[98]

SCHOLARLY RESISTANCE TO A PROCESS NONVIOLENCE?

The overt path of karmic reduction in Jainism illuminates the parallel dipolar aim of nonviolence that has been too often overlooked, or trapped in conceptual metaphor, by those engaging Whitehead's cosmology. Whitehead's "metaphysics of morals"[99] are often referred to as an aesthetics,[100] intensity of feeling,[101] or a pursuit of axiological value,[102] phrases that abstract from contextual guidance for action. If the telic aim of the entire universe—and not just "the human"—is toward minimized obstruction, indeed toward love—to "Be Multiplicity!"[103]—why does there seem to be a reticence among process thinkers to characterize this telos as nonviolence? Why, for example, would Cobb privilege ecology *over* nonviolence if the universal aim of the creative advance aim is toward minimized obstruction? Here, I would like to make a few speculative suggestions, drawing upon three contemporary criticisms of the concept of "nonviolence" that might be operative among process philosophers, in hopes of making space for a bold endorsement of a Whiteheadian ethics of nonviolence.

The first criticism is that nonviolence may be unnatural or anti-ecological. This view is often fueled by cultural factors that retain violence as essential to biological and social evolution, ranging from Thomas Hobbes' sixteenth-century view of violence as the "natural condition of mankind" based on resource competition[104] to Tennyson's depiction of nature "red in tooth and claw."[105] Of course, most process thinkers would dispute such accounts of natural violence as overstated; outgroup aggression and resource pressures have also produced collaboration as noted in the many positive process-based analyses of deep ecology, ecofeminism, and environmental action groups such as the Whiteheadian-inspired global Institute for Ecological Civilization.[106] As Darwin noted in *On the Origin of Species*, the "Struggle for Existence . . . [includes] dependence of one being on another" as well as variations "advantageous to the community," and not only the individual.[107]

The larger concern within process camps seems to be that a thorough-going nonviolent ethic might flatten out all ontological differences. Shaviro points out that "Whitehead writes without embarrassment of the feelings and satisfaction of a plant, an inorganic object like Cleopatra's needle, or even an electron,"[108] yet as Brian Henning has argued, people still need a way to conceive of "hierarchy without anthroparchy."[109] Taxonomic distinctions, according to Shaviro, are "of the greatest pragmatic importance. I should not treat a human being the way that I treat a stone," even though a human and stone, from a process perspective, both share the same ontological features of creative becoming.[110]

But aren't these concerns over losing one's discernment of relational difference counter to the deconstructive shock of Whitehead's metaphysics? If the subject has been replaced with event multiplicities, then each actual entity, nexus, and society is liberated from any fixed location in a taxonomy. This is not to say that one does not consume to live; "Life is robbery," asserts Whitehead,[111] but the claim of robbery is not opposed to nonviolence. Rather, robbery is the realist acknowledgement of embodied existence that motivates a process aim toward nonviolence. Whitehead states, "The dominance of societies, harmoniously requiring each other, is the essential condition for depth of satisfaction," affirming that obstruction is not a tragic necessity, but the "first step towards another temporal order seeking to minimize obstructive modes."[112] Including a stone—or any existent entity—in relations of harmonious requirement rather than violent disruption, whenever possible (even if not always possible), is still an act of novel and expansive satisfaction within *that* becoming.

The prevalence of harm-reducing social practices within ecological relations is evident in the flood of research on the social altruism and cooperation of other animals which contradicts any cultural assumptions of "animalized" or naturalized violence.[113] Beyond the bonobos, often touted as our "peaceful primate" kin, even Jane Goodall concluded her analysis in *The Chimpanzees of Gombe* by warning against a mistaken impression of inherent chimp violence. "In actuality, peaceful interactions are far more frequent than aggressive ones," she writes, "mild threatening gestures are more common than vigorous ones; threats per se occur much more often than fights; and serious, wounding fights are very rare compared to brief, relatively mild ones."[114] As American biologist and ecological behaviorist Marc Bekoff summarizes:

> People who claim nonhuman animals are inherently aggressive and warlike are wrong. So, when they use information from animal studies to justify our own cruel, evil, and warlike behavior, they're not paying attention to what we really know about the social life of animals. Do animals fight with one another? Yes. Do they routinely engage in cruel, warlike behavior? Not at all. Numerous species display wild justice and carefully negotiate their social relationships so that fairness, cooperation, compassion, and empathy are quite common.[115]

Even in the political sphere, Jean-Jacques Rousseau challenged Hobbes's view of natural aggression, instead attributing violence to culture and civilization, while John Locke emphasized the evolution of mutual obligations in society. One who rejects nonviolence on the basis, as Hobbes suggested in *Leviathan*, that the state of nature and the life of *homo sapiens*, is "solitary, poor, nasty, brutish, and short," must yet grapple with ongoing evidence that societies also evolved based on compassion, ethical action and cooperation.[116] Consequently, process philosophers have ample support in thinking ecology, including pragmatic hierarchies, together with experimental practices of nonviolence.

The second critique of nonviolence is exemplified in Frantz Fanon's claim in *The Wretched of the Earth* that nonviolent strategies preserve the colonial and capitalist powers of the state.[117] This critique has at least two facets: first, erroneously applying a Eurocentric view of the ideal "human" to non-Western cultures, and second, equating nonviolence with private attitudes rather than public action.

American Indian scholar Tink Tinker describes the "irrelevance" of nonviolence for indigenous peoples as "a contemporary euro-christian cultural signifier that is too easily presumed by the 'liberal colonizer' as some sort of universal that all people should automatically affirm."[118] Indeed, the field of peace studies is undergoing critical post- and de-colonial engagement to challenge the less-visible epistemic, geopolitical, and methodological assumptions of the discipline.[119] These critiques, however, do not necessarily advocate abandoning theories and practices of nonviolence but rather resisting a one-size-fits-all application. In reading Tinker's critique more closely, for example, Eurocentric nonviolence is deemed irrelevant because it frames violence only as a human-on-human phenomenon that overlooks inherent multispecies relations. Tinker writes, "Indian peoples could never limit the category of violence only to acts perpetrated on other human beings."[120] To eat, for example, is a primary aspect of violence, according to Tinker, overlooked in dominant concepts of nonviolence; even ecological nonviolence in the Eurocentric frame is primarily justified based on sustaining humanity.[121] Tinker rejects the imposition of nonviolence, advocating instead for an approach to cosmic relational balance with an aim to "disrupt [the world] as little as possible."[122] If we consider nonviolence through a Jain and process perspective, however, we can see a universal mode of affirming the cost of life, especially through consumption, alongside a metaphysical aim to minimize that obstruction whenever possible, sidestepping Tinker's critique of a Eurocentric and isolated human.

When considering how nonviolence protects the state, we do find many examples of minority or separatist individuals and communities utilizing nonviolence when their own impact on dominant powers is limited. This can be exemplified in naive readings of particular religious commitments to nonviolence or pacifism, such as Jesus's command to followers, in the face of harassment and assault by Roman soldiers, to "offer the other cheek also" (Matthew 5.38–40) or isolationist monastic practices of nonviolence—whether among Christian, Eastern Orthodox, Hindu, Buddhist, or Jain monks and nuns. Likewise, more nuanced reading are required of secular civil disobedience to state and military violence such as Tolstoy or Thoreau's insistence on following the claims of conscience over governmental laws.

Yet, even long-standing critics of nonviolence, including Fanon, as well as contemporary authors such as Ward Churchill[123] and Peter Gelderloos[124] are primarily arguing against a strategy of *pure* nonviolence. That is, they acknowledge that "peaceful tactics" have also been part of combative or "diverse resistance tactics."[125] That is, the practices of nonviolence—even at the civic level, to say nothing of the ecological level—are not always of one kind. In *Nonviolence: The History of a Dangerous Idea*, Mark Kurlansky argues that:

> Some advocates of nonviolence believe that boycotts and embargoes that cause hunger and deprivation are a form of violence. Some believe that using the less lethal means of force, rock throwing or rubber bullets, is a form of nonviolence. But the central belief is that forms of persuasion that do not use physical force, do not cause suffering, are more effective.[126]

Gandhi, for example, was known to use fasting (*dhāraṇa*) as a coercive tool to convince opponents to consider protestors' demands in the Indian freedom struggle. Walter Wink explores the "third way of Jesus," not as passive nonresistance, but of "assertive nonviolence" or "nonviolent resistance"[127] to state power, religious authorities, and mob mentality. Martin Luther King, Jr., similarly, described strategies of nonviolent resistance dependent on "some form of constructive coercive power,"[128] sidestepping any clean dichotomy between persuasion and coercion.[129] Consider King's statement in "Letter from a Birmingham Jail," that "The purpose of our direct action program is *to create a situation so crisis packed* that it will inevitably open the door to negotiation."[130] These are neither passive nor private modes of engagement, nor are they idealized forms of pure nonviolence. As Kurlansky put it, "not passivity, but activism by other means—nonviolence."[131] From a process perspective, any event, whether obstructive or not, has a three-fold private and public dimension of both inclusion and exclusion as (a) emerging "superject" from the world, (b) as a subject of novel creation, and (c) becoming an object for others.[132] The nonviolence aim of process and Jain metaphysics provides a concept of the "polis" that includes and exceeds the limited scope of human politics.

Third, and finally, nonviolence is often dismissed when it's viewed an all-or-nothing ethical absolute. Yet, as I have gestured to above, nonviolence is not a universal, flat application in theory or practice. It is better understood as contextual practice of enlarging what is included in creative becoming; a contextual practice of minimizing obstructive modes. Nonviolence characterizes diverse tactics for personal and social resistance, with coercive and persuasive elements and ideals. As seen in the various Jain practices aimed to minimize karmic intake, doing and nondoing are both aspects of nonviolence, such as positively cultivating knowledge about less visible beings or challenging accepted vocations of harm, as well as avoiding killing life or refraining from eating or traveling.[133] Nonviolence is also not adequately understood as passive separatism, though this may be a feature in certain contexts. Nonviolence need not be limited to an isolated "human" or to intra-human conflicts, but as is evidenced in the Jain assessment of inherent cost and the process affirmation of universal "robbery," nonviolence must be examined in multispecies relations such as food consumption as a way of minimizing disruption to other beings. As Judith Butler describes in *The Force of Nonviolence*, "we can think of nonviolence not simply as the absence of violence, or as the act of refraining from committing violence, but as a sustained commitment, even a way of rerouting aggression for the purposes of affirming ideals of equality and freedom" among all living beings who existence is relationally conditioned.[134]

PROCESS NONVIOLENCE: BE MULTIPLICITY(!) WITHOUT TYRANNY

The dipolar frameworks of process and Jainism both provide speculative philosophical tools demonstrating how violent oppositions can be transformed in each moment into harmonies of contrast; from tyranny to ever-enlarging multiplicities.

Several process thinkers recognize an urge toward harmony in Whitehead's metaphysics, often through his use of the term "Peace," described in the final chapter of *Adventures of Ideas* as the modulating factor of Truth, Beauty, Adventure, and Art enabling the advance of "civilization." Much as the *guṇa-sthāna*s framework maps the removal of karmic bondage associated with obscured understandings of self and others enabling the *jīva*'s full omniperception (momentarily glimpsed at the fourth step of *samyak darśana*), Whitehead describes the first effect of peace as "removal . . . [of] the soul's preoccupation with itself . . . [as] the surpassing of personality." Whitehead continues:

> Peace is the removal of inhibitions . . . [and] results in a wider sweep of conscious interest. . . . Peace is self-control at its widest,—at the width where the 'self' has been lost, and interest has been transferred to coördinations wider than personality. . . . It is a barrier against narrowness.[135]

David L. Hall, in his 1973 book *The Civilization of Experience*, clarifies the close link between Whitehead's concepts of "peace" as the aim of "civilization," such that peace "involves a broadening of the sense of Eros to include all individuals,"[136] a kind of "ecstatic individuality" according to Judith Jones, that rejects "subjectivism, solipsism, or egoism."[137] Whitehead elaborates in *Modes of Thought*, "Everything has some value for itself, for others, and for the whole. This characterizes the meaning of actuality."[138] More strongly, Whitehead describes this as "aims that lie beyond any personal satisfaction" and "beyond personality" or "self-regarding motives."[139]

In one sense, such a sense of peace seems like an elusive conceptual ideal, abstracted from embodied living. Yet Whitehead's concept of civilization as exemplified by peace is not merely abstract. Civilized order, he writes, is "the victory of persuasion over force," expressed tangibly in the "gradual purification of conduct."[140] Henning asserts that the concept of peace plays a practical, and not merely conceptual, role "by inspiring us to reject our own narrow, egoistic commitments and strive instead for the most inclusive perspective possible."[141] He writes, "One's bias must always and everywhere be toward inclusiveness," within both human conflicts and relations with the more-than-human world.[142]

Belgian chemist Isabelle Stengers further concretizes Whitehead's vision as a "peace-producing philosophy" that must move "beyond conversation" to embrace significant risks.[143] Peace is enacted she argues, drawing upon Deleuze and William James, in new "experimental habits, be they moral, aesthetic, or conceptual" of "propositional togetherness" pertaining well beyond any narrow "human."[144] Such efforts to "turn an opposition into a possible matter of contrast" are attempted, according to Stengers, in an empirical "middle ground for testing" in order to "fabricate the possibility of peace" verified "through it practical effects" that minimize negative prehensions.[145] In his dissertation exploring a process theology of nonviolent resistance, Alonzo Valentine describes nonviolence as a risk-taking unity of God's dipolarity fusing the "primordial" ordering of justice with the "consequent" receptivity of love; "nonviolent actions," he writes, "are those that place the resister at more risk than those to whom the act is directed."[146]

In one of the few explicit analyses of nonviolence and process philosophy, Robert L. Kinast argues that "peace is such that it brings one into a conformal relationship with the way God functions for the world."[147] He further states that nonviolence is one of the ways in which the eternal object of peace may be experienced, not as passivity but as a "harmony of harmonies which calms destructive turbulence and completes civilization."[148] Kinast describes nonviolence as a subjective form "according to which contrast is introduced which is compatible with creative becoming" beginning with the individual and expanding in intensity to societies and civilizations. According to Kinast:

> Nonviolence may be seen as a total lifestyle which seeks the highest satisfaction of every occasion. In a negative sense, it does this by excluding violent/coercive conflict; in a positive sense, it does this by including compatible contrast and an appeal to creative becoming. . . . Nonviolence is the most compatible and appropriate forms of feeling this eternal object [of peace]. Indeed they are virtually identical; but in any event nonviolence allows peace to be what it is more fully than any other way of prehending peace can.[149]

Roland Faber, in his recent text *Cosmic Spirit* explicitly connects Whitehead's "peace" with the South Asian concept of nonviolence through "compassion." He describes the ideal of "disinterested" compassion—expressed notably in Buddhism, but also in Jainism—as a nexus, "free from any biases and agendas, and can, thereby, sympathetically and inherently relate 'other lives' *as our own* but without possessive implications."[150] The "universal tendency" toward compassion, according to Faber, "will be towards non-violence and Peace, as the ideal realization of any society that deserves . . . the name of Civilization," referring not only to human societies, but toward the totality of the multiverse as "Civilized Cosmos." "Peace is not an empty idea, an empty promise, or an idealistic illusion," writes Faber, "but the real, activating, teleological *attractor*" inherent in the cosmos of becoming "in which no opposition is ultimate and no alternative destructive."[151]

Nonviolence is not a one-size solution to any conflict, but if exclusionary loss is the problematic cost of existence, both Jainism and Process construct an ideal of univocal multiplicity without loss, luring every embodied activity toward less obstruction in its own context. To "be multiplicity" without tyranny is a personal, social, and cosmic ethic striving toward the ultimate aim of nonviolence.

Nonviolence is not just an abstract ideal but is manifest in activity. As Whitehead states in *Aims of Education*, "What is proved should be utilized, and that what is utilized should—so far as it is practicable—be proved."[152] In this chaper I have attempted to show, to prove, that Jainism and Process philosophy reflect a common metaphysical aim toward multiplicity without loss, though only Jainism has clearly translated this vision to an embodied ethic of nonviolence. May it be that the future of process philosophy can boldly assert Whitehead's pragmatic relevance as a rare ethical metaphysics of nonviolence aimed toward the implausible hope of cosmic thriving.

NOTES

1. Certain sections in this analysis draw from ideas in my previous work, especially *Creaturely Cosmologies* and "Immortality and Omniscience."
2. Whitehead, *Process and Reality*, 41–42.
3. Ibid., 105–6.
4. Jaini, *The Jaina Path of Purification*, 284.
5. Whitehead, *Process and Reality*, 340.
6. Ibid., emphasis added.
7. Ibid.
8. Ibid., 105.
9. Whitehead, *Adventures of Ideas*, 268.
10. Whitehead, *Process and Reality*, 244.
11. Faber, *The Garden of Reality*, 207ff.
12. Whitehead, *Process and Reality*, 7.
13. A notable exception is Jeffery D. Long's dissertation "Plurality and Relativity," some of which was included in his book *Jainism*, chs. 5–6. For other Jain/Process comparisons, see Chapple "Life All Around"; Donaldson "'They'll Know We Are Process Thinkers,'" "Immortality and Omniscience," and *Creaturely Cosmologies*; Long, "The Paradoxes of Radical Asceticism"; and Schwartz, *The Metaphysics of Paradox*.
14. WRD uses four sources to estimate global religions: (1) the World Christian Database, (2) censuses in which a religious question is asked, (3) censuses in which an ethnicity or language question is asked, and (4) other surveys and polls. For a full description of each of these sources, as well as the wider methods of the WRD, see Johnson and Grim, *World Religion Database*.
15. Johnson and Grim, *World Religion Database*.
16. Long also completed his dissertation on religious pluralism in Whitehead and Jainism. See fn. 2.
17. Long, *Jainism*, xviii–xix, emphasis his.
18. Donaldson, "Jainism, Process Thought, and Animal Ethics."
19. Jackson, "The Eros and Tragedy of Peace," 100.
20. Cobb, "Organisms as Agents of Evolution," 241.
21. Henning, *The Ethics of Creativity*, 171.
22. Whitehead, *Process and Reality*, 340.
23. Tatia, *Tattvārtha Sūtra* 1.1.
24. Riepe, *The Naturalistic Tradition in Indian Thought*, 6.
25. Barad, *Meeting the Universe Halfway*, 90.
26. Scholars list several sources for the *guṇa-sthāna* framework. For a summary, see Donaldson and Bajželj, *Insistent Life*, 54ff.
27. Tatia, *Tattvārtha Sūtra*, 5.1–5.2.
28. Jaini, *The Jaina Path of Purification*, 90
29. Balcerowicz, *Early Asceticism in India*, 174.
30. Barbato, *Jain Approaches to Plurality*; Long, *Jainism*, 117–72.
31. Whitehead, *Process and Reality*, 6.

32. Ibid., 104.
33. Jaini, *The Jaina Path of Purification*, 91.
34. Upadhye, *Śrī Kundakundācārya's Pravacanasāra*, 2.76
35. For an overview of the changing meanings of karma in the Jain texts, see Ohira, *A Study of the Bhagavatīsūtra*, 174ff.
36. Wiley, "Aghātiya Karmas." For an overview of life and nonlife in the Jain taxonomy of existence as determined by these two kinds of karmas, see Donaldson and Bajželj, *Insistent Life*, ch. 2.
37. Tatia, *Tattvārtha Sūtra*, 2.8–25.
38. Williams, *Jaina Yoga*, 65–56.
39. Ibid.
40. Cort, *Jains in the World*, 25.
41. Jaini, *The Jaina Path of Purification*, 147–49.
42. Glasenapp, *Doctrine of Karman in Jain Philosophy*, 82.
43. Ibid., 225.
44. Cakravarti, *Ācārya Kundadunda's Samayasāra*, 1.3
45. Jacobi, *Gaina SÛtras*, 5.121, emphasis added.
46. Donaldson and Bajželj, *Insistent Life*, 39–42.
47. Deleuze and Guattari, *A Thousand Plateaus*, 40.
48. See Donaldson "Immortality and Omniscience," 131ff; *Creaturely Cosmologies*, 78–82.
49. Cakravarti, *Ācārya Kundadunda's Samayasāra*, 7–9.
50. For example, of nonhuman restraints, see Vallely, "Being Sentiently with Others" and "Vulnerability, Transcendence, and the Body" and Appleton, *Narrating Karma and Rebirth*, 24ff.
51. Williams, *Jaina Yoga*, 116.
52. For additional information on occupational restraints, including Jain textual sources, see Donaldson and Bajželj, *Insistent Life*, 70–73.
53. Jacobi, *Gaina SÛtras*, 24.1–18
54. Williams, *Jaina Yoga*, 108–9.
55. The *Āvaśyaka-sūtra* (2nd–3rd c. CE) states obligatory practices for mendicants: (1) attaining a state of mental calm (*sāmāyika*), (2) venerating the *tīrthaṅkara*s (*caturviṃśati-stava*), (3) honoring mendicants (*guru-vandana*), (4) confessing one's daily harms (*pratikramaṇa*), (5) meditating without thought to the body (*kāyotsarga*), and (6) temporary fasting from certain foods or activities (*pratyākhyāna*). Williams, *Jaina Yoga*, 185.
56. Settar, *Pursuing Death*.
57. Whitehead, *Process and Reality*, 18.
58. Whitehead, *Modes of Thought*, 151.
59. Whitehead, *Process and Reality*, 134, 7.
60. Faber, "De-Ontologizing God," 7.
61. Whitehead, *Process and Reality*, 7.
62. Ibid., 207.
63. Ibid., 344–45.
64. Ibid., 350.

65. Shaviro, "Deleuze's Encounter With Whitehead," 8; Whitehead, *Process and Reality*, 346.

66. Shiviro borrows the term "virtual" for his description of God's consequent nature from Deleuze, *The Fold*, 51ff. I also use the term "virtual" to describe the siddha as similar to a Deleuzian "body without organs" earlier in this chapter. Both the Jain *siddha* and process God/Receptacle/Infinite/World of Value are "virtual" nonmaterial, nontemporal states that are nonetheless real when provocative or enfolded into actual happenings.

67. Whitehead, *Adventures of Ideas*, 134.

68. Whitehead, *Religion in the Making*, 115.

69. Whitehead, *Adventures of Ideas*, 178.

70. Whitehead, *Process and Reality*, 345.

71. Whitehead, *Essays in Science and Philosophy*, 61.

72. Whitehead, *Process and Reality*, 345.

73. Ibid., 343.

74. Ibid.

75. Ibid., 244.

76. Ibid., 349.

77. Ibid.

78. Whitehead, *Adventures of Ideas*, 134.

79. Whitehead, *Science and the Modern World*, 192.

80. Whitehead, *Process and Reality*, 348–49.

81. Whitehead, *Essays in Science and Philosophy*, 64–65.

82. Cobb, "Organisms as Agents of Evolution," 230.

83. Ibid., 231.

84. Whitehead, *The Function of Reason*, 4, 8.

85. Tatia, *Tattvārtha Sūtra*, 1.1.

86. Whitehead, *Essays in Science and Philosophy*, 81.

87. Ibid., 86, 81.

88. Whitehead, *Essays in Science and Philosophy*, 62.

89. Whitehead, *Process and Reality*, 346.

90. Ibid., 340.

91. Ibid., 346, 348.

92. Whitehead, *Adventures of Ideas*, 268.

93. Loomer, "Process Theology," 252.

94. Whitehead, *Adventures of Ideas*, 285.

95. Whitehead, *Process and Reality*, 342.

96. Faber, "De-Ontologizing God," 202.

97. Whitehead, *Process and Reality*, 351, 340.

98. Ibid., 343.

99. Henning, "On the Possibility of a Whiteheadian Aesthetics of Morals."

100. Kaplický, "Aesthetics in the Philosophy of Alfred North Whitehead"; Odin, *Tragic Beauty in Whitehead and Japanese Aesthetics.*

101. Jackson, "The Eros and Tragedy of Peace in Whitehead's Philosophy of Culture"; Jones, *Intensity*; Lango, "Does Whitehead's Metaphysics Contain an Ethics?"

102. Henning, *The Ethics of Creativity*, 41ff.
103. Faber, *The Garden of Reality*, 207ff.
104. Hobbes, *Leviathan*.
105. Tennyson, *In Memoriam*.
106. https://ecociv.org.
107. Darwin, *On the Origin of Species*, 62, 238–39.
108. Shaviro, "Deleuze's Encounter With Whitehead," 9.
109. Henning, "Hierarchy Without Anthroparchy," 107.
110. Shaviro, "Deleuze's Encounter With Whitehead," 9.
111. Whitehead, *Process and Reality*, 105.
112. Ibid., 93, 340.
113. Bekoff and Pierce, *Wild Justice*; de Waal, *The Age of Empathy*; Horgan, "Quitting the Hominid Fight Club" and *The End of War*; Kokko, "Conflict and Restraint in Animal Species"; Sussman and Cloninger, *Origins of Altruism and Cooperation*; Verbeek, "An Ethological Perspective on War and Peace."
114. Goodall, *The Chimpanzees of Gombe*, 114.
115. Bekoff, "Demonic, Warlike Animals Are the Rare Exception, Not the Rule."
116. Keltner, *Born to Be Good*.
117. Fanon, *The Wretched of the Earth*, 23.
118. Tinker, "The Irrelevance of Euro-Christian Dichotomies," 215.
119. Brunner, "Conceptualizing Epistemic Violence;" Byrne et al., "Colonialism and Peace and Conflict Studies;" Jabri, "Post-Colonialism."
120. Tinker, "The Irrelevance of Euro-Christian Dichotomies," 215.
121. Ibid., 216.
122. Ibid.
123. Churchill, "Pacifism as Pathology."
124. Gelderloos, *How Nonviolence Protects the State*.
125. Ibid., 13, 14–15.
126. Kurlansky, *Nonviolence*, 6–7.
127. Wink, *The Powers That Be*, 103, 142.
128. King, Where Do We Go From Here?, 137.
129. Delmas and Brownlee, "Civil Disobedience."
130. King, "Letter from Birmingham Jail," emphasis added.
131. Kurlansky, *Nonviolence*, 8.
132. Whitehead, *Process and Reality*, 87.
133. Aristarkhova, "Thou Shall Not Harm All Living Beings," 639.
134. Butler, *The Force of Nonviolence*, 27.
135. Whitehead, *Adventures of Ideas*, 285–86.
136. Hall, *The Civilization of Experience*, 155.
137. Jones, *Intensity*, 71, 180.
138. Whitehead, *Modes of Thought*, 111.
139. Whitehead, *Adventures of Ideas*, 288.
140. Ibid., 25.
141. Henning, *The Ethics of Creativity*, 119.
142. Ibid., 154.

143. Stengers, "Beyond Conversation," 244.
144. Ibid., 241, 238.
145. Ibid., 236, 237, 243, 245, 251.
146. Valentine, "Love Seeking Justice," 7. See ch. 3 for the analysis of nonviolence as unification of God's dipolarity.
147. Kinast, "Nonviolence in a Process Worldview," 279.
148. Ibid., 280–81.
149. Ibid., 282–83.
150. Faber, *The Cosmic Spirit*, 279. See also Faber, "Space, Time, and the Deity of Peace," 33–60 and Faber, "Must 'religion' always remain as a synonym for 'hatred?'"
151. Faber, *The Cosmic Spirit*, 278–79.
152. Whitehead, Alfred North. *The Aims of Education* (New York: The Free Press 1967 [1929]).

BIBLIOGRAPHY

Appleton, N. *Narrating Karma and Rebirth: Buddhist and Jain Multi-life Stories.* New York: Cambridge University Press, 2014.

Aristarkhova, I. "Thou Shall Not Harm All Living Beings: Feminism, Jainism, and Animals." *Hypatia* 27, no. 3 (2012): 636–50.

Balcerowicz, P. *Early Asceticism in India: Ājīvikism and Jainism*. New York: Routledge, 2016.

Barad, K. *Meeting the Universe Halfway: Quantum Physics and the Entanglement of Matter and Meaning.* Durham: Duke University Press, 2007.

Barbato, Me. *Jain Approaches to Plurality, Identity as Dialogue*. New York: Brill, 2017.

Bekoff, M. "Demonic, Warlike Animals Are the Rare Exception, Not the Rule: Wild Justice in Animals Redux." *Psychology Today*, July 5, 2010. Accessed September 1, 2021. https://www.psychologytoday.com/us/blog/animal-emotions/201007/demonic-warlike-animals-are-the-rare-exception-not-the-rule-wild-justice.

———. "'What Were Wars?' Don't Blame Other Animals For Human Violence." *Psychology Today*. January 22, 2012. Accessed September 2, 2021. https://www.psychologytoday.com/us/blog/animal-emotions/201201/what-were-wars-dont-blame-other-animals-human-violence.

Bekoff, M., and J. Pierce. *Wild Justice: The Moral Lives of Animals*. Chicago: University of Chicago Press, 2010.

Brunner, C. "Conceptualizing Epistemic Violence: An Interdisciplinary Assemblage for International Relations." *International Politics Reviews*, Vol 9 (2021): 193–212. https://doi.org/10.1057/s41312-021-00086-1

Butler, J. *The Force of Nonviolence: An Ethico-Political Bind*. New York: Verso Books, 2020.

Byrne, S., et al. "Colonialism and Peace and Conflict Studies." *Peace and Conflict Studies Journal* 25, no. 1 (2018). Accessed November 2, 2021. https://nsuworks.nova.edu/pcs/vol25/iss1/1/.

Cakravarti, A., trans. [Kundakunda, *Samayasāra*] Ācārya Kundadunda's Samayasāra. New Delhi: Bharatiya Jnanpith, 2000.

Chapple, C. K. "Life All Around: Soul in Jainism." In *Panentheism Across the World's Traditions*, edited by L. Biernacki and P. Clayton, 100–22. Oxford: Oxford University Press, 2014.

Churchill, W. "Pacifism as Pathology: Notes on an American Pseudo-Praxis: I." *Issues in Radical Therapy* 12, no. 1 (1986): 36–39, 55–69.

Cobb, J. B., Jr. "Organisms as Agents of Evolution." In *Back to Darwin: A Richer Account of Evolution*, edited by J. B. Cobb, Jr., 215–41. Grand Rapids: William B. Eerdmans Publishing, 2008.

Cort, J. E. *Jains in the World: Religious Values and Ideology in India*. New York: Oxford University Press, 2001.

Darwin, C. *On the Origin of Species*. Introduction by Ernst Mayr. Cambridge, MA: Harvard University Press, 1998.

de Waal, F. *The Age of Empathy: Nature's Lessons for a Kinder Society*. New York: Crown, 2010.

Deleuze, G., and F. Guattari. *A Thousand Plateaus: Capitalism and Schizophrenia*. Trans. Brian Masumi. Minneapolis: University of Minneapolis Press, 2011.

Delmas, C., and K. Brownlee. "Civil Disobedience." *The Stanford Encyclopedia of Philosophy*, edited by E. N. Zalta, 2021. Accessed November 2, 2021. https://plato.stanford.edu/archives/win2021/entries/civil-disobedience/.

Donaldson, B., and A. Bajželj. *Insistent Life: Principles for Bioethics in the Jain Tradition*. Oakland: University of California Press, 2021.

Donaldson, B. "Jainism and Nonviolence: Variable Practices of an Ultimate Principle." *Nonviolence In the World's Religions*, edited by J. D. Long and M. J. Long, 24–38. New York: Routledge, 2021.

Donaldson, B. "Jainism, Process Thought, and Animal Ethics," Lecture for Cobb Institute, May 11, 2021.

Donaldson, B. "'They'll Know We Are Process Thinkers by Our . . . ': Finding the Ecological Ethic of Whitehead through the Lens of Jainism and Ecofeminist Care." In *Polydoxy: Theologies of the Manifold*, edited by C. Keller and L.C. Schneider, 203–16. New York: Routledge, 2010.

Donaldson, B. "Immortality and Omniscience: Conceptual Bridges to Reworlding in Process and Jainism." In *Beyond the Bifurcation of Nature: A Common World for Animals and the Environment*, edited by B. Donaldson, 117–40. Newcastle upon Tyne: Cambridge Scholars Publishing, 2014.

Donaldson, B. *Creaturely Cosmologies: Why Metaphysics Matters for Animal and Planetary Liberation*. Lanham, MD: Lexington, 2015.

Faber, R. "De-Ontologizing God: Levinas, Deleuze and Whitehead." In *Process and Difference: Between Cosmological and Poststructuralist Postmodernisms*, edited by C. Keller and A. Daniels, 209–34. New York: SUNY Press, 2002.

Faber, R. "Organic or Orgiastic Metaphysics? Reflections on Whitehead's Reception in Contemporary Poststructuralism." The Japan Society for Process Studies, Chuo University, October 24, 2010. Accessed October 12, 2021. https://drive.google.com/file/d/1mcYp1ZvbmaPirGok4xCCFzNUyz_AvHYg/view.

Faber, R. "Space, Time, and the Deity of Peace. In *Propositions in the Making: Experimentations in a Whiteheadian Laboratory*, edited by R. Faber, M. Halewood, and A. M. Davis, 33–60. Lanham, MD: Lexington, 2020.

Faber, R. "Must 'religion' always remain as a synonym for 'hatred?'": Whiteheadian Meditations on the Future of Togetherness." In *Living Traditions and Universal Conviviality: Prospects and Challenges for Peace in Multireligious Communities*, edited by R. Faber and S. Slabodsky, 167–82. Lanham, MD: Lexington.

Faber, R. *The Cosmic Spirit: Awakenings at the Heart of All Religions, the Earth, and the Multiverse.* Eugene: Cascade Books, 2020.

Faber, R. *The Garden of Reality: Transreligious Relativity in a World of Becoming.* Lanham, MD: Lexington, 2018.

Fanon, F. 2004. *The Wretched of the Earth.* Trans. Richard Philcox. New York: Grove Press, 2004.

Gelderloos, P. *How Nonviolence Protects the State.* Boston: South End Press, 2007.

Gelderloos, P. *The Failure of Nonviolence: From the Arab Spring to Occupy.* St. Louis, MO: Left Bank Books, 2013.

Gilles, D. *The Fold: Leibniz and the Baroque*. Trans. T. Conley. Minneapolis: University of Minnesota Press, 1993.

Glasenapp, H. *Doctrine of Karman in Jain Philosophy*, trans. G. B. Gifford. Varanasi: Parshwanath Vidyapeeth Research Institute, 1991.

Gómez, J. M. et al. "The Phylogenetic Roots of Human Lethal Violence." *Nature* 538 (2016): 233–37. https://doi.org/10.1038/nature19758.

Goodall, J. *The Chimpanzees of Gombe: Patterns of Behavior*. Cambridge, MA: Belknap Press, 1986.

Hall, D. L. *The Civilization of Experience: A Whiteheadian Theory of Culture*. New York: Fordham University Press, 1973.

Henning, B. "On the Possibility of a Whiteheadian Aesthetics of Morals." *Process Studies* 31, No. 2 (2002): 97–114.

Henning, B. "Hierarchy Without Anthroparchy." In *Beyond the Bifurcation of Nature: A Common World for Animals and the Environment*, edited by B. Donaldson, 107–17. Cambridge: Cambridge Scholars Publishing, 2014.

Henning, B. *The Ethics of Creativity: Beauty, Morality, and Nature in a Processive Cosmos*. Pittsburgh: University of Pittsburgh Press, 2005.

Hobbes, T. *Leviathan*. Edited by C. B. Macpherson. Harmondsworth: Penguin Publishers, 1968.

Horgan, J. "Quitting the Hominid Fight Club: The Evidence is Flimsy for Innate Chimpanzee—Let Alone Human—Warfare." *Scientific American*. June 29, 2010. Accessed September 1, 2021. https://blogs.scientificamerican.com/cross-check/quitting-the-hominid-fight-club-the-evidence-is-flimsy-for-innate-chimpanzee-let-alone-human-warfare/#:~:text=Cross%2DCheck-,Quitting%20the%20hominid%20fight%20club%3A%20The%20evidence%20is%20flimsy%20for,let%20alone%20human%2D%2Dwarfare&text=Extraordinary%20claims%2C%20Carl%20Sagan%20liked%20to%20say%2C%20require%20extraordinary%20evidence

———. *The End of War.* San Francisco: McSweeney's Publishing, 2014.

Jabri, V. "Post-Colonialism: A Post-Colonial Perspective on Peacebuilding." In *The Palgrave Handbook of Disciplinary and Regional Approaches to Peace*, edited by O. P. Richmond, S. Pogodda, and J. Ramović, 154–67. New York: Springer, 2016.

Jackson, M. M. "The Eros and Tragedy of Peace in Whitehead's Philosophy of Culture." *Essays in the Philosophy of Humanism* 23, no. 1 (2015): 93–112. https://doi.org/10.1558/eph.v23i1.26923.

Jacobi, H., trans. [Bhadrabahu, Kalpa-sūtra] *Gaina Sûtras, Part 1, The Kalpa Sûtra.* The Sacred Books of the East, vol. 22. Oxford: Clarendon Press, 1884.

———. [Uttarādhyayana-sūtra]. *Gaina Sûtras, Part 2, The Uttarâdhyayana Sûtra.* The Sacred Books of the East, vol. 45. Oxford: Clarendon Press, 1895.

Jaini, P. S. "Fear of Food: Jaina Attitudes on Eating." In *Collected Papers on Jaina Studies*, 2nd ed., 281–96. Delhi: Motilal Banarsidass, 2010.

———. *The Jaina Path of Purification.* Delhi: Motilal Banarsidass, 2001.

Johnson, T. M., and B. J. Grim, eds. *World Religion Database.* Leiden, The Netherlands: Brill, 2020.

Jones, J. A. *Intensity: An Essay in Whiteheadian Ontology.* Nashville: Vanderbilt University Press, 1998.

Kaplický, M. 2011. "Aesthetics in the Philosophy of Alfred North Whitehead." *Estetika* 48, no. 2 (2011): 157–71. https://doi.org/10.33134/eeja.82.

Keltner, D. *Born to Be Good: The Science of a Meaningful Life.* New York: W. W. Norton and Co., 2009.

Kinast, R. L. 1981. "Nonviolence in a Process Worldview." *Philosophy Today* 25, no. 4 (1981): 279–85. https://doi.org/10.5840/philtoday19812541.

King, M. L., Jr. "Letter from Birmingham Jail." In *Civil Disobedience in Focus*, edited by Hugo Adam Bedau, 68–84. London: Routledge, 1991.

———. *Where Do We Go From Here? Chaos or Community?*. New York: Harper & Row, 1968.

Kokko, H. "Conflict and Restraint in Animal Species." In *War, Peace, and Human Nature: The Convergence of Evolutionary and Cultural Views*, edited by D.P . Fry, 38–53. Oxford: Oxford University Press, 2013.

Kurlansky, M. *Nonviolence: The History of a Dangerous Idea.* New York: Modern Library, 2008.

Lango, J. W. "Does Whitehead's Metaphysics Contain an Ethics?" *Transactions of the Charles S. Peirce Society* 37, no. 4 (2001): 515–36.

Long, J. D. "Plurality and Relativity: Whitehead, Jainism, and the Reconstruction of Religious Pluralism." PhD diss., University of Chicago, 2000.

———. "The Paradoxes of Radical Asceticism: Jainism as a Therapeutic Paradigm." In *The Roar of Awakening: A Whiteheadian Dialogue Between Western Psychotherapies and Eastern Worldviews*, edited by G. Derfer, Z. Wang and M. Weber, 71–84. Boston: DeGruyter, 2009.

———. *Jainism: An Introduction.* New York: I. B. Tauris, 2009.

Loomer, B. "Process Theology: Origins, Strengths, Weaknesses." *Process Studies* 16, no. 4 (1987): 245–54.

Odin, S. *Tragic Beauty in Whitehead and Japanese Aesthetics.* Lanham, MD: Lexington, 2016.

Ohira, S. *A Study of the Bhagavatīsūtra: A Chronological Analysis*. Vol. 28. Prakrit Text Series. Ahmedabad: Prakrit Text Society, 1994.

Riepe, D. *The Naturalistic Tradition in Indian Thought*. Delhi: Motilal Banarsidass, 1996.

Schwartz, A. W. *The Metaphysics of Paradox: Jainism, Absolute Relativity, and Religious Pluralism.* Lanham, MD: Lexington, 2018.

Settar, S. *Pursuing Death: Philosophy and Practice of Voluntary Termination of Life*, revised ed. Delhi: Primus Books, 2017.

Shaviro, S. "Deleuze's Encounter With Whitehead," 2007. Accessed July 20, 2021. http://shaviro.com/Othertexts/DeleuzeWhitehead.pdf.

Stengers, I. "Beyond Conversation: The Risks of Peace." In *Process and Difference: Between Cosmological and Poststructuralist Postmodernisms*, edited by C. Keller and A. Daniels, 235–56. New York: SUNY Press, 2002.

Sussman, R. W. and C. R. Coninger, eds. *Origins of Altruism and Cooperation*. New York: Springer, 2013.

Tatia, N., trans. [Umāsvāti, *Tattvārthasūtra*] *Tattvārtha Sūtra: That Which Is*. New Haven: Yale University Press, 2011.

Tennyson, A. L. *In Memoriam* LVI *A. H. H. OBIIT MDCCCXXXIII.* London: Edward Moxon, Dover Street, 1850.

Tinker, T. "The Irrelevance of Euro-Christian Dichotomies for Indigenous Peoples: Beyond Non-violence to a Vision of Cosmic Balance." In *Peacemaking and the Challenge of Violence in World Religions*, edited by I. A. Omar and M. K. Duffey, 206–29. Hoboken, NJ: Wiley-Blackwell, 2015.

Upadhye, A. N., trans. [Kundakunda, *Pravacanasāra*] *Śrī Kundakundācārya's Pravacanasāra.* Agas: Parama-Śruta-Prabhāvaka Mandal, Shrimad Rajachandra Ashrama, 1984.

Valentine, A. M. "Love Seeking Justice: A Process Theology of Nonviolent Resistance." PhD diss., Emory University, 1989.

Vallely, A. "Being Sentiently with Others: The Shared Existential Trajectory among Humans and Nonhumans in Jainism." In *Asian Perspectives on Animal Ethics: Rethinking the Nonhuman*, edited by N. Dalal and C. Taylor, 38–55. New York: Routledge, 2014.

Vallely, A. "Vulnerability, Transcendence, and the Body: Exploring the Human/Nonhuman Animal Divide within Jainism." *Society and Animals* 28 (2020): 550–66. https://doi.org/10.1163/15685306-12341520.

Verbeek, P. "An Ethological Perspective on War and Peace." In *War, Peace, and Human Nature: The Convergence of Evolutionary and Cultural Views*, edited by D. P. Fry, 54–77. Oxford: Oxford University Press, 2013.

Whitehead, A. N. *Adventures of Ideas.* New York: The Free Press, 1967.

———. *Essays in Science and Philosophy.* New York: Philosophical Library, 1948.

———. *Modes of Thought*. New York: The Free Press, 1966.

———. *Process and Reality: An Essay in Cosmology*. Corrected Edition, D. R. Griffin and D. W. Sherburne. New York: The Free Press, 1978.

———. *Religion in the Making.* New York: Fordham University Press, 1996.

———. *Science and the Modern World.* New York: The Free Press, 1967.

Wiley, K. "Aghātiya Karmas: Agents of Embodiment in Jainism." PhD diss., University of California, Berkeley, 2000.
Williams, R. *Jaina Yoga: A Survey of the Mediaeval Śrāvakācāras*. Delhi: Motilal Banarsidass, 1991.
Wink, W. *The Powers That Be: Theology for a New Millennium*. New York: Galilee Doubleday, 1998.

Chapter 13

Extending the Noosphere into Intergalactic Life

Teilhard de Chardin and the Third Axial Age

Ilia Delio

The Jesuit scientist Pierre Teilhard de Chardin was an intuitive thinker with a devotion to science, He speculated that the principles of earthly life, including theological principles, could extend to extraterrestrial life. His broad vision of theistic evolution, marked by the rise of complexity-consciousness, included the search for extraterrestrial life. The openness of matter to consciousness, and the transcendence of conscious matter toward more consciousness, suggests that terrestrial life cannot contain the infinite capacity of consciousness to complexify. As one who lived at the dawn of the computer age, Teilhard anticipated the internet and a new level of shared mind which he called "noogenesis." He predicted that computer-linked minds would create a process of hyperpersonalization and thus a new age of complexified consciousness. The dawning of ultrahumanity on the level of noogenesis suggests that the ongoing need to complexify cannot be exhausted by humanity alone; hence the need to extend conscious intergalactically.

Theologically, the law of increasing complexity-consciousness coincides with Teilhard's notion of theogenesis, that is, God seeks to complete Godself in creation by self-actualizing through higher levels of consciousness. In this respect, the pursuit of extraterrestrial life is not merely a scientific endeavor but a theological endeavor as well. The search for meaning and for new levels of conscious interaction suggest that the search for extraterrestrial life is inherent to the process of evolution. This paper will examine Teilhard's notion of evolution as a process of theogenesis, with an emphasis on the principle of complexity-consciousness, and the emergence of hyperpersonal and intergalactic conscious life.

A BRIEF HISTORY

While advances in astronomy and space technology, as well as movies like *Contact*, render the modern person attuned to the possibility of extraterrestrial [ET] life, speculation on ET life is not entirely new. The ancient Greeks wondered if there were many worlds or only one world. Among atomistic philosophers, Epicurus (341–270 BC) and later Lucretius (98–54 BC), suggested that because there are an infinite number of atoms, there could be an infinite number of worlds. They even asserted the existence of plants and creatures on other worlds.[1] The champion of Greek philosophy, Aristotle, rejected the argument of the atomists on the belief that all things naturally seek their proper place. He posited a cosmological vision of a single reality with the earth at its center and concluded, since there is only one center, there is only one world. Although Aristotle's argument prevailed over the atomists, the doyen of scholastic theology, Thomas Aquinas, did not quite adhere to the geocentric argument. Instead, he argued that perfection is found in unity; hence, an omnipotent God would best express himself in a single perfect world.[2] Subsequent thinkers, however, did not limit God's perfection to a single world. John Buridian (1295–1358), Nicole Oresme (1320–1382) and Nicholas Cusa (1401–1464) all held to a "many worlds" idea, and some, like Cusa, even maintained belief in ET life. Ted Peters writes: "During the formidable period of medieval scholasticism, despite the forceful impact of Aristotelian philosophy, Christian theology was by no means wedded to the idea that God created only one world. . . . Not only did some of our best minds affirm the idea of multiple worlds, some even spoke positively regarding the existence of extraterrestrial life."[3]

In our own time, the late Carl Sagan described an attitude of "earth chauvinism" which makes the earth the standard for everywhere.[4] Andrew Burgess identified this chauvinism as a "cosmic hubris," believing that the Creator of the universe was incarnate on our own bit of interstellar debris, and maintains a special relationship to the human species.[5] As Burgess and others have pointed out, the Christian story is based on a pre-Copernican cosmology. He writes:

> As long as someone is thinking in terms of a geocentric universe and an earth-deity, the story has a certain plausibility. . . . As soon as astronomy changes theories, however, the whole Christian story loses the only setting

> within which it would make sense. With the solar system no longer the center of anything, imagining that what happens here forms the center of a universal drama becomes simply silly."[6]

Burgess identified a relationship that was known for centuries among patristic and medieval writers, namely, theology is integrally related to cosmology. Once cosmology changes, so too does theology, whether implicitly or explicitly.

There is no doubt that advances in modern astronomy are changing our view of life in the universe. According to leading scientists, there is mounting evidence that Earth might not be the only oasis in the cosmos, or even in our solar system. Researchers have long known that tiny fragments of one planet, such as meteors, can crash onto another planet's surface. This gives life the potential to jump from place to place in our solar system. Recent research suggests that the ingredients necessary for life as we know it to evolve from scratch are present in many places besides Earth. The ingredients include liquid solvents, such as water; a variety of elements, such as carbon, nitrogen and phosphorus for constructing organic molecules; and compounds that store energy. According to Bruce M. Jakosky, director of the Center for Astrobiology at the University of Colorado in Boulder, the planet Mars and Jupiter's moon Europa offer tantalizing combinations of these ingredients right in our celestial neighborhood.[7] Lewis Ford writes: "Even if only one planet out of every 150,000 contained life, there would be one million life-worlds in our galaxy, some of which we can reasonably assume contain intelligent life, for whom, we presume, God would also be concerned."[8]

Almost fifty years ago, J. Edgar Burns viewed the widening horizon of the universe through space exploration as the emergence of a new "space-faith" or what he called "cosmolatry." He posited, somewhat prophetically, that "what seems to have received scant attention is the appearance of a phenomenon which has all the earmarks of a new religion."[9] By "new religion" it is not entirely clear whether Burns meant an annihilation of the past or a breakthrough in religious consciousness that renders our "truth claims" obsolete. What the term does connote, however, echoes an insight of Pierre Teilhard de Chardin, namely, Christianity is reaching the end of one of the natural cycles of its existence.

MIND, MIND AND EVOLUTION

Science tells us that our universe was born out of the Big Bang about 13.8 billion years ago. According to Teilhard, evolution is the term that describes cosmological as well biological movement towards greater unity and complexity. Evolution is not merely an explanation of physical life; rather, evolution ushers in a whole new *weltanschauung* that affects every aspect of created life. Teilhard described evolution as a movement toward more complexified life forms which, at critical points in the evolutionary process, qualitative differences emerge.[10] At some point, evolution reaches a reflexive state which generates the idea of evolution: "There is only one real evolution, the evolution of convergence, because it alone is positive and creative." Teilhard's views concerning the energy that propels the evolutionary process are conceptualized within the framework of an entropic universe, yet one in which the diminishment of usable physical energy is counterbalanced by negentropic energy that is psychic or spiritual energy. The anti-entropic phenomenon of human emergence is a movement arising "upstream against the flow of entropy."[11]

Governing the complexifying process of evolution is a centrating factor that holds the entire process together and moves it forward toward greater complexity and unity. Teilhard called this principle of unity "Omega" because it is the point of unity toward which complexification aims. The Omega point emerges from seeing that the universe is psychically convergent, a phenomenon which cannot be accounted for by the increasing entropy of emergent life. Hence, the Omega point must be independent of the collapse of forces with which evolution is woven. Evolution proceeds toward greater consciousness because something in the cosmos escapes entropy and does so increasingly. Multiplicity is dependent on unity and on some final unity which does not need any principle beyond itself to unify it, since it is "already One." This principle of absolute unity is Omega. To empower evolution towards ultimate unity, however, Omega must be supremely present and supremely personal, a universal center that influences unification of individually centered human personalities without annihilating those centers.[12] Patrick Byrne states: "What is deepest in the center of every human being is the very thrust of evolution toward ultimate fulfillment that produces every evolved being."[13] However, Omega cannot be merely the point that would eventually be reached by the process of evolution itself. Rather, Omega must be accepted as "already in existence and operative at the very core" of evolution.[14] Science cannot prove the existence of Omega; what is needed is faith in the existence of Omega, that is, a distinct kind of faith in the existence of an organized whole seeking greater wholeness in which conscious and intelligence play a fundamental role. As the principal form of evolution's dynamic and emergent process, Omega undergirds the rise of greater complexity-consciousness, that is, an increase of intention, of acting with a goal.

Although evolution is a terrestrial phenomenon, the idea that we would be alone in this universe became increasingly untenable for Teilhard. How can all this tremendous expenditure of cosmic energy be explained except as a necessary precondition and/or concomitant of the emergence of thought? And, if so, how can one explain the emergence of thought on only one planet among millions over billions of others? On the other hand, if the universe turns out to be quite otherwise, to be eternal (as Einstein once thought) or perhaps self-renewing, then everything would appear in quite a different light. But an eternal universe or one that is self-renewing has no real need for evolution, a universe with no higher destiny. For Teilhard, on the other hand, this universe has a purpose, a goal, an "end" which is the Pleroma, the completion of Omega, in which the one God and the many aspects of evolved reality are united in a cosmotheandric wholeness, a quantitative repletion and qualitative consummation of all things.[15]

THE HUMAN PHENOMENON

In 1940 Teilhard completed his most important work, *The Phenomenon of Man*, in which he described the fourfold sequence of the evolution of galaxies, Earth, life, and consciousness. The human person is not a ready-made fact but the outflow of billions of years of evolution, beginning with cosmogenesis and the billions of years that led to biogenesis. To realize that humans are part of larger process which involves long spans of developmental time brings a massive change to all of our knowledge and beliefs. Evolution, he claimed, is ultimately a progression towards consciousness; the material world contains within it a dynamism toward spirit. Matter and spirit are not two separate entities but two forms of energy: spirit is the peak reflection of matter. Teilhard recognized that at all known levels of the universe there are units or "grains" which include stars, atoms, molecules, cells, people, etc. Sometimes the particles are gathered together in aggregations (for example, a pile of sand or a galaxy of stars); sometimes they are linked together to form a crystal. While a crystal builds up through indefinitely repeating the same molecular pattern, life builds up unities by structures ever more complex and intricately related. This "center to center" bonding of complexity which is characteristic of all living things captured Teilhard's attention.[16]

Consciousness is intrinsic to the process of evolution, according to Teilhard. He held a dual-aspect monist position to explain evolution. Life, he wrote, is "a specific effect of matter turned complex; a property that is present in the entire cosmic stuff."[17] He considered matter and consciousness not as "two substances" or "two different modes of existence, but as two aspects of the same cosmic stuff."[18] From the Big Bang onward there is a "withinness" and "withoutness," or what he called, radial energy and tangential energy.[19] Consciousness is, in a sense, the withinness or "inside" of matter, and attraction is the "outside" of matter; hence, matter is both attractive (tangential) and transcendent (radial). The complementarity of mind and matter explains both the rise of biological complexity and the corresponding rise of consciousness. Teilhard's position on conscious matter corresponds to the renewal of panpsychism. In the early twentieth century physicist Max Planck spoke of consciousness as fundamental to matter, that is, we cannot consider matter apart from consciousness. He wrote:

> All matter originates and exists only by virtue of a force which brings the particle of an atom to vibration and holds this most minute solar system of the atom together. We must assume behind this force the existence of a conscious and intelligent mind. This mind is the matrix of all matter.[20]

Physicist Erwin Schrödinger, like Planck, thought that consciousness is absolutely fundamental to matter and always experienced in the singular; everything begins with consciousness which itself is immaterial.[21] The philosopher Bertrand Russell said: "We know nothing about the intrinsic quality of physical events except when these are mental events that we directly experience."[22] These insights have led to "the hard problem of matter," namely, we cannot talk about matter apart from consciousness.[23] In the 1950s astrophysicist James Jean wrote: "The universe looks more like a great thought than a great machine. Mind no longer appears as an accident intruder into the realm of matter. . . . The quantum phenomena make it possible to propose that the background of the universe is mindlike."[24] Since consciousness is absolutely fundamental to matter, everything begins with consciousness, which itself is immaterial.[25] No one really knows what consciousness is or how it arises; what is known is that it is fundamental to existence itself. Nothing can be said about anything apart from consciousness.

Teilhard indicated that intelligent life cannot be considered in the universe any longer as a superficial accident but must be considered to be under pressure everywhere—ready to burst from the smallest crack no matter where in the universe—and, once actualized, it is capable of using every opportunity and means to arrive at the extreme of its potentiality, externally of complexity, and internally of consciousness.[26] The universe orients itself toward conscious intelligent life. In light of evolution, Teilhard did not see the human person as a chance arrival but an integral element of the physical world. He described human distinction in three ways: 1) The extreme physic-complexity (apparent in the brain) marks the human person as the most highly synthesized form of matter known to us in the universe; 2) in light of this complexity, the human is the most perfectly and deeply centered of all cosmic particles within the field of our experience; and 3) the high degree of mental development (reflection, thought) places the human person above all other conscious beings known to us.[27] Thus the human person is integrally part of evolution in that we rise from the process but in reflecting on the process we stand apart from it. Teilhard defines reflection as "the power acquired by a consciousness to turn in upon itself, to take possession of itself *as an object* . . . no longer merely to know, but to know that one knows."[28] He quotes a phrase of Julian Huxley, the human person "is nothing else than evolution become conscious of itself." To this idea he adds, "the consciousness of each of us is evolution looking at itself and reflecting upon itself."[29] Thus the human person emerges from the evolutionary process and is integral to evolution. She/he is "the point of emergence in nature, at which this deep cosmic evolution culminates and declares itself."[30]

THE THIRD AXIAL AGE

Human life in evolution is an ongoing and creative biological development of culture, community, meaning and purpose. The German philosopher Karl Jaspers coined the phrase "Achsenzeit" ("Axial Age" or "Axis age" in English) in 1949 to describe a time between approximately 800–200 BCE when the spiritual foundations of modern humanity were established. As a pivotal point of human consciousness, the axial age provides a framework to assess the emergence of the human person in evolution. The preaxial period is the age of the tribe or the collective which could extend as far back as the Neanderthals to 64,000 BCE, marked by a level of religious-mythic consciousness that was cosmic, collective, tribal, and ritualistic. Ancient civilizations looked at the physical and human worlds as interdependent. An imbalance in one sphere could result in an imbalance in the other. Ewert Cousins notes that the preaxial consciousness of tribal cultures was located in the cosmos and in fertility cycles of nature.[31] These early humans or "first earth persons" "mimed" nature and venerated nature in which nature appeared as a sacred reality determining one's destiny. The new "axial age" differed from preaxial consciousness in that it was marked by the rise of the individual and religious cultures. Axial persons were in possession of their own identity but they lost their organic relationship to nature and community, severing the harmony with nature and the tribe. Jaspers notes that the axial age gave rise to "a new departure within mankind," "a kind of critical, reflective questioning of the actual and a new vision of what lies beyond."[32] He writes: "Man becomes conscious of Being as a whole, of himself and his limitations. He experiences the terror of the world and his own powerlessness. He asks radical questions. Face to face with the void he strives for liberation and redemption. *By consciously recognizing his limits* he sets himself the highest goals. *He experiences absoluteness in the depths of selfhood and in the lucidity of transcendence*.[33] William Thompson states, "what makes this period the 'axis' of human history, even our own history today, is the fact that humans emerged as 'individuals' in the proper sense."[34] Axial consciousness generated a new self-awareness, including awareness of autonomy and a new sense of individuality. The human person as *subject* emerged.

A new axis of consciousness dawned in the twentieth century with the rise of big bang cosmology, evolution and quantum physics. Ewert Cousins and Thomas Berry called this new period of consciousness the "Second Axial Age."[35] Like the first period, this new axial age has been developing for several centuries, beginning with the rise of modern science. And like the first, it is effecting a radical transformation of consciousness. While the first axial period produced the self-reflective individual, the second axial period is giving rise to the hyperpersonal or hyperconnected person. Technology has fundamentally altered our view of the world and ourselves in the world. The tribe is no longer the local community but the global community which can now be accessed immediately via television, internet, satellite communication and travel. "For the first time since the appearance of human life on our planet," Ewert Cousins wrote, "all of the tribes, all of the nations, all of the religions are beginning to share a common history."[36] People are becoming more aware of belonging to humanity as a whole and not to a specific group.

Teilhard's writings undergird the second axial age; however, his insights on evolution as an ongoing process of complexifying consciousness do not terminate here but open up to the possibilities of life on other planets and, hence, a third axis of consciousness, which is intergalactic in nature. To appreciate the strong current of orthogenesis in Teilhard's thought, it is worthwhile to point out that he was acutely aware of internal forces that could thwart the direction of evolution toward the Omega Point. In an article on "The Directions and Conditions of the Future" he had begun to worry whether or not an expanding population would be able to live amiably and in peace with each other under conditions which he no longer described as "convergence" but "external compression."[37] In other words, the forces of history acting on humanity must either impel toward complexification and deeper wholeness or force it to wither.

Teilhard saw the choice, in his own day, between political totalitarianism or some new breakthrough into a new state of human interaction, the emergence of an "ultra-humanity."[38] As he saw it, the *first* phase happened long ago when our ancestors first emerged into the state of reflective consciousness. Man is psychically distinguished from all other animals, he wrote, by the fact that *he not only knows, but knows that he knows.*"[39] The *second* phase of human evolution involved not just this reflective consciousness, but what Teilhard termed a "*co-consciousness,*" a collective awareness brought about by the convergence of human beings (the noosphere) over the surface of the earth. Finally, he envisioned, fast approaching on the horizon, a *third*, even more critical phase facing humanity which he described as "a completely new mode of phylogenesis," a "mega-synthesis"[40] elaborated in terms of a "planetization."[41] Hence, evolution is not only a process of increasing consciousness; rather, it is a new phase of life in the universe toward unification of mind by which the whole cosmic evolution progresses toward greater unity.

THEOGENESIS

Teilhard did not see technology merely as a tool; rather, technology can further the religious spirit which is the heart of evolution. As he wrote, religion begins with the genesis of the universe, as he wrote: "Religion, born of the earth's need for the disclosing of a god is related and coextensive with not the individual man but the whole of humankind."[42] Ursula King states: "Teilhard is one of the few modern thinkers on religion for whom evolution provided the dominant note of his entire work."[43] The basic structure of an evolutionary perspective underlies all his thought on religion. In 1916 claimed

> Religion and evolution should neither be confused nor divorced. They are destined to form one single continuous organism, in which their respective lives prolong, are dependent on, and complete one another, without being identified or lost. . . . Since it is in our age that the duality has become so markedly apparent, it is for us to effect this synthesis.[44]

Religion, therefore, is not strictly a personal matter of salvation and personal belief but is integral to nature itself: "The religious phenomenon taken as a whole is simply the reaction of the universe as such, of collective consciousness and human action in process of development."[45]

Religion is integral to the future of the earth because it "is biologically the necessary counterpart to the release of the earth's spiritual energy." In his essay on "The Spirit of the Earth" (1931), Teilhard wrote that the true function of religion, not always clearly perceived in the past, is to sustain and spur on the progress of life. Thus, the religious function increases in the same direction and to the same extent as "hominization," that is, the emergence and growth of the human: "Religion grows 'continuously with the human being itself' by taking on a new and more closely determined form with each new phase of humankind."[46] The purpose of religion is to sustain the human zest for life. On the human level, religion is primarily on the level of consciousness and human action, rather than on the level of institutions or belief systems. This led Teilhard to realize that no one religion can satisfy the religious spirit of the earth because religion is directly concerned with the universe and its evolution towards Omega. The religions of the world represent the highest development of human consciousness in that direction. For Teilhard, the end is Omega, the total unification and complexification of being-in-love which is the full flowering of consciousness. In a universe that will continue on for billions of years, humans are not the center of life but "the arrow pointing the way to the final unification of the world in terms of life."[47]

Teilhard developed a doctrine of theogenesis (literally, the birthing of God) based on the rise of consciousness in evolution. If God is love and love is relational, God can only be the fullness of love through the deepening of relationships in evolution. He wrote: "As a direct consequence of the unitive process by which God is revealed to us, he in some way 'transforms himself' as he incorporates us."[48] As we come to a higher consciousness of a point of unity, God rises up in us; God *becomes* God in us. This is the meaning of incarnation; God "enters into" matter by rising up in matter as the unitive power of love. God is in us and we are in God without collapsing or merging these two realities, since they form a single reality. It is not enough to simply believe in God, Teilhard said; rather we are to incarnate God and help God become God, if we are to realize the potential of created existence.[49] As God emerges through higher levels of consciousness, the human evolves from an incomplete whole to a new level of completion and thus a new vision, a new knowing, and a new way of acting in the world. Peter Todd writes:

> Like Jung, Teilhard thinks God needs humankind to become both whole and complete. The implication is that God and humanity are in an entangled state and that the individuation of each is inextricably bound with the other. This entanglement of God and world is symbolized by the concept of Omega. Teilhard develops an understanding of personalization whereby God becomes God in union with another because only in union with another can one's true personality be found.[50]

Since God is love and love is personal—center to center attraction—God is most deeply actualized in personal love. God loves in and through our love for one another. Only in actualization is love experienced on a personal level, in the attraction and relations between one's deepest center, drawn to another center. The universe may be understood as God's actualization in deepening and personalizing love. As the most conscious point of the universe, carbon-based intelligent life is the basis for divine realization and interpersonal love. The full flowering of conscious being-in-love is symbolized by Omega.

NOOSPHERE AND CO-REFLECTION

Teilhard's vision is like an unfinished symphony in which the various players are constantly interacting to produce a glorious sound. Evolution describes the process, consciousness is the conductor, and the rising sound is intelligent life giving voice to the emergence of God. The process of theogenesis is the outflow of human evolution which continues today, primarily through technology and the various ways technology shapes culture. It is this process, as well, that undergirds the search for extraterrestrial life. As Teilhard realized, technology extends biology.

Teilhard lived at the dawn of the computer age and was fascinated by the thinking machine as a new level of interconnecting mind. He insisted that humankind does not dissipate itself but continually concentrates upon itself, that is, evolution is progression toward more being.[51] He wrote:

> It is not *well being* but a hunger for *more-being* which, of psychological necessity, can alone preserve the thinking earth from the *taedium vitae*. . . . It is upon its point (or superstructure) of spiritual concentration, and not upon its basis (or infra-structure) of material arrangement, that the equilibrium of Mankind biologically depends.[52]

Teilhard distinguished "more being" from "well-being" by saying that materialism can bring about well-being but spirituality and an increase in psychic energy or consciousness brings about more being.[53] Technology is a means of convergence and the noosphere is the evolutionary convergence of mind through technology; humankind does not dissipate itself but continually concentrates upon itself.[54] He saw that computer technology can advance co-reflective consciousness and hyperpersonalization and the emergence of a new global entity that opens up to a new cosmic unity. This new level, he said, is a new stage of convergence in evolution, the formation of what he called "the noosphere," a new level of coreflective thought and action.[55] The noosphere is a psychosocial process, a planetary neo-envelope *essentially linked with the biosphere* in which it has its root yet distinguished from it. It is a new stage for the renewal of life and not a radical break with biological life. If there is no connection between noogenesis and biogenesis, Teilhard said, then the process of evolution has halted and man is an absurd and "erratic object in a disjointed world."[56] Hence the noosphere is a superconvergence of psychic energy, a higher form of complexity in which the human person does not become obsolete but rather acquires more being through interconnectivity with others. In this respect the Noosphere is not the realm of the impersonal but conversely it is the realm of the *deeply personal* through *convergence* or the bringing together of diverse elements, organisms, and the currents of human thought. Teilhard wrote: "It is a mistake to look for the extension of our being or of the Noosphere in the impersonal. The Future universal cannot be anything else but the *hyperpersonal,* a new level of evolution and the expansion of community through greater unity. [57] He envisioned the noosphere like "the living membrane which is stretched like a film over the lustrous surface of the star which holds us."[58] It is a new layer of thinking that clothes the surface of biological life. "Just as Earth once covered itself with a film of interdependent living organisms which we call the biosphere, so mankind's combined achievements are forming a global network of collective mind."[59] Hence, the noosphere is a sphere of collective consciousness which preserves and communicates everything precious, active and progressive contained in this earth's previous evolution.[60] It is the natural culmination of biological evolution and not a termination of it, an organic whole, irreducible to its parts, destined for some type of superconvergence and unification.[61]

The formation of the noosphere begins with a global network of trade, communications, accumulation and exchange of knowledge, cooperative research, mixture of populations and production of energy—all go into the weaving of the material support for a sphere of collective thought. Although mass communication technology was just beginning to develop in Teilhard's time, he appreciated the role of machines in the emergence of the noosphere. In his *Future of Man* he wrote of "the extraordinary network of radio and television communications which, perhaps anticipating the direct intercommunication of brains through the mysterious power of telepathy, already link us all in a sort of 'etherized' universal consciousness."[62] Teilhard predicted the evolution of the computer as the "brain" behind the noosphere and thus the catalyst for the next step of evolution. His anticipation of what computers would do for us was twofold: first, they would complete our brains through instantaneous retrieval of information around the globe so that what one person lacks is immediately provided by another, and second, they would improve our brains by facilitating processes more quickly than our own resources can achieve them.[63] His vision of the noosphere as cybernetic mind anticipated the emergence of cyberspace as a field of global mind through interconnecting computer pathways. With the rise of technology, he saw a forward movement of spiritual energy, a maximization of consciousness and a complexification of relationships.

Teilhard saw evolution proceeding to a greater unification of the whole in and through the human person who is the growing tip of the evolutionary process. In his introduction to Teilhard's *Phenomenon of Man* Julian Huxley wrote: "We should consider inter-thinking humanity as a new type of organism whose destiny it is to realize new possibilities for evolving life on this planet."[64] Both Huxley and Teilhard saw this new type of person as a hyperpersonalizing person on a new level of "cooperative interthinking."[65]

Just as human persons develop a complex brain, Teilhard saw that the earth is developing a "planetary brain," a global complex brain, made possible by computer-mediated interconnected minds. He posited a new type of person to embody this new type of brain, an "ultra" human, whereby thought is no longer on the level of the individual but on the level of the convergent and collective. The hybridization of human and machine intelligence would complete the material and cerebral sphere of collective thought.[66] His hopeful vision was a richer and more complex domain of matter and mind through the development of technology, a way of constructing or joining all minds together in a collective or global mind for the forward movement of cosmic evolution. In his *Heart of Matter* he wrote: "How can we fail to see that the process of convergence from which we emerged, body and soul, is continuing to envelop us more closely than ever, to grip us, in the form of . . . a gigantic planetary contraction?"[67] The individual human person, he thought, will be surpassed by a collective convergence of consciousness giving rise to the ultrahuman, a new person who is part of the new planetary consciousness.

The ultrahuman represents a new collective consciousness that transcends individual consciousness and evokes a new type of person whose body now extends to the whole electronically connected planet. Hominization continues in and with new lines of shared information. What is "staring us in the face," according to Teilhard, is the rise of a "collective reflection" which is now realized by the internet; the concept of personhood is changing in accord with the new consciousness. He anticipated that each ego will be "forced convulsively beyond itself into some mysterious *super ego*."[68] The individual is coming to an end and a new "hyper-personal" person is emerging in evolution. Only when the noosphere is aligned with the whole, the cosmos/universe, can it facilitate the *deeply personal* through *convergence* by bringing together consciousness, person and creativity. He wrote: "The future universal cannot be anything else but the *hyperpersonal*."[69] This hyperpersonal for Teilhard is a "folding in" of consciousness, as if the lines of consciousness are merging together into one great complexified brain of planetary thought and planetary thought is giving rise to a new planetary body. This "grand enigma presented by the human phenomenon is not the question of knowing how life was kindled on earth, but of understanding how it might be extinguished on earth without being continued elsewhere. Having once become reflective it cannot abandon itself or totally disappear without biologically contradicting itself."[70] Teilhard realized that even if computer technology unites us into a fully global mind, a "planetization," this will not be the end of things but the next step in our passage to another sphere of the universe.

EXTRATERRESTRIAL LIFE

Sociologist and historian John Torpey advances the idea that there are in fact three "Axial Ages," instead of one original Axial Age and several subsequent, smaller developments. Each of the three ages contributed decisively to how humanity lives, and the difficulties it faces. The earliest, or original, Axial Age was a moral one; the second was material and revolved around the creation and use of physical objects; and the third is chiefly mental and focused on the technological. While there are profound risks and challenges, Torpey shows how a worldview that combines the strengths of all three ages has the potential to usher in a period of exceptional human freedom and possibility.[71] Torpey's thesis is interesting; however, I do not think it sufficiently considers the open process of evolving consciousness and complexity. I would suggest that Teilhard's level of the noosphere and the emergence of a third axial age of consciousness opens up to greater levels of complexification and consciousness. The noosphere is *not the final stage* of evolution but a node of consciousness in the ongoing transcendence toward more new life, and hence in the direction of intergalactic consciousness and the pursuit of intelligible life on other planets.

While Teilhard believed that evolution was teleologically oriented, he never assumed that life on planet earth could be its sole intelligent product. In a 1950 essay, he wrote about the increasing probability of other thinking planets[72]; by 1953 he had come to the inevitable conclusion that it is "practically certain" that millions of other galaxies exist in the universe in which there are 'thousands of millions' of solar systems . . . in which life has an equal chance of being born and becoming hominized.[73] Given the current estimate of the number of galaxies in the observable universe now being around 125 billion, the possibility of life elsewhere far outstrips Teilhard's earlier estimates. However, Teilhard was convinced of a general noogenic and hence orthogenic direction of evolution, rendering the search (and possibility) of extra-terrestrial intelligent life the next stage of the third axial period.

There are two aspects to Teilhard's thought that render the search for extraterrestrial life not only a matter of science and discovery but one of spirituality and meaning. The first is laid out in his 1953 essay on the plurality of inhabited world in which he indicates that matter tends to group itself into as large molecules as possible. He explains: "Life continues naturally to become both complexified externally and more conscious internally . . . the appearance of man on earth in the Pliocene is simply the normal . . . property common to all . . . matter. There are millions of galaxies in the universe, in each of which matter has the same general composition and is going through essentially the same evolution as that inside our own Milky Way." This leads to a set of propositions:

1. If it is true that the proteins appear in the universe as soon as it is possible for them to do so and where it is possible
2. And if when life is once taken hold on a star, it not only propagates itself on that star but carries itself as far and to as high a degree as possible [that is, hominization, if possible]
3. And if in addition there are thousands of millions of solar systems in the world in which life has equal chances of being born and becoming hominized
4. Then our minds cannot resist the inevitable conclusion that were we by chance to possess plates that were sensitive to the specific radiation of the noosphere scattered throughout space, it would be practically certain that what we saw registered on them would be a cloud of thinking stars.[74]

Hence, Teilhard accepts, almost as a given, that the search for other forms of intelligent life will eventually be discovered/found. Richard Kopf sums up Teilhard's position on extraterrestrial life by saying, "Teilhard's orthogenesis amounts to an 'anthropic principle' in which *anthropos* (that is, humanity) itself may be but an infinitesimally small but typical specimen in a universe teeming with life."[75]

While Teilhard's Omega-driven evolution naturally yields to the quest for extraterrestrial life, the real impetus behind such a quest is the completion of God, that is, the fulfillment of Omega. His essay on "The Heart of Matter" clearly affirms that matter is God-filled and thus rootless and restless. Conscious material life will continuously search for more life and consciousness, primarily because divinity seeks completion in intelligent life. The rise of the ultrahuman is still frightening to those who still shrink from entering into the "great ocean of Matter," Teilhard states, precisely because fear prevents seeing God in a new way; yet, divinity constantly seeks to "pleromize itself" by enfolding the universe into its own life. Elsewhere he writes that as God "metamorphizes the World from the depths of matter to the peaks of Spirit," so the world "endomorphizes" God.[76] This entanglement of God and universe is not limited to terrestrial life but to all conscious life in the universe in which God and universe form a cosmotheandric whole: "As a direct consequence of the unitive process by which God is revealed to us, he in some way 'transforms' himself as he incorporates us."[77] Just as the infinite universe stretches into new life by the expansion and hence convergence of consciousness, so too the infinite God is not limited by terrestrial life but continues to seek fulfillment in all possible forms of intelligent life. Teilhard uniquely brings together theology and extraterrestial life through the evolution of complexity-consciousness. His brilliant synthesis of science and religion shows that the quest for intergalactic life is not a fringe benefit for the wealthy or a platonic ideal but the necessary next phase of evolution toward Omega. The search for extraterrestrial life is not only a scientific pursuit but a theological mandate. God will not rest until God is all in all.

CONCLUSION

We are in the midst of the second axial age of consciousness, but the rapid development of computer technology and the quest for extraterrestrial life is quickly ushering in a third age of consciousness. Teilhard's theological vision, governed by the process of evolution, is one that orients itself to intergalactic life. The human phenomenon marked by the capacity for self-reflection and co-reflection does not terminate on the level of terrestrial life because the human person is transcendent by nature and constantly seeks the perfectability of knowledge. Matter is driven by consciousness and the infinite power of Omega; hence, it will not rest until all forms of intelligent life are gathered into the unity of intelligible love. From matter to spirit, conscious material life in the universe is an ever-emergent dynamism of divine love. The search for extraterrestrial life reflects nothing less than the ongoing emergence of God.

NOTES

1. Ted Peters, *Science, Theology, and Ethics* (Burlington: Ashgate, 2003), 122.
2. Ibid., 123.
3. Ibid., 125.
4. Andrew J. Burgess, "Earth Chauvinism" *Christian Century* 93 (1976), 1098.
5. Ibid.
6. Ibid., 1100.
7. Christen Brownlee, "Is Anybody Out There?" *Science News* 169.3 (2006), p. 42. Online at http://www.sciencenews.org/articles/20060121/bob10.asp.
8. Lewis S. Ford, *The Lure of God: A Biblical Background for Process Theism* (Philadelphia: Fortress Press, 1978), 56.
9. J. Edgar Burns, "Cosmolatry," *The Catholic World* 191 (1960): 284.
10. Robert North, *Teilhard and the Creation of the Soul,* 49.
11. Pierre Teilhard de Chardin, *The Phenomenon of Man*, trans. Bernard Wall (New York: Harper & Row, 1959), 290.
12. Teilhard de Chardin, *The Phenomenon of Man,* 261–62.

13. Patrick H. Byrne, "The Integral Visions of Teilhard and Lonergan: Science, the Universe, Humanity, and God," in *From Teilhard to Omega: Cocreating an Unfinished Universe,* ed. Ilia Delio (New York: Orbis, 2014), 88.

14. Teilhard de Chardin, *The Phenomenon of Man,* 289.

15. Pierre Teilhard de Chardin, *The Divine Milieu,* trans. Sion Caldwell (New York: Harper, 2001), 100.

16. Thomas M. King, *Teilhard's Mysticism of Knowing* (New York: Seabury Press, 1981), 32.

17. Pierre Teilhard de Chardin, *Man's Place in Nature,* trans. Noel Lindsay (New York: Collins, 1966), 34

18. Teilhard de Chardin, *Phenomenon of Man*, 56–64.

19. Teilhard de Chardin, *Phenomenon of Man*, 56–64.

20. Susan Borowski, "Quantum Mechanics and the Consciousness Connection," *AAAS* (July 16, 2010) https://www.aaas.org/quantum-mechanics-and-consciousness-connection

21. Erwin Schrodinger, *What is Life?* Trans. Verena Schrodinger (Cambridge: Cambridge University Press, 2012, reprint edition), 93–5.

22. Bertrand Russell, "Mind and Matter," 1950 https://russell-j.com/19501110_Mind-Matter.HTM

23. Gaylen Strawson, "Consciousness Isn't a Mystery. Its Matter," *New York Times* (May 16, 2016) https://www.nytimes.com/2016/05/16/opinion/consciousness-isnt-a-mystery-its-matter.html

24. James Jeans, *The Mysterious Universe* (New York: Macmillan, 1931), 158.

25. Erwin Schrödinger, *What is Life?* Trans. Verena Schrödinger (Cambridge: Cambridge University Press, 2012, reprint edition), 93–5.

26. Teilhard de Chardin,"The Position of Man in Mature and the Significance of Human Socialization," in *The Future of Man*, trans. Norman Denny (New York: Harper & Row, 1964), 211–17.

27. Teilhard de Chardin, *Future of Man*, 90.

28. Teilhard de Chardin, *Phenomenon of Man*, 165.

29. Teilhard de Chardin, *Phenomenon of Man*, 221.

30. Pierre Teilhard de Chardin, *Human Energy*, trans. J. M. Cohen (New York: Harcourt Brace Jovanovich, 1969), 23.

31. Ewert H. Cousins, *Christ of the 21st Century* (Rockport, MA: Element Books, 1992), 5.

32. Benjamin I. Schwartz, "The Age of Transcendence," *Daedalus* 104 (1975): 3.

33. Karl Jaspers, "The Axial Period," in Karl Jaspers, *The Origin and Goal of History* (New Haven: Yale University Press, 1953), 2; cf. *The Origin and Diversity of Axial Age Civilizations,* ed. S. N. Eisenstadt (Albany, NY: State University of New York Press, 1986); Karen Armstrong, in *The Great Transformation: The World in the Time of Buddha, Socrates, Confucius and Jeremiah* (London: Atlantic Books, 2006).

34. William M. Thompson, *Christ and Consciousness: exploring Christ's contribution to human consciousness* (New York: Paulist, 1977), 21.

35. Cousins, *Christ of the 21st Century,* 7–8. Thomas Berry used the term "Second Axial Ages" to refer to the convergence of world religions, leading to a new

phase of human culture and civilization. See Mary Evelyn Tucker, John Grim and Andrew Angyal, *Thomas Berry: A Biography* (New York: Columbia University Press, 2019), 93.

36. Cousins, *Christ of the 21st Century*, 7–10.

37. Teilhard de Chardin, *Future of Man*, 235.

38. Teilhard, *Future of Man*, 270–80.

39. Teilhard, *Future of Man*, 293.

40. Teilhard, *Phenomenon of Man*, 243–44.

41. Teilhard de Chardin, *Future of Man*, 124–39; ibid., *Phenomenon of Man*, 243.

42. Pierre Teilhard de Chardin, *Christianity and Evolution*, trans. René Hague (New York: Harcourt, Brace, Jovanovich 1971, 119.

43. Ursula King, *Teilhard de Chardin and Eastern Religions: Spirituality and Mysticism in an Evolutionary World* (Mahwah, NJ: Paulist Press, 2011), 179.

44. King, *Teilhard and Eastern Religions,* 179–180.

45. Teilhard de Chardin, *Christianity and Evolution,* 118.

46. King, *Teilhard and Eastern Religions,* 186.

47. Teilhard de Chardin, *Phenomenon of Man*, 224; W. Henry Kenny, S.J., *A Path Through Teilhard's Phenomenon* (Dayton, OH: Pflaum Press, 1970), 99.

48. Pierre Teilhard de Chardin, *The Heart of Matter*, trans. René Hague (New York: Harcourt, Inc., 1978), 53.

49. Teilhard de Chardin, *Heart of Matter,* 53.

50. Peter Todd, "Teilhard and Other Modern Thinkers on Evolution, Mind and Matter," *Teilhard Studies* 66 (2013): 1–18 at p. 5.

51. Teilhard de Chardin, *Future of Man*, 316.

52. Teilhard de Chardin, *Future of Man*, 317.

53. Joseph A. Grau, *Morality and the Human Future in the Thought of Teilhard de Chardin A Critical Study* (Cranbury, NJ: Associated University Presses, Inc., 1976), 275.

54. Teilhard de Chardin, *Future of Man*, 316.

55. Teilhard de Chardin, *Future of Man,* 204. "In the 1920s Teilhard coined the word *noosphere* in collaboration with his friend Edouard Le Roy. The noosphere (sometimes noösphere) is the sphere of human thought. The word derives from the Greek νοῦς (nous "mind") and σφαῖρα (sphaira "sphere"), in lexical analogy to "atmosphere" and "biosphere." It may have been introduced by Pierre Teilhard de Chardin in 1922 in writings on cosmogenesis. Another possibility is the first use of the term by Édouard Le Roy (1870–1954), who together with Teilhard was listening to lectures of Vladimir Ivanovich Vernadsky at the Sorbonne; cf. Ursula King, "One Planet, One Spirit: Searching For an Ecologically Balanced Spirituality," in *Pierre Teilhard de Chardin on People and Planet*, ed. Cecelia Deane-Drummond (London: Equinox, 2008), 82.

56. Robert J. O'Connell, *Teilhard's Vision of the Past: The Making of a Method* (New York: Fordham University Press, 1982), 145.

57. Teilhard de Chardin, *Phenomenon of Man*, 260.

58. Blanche Gallagher, *Meditations with Teilhard de Chardin* (Santa Fe: Bear & Co., 1988), 39.

59. Michael H. Murray, *The Thought of Teilhard de Chardin* (New York: Seabury Press, 1966), 20–1.

60. Murray, *The Thought of Teilhard de Chardin*, 20–1.

61. Kenny, *A Path Through Teilhard's Phenomenon,* 110.

62. Lawrence Hagerty, *The Spirit of the Internet: Speculations on the Evolution of Global Consciousness* (Matrix Masters, 2000), 33.

63. Pierre Teilhard de Chardin, *Man's Place in Nature*, trans. Bernard Wall (New York: HarperCollins, 2000), 111.

64. Teilhard de Chardin, *Phenomenon of Man,* 20.

65. Teilhard de Chardin, *Phenomenon of Man*, 21.

66. See Eric Steinhart, "Teilhard de Chardin and Transhumanism," *Journal of Evolution and Technology* 20.1 (2008): 22. http://jetpress.org/v20/steinhart.htm. Steinhart begins his paper by saying that "Teilhard was one of the first to articulate transhumanist themes. Transhumanists advocate the ethical use of technology for human enhancement. Teilhard's writing likewise argues for the ethical application of technology in order to advance humanity beyond the limitations of natural biology." However, Teilhard's transhumanism is transcendent evolution on the level of mind. His ideas do not fall into the philosophy of transhumanism per se but combine aspects of transhumanism and posthumanism.

67. Teilhard de Chardin, *Heart of Matter*, 36.

68. Teilhard de Chardin, *Heart of Matter,* 38.

69. Teilhard de Chardin, *Phenomenon of Man,* 260.

70. Teilhard de Chardin, *Future of Man,* 297–98.

71. John Torpey, *The Three Axial Ages: Moral, Material, Mental* (New Brunswick, NJ: Rutgers University Press, 2017).

72. Teilhard de Chardin, *Heart of Matter*, 56.

73. Teilhard de Chardin, *Christianity and Evolution*, 230–31.

74. Teilhard de Chardin, *Christianity and Evolution*, 231.

75. Richard Kopf, "Teilhard de Chardin's Vision of Ultimate Reality and Meaning in the Light of Contemporary Cosmology," *Toronto Journal of Theology* vol. 3.3 (Sept 2012): 238–59 at p. 249.

76. Teilhard de Chardin, *Heart of Matter*, 52–3.

77. Teilhard de Chardin, *Heart of Matter*, 52–3.

REFERENCES

Armstrong, K. *The Great Transformation: The World in the Time of Buddha, Socrates, Confucius and Jeremiah*. London: Atlantic Books, 2006.

Borowski, S. "Quantum Mechanics and the Consciousness Connection," *AAAS* (2010). Accessed March 30, 2023. https://www.aaas.org/quantum-mechanics-and-consciousness-connection.

Brownlee, C. "Is Anybody Out There?" *Science News*169, no. 3 (2006). Accessed March 30, 2023. http://www.sciencenews.org/articles/20060121/bob10.asp.

Burgess, A.J. "Earth Chauvinism." *Christian Century* 93, no. 4 (1976): 1098–1102.

Burns, J. E. “Cosmolatry.” *The Catholic World* 191 (1960): 284.
Byrne, P. H. “The Integral Visions of Teilhard and Lonergan: Science, the Universe, Humanity, and God.” In *From Teilhard to Omega: Cocreating an Unfinished Universe*, edited by I. Delio, 83–110. Maryknoll, NY: Orbis, 2014.
Cousins, E. H. *Christ of the 21st Century*. Rockport, MA: Element Books, 1992.
Eisenstadt, S. N. *The Origin and Diversity of Axial Age Civilizations*. Albany: SUNY Press, 1986.
Ford, L. S. *The Lure of God: A Biblical Background for Process Theism*. Philadelphia: Fortress Press, 1978.
Gallagher, B. *Meditations with Teilhard de Chardin*. Santa Fe: Bear & Co., 1988.
Gaylen S. “Consciousness Isn’t a Mystery. Its Matter,” *New York Times* (2016). Accessed March 30, 2023. https://www.nytimes.com/2016/05/16/opinion/consciousness-isnt-a-mystery-its-matter.html
Grau, J.A. *Morality and the Human Future in the Thought of Teilhard de Chardin.* Cranbury, NJ: Associated University Presses, Inc., 1976.
Hagerty, L. *The Spirit of the Internet: Speculations on the Evolution of Global Consciousness*. Matrix Masters, 2000.
Jaspers, K. “The Axial Period.” In *The Origin and Goal of History*. New Haven: Yale University Press, 1953.
Jeans, J. *The Mysterious Universe*. New York: Macmillan, 1931.
Kenny, W. H. *A Path Through Teilhard’s Phenomenon*. Dayton: Pflaum Press, 1970.
King, T. M. *Teilhard’s Mysticism of Knowing*. New York: Seabury Press, 1981.
King, U. “One Planet, One Spirit: Searching For an Ecologically Balanced Spirituality.” In *Pierre Teilhard de Chardin on People and Planet*, edited by C. Deane-Drummond, 74–95. London: Equinox, 2008.
———. *Teilhard de Chardin and Eastern Religions: Spirituality and Mysticism in an Evolutionary World*. Mahwah, NJ: Paulist Press, 2011.
Kopf, R. “Teilhard de Chardin’s Vision of Ultimate Reality and Meaning in the Light of Contemporary Cosmology.” *Toronto Journal of Theology* 3, no. 3 (2012): 238–59. https://doi.org/10.3138/uram.23.3.238.
Murray, M. H. *The Thought of Teilhard de Chardin*. New York: Seabury Press, 1966.
North, Robert. *Teilhard and the Creation of the Soul*. Milwaukee: Bruce Publishing Co., 1967.
O’Connell, R. J. *Teilhard’s Vision of the Past: The Making of a Method*. New York: Fordham University Press, 1982.
Peters, Ted. *Science, Theology, and Ethics*. Burlington: Ashgate, 2003.
Russell, B. “Mind and Matter.” 1950. Accessed March 30, 2023. https://russell-j.com/19501110_Mind-Matter.HTM.
Schrodinger, E. *What is Life?* Translated by Verena Schrodinger. Cambridge: Cambridge University Press, 2012.
Schwartz, B. I. “The Age of Transcendence.” *Daedalus* 104, no. 2 (1975): 1–7.
Steinhart, Eric. “Teilhard de Chardin and Transhumanism.” *Journal of Evolution and Technology* 20, no. 1 (2008): 1–22. http://jetpress.org/v20/steinhart.htm.

Teilhard de Chardin, P. "The Position of Man in Mature and the Significance of Human Socialization." In *The Future of Man*. Translated by N. Denny, 211–17. New York: Harper & Row, 1964.

———. *Christianity and Evolution*. Translated by R. Hague. New York: William Collins & Sons, 1971.

———. *Human Energy*. Translated by J. M. Cohen. New York: Harcourt Brace Jovanovich, 1969.

———. *Man's Place in Nature*. Translated by B. Wall. New York: HarperCollins, 2000.

———. *The Divine Milieu*. Translated by S. Caldwell. New York: Harper, 2001.

———. *The Heart of Matter*. Translated by R. Hague. New York: Harcourt, Inc., 1978.

———. *The Phenomenon of Man*. Translated by B. Wall. New York: Harper & Row, 1959.

Thompson, W. M. *Christ and Consciousness: Exploring Christ's Contribution to Human Consciousness*. New York: Paulist, 1977.

Todd, P. "Teilhard and Other Modern Thinkers on Evolution, Mind and Matter." *Teilhard Studies* 66 (2013): 1–18.

Torpey, J. *The Three Axial Ages: Moral, Material, Mental*. New Brunswick, NJ: Rutgers University Press, 2017.

Tucker, M. E., J. Grim, and A. Angyal. *Thomas Berry: A Biography*. New York: Columbia University Press, 2019.

Chapter 14

Astrobiology without Biology

Will AI Be Our Emissary or Our Bottleneck?

Noreen Herzfeld

The search for extraterrestrial intelligence (SETI) project has been scanning the heavens for a signal, some sign of intelligent life, for almost half a century. So far, radio silence. Despite the claims of UFO enthusiasts, self-professed alien abductees, and supermarket tabloids, we have seen no verifiable evidence of intelligent extraterrestrial life. On statistical grounds, this is odd. There are at least 100 billion galaxies in the observable universe. Our own relatively small Milky Way contains 100 billion stars. We have located more than four thousand planets orbiting stars in our galaxy, and this number represents only planets large enough for their gravitational pull to perturb the light of their star. Using an expanded version of an equation developed by Frank Drake in 1961, astronomers Frank and Sullivan concluded that the odds of our planet being the only one on which life evolved are somewhere in the vicinity of one in ten billion trillion (10^{-24}). Drake suggests, "Think of it this way. Before our result you'd be considered a pessimist if you imagined the probability of evolving a civilization on a habitable planet were, say, one in a trillion. But even that guess, one chance in a trillion, implies that what has happened here on Earth with humanity has in fact happened about a 10 billion other times over cosmic history!"[1]

This raises the famous question posed by physicist Enrico Fermi at a lunch discussion at Los Alamos Labs in 1951: "Where is everybody?" If the probability of intelligent life is that high, then why the great silence? Someone, somewhere, should have by now colonized large portions of the galaxy, either in person, or with their machines. While we all admit that space travel is difficult for biological beings, recent advances in AI technology have led to predictions that superintelligent computers might exist in the near future. Surely these could be the emissaries of an intelligent race. Yet we see no sign of intelligent life beyond our planet. Why not? I contend that three factors—the hostility of space, the nature of AI, and our own nature as evolved creatures—make it unlikely that we will hear from anyone any time soon, if ever.

THE HOSTILITY OF SPACE

If we are to contact intelligent biological life from another planet or they us, it is unlikely to happen in person. Space travel is difficult for biological carbon-based creatures. NASA has identified multiple issues that make travel far from earth prohibitive for humans. A number of these are physical. The lack of gravity in space causes problems with fluid distribution throughout the body, loss of bone mass and density, difficulties with our balance system, and even issues with astronauts' eyesight. Exposure to radiation in space can lead to cell mutations that can result in a variety of cancers. It also causes long-term brain damage, resulting in dementia, memory loss, anxiety, and impaired decision-making. Further deterioration of astronauts' mental health can come from extended periods in cramped quarters, bad food, and loss of privacy. Additionally, sufficient supplies of food and water are hard to come by in space because spacecraft have limited room for supplies and gardening requires soil and water.[2]

More than these, there is the magnitude of space and the brevity of life. It takes a long time to get to even those planets in our own solar system with current technology, let alone those revolving around distant stars. The closest star to Earth, Proxima Centauri, is more than four light-years, or approximately 25 trillion miles, away. Currently, our fastest spacecraft, the Parker Solar Probe, has a top speed of 450,000 mph. At that speed, it would take 6,633 years to reach any planets in our nearest neighboring system.[3] To reach planets at any distance, we would need to travel faster than light, a physical impossibility (the warp drive envisioned in *Star Trek* and other science fiction would require either negative energy or negative mass—definitely not suited to transporting human beings).

Given the physical, mental, and theoretical limitations of space travel for biological beings, it is likely that our first encounter with another intelligence would be technological. Physicist Paul Davies writes, "I think it very likely—in fact inevitable—that biological intelligence is only a transitory phenomenon, a fleeting phase in the evolution of intelligence in the universe. If we ever encounter extraterrestrial intelligence, I believe it is overwhelmingly likely to be post-biological in nature."[4] This raises the question of whether an AI would be our best bet for colonizing other planets. With the Rover series exploring Mars, we have taken the first steps in sending our machines to places it is difficult for us to go. Fermi quite expected that an intelligent society with artificially intelligent machines, would colonize large parts of the galaxy, first sending machines to nearby planets, then propagating those machines from solar system to solar system.

Robots obviate many of the difficulties inherent in human space travel. They can withstand extreme temperatures, loss of gravity, need no atmosphere, and can be more easily shielded from radiation. They do not suffer from boredom, anxiety, depression, or the close confines of space travel. If a mission fails, we would not grieve the loss of a robot as we would the death of a human being. Robots require few supplies—they do not eat, sleep, breathe or go to the bathroom. They excel at collecting and assessing data, making predictions and categorizations much more quickly than a human could. And while they do not last forever, their functional span is much longer than ours.

However, while machines are more resistant to the rigors of space travel and hostile conditions on planets unprotected by an atmosphere, they are not impervious to them. Radiation can still play havoc with sensitive electronics. The constant bombardment of high-energy particles in space causes electronic equipment to become unstable and fail over time. Meteorites and other space debris can cause damage to a probe's physical structure and its electronics. However, these difficulties pale in comparison to the vastness of space, which remains the greatest barrier to interstellar travel, even for AI. While we can propel a machine into space at much greater velocities and with far less energy consumption than is needed for human space travel, we remain limited to speeds well below the speed of light. Voyager 1 has left our solar system, but at its speed of 62,000 km per hour, it would take it 77,000 years to reach our nearest star. Both it and we will likely fail long before then. Thus, travel, even for machines, remains a prohibitively slow business relative to the vastness of space. Ultimately, most of our galaxy, let alone other galaxies, is basically out of reach. Even a lineage of AIs that propagates steadily would not get far beyond the home solar system, or perhaps the closest stars, in a practical time frame.

THE NATURE OF AI

The propagation of a lineage of AIs, even on a local level, would not be an easy task. For AI technology to leapfrog from planet to planet without human intervention it would need to establish mining, smelting, and the building of intricate computer chips on each planet. Currently, production of such chips requires a sterile environment, something that would also have to be established. Theoretically, this is all possible. In practice, it would be exceedingly difficult.

Currently, AI functions as a tool, one that extends and amplifies human capabilities. The Mars rovers calculate their own best landing site. They collect measurements and data, take pictures, and transmit them back to humans on earth. But they are nothing without their human designers, programmers, and handlers. They do not, and cannot, think or plan on their own. While humans adapt to changing situations, current AIs execute their programming or call on their nearby human programmers for help. Programming may be flawed, as was the case for the ill-fated Mars Climate Orbiter, which crashed into the Martian surface because two different programming teams had used different measurement systems, or it may simply be insufficiently adaptable to unforeseen conditions. To function as our agents in space, AIs would need to make fully autonomous decisions, interact with the space environment, each other, and any other intelligences they encounter, and determine their own health status.[5] A tall order.

When futurists posit large-scale colonization of space by AI, they are not talking about anything we currently know of as AI; instead, they are predicting something quite different from current space probes or rovers—a superintelligent AI, one that can function independently from human guidance, that can learn through observation, and alter its own programming at will. Only such an automated general intelligence (AGI), would have the autonomy and adaptability outlined above.

Kurzweil, Bostrom, Hawking, and others have predicted that just such a machine is right around the corner. However, we should take such predictions with a grain of salt, considering past predictions for the field, which have tended to be long on enthusiasm and woefully short of accuracy. Herbert Simon suggested in 1965 that "machines will be capable, within twenty years, of doing any work that a man can do."[6] Marvin Minsky, in 1970, was even more optimistic:

> In from three to eight years, we will have a machine with the general intelligence of an average human being. I mean a machine that will be able to read Shakespeare, grease a car, play office politics, tell a joke, have a fight. At that point, the machine will begin to educate itself with fantastic speed. In a few months, it will be at genius level, and a few months after that, its power will be incalculable.[7]

Futurists such as Ray Kurzweil and Nick Bostrom continue the trend. Kurzweil predicts that by 2045 AI will exceed human intellectual capabilities, bringing about "a rupture in the fabric of human history," an event that Vernor Vinge popularized in 1993 as "the Singularity."[8] While we call programs that are new and exciting "AI," the ultimate goal, to produce an AGI that can function in a wide variety of domains and transfer learning from one arena to another, as humans do, still seems to be, as it was for Simon, about twenty years in the future.

How we will reach this goal is a matter of conjecture. Kurzweil claims it is the inevitable result of a continued exponential increase in computing power, an increase dubbed Moore's Law. Not everyone agrees. Gordon Moore himself has suggested that computing power is likely to level off in the near future, as the miniaturization of circuits reaches its nadir. Moore's Law also only applies to computer hardware or processing power, not to software.

Psychologist Steven Pinker notes: "There is not the slightest reason to believe in a coming singularity. The fact that you can visualize a future in your imagination is not evidence that it is likely or even possible. . . . Sheer processing power is not a pixie dust that magically solves all your problems."[9] Most futurists who posit a coming Singularity also posit intelligence as an emergent property of complexity, specifically the complexity of the human brain. They expect that, as we learn more about the neural structure of the brain, we might soon be able to model that structure in silicon. But that does not mean such a model would necessarily be functional. According to Oxford physicist David Deutsch, "no brain on Earth is yet close to knowing what brains do in order to achieve . . . functionality. The enterprise of achieving it artificially—the field of 'artificial general intelligence' or AGI—has made no progress whatever during the entire six decades of its existence."[10] He posits that any AGI would have to be "qualitatively, not quantitatively, different from all other computer programs. . . . Expecting to create an AGI without first understanding in detail how [the brain] works is like expecting skyscrapers to learn to fly if we build them tall enough."[11]

Despite his skepticism regarding current AI technology, Deutsch believes an AGI is theoretically possible, since the brain is composed of atoms, each of which obeys the laws of physics. Yet computers, as we know them, according to philosopher John Searle, "have, literally . . . , no intelligence, no motivation, no autonomy, and no agency. We design them to behave as if they had certain sorts of psychology, but there is no psychological reality to the corresponding processes or behavior. . . . The machinery has no beliefs, desires, [or] motivations."[12] Perhaps the problem here is that computer scientists have, heretofore, held a too limited understanding of intelligence.

The field of AI began with a model that sought to reproduce deductive reasoning, assuming intelligence meant problem-solving or the application of symbolic logic. This model met with initial success in fields such as mathematical problem solving or game playing, successes that led to Simon and Minsky's optimistic predictions. However, it soon became clear that not all human intelligence was deductive or represented by a rule-based system. The

1980s saw the rise of expert systems, computer programs that mimicked the inductive reasoning of an expert in a particular field. Inductive reasoning in AI has received a big boost since the turn of the century through the rise of machine learning, computer programs that make statistical inferences based on large data sets or training trials. While we call these programs "artificial intelligence" they do not differ from other computer programs in kind. They solve previously unsolved problems primarily through their speed and ability to handle amounts of data that no human could ever handle.

They are not, however, exceedingly adaptable. Deep Blue beat then reigning champion Gary Kasparov at chess; since then, Kasparov has gone on to play a significant role in Russian politics. Deep Blue was dismantled before Kasparov could demand a rematch. In fact, it later came to light that Deep Blue's programmers were pulling strings behind the scenes, altering Deep Blue's programming between games. This illustrates a limitation of almost all current AI. It depends deeply on its human trainers or on the work of countless human beings hidden behind the scenes. While Google Translate looks like a linguistic expert, it relies on the millions of extant translations already made by human beings. Philosopher Daniel Dennett notes that "AI in its current manifestations is parasitic on human intelligence. It quite indiscriminately gorges on whatever has been produced by human creators and extracts the patterns to be found there."[13] An AI in space, unconnected to its human masters, would lack the adaptability we currently assume as one of AI's features.

In a similar vein, computers also lack resiliency. Remove a single transistor from a computer chip and it crashes, while the human brain remains quite functional, even when large parts of it have been removed. Remove or corrupt part of a computer's memory and whatever data was stored there is gone for good, while human memories seem to be diffusely stored and distributed about the brain. Our brains are extremely plastic, engaged in a continual process of establishing new neuronal connections, strengthening some and weakening others, new neurons coming on-board as old cells die. A truly adaptable AI, one capable of what we consider intelligent thought and expression over a variety of fields is unlikely to arise from either the deductive or inductive approach. Current hopes for an AGI lie in the creation of ever larger neural networks, layered networks of nodes that change the strength of pathways between nodes each time they make a right, or wrong, decision. These networks can, in essence, program themselves through trial and error. They underlie recent successes such as that of AlphaGo, a program that used neural networks and both self and external reinforcement to master the game of Go, considered much more difficult than chess. While, like Deep Blue, AlphaGo only knows how to play Go, some commentators see it as a precursor to developing an AGI, in that AlphaGo, and its cousins, are able to self-correct once they know a desired outcome.

Producing an AGI with the power and flexibility of the human brain is a tall order. Several projects, including the US funded Brain Research through Advancing Innovative Neurotechnologies (BRAIN) or the European Human Brain Project hope to decode the neural connectome of the human brain, much as we decoded the human genome and use this understanding to reverse engineer a brain much like ours. This is, however, a much more difficult task that understanding the genome. So far, we understand how individual neurons connect with and signal one another. We can also use technologies such as MRI scanning to identify patterns of blood flow and thus, the stimulation or quiescence of large segments of the brain. It is the middle ground that remains unknown—how we store a particular memory, or what makes up a thought. A second problem is that of scale. The human brain has over a hundred billion neurons, each of which is capable of thousands of different connections which, as we have already noted, are constantly changing. Theoretically, a copy of a human brain is possible. So far, we have managed to map the brain of a nematode, which has only 302 neurons and 7000 synapses.[14]

THE NATURE OF HUMAN INTELLIGENCE

Even should we succeed in mapping the human brain and building a model, would it be functional? More specifically, could an AI become conscious? Conscious computers have long been a staple of science fiction, and figure largely in the future prognostications of several scientists, ranging from Nick Bostrom to the late Stephen Hawking. Yet the possibility of a conscious computer remains highly controversial.

We do not know what consciousness is or how it arises in human and other beings. Philosopher David Chalmers has called the question of consciousness, as distinguished from cognitive functioning, the quintessential "hard problem."[15] One of the roots of this problem is that we lack a definition for consciousness. Chalmers has catalogued more than twenty thousand papers on the subject, but they reach no consensus. Let us consider the definition put forward by physicist Michio Kaku, that consciousness is "the process of creating a model of the world using multiple feedback loops . . . in order to accomplish a goal."[16] Modeling the environment and accomplishing a goal seem to be two things that would be necessary for any competent space emissary.

Creating models and setting up feedback loops are standard tasks, and eminently doable, particularly by neural networks, which are based on feedback loops. It is having a goal that presents the stumbling block. Computers have no goal of their own. They can only be given goals by us. AlphaGo does not "want" to play Go, or even, once playing, to win the game. Would an AI traveling into distant space need its own goals? As Viking I heads out of our solar system, it has no goals of its own. But for AI to truly be our emissary to other intelligences it would need to adapt to their world conditions, their environment, their presentation and needs. Without the ability to formulate goals this would be rudimentary, at best.

From where do we get our goals? We like to think our goals are the product of rational deductive thought. But they are generally not. Social scientists have noted that we tend to make decisions first, then use deductive reasoning to justify them. So where do these decisions come from? Neuroscientist Antonio Damasio found that when the link between the prefrontal lobe, which governs reasoning, and the limbic system, which governs emotion, is damaged, patients are unable to make even the simplest decisions, such as what to have for lunch.[17] To have a goal we need to want something, and to want something we need emotions.

Psychologist Jerome Kagan notes that the ancients described emotion as an appraisal of a change in feeling. This definition divides emotion into two parts: a change in feeling and a mental appraisal of that change. Setting aside for a moment the question of what a feeling is, the appraisal of that feeling would seem to make emotion a mental calculation. So far so good for the proponents of AI, who would argue that the result of any such calculation correlates to a state of the brain which could, in theory, be modeled. Kagan notes, however, two difficulties with correlating emotion with a brain state. The first is that the phenomenon of emotion is "underdetermined by a brain state because each brain profile can give rise to an envelope of emotions."[18] Which emotion emerges depends on our interpretation, an interpretation based on context, memory, and a person's biology. An AI could bring both context and memory to bear when forming an interpretation. Biology, however, raises a complication.

This complication is also present in the first part of emotion's definition, that it begins as a feeling. In an 1884 article in *Mind*, philosopher William James already noted this: "Our natural way of thinking about . . . emotions is that the mental perception of some fact excites the mental affection called the emotion, and that this latter state of mind gives rise to the bodily expression. My thesis on the contrary is that *the bodily changes follow directly the perception of the exciting fact, and that our feeling of the same changes as they occur IS the emotion* (italics original)."[19]

Kagan elaborates on this definition, breaking emotion into a four-step process: 1) a change in brain activity due to a stimulus, 2) a perceived change in feeling that is sensory and may contain an involuntary motor response, 3) a cognitive appraisal of that feeling, and 4) a preparedness toward or display of a response.[20] Clearly computers are capable, to some degree, of parts one, three, and four. They can note a stimulus, appraise that stimulus, and calculate an appropriate response. However, the second step, a change in feeling that is sensory, returns us to the question of what is a feeling? For Kagan, "feelings refer to detected sensory changes that originate in spontaneous brain processes or activity in the gut, muscles, genitals, and vestibular apparatus."[21] He notes that most environmental stimuli, particularly if not intense or unusual, are "felt" but do not rise to a level that evokes an involuntary motor response. Those that do, however, require cognitive appraisal. Consider the rapid heartbeat, flushed face, and weak knees of anxiety or fear, the way the stomach turns in disgust, or the warm relaxation of love. These physical responses precede conscious recognition of the emotion. They come unbidden. We involuntarily wince when we see pain being inflicted on another; our heart speeds up long before our consciousness tells us that we are afraid. These physical responses, in turn, influence our process of cognitive appraisal. Brain influences body, which in turn influences brain.

While conventional wisdom has it that since feelings are registered in the brain, they are nothing more than a product of the brain, Damasio notes that the brain cannot be detached from the rest of the body with which in interacts, not only though the neural pathways distributed throughout the body, but also through chemical signals in the blood: "The circularity of these operations is remarkable. On the face of it, mind and brain influence the body proper just as much as the body proper can influence the brain and the mind."[22] Thus, emotions happen in the body as much as they do in the brain. Consider, for example, the old sayings to "trust one's gut feelings" or "I was simply gutted by his criticism" hold a great deal of truth. The enteric nervous system is huge, comprised of 100 to 600 million neurons. This neural structure is independent of the brain, yet intimately connected to the brain through the vagus nerve. Some have referred to it as our "second brain."[23] Recent experiments have shown that this system plays a major role in regulating our feelings and mood, in part, by producing 95 percent of the body's serotonin. Our moods are also influenced by the billions of bacteria that reside in our gut, though we do not yet know the exact mechanism for this. Yet it seems increasingly clear that our emotions are a product of much more than our brain.

Can a person live without feelings and emotions? Saint Jerome assumed to do so one must either be like a rock or like God.[24] What would it mean for a human being to lack the feeling part of an emotion? Regrettably, some do. They recognize stimuli in certain situations and calculate what would be the normal response, yet they do not experience the gut-wrenching or heart-rending feelings that set up the feedback loop that influences those responses. In *Confessions of a Sociopath*, M. E. Thomas writes "I am generally free of entangling and irrational emotions. I am strategic and canny, intelligent and confident."[25] According to psychologist Simon Baron-Cohen, the defining feature of a sociopath is just such an incomplete circuit. A computer may use facial recognition to identify an emotion and can also exhibit an emotional or helpful response. However, hidden between these two movements is the third—we must respond out of an inner feeling. While we can turn the existence of a feeling into information, we cannot digitize the feeling itself. Without that, an AI's response would ultimately feel manipulative or empty, as do the responses of an intelligent sociopath. Perhaps not our greatest emissary.

Given this limitation, an intelligent race might well send an AI out into space to do precisely, and only, what computers do well—to observe and gather information. Perhaps some of the unexplainable objects we have encountered are doing precisely that—observing us and hightailing it out of here when they note that they themselves have been observed. A second reason an AI emissary might simply and stealthily observe is posed by ethicist Brian Green. Green suggests a possible reason for the Fermi paradox, in the face of evil and sin, may be that all intelligences "are hiding, trying to escape detection, lest they be targeted and destroyed. Like a sort of nightmarish galactic 'Central Park' at night, the only sounds ever heard are occasional pleas for help and death-screams, or perhaps a bumbling ignoramus, new to the neighborhood and thinking only the best of people, who just walks through whistling happily, not knowing what lies ahead."[26] The rapid development of autonomous weapons here on earth demonstrates the dark side of AI, and of at least one species that wields it.

AN EVOLUTIONARY BOTTLENECK?[27]

So, if there are AIs out there, they may be programmed specifically not to make contact with any other intelligences they encounter. This provides one answer to Fermi's famous question, "Where is everybody?" However, there is another, equally dark, possibility—that they simply are not out there.

Let's take a closer look at Drake's equation for the possibility of intelligent life:

$$N = R * f_p * n_e * f_l * f_i * f_c * L$$

where:

N = the number of civilizations in the observable galaxy

R = the average rate of star formation in our galaxy

f_p = the fraction of those stars that have planets

n_e = the average number of planets that can potentially support life per star that has planets

f_l = the fraction of planets that actually develop life

f_i = the fraction of planets that develop intelligent life

f_c = the fraction of civilizations that develop a technology that releases detectable signs

L = the length of time such civilizations release detectable signals

The one variable whose value we cannot know is L, the length of time a civilization might release detectable signals or machines into space. How long does a technological civilization last? Futurist Nick Bostrom has noted that the great silence we observe means technologically capable intelligent life is probably rare, if not unique. He suggests the existence of a "Great Filter," an evolutionary step which is exceedingly difficult for life to transcend. This filter might occur early in the evolutionary process. Perhaps life is rare. Bostrom hopes this is indeed the case. But if life *has* evolved on multiple planets, then Bostrom's "Great Filter" lies ahead of us. The most likely candidate is a technological bottleneck, a point at which technology outstrips a culture's ability to control it and use it wisely.

Key properties of evolution make such a bottleneck likely. The natural environment of an organism is the proving ground that selects for specific traits. Individuals that better fit their environment have a better chance for survival and reproduction. At first, technologies are remarkable at improving a species' ability to cope with the dangers and difficulties a given environment might afford. The harnessing of fire, the development of tools, the design of better clothing and shelter, the development of agriculture—each allowed our species to flourish in number and expand our range. It is easy to think technology brings only advantages for the species that wields it. Microsoft founder Bill Gates expressed this view to world leaders at Davos in 2008: "Advances in technology, have brought us to a high point in human welfare. We're really just at the beginning of this technology-driven revolution in what people can do for one another. In the coming decades, we'll have astonishing new abilities: better software, better diagnosis for illness, better cures, better education, better opportunities and more brilliant minds coming up with ideas that solve tough problems."[28]

But technology does not always alter the environment for the better. Jared Diamond, in his monumental study *Collapse*, chronicles several human societies that altered their environment beyond its capacity to rejuvenate. Process such deforestation, soil erosion, loss of water, overhunting, and population growth led these early societies to wars, famines, and loss of social and political complexity. The collapse of Diamond's societies was local. A true technological bottleneck would be global. Unfortunately, we have come to that point. Diamond writes: "Globalization makes it impossible for modern societies to collapse in isolation, as did Easter Island and the Greenland Norse in the past. Any society in turmoil today, no matter how remote . . . can cause trouble for prosperous societies on other continents and is also subject to their influence (whether helpful or destabilizing). For the first time in history, we face the risk of a global decline.[29]

Diamond speculates as to what might have gone through the mind of the man who cut down the last tree on Easter Island. Why do we cut down the last tree, hunt a species to extinction, or build machines that might destroy ourselves? The problem is that evolution functions on an individual level. In other words, it selects individuals who show a reproductive advantage, thus inevitably introducing competition, both between species and within a species. Theologian Reinhold Niebuhr warns us of the pitfalls lurking in interspecies competition. He notes Albrecht Ritschl's observation that humans experience themselves as both a part of nature and as dominating nature. The domination of nature is, of course, the first goal of technology. However, Niebuhr notes that we find ourselves unable to avoid overreaching and disturbing the harmony of nature. Niebuhr ascribes this to our limited knowledge regarding the effects of our actions and our tendency to overestimate our capabilities. He notes that, "man is a finite spirit, lacking identity with the whole, but yet a spirit capable in some sense of envisaging the whole, so that he easily commits the error of imagining himself the whole which he envisages."[30] This separates us from a nature on which we depend. We dream of living in a self-constructed world, a vision which underlies cyberspace and many of our transhuman and posthuman dreams.

Evolution also rewards those who reproduce. This leads to intraspecies competition—we turn our technologies against one another. Foreign policy analyst Martin van Creveld notes, "War and technology have always been linked very closely. Indeed, without technology, there would probably have been no war. After all, without technology, if only in the form of sticks and stones, man's ability to kill his own kind is extremely limited."[31] This link goes both ways. While technology has shaped war, so has war provided the impetus for the development of numerous technologies, not limited to armaments. Transportation systems, new materials, radar, satellite and missile technology, computer technology, the Internet—all have been advanced primarily with military uses in mind.

While we can imagine the evolution of a species devoted to intraspecies harmony, the laws of evolution make this unlikely. So do the laws of social cohesion. Niebuhr notes that, while we as individuals are susceptible to the overreach of pride in our drive for dominance and security, we are far more susceptible as a group. Thus, we engage in competition in two venues, between ourselves and the rest of nature and among ourselves, and on two levels, as individuals and as groups. We develop and deploy technologies to aid us in the battle and our battles, to a large extent, determine the shape of those technologies. As these technologies become more and more powerful and wider in scope, they become more and more dangerous. Nuclear weapons were the first "doomsday" technology humans invented, able to destroy our entire species. Is a superintelligent AGI another such doomsday technology? R. Martin Chavez, senior director and former global head of securities for Goldman Sachs thinks we should fear more the AI we already have. Considering the massive data centers of firms such as Facebook and Google, Chavez notes that AI is already "systematically exploiting weaknesses in human psychology: our tribalism, our gullibility, or wanting to be told what to believe, our wanting to be liked, our wanting to be told that we're right. And they're exploiting it to the end of maximizing advertising revenue."[32] Should misinformation or the tribal bubbles that form on social media bring down our functioning democracies, our heavily armed world will become a far more dangerous place.

We do not need a malevolent AGI for AI to bring us to our own technological bottleneck. The single-mindedness of an AI that does exactly what we programmed it to do could bring a similarly tragic result. Computers lead us to treat people as data, overwhelming us with too much information, separating us by catering to our preferences, and providing an all too tempting diversion. [AQ4]The late diplomat Henry Kissinger warned that the emphasis on speed engendered by computers limits our time for reflection. After nodding to the possibilities of "extraordinary benefits" in medical science, clean-energy provision, and other environmental issues, Kissinger warned of AI's potential for unintended consequences, especially those that may arise from the inability of an AI to contextualize. Kissinger asked, "Can we, at an early stage, detect and correct an AI program that is acting outside our framework of expectation? Or will AI, left to its own devices, inevitably develop slight deviations that could, over time, cascade into catastrophic departures?"[33] The latter is, perhaps, what should worry us most. As Sir Nigel Shadbolt, professor of computer science at Oxford, recently noted, "The danger is clearly not that robots will decide to put us away and have a robot revolution . . . If there [are] killer robots, it will be because we've been stupid enough to give it the instructions or software for it to do that without having a human in the loop deciding."[34] Recall the game-playing program AlphaGo, programmed only to win. Just as Go can be reduced to "winning," so too in other areas, the single-minded efficiency of AI might go for the "win" at too great a cost.

There is a classic story from the early days of machine learning of a program devised by the US Department of Defense to locate hidden tanks. The machine quickly became quite proficient at identifying tanks in its initial set of pictures, but when given a new set, it totally failed. It turned out that the photos in the training set harboring hidden tanks were all taken on cloudy days. The machine had learned nothing about tanks but knew how to distinguish a cloudy from a sunny day. Whether true or apocryphal, this story illustrates how machine learning can reach conclusions we do not understand. This makes AI as inscrutable as the spells in a sorcerer's magic book. We will know it works, but we won't know how.

So, what are we to do? First, we must avoid the category error of personifying AI. A computer has no consciousness, no emotions, no will of its own and, despite all predictions, these are not "right around the corner." The idea, often espoused by computer futurists such as Kurzweil or Musk, that just a little more complexity will suddenly cause consciousness to emerge is, in my opinion, risible. We are still a long way from knowing what consciousness is or where it comes from. In the meantime, we would do well to think of AI, then, as a means rather than an end. Like all technologies, AI is a tool. As such, it is not just a means of power over nature, but primarily a means of power by some persons over other persons. As Weiner so presciently pointed out, the source of our undoing has always been and will remain the human use of other human beings.

Jaron Lanier calls our dreams of a superintelligent AI and a posthuman future a "death-defying religion" but not a spiritual path. We need to remember that "computer scientists are human, and are as terrified by the human condition as anyone else. We, the technical elite, seek some way of thinking that gives us an answer to death . . . What we are seeing is a new religion, expressed through an engineering culture."[35] But that engineering is most likely to fail and could be our total undoing, unless it is coupled with humility, unless we pause to ask if we are, metaphorically, about to chop down the last tree on Easter Island.

In theological terms, AI can be viewed as a millenarian project, one suited to the twenty-first century. Groups predicting the immanent second coming of Christ, or of some other form of salvation, such as found in cargo cults, have flourished in times of societal stress. Millenarian thinking has typically risen among groups challenged by demographic change or climatic stressors. It has found purchase in agrarian cultures and in cosmopolitan ones, such as Renaissance Munster. It promises a solution to current problems that is delivered from outside, rather than from the society itself. AI promises the same. Dreams of a salvation that is always coming but not yet here, a salvation at the hands of something or someone other. These dreams remove the onus for solving society's ills ourselves, and soothe those who feel powerless in the face of what look like unsurmountable obstacles. Help is just around the corner. When the corner is reached and either God or AI have failed to deliver, millenarians simply recalibrate their timeline, or suggest that we misunderstood the extent of their promises.

This millennarian orientation has allowed both researchers and the general public to project a variety of eschatological hopes, including the hope that AI will help us "explore new world, seek out new life and new civilization, [and] boldly go where no man has gone before." Apocalyptic scenarios involve the creation of an AI that outstrips human intelligence and control, one that could turn on its creators or be used as a weapon by those creators. However, we do not need conscious or superintelligent AIs to produce programs that can be quite damaging to the fabric of human society, a fact obvious in today's world of ransomware attacks, autonomous weapon development, social media dispersal of falsehoods, and machine learning programs with inherent biases. Such mundane uses of computers, coupled with our natural tendencies toward competition and domination may keep us, and may have kept all other technologically developed societies, from anything more than the most tenuous steps into space.

NOTES

1. Sierra, "Are We Alone in the Universe?"
2. See NASA Headquarters Library, "Long-Term Challenges to Human Space Exploration"; Body, "From Shrinking Spines to Space Fungus."
3 Borunda, "Warp Drives and Negative Energy."
4. Davies, *The Eerie Silence*, 159.
5. Girimonte and Izzo, "Artificial Intelligence for Space Applications," 236.
6. Quoted in Dreyfus, "Making a Mind," 19.
7. Quoted in Darrach, "Meet Shakey," 58D.
8. Kurzweil, "The Law of Accelerating Returns," 507.
9. Pinker, "Tech Luminaries Address Singularity."
10. Deutsch, "Creative Blocks."
11. Ibid.
12. Searle, "What Your Computer Can't Know," 54.
13. Dennett, "What Can We Do?," 48.
14. Kaku, *The Future of the Mind*, 253.
15. Chalmers first used the phrase "hard problem" in a talk at "Toward a Scientific Basis of Consciousness," Tucson, 1994.
16. Kaku, *The Future of the Mind*, 43.
17. Damasio, *Self Comes to Mind*, 108–29.

18. Kagan, *What Are Emotions?*, 4.
19. James, "What Is an Emotion?," 189–90.
20. Kagan, *What Are Emotions?*, 23.
21. Ibid., 44.
22. Damasio, *The Strange Order of Things*, 117.
23. Ibid., 134.
24. Rasmussen, "Like a Rock or like God?," 147.
25.Thomas, "Confessions of a Sociopath," 52–61.
26. Green, "Astrobiology, Theology, and Ethics," 349.
27. Much of this section has been published previously in "Where is Everybody? Fermi's Paradox, Evolution, and Sin."
28. Gates, "A New Approach to Capitalism."
29. Diamond, *Collapse*, 23.
30. Niebuhr, *The Nature and Destiny of Man*, 181.
31. van Creveld, "War and Technology."
32. Basak, "Wall Street Visionaries Provide Chilling Views on Next Big Risk."
33. Kissinger, "How the Enlightenment Ends."
34. Devlin, "Killer Robots."
35. Lanier, "The First Church of Robotics."

BIBLIOGRAPHY

Basak, S. "Wall Street Visionaries Provide Chilling Views on Next Big Risk." Bloomberg News (2021). Accessed March 31, 2023. https://www.bloomberg.com/news/articles/2021-01-12/what-do-wall-street-leaders-think-is-the-next-big-risk?

Body, J. "From Shrinking Spines to Space Fungus: The Top Five Dangers of Space Travel." *Science* 2 (2016). Accessed March 31, 2023. https://www.science.org/content/article/shrinking-spines-space-fungus-top-five-dangers-space-travel

Borunda, M. "Warp Drives and Negative Energy: Physicists Give Chances of Faster-Than-Light Space Travel a Boost." *SciTechDaily* (2021). Accessed March 31, 2023. https://scitechdaily.com/warp-drives-and-negative-energy-physicists-give-chances-of-faster-than-light-space-travel-a-boost/

Damasio, A. *Self Comes to Mind: Constructing the Conscious Brain*. New York: Pantheon, 2010.

———. *The Strange Order of Things: Life, Feeling, and the Making of Cultures*. New York: Pantheon, 2018.

Darrach, B. "Meet Shaky, the First Electronic Person: The Fascinating and Fearsome Reality of a Machine with a Mind of Its Own." *Life* (1970): 58B.

Davies, P. *The Eerie Silence: Reviewing Our Search for Alien Intelligence*. New York: Houghton Mifflin Harcourt, 2011.

Dennett, D. "What Can We Do?" In *Possible Minds: Twenty-Five Ways of Looking at AI,* edited by J. Brockman, 41–53. New York: Penguin Books, 2020.

Deutsch, D. "Creative Blocks." *Aeon* (2012). Accessed March 31, 2023. https://aeon.co/essays/how-close-are-we-to-creating-artificial-intelligence.

Devlin, H. "Killer Robots Will Only Exist if We Are Stupid Enough to Let Them." *The Guardian* (2018). Accessed March 31, 2023. https://www.theguardian.com/technology/2018/jun/11/killer-robots-will-only-exist-if-we-are-stupid-enough-to-let-them/.

Diamond, J. *Collapse: How Societies Choose to Fail or Succeed.* New York: Viking, 2005.

Dreyfus, H., and S. Dreyfus, "Making a Mind Versus Modeling the Brain: Artificial Intelligence Back at a Branchpoint." *Daedalus* 117, no. 1 (1988): 15–43.

Gates, B. "A New Approach to Capitalism in the 21st Century." World Economic Forum, Davos, Switzerland (2008). Accessed March 31, 2023. https://www.networkworld.com/article/2282669/microsoft-s-bill-gates---a-new-approach-to-capitalism-in-the-21st-century-.html

Girimonte, D., and D. Izzo. "Artificial Intelligence for Space Applications." In *Intelligent Computing Everywhere*, edited by A. Schuster, 235–54. London: Springer, 2007.

Green, B. "Astrobiology, Theology, and Ethics." In *Anticipating God's New Creation: Essays in Honor of Ted Peters*, edited by C. R. Jacobsen and A. W. Pryor, 339–50. Minneapolis: Lutheran University Press, 2015.

Herzfeld, N. *The Artifice of Intelligence: Divine and Human Relationships in a Robotic Age*. Minneapolis: Fortress, 2023.

James, W. "What Is an Emotion?" *Mind* 9, no. 34 (1884): 188–205.

Kagan, J. *What Are Emotions?* New Haven: Yale, 2007.

Kaku, M. *The Future of the Mind: The Scientific Quest of Understand, Enhance, and Empower the Mind.* New York: Doubleday, 2014.

Kissinger, H. "How the Enlightenment Ends." *The Atlantic*, June 2018. Accessed March 31, 2023. https://www.theatlantic.com/magazine/archive/2018/06/henry-kissinger-ai-could-mean-the-end-of-human-history/559124/.

Kurzweil, R. "The Law of Accelerating Returns." *Nature Physics* 4, no. 507 (2001): https://doi.org/10.1038/nphys1010.

Lanier, J. "The First Church of Robotics." *New York Times* (2010). Accessed March 31, 2023. http://www.nytimes.com/2010/08/09/opinion/09lanier.html?pagewanted=1/

NASA Headquarters Library. "Long-Term Challenges to Human Space Exploration," 2020. Accessed March 31, 2023. https://www.nasa.gov/centers/hq/library/find/bibliographies/Long-Term_Challenges_to_Human_Space_Exploration.

Niebuhr, R. *The Nature and Destiny of Man: A Christian Interpretation, Vol I: Human Nature*. New York: Scribner's, 1941.

Pinker, S. "Tech Luminaries Address Singularity." *IEEE Spectrum* (2008). Accessed March 31, 2023. https://spectrum.ieee.org/tech-luminaries-address-singularity.

Rasmussen, M. "Like a Rock or like God? The Concept of Apatheia in the Monastic Theology of Evagrius of Pontus." *Studia Theologica* 59, no. 2 (2005): 147–62. https://doi.org/10.1080/00393380500322285.

Searle, J. R. "What Your Computer Can't Know." *New York Review of Books* (2014). Accessed March 31, 2023. https://www.nybooks.com/articles/2014/10/09/what-your-computer-cant-know/.

Sierra, L. "Are we alone in the universe? Revisiting the Drake equation." *NASA Exoplanet Exploration*, May 19, 2016. Accessed March 31, 2023. https://exoplanets.nasa.gov/news/1350/are-we-alone-in-the-universe-revisiting-the-drake-equation/.

Thomas, M. E. "Confessions of a Sociopath." *Psychology Today* 46, no. 3 (2013): 52–61.

van Creveld, M. "War and Technology." Foreign Policy Institute (2007). Accessed March 31, 2023. https://www.fpri.org/article/2007/10/war-technology-2/

Chapter 15

A Darker Forest?

The Fermi Paradox and Extraterrestrial Spiritual Life (ETS)

Roland Faber

A CONJECTURE

What follows is a conjecture. It is as good as any, given that we don't have the facts confirming them either way. Yet this conjecture is based on multiple long-standing and tested sources and meant to create the space for considerations on an alternative to certain assumptions that underly presuppositions opposing it. What is this conjecture and what are the opposing presuppositions? What are the sources supporting it? And why may the conjecture be relevant? Those are the questions that will structure this chapter.

What is this conjecture about? It is about the ability to understand the reasons why we should consider that we and this Earth are expressions of a universe, even a multiverse,[1] in permanent becoming, infinite and multifaceted beyond imagination, but restrained in its development, allowing beings like us to exist with feelings and thought, consciousness and conscience, a sense of beauty and of the mysterious nature of existence, yet also with destructive drives and self-negating impulses, grave predicaments and uncontrolled powers of oppression and annihilation—evolving, as it were, into new and unchartered, only envisioned and poetically sung about lands of a cosmic community of existence.[2] Without such a transformation, which can be preliminarily called "spiritual" in nature, so the conjecture, humanity will not survive and, in the wider cosmic context, may not be the only planetary civilization to disappear, but contribute to a cosmic graveyard of failed evolutions, if they ever happened—a failure of evolution like that which happened to so many species that cosmic forces and humanity have erased on Earth.[3]

In fact, maybe humanity is only meant to be a precursor of such a transformation or for the rise of other evolutionary branches on Earth[4] or somewhere else in the expanse of the universe that *will* overcome the threshold of potential self-destruction generated by the bifurcation of their existence into nature and culture,[5] by engaging chaotic possibilities and malleable orders of transition in order to shift from old forms of life, defined by exploitation and decay, to new ones that will cross a boundary beyond which another, different, and much more exciting evolution awaits the adventure that is the universe.[6]

Is there any evidence for this? Perhaps, a sign of that which may be awaiting the long journey of universes in reaching meaning beyond the effects of destructive connotations can be gleaned from the flickering through human history of undeniable modes of integration, transformation and activation of such ultimate motives and forces that humanity has practiced, although only with glimpses of hope, in its own existence: forces of composition and harmonics of love, mutuality and goodness—desiring an existence *within* becoming and perishing, but *beyond* their toxic noise, remaining its downfall. What shows in these motives and forces—really feelings of the heart and imaginations of the mind—is a *spirit of life* that, indeed, tends toward intensity and harmony, adventure and zest, love and compassion, wisdom and freedom from oppression and the enslavement by forces of domination.[7] It may be this spirit that holds the very key for the activation of these abilities that readily are called *spiritual*,[8] and that throughout the history of human existence manifested in the form of visions of prophets and ideas of philosophers, imaginations of religions and social experiments, dreaming of another world—a world in which to feel at home has become a reality and that, beyond the adverse contingencies of natural and cultural happenstances, is ready to venture out into the vastness of empty space in search for companionship, perhaps following the traces of other such beings in their exploration of meaning and curious of the wisdom of their new evolutionary potential, instead of being driven into a dark pit by mere survival and with the intention of conquistadores for the exploitation of its assumed riches.[9]

The conjecture I am pondering is as much about the future of humanity and the Earth it inhabits as it is about the nature of the universe allowing for such inhabitation, and with the assumption that we and the Earth that birthed us are not alone in this community of existence. This assumption is not so much about the question *whether* such other beings with life and intelligence and of such a spirit of life exist or *how* it would be and feel when we, in fact, encounter their reality—we will probably find out eventually one way or the other if we allow for certain conditions to be realized[10]—as it is about the *conditions* under which we can expect such a contact with alien life-forms with the ability to communicate with us (or us with them) to happen in the first place.[11]

Not that these conditions decide their facticity, but they drive the preconceptions and, perhaps, preparations for such an encounter; they have had their impact on the visions for an eventual contact, be they fed by religious, philosophical, scientific or literary imaginations and computations; they feed the financial and economic resources, provided for ideological and technological endeavors in preparing for its happening, either conforming their existence or their intentions; and they are the stuff that our dreams, in projecting our own future and its individual and social potentials, have used, capturing our attention for centuries and maybe as far back as time allowed humanity to arrive on this planet.

In short, the *conditions* for assuming extraterrestrial life (ETL) and extraterrestrial intelligence (ETI) that I mean, here, are these three: First, that we expect alien life-forms with the ability of intellect and communication across the cosmic expanse to be projections of our own current state of affairs on Earth: sentient and with intellect, but in a state of awry; or even worse: expressions of their extremes: being angels or devils—yet the more rational in-between versions also not being necessarily more than a form of wolves in sheep's clothes.[12] Second, that we expect our own intellectual abilities for researching the conditions of life and intellect to follow the same ideological restrictions that have accompanied scientific knowledge since the antireligious sentiment of the Enlightenment and the beginning of empirical research with its guiding and overarching materialism, physicalism, reductionisms, and foundationalism: reducing characteristics of living beings to biochemical processes and relegating consciousness and conscience to matters of determinism by, or emanentism from, closed causal loops.[13] Third, that we expect the evolution of life and intelligence always only to follow the rules of Darwinian mechanisms of the production of self-replicating organisms with a drive for survival in hostile environments of limited resources and by elimination of the weak by selection.[14]

Not that these assumptions are not true or worth considering in relation to alien life-forms of intelligence that humanity wants to find and with which it desires to communicate. Not that these perspectives are not valuable in their own right in understanding ourselves and life on Earth, or meaningless by projecting them onto the grand scale by the measure of principles such as mediocrity or universal laws of nature and emergence.[15] Not that we should not be careful in avoiding any other ideological basis to be *as* problematic, or even more so, in our determination of the parameters of life and intelligence in general.[16] All that, yes, but also consider a self-critical review of their potential limitations, given the *other* side of life and intelligence, and the long-standing reflection on them by philosophical endeavors to match our experience and thought with the fullness of their expressions: aesthetic experience, the indeterminateness of rationality, the sense of mystery, the poetic side of encounters with nature and the search for meaning, the joy of living, the suffering of their lack, and the polyvalency of all conceptual determinations.[17]

The use of *such* sources may create a countertheme. They can be found in the history of human thought and art, literary expressions, and cultural diversities, but also in religious visions and philosophical experiments with the freedom of thought and the organization of societies. In other words: the *spiritual* adventure of humanity may provide a different background for considering life and intelligence in the universe; and it may be a way for supplementing and changing the formulation of the very conditions under which we cannot only reach out across the dark blanket in order to find life and intelligence, but may also reflect how we understand *ourselves* as life and intelligence venturing beyond our home.[18] Hence, I have chosen to address these conditions with the help of selected religious, philosophical, and literary sources by following up on a specifically extreme proposition employing those named conditions: that we live in a "dark forest," dangerous and endangered, demanding silence and hiding from the grand scheme of the universal mechanisms of destruction that allegedly populates the universe with viruses and alligators, hyenas and sharks, striving for survival at the expense of all encountered life-forms—and that we are unfit for the full recognition of this bleak reality.[19]

THE PROBLEM

What if we are not alone?[20] More symbolically than numerically, but still as a matter of probability, the Drake equation, devised by the astronomer Francis Drake (1930–2022) in 1961, has enkindled our excitement for the rather large potential of the flourishing of life and intelligent civilizations in the universe.[21] Assuming certain convergencies by which any life and mind may develop across star systems and galaxies,[22] we may expect several millions of such civilizations in our galaxy alone![23] The more pessimistic recalculation of this probabilistic formula, in recent times, has stiffened this enthusiasm to some extent[24]—many of its variables are still not seriously enough constrained in their range. But to assume that it tends toward zero or miraculously points to Earth as the only habitable planet in a habitable zone of planets around a sun favorable to the emergence of life, is as pessimistic as it gets and, in this sense, perhaps itself improbable.[25] In any case, the Drake equation has empowered programs like SETI and is still in use for categorizing and analyzing contributing factors, matching them with ever new insights regarding their factual numbers provided by diverse scientific inputs from physics and biology to astronomy and evolutionary theory and the like.[26]

The analysis of its diverse factors is beyond the scope of the current considerations. I will assume the search for alien life and intelligence to be, in any case, built on some assumptions that underly this formula: that there are planets (or other habitats) in which life and intelligence *can* eventually develop[27]—otherwise any search would be meaningless, except in demonstrating the futility of this enterprise for which we may not have the temporal means to even probe all heavenly bodies of *our* own galaxy alone in a time frame that would be more limited than the expansion of our sun, swallowing the Earth in the next five billion years.[28] And while other factors or fractions of factors may fine-tune the formula further, the last two of its original six factors merit further consideration in our context: In order to be in *contact* with such an assumed intelligent civilization, the Equation assumes that such a civilization must develop the technological ability of communicating cross spacetime (Fc) and exist for such a length of time that allows a civilization to use these technologies with meaningful results (L).[29]

Indeed, regarding the first factor (Fc), the length of time a technological civilization is *able* to send signals into space may be limited because of internal problems in the use of the advanced technology of sending and receiving signals, or, in the case of devising AIs and nanotechnology for contact, the ability of creating and maintaining such tools.[30] Such a civilization may also have been hampered by emergencies in the wake of developing such technologies or by natural disasters of the magnitude of extinction-level events. And such a civilization may, especially by its technological abilities, concomitant to the ability to communicate across the universe, have fallen prey to self-destructive tendencies inherent in their specific civilizational use of technologies on the grand scale of their habitat.[31]

We can add that such a civilization may also be limited by the length it is *willing* to send or receive signals or produce tools for communication and exploration.[32] It may have come to the insight that, besides the theoretical possibility, searching and finding companions in the nearer or farther cosmic environment is futile, assuming the limitations for technologically tricking the spacetime expanse into any FTL-communication or -travel beyond the limitations of the Einstein universe.[33] Or such a civilization may, indeed, have come to the conclusion that it would be too dangerous to use its technology for such matters of interstellar communication or contact because of the fear of occupation or destruction of its integrity by alien encounters or even their very existence.[34]

Regarding the second factor (L), it is a fundamentally open question how long such a civilization with technological abilities for interstellar contact may *actually exist* in *such* a technological state.[35] Natural or artificial limitations may restrict this ability of maintaining its technological state, if not its existence as such, given that the universe is a dangerous place in which internal catastrophes, such as the economic exhaustion of resources or ecological destruction, or cosmic cataclysms, such as supernovae, freewheeling black holes, comets or other cosmic debris, may destroy its habitat, despite its technological abilities.[36] Furthermore, given the temporal expansion of space, we cannot even be sure whether we may be the first or last of such civilizations[37]—if others ever existed—to be able to communicate across the expanse in a mutually related time frame, either allowing for an encounter or even only for the recognition of remnants of fallen and extinguished civilizations.[38]

To all of these uncertainties (and that of the other factors in the Drake equation), we must also add the disappointing fact that we don't have any conclusive evidence *yet* of any such alien civilizations to be extant: No nonambiguous finding has yet confirmed ETL or ETI. While signs for ETL and ETI are researched—astrobiology[39] and SETI stand for this endeavor—no factual conclusions have been reached. "Unexplained phenomena" on Earth are similarly, if not more so, ambivalent and inconclusive.[40] We are, in this current situation, bound by "Fact One,"[41] that is, *that* fact that we have no clarity on the very existence of ETL and ETI, despite the ongoing conversation that lives from a long history of human imaginations of such other life and intelligence.[42] Despite the cosmological, philosophical and religious weight of arguments and assumptions, there is no evidence that supports anything tangible beyond Fact One.

All of this is to say that despite the nonzero probability of the Drake equation, the most important fact is stated with the simple question: If they exist, where are they? This interruption of fantasies and imaginations, as devised by the physicist Enrico Fermi (1901–1954), originating during a lunch conversation at Los Alamos with Emil Konopinski, Edward Teller, and Herbert York in 1950, has become the most fantastic folio for speculations on our question ever since.[43] I cannot go into the details of the occupation with sophisticated arguments answering this Fermi paradox, here,[44] but I will at least list a few obvious and more exquisite rebuttals: Sure, the simplest explanation of this paradox between probability and fictionality is to assume that neither ETL nor ETI have ever evolved beyond the sole exemplar of Earth.[45] This would make SETI irrelevant, but not necessarily astrobiology, as we need not equate the development of biological mechanisms of life in the cosmos with that of something equal to human mind and subjectivity.[46] Hence, another quite simple answer assumes that ETL may exist, but it has never given rise to ETI. So, if we find creditable evidence of ETL, nothing regarding ETI is decided or implied thereby.[47] A more informed answer, with the same intention of heightening the improbabilities of encountering ETL and ETI, would concede that both are not only possible, but also connected with one another to the extent that, if one arises, we have good reasons to assume that the other will eventually also emerge, *but* that the factors Fc and L (or similar factors) in the Drake equation demonstrate insurmountable improbabilities for any contact.[48]

Another host of directions for answering or confirming the implications of the Fermi paradox, namely of the improbability of ETI to *actually* have arisen beyond Earth, surround the question of technology: either that we have not yet the technological means for even registering extraterrestrial civilizations (with or without matching their technological abilities), as even their past existence may escape our recognition and artifacts of such civilizations may forever remain in the dark since no technology may be able to confirm ambivalent signs in one or the other direction if we don't, at the same time, develop the technological means to follow up on them with actual *visits*.[49] Even if ETI exist and are liberated from the constraints we must recognize for the factors Fc and L in the Drake equation in order to actually communicate, they may decide to do so only by AIs and nanotechnological probes that, for reasons of cautiousness, or for the conviction of a non-interventionalist stance, hide rather than seek communication.[50] Their means of investigation may be invisible to us—anti-Bracewell probes, as it were.[51]

Another direction of thought may employ the vastness of the universe, even a multiverse, in which the probability of developing ETI may be large, even infinite, but still distributing them in such a way that no encounter can possibly ever be achieved: our galaxy may have birthed only *one* such civilization while every other of the trillion galaxies may have done the same; or an infinite multiverse may have allowed infinitely many universes to develop such a rare civilization only once, but in an infinite array of possibilities still infinitely many times.[52] In such an infinite universe or multiverse with infinitely many civilizations, we may still be a rare Earth.[53]

There is, however, a particular answer to the Fermi paradox that assumes a wild proliferation of ETL and ETI in any given galaxy of the visible universe—not to speak of the whole universe beyond observation and measurement—but under the constraint that no civilization that is intelligent enough to communicate and even technologically so advanced as to follow up on such a conscious intent (METI)[54] would dare to do so without being able to defend itself with utmost power against the potential adversarial effects of such a contact.[55] In other words, if such a civilization exists, it would not communicate at all out of the fear that if it does so it will become subject to extinction by the contamination of another civilization that may be presumed to be even more powerful and driven by the same motivations and insights that haunt its own struggles. Such a civilization would hide.[56] This is a particularly hostile and dark but fascinating answer to Fermi's Paradox, to which I will now turn.

THE DARK FOREST HYPOTHESIS

In the image of the "dark forest," in one of the most pessimistic series of sci-fi novels—*The Three Body Problem* (2008), *The Dark Forest* (2008) and *Death's End* (2010)—on the future of life in the universe, Cixin Liu (b. 1963), in the center of this trilogy, has devised the following scenario:

> The universe is a dark forest. Every civilization is an armed hunter stalking through the trees like a ghost, gently pushing aside branches that block the path and trying to tread without sound. Even breathing is done with care. The hunter has to be careful, because everywhere in the forest are stealthy hunters like him. If he finds other life—another hunter, an angel or a demon, a delicate infant or a tottering old man, a fairy or a demigod—there's only one thing he can do: open fire and eliminate them. In this forest, hell is other people. An eternal threat that any life that exposes its own existence will be swiftly wiped out. This is the picture of cosmic civilization. It's the explanation for the Fermi paradox.[57]

Again, without being able to venture into the fascinating details of the whole background for this assumption in Cixin's trilogy, which defines the appearance of life in the universe as such a catastrophic event[58] that it continuously and terminally dooms all existence to extinction, at least the following assumptions can be deduced from its Dark Forest hypothesis. It claims the ability of technological civilizations to send and receive signals to and from other civilizations; it claims the ability of an alien civilization of bridging the spacetime expansions, following up on such signals; it claims that any such knowledge of a civilization of others' existence *would*, under these two conditions and abilities, lead to the destruction of the encountered civilization. Further, it claims that Fact One (that we don't have any evidence) is explained in two ways: by the nonexistence of other civilizations, as they have gone through a life cycle that ended in mutual destruction; and by the mutual hiding of potent civilizations that survived, or at least the ones that have come to the insight of the devastating nature of any potential mutual encounter from exposure in order to avoid destruction. Hence, in this explanation both the optimistic perspective on the Drake equation *and* the pessimistic nature of the Fermi paradox are equally affirmed.

It is worthwhile to dive deeper into the fundamental presuppositions of this hypothesis. What makes us think that Cixin's hypothesis is in any way reasonable in the sense that it lets us pause and ponder its viability; at least, that it gives us the chills, raises real fears, and paints a monster that is not beyond imagination? And what are the hidden assumptions that make this proposal not only fascinating, but powerful, perhaps gripping us as a catalysmic thought experiment that deals with real human fears?[59] Sure, if human history is our teacher, we will not have to think long and deep in order to uncover immediate reasons for its impactful relevance: humanity has demonstrated its warring character for as long as the collective memory and, in many cases, the immediate experience serves: occupation, subjection, exploitation, exclusion, denigration, genocide, holocaust, colonialization, xenophobia, racism, environmental irresponsibility, and whatever else such a list of the human lack of feelings and assumptions of the Good, humaneness, compassion and wisdom may provide—they all are a real background for a cosmic scenario in which interspatial encounters may make us expect more of the same.[60]

There are some improbabilities involved, although these may be a matter of further fact finding and the refinement of tools of observation: that we would be able to find signs of such destructive events;[61] that we should, even in the case of AIs, rampaging through the universe, find artofacts of such a destruction or of an occupation, for instance, of von Neumann machines;[62] and that it is virtually impossible to encounter such destructive civilizations if the Einstein Universe prevails and FTL is impossible, because they might be far and between, in other galaxies or universes, or not yet interested in the Earth (the rancher waits until cattle is ready to be slaughtered).[63] We may still be the first or, per happenstance, the last of such civilizations, on the verge of projecting destructiveness into the cosmos[64]—so it is worth reverting the perspective and, in fact, look at ourselves as *this* kind of civilization: Would we, like any other civilization to be encountered by us in our state of turbulence and in prolongation of our barbarism, act differently?[65]

However, the deeper presupposition that carries all others, I propose, is precisely based on our self-experience as barbaric humanity—and it assumes this to be the default by which the future is bound to confirm itself in case we ever became star-faring: *It is the understanding of the evolution of Life and Intelligence exclusively in terms of reductionist interpretations of the Darwinian mechanism of self-preservation and -expansion and the survival of the fittest.*[66] Social Darwinism is an explanation of the human history of destructiveness, here, *and* a prescriptive imperative for the future of humanity and all evolution of Life and Intelligence in the cosmos.[67] Perhaps, Cixin's political and ideological views are inscribed in this vision; but however strange those may be, we cannot escape the horrific implications and projections of his vision by reducing it to party politics and ideological idiosyncrasies of one author.[68] Rather, we are confronted with a much more fundamental question: Whether we believe—and it is not a matter of science or philosophy, but of *belief*—that the factual human experience of itself and the scientific exploitations of our mind will eternally limit the cosmos to conform to the conditions we determine for our past and present.[69] And what would we hold against such an assumption? Yet before I venture into answering this question, a few reflections on the viability of the assumption of Life and Intelligence in the cosmos beyond Earth are in order—including the Spirit in which we may counter the devastating continuation of the Dark Forest hypothesis.

EXO-LIFE, INTELLIGENCE, AND SPIRIT

The general assumption of Life beyond Earth, even the development of ETI, can be summarized by employing one or more of the following principles, being themselves assumptions of logic *and* aesthetics, that is, of human intelligence *and* spirit.[70]

Assumption 1: *Mediocrity*. This principle claims that Life and Mind are not exceptional throughout the cosmos—also known as the Copernican principle.[71] It is based on the logic that for the Earth and humanity to appear among a vast array of possible worlds and alien habitats in trillions of galaxies' habitable zones around uncounted stars, under similar conditions and physical, chemical, and biological processes, is itself much more reasonable than the opposite, namely, that we are an expression of an extraordinary and rare state of things.[72] Mediocrity is the by far more probable state of affairs to appear and in which to find oneself than a boundary-state: to be one of many in the middle of appearances may have special conditions that constrain mediocracy, but that would not upset its far greater probability. Mediocrity may inherently be more anthropomorphic and -centric than expected, as it assumes similarities to be the default option and repetition to grant the greatest probability for the proliferation of Life[73]—if not heightened by some version of the Anthropic principle.[74]

Assumption 2: *Pleroma*. This argument is ancient, can be found in many prescientific philosophies, and is fundamental to aesthetics, both religious or materialist, for instance the *Rig Veda*, Plotinus, and the Stoics.[75] Today it is known as the Principle of Plenitude.[76] It assumes that the world is in becoming, and that such a becoming needs, in order not to decay into states of uniformity and stagnation in mere repetition, some kind of creative drive toward novelty, forming a manifold that always overflows any given state of affairs and all limitations arising from the rehearsal of the same. Despite the Darwinian connotation of the pressure of evolutionary differentiation, being connected to scarcity of environmental resources for the self-sustenance of the population of a species, in order for natural selection to work its magic of diversification it needs a pleroma of populations to work on.[77] Anyway, pleroma is, in difference from mediocrity, less anthropocentric or anthropomorphic, as it allows the diversification of Life and Intelligence to be the driving force of their evolution.[78]

Assumption 3: *Large Numbers and Statistics*: This mathematical assumption views the development of trillions of galaxies with billions of stars, and a great amount of them with habitats such as planets, as similar and diverse enough so as to reduce the probability of the singularity of Life and Intelligence to only one place to virtually zero.[79] Uncertainty is not a hinderance, but an ally.[80] As a mathematical tool of computation of large numbers, it does not yet imply other restrictions that may reduce the numbers, but, in any case, will not reduce them toward naught, either.[81] It is independent from evolutionary considerations of similarity or dissimilarity of the rising of life-forms and modes of similarities and diversities of intelligence, but does not hinder considerations in either direction.[82] In this sense, this tool does not yet decide on the ability of communicating with, or even recognizing, alien life-forms because of a human limitation of senses, of scientific expansions of our ability of recognizing other physical states or of deciphering potential means of communications that may be beyond our imaginations, yet not physically impossible.

As a means of countering mediocrity, but of upholding plenitude, this principle may even allow for a rare Earth hypothesis—that we may be the only appearance of Life and Intelligence in the whole universe—if we assume an infinity of other universes with their own rules and variety such that even under the condition of their singularity for one universe infinitely others' may carry more exemplifications of Life and Intelligence. What is more, even if in any given universe Life and Intelligence appear only once, in the infinity of the multiverse, their existence would still be infinite, and in this sense even mediocrity would be sustained.

Assumption 4: *Convergence of Evolution.* Closer to home, this assumption is a constraint on the other three principles and antagonistically discussed. Paradigmatically, it is, among others, formulated for the development of evolution of Life and Mind on Earth by the evolutionary biologist Simon Conway Morris (b. 1951), and, to a certain extent, opposed by Steven J. Gould (1941–2002).[83] One of the central empirical arguments for the convergence of evolution arises from the evidence for the development of similar evolutionary solutions to environmental encounters of organisms, for instance, the development of senses, from vastly diverse starting points and functions of an already diversified life-form, and even repeatedly and independently in the course of the evolutionary time on Earth.[84] The theoretical argument adds the hypothesis that this is not just how evolution works on Earth but demonstrates a universal mechanism of the emergence of life-forms in a probability space that is restrained by its implied mutuality of the coordination of conditions of becoming, physically, chemically, biologically, in the cosmic environment—much like islands of stability in the appearance of proven or hypothetical superheavy elements on the Periodic Table.[85]

While Gould's counterargument from divergent evolution, that is, of the singularity of every evolutionary process, should it appear in a different cosmic habitat, even if it would reappear on Earth after the destruction of all life that inhabits it now, that will never repeat the same pathways and forms, is closer to the principle of plenitude than Conway Morris's paradigm of convergence, but rather removed from the principle of mediocrity; yet in light of the principle of large numbers, even such a singularity and divergence of evolutionary pathways would still be an exemplification of the evolution of *Life* and, perhaps, *Intelligence*, that is, appear in the same universe, and cannot be considered as being beyond any kind of communication on at least the physical, chemical, and biological basis of the category of evolution,[86] if not in appearance and the potential for mutual contact.[87] And if they are *anything* like Earthly evolution suggests, regarding their dimensions, not necessarily the forms, they would not only exhibit intelligence but some aesthetic and spiritual qualities as well.[88] This is neither an anthropomorphic nor an anthropocentric assumption.

Assumption 5: *Decentralization.* The avoidance of exceptionalism of Life and Mind, as well as Spirit, comes also from this line of argumentation: that their development in an infinite universe must be beyond concentric orderings of space and time.[89] It is also referred to as the Cosmological Principle.[90] This anti-Aristotelian way of conceptualizing the universe has been formulated on mathematical as well as philosophical and theological bases.[91] While many of the mythological cosmologies of bygone times are Earth-centered, with humanity as the qualitative center of the Earth, embedded in a concentric multiplicity of spheres between Earth, the Planets, the Heavens of fixed stars, and beyond that, a divine sphere, the decentralization of this layered universe—in some kind of pre-postmodern eradication of levels of grounding[92]—assumes the universe as an infinite circle that has, accordingly, in this thought experiment, neither circumference nor center; rather, in this universe, center and circumference coincide at *any* point of its appearance, which is itself, hence, an infinite multiplicity.[93] This flat ontology was devised by Nicolas of Cusa (d. 1464), virtually against all other ancient and contemporary Renaissance cosmological models,[94] and it is in its paradigmatic radicality not matched until the heliocentric cosmology gave way to a multiverse of infinite variations of universes in more recent times.[95]

One of its most interesting implications that I cannot go deeper into here, but that influences the whole discussion on alien life and intelligence, especially in contemporary considerations, but in its variability pervades the discussion as such, is this: *spiritual altitude* and *spatial distance* become transmutable into one another.[96] What I mean by these terms is this: While ancient religious and mythological worlds, populated by angels and demons or some kind of higher and lower beings, less material, more spiritual in nature, depended of layers of existence, that can, while under certain conditions, become present in this physical world, they really exist on a spiritual altitude that transcends this physical plane; however, in the decentralized universe, all levels are mutually not only inherent but coinhabit the same *one* sphere of existence.[97] This allows a bifurcation or mutual transformation between two movements: One movement maps the diversified altitude of spiritual spheres onto the physical spatiotemporal distance—assuming angels or higher beings to be physically developed beings like us, but more advanced, occasionally showing up on our physical space.[98] With the other movement, spatial differentiation can now also be mapped onto spiritual levels—assuming alien beings not to be like us, but of higher nature, for our salvation or detriment.[99] UFO religions assume the latter, while some UFO research assumes the former for the mentioning of higher beings in ancient religious texts, such as witnessed by the famous visionary encounter of the prophet Ezekiel.[100]

In any case, this argument from decentering allows us to implicate all other principles to the extent that we should be theoretically open not only for the multiplicity of life-forms with intelligence and aesthetic abilities, as well as spiritual sensitivities, as they manifest in human beings and other sentient beings on Earth, but also to their vast diversity, without, however giving up on the mutual communicability of their existence.[101]

With this background in mind and the openness that it implies for experimental tests that don't exclude possibilities of such life-forms, I can now add that the wisdom of human visionary imaginations on these matters of exo-life, mind, and spirit, in philosophies and religions, has, in fact, employed one or the other, or all of these principles, in assuming, as historical research shows, at least at one point of their development, the faith in, reflection on, and even experience of, alien life-forms,[102] with the *twist*, that they are understood to be expressions of a fundamentally *spiritually infused* universe.[103] The intricate details of their explorations are most interesting, but not relevant at this point. What is important, however, is the assumed diversity of dimensions that may or may not, in any particular rendering of this immensely populated universal space, in diverse ways be a mixture of spiritual altitude and spatial distance: either as levels of existence or as phases of cosmic becoming; either as lower (demonic) and higher (angelic) beings or as biological life like on Earth;[104] either in this cosmic epoch (or region) or in another; either as a projection into a future or as a fall from the past; either as local organisms or as cosmic forces or regions (superorganisms); in the best case, with sentience and consciousness, feeling, reason, and technology; in the worst case, as mere predators or mere algorithmic programs.

The important dimensional expansion, in all of these explorations, and especially in the diverse philosophical and religious reflection on them, is that humanity—if not all invoked beings—ought to be understood to exhibit a vast array of elements that define them not only as living and intelligent, but as *spiritual* in a wider sense. Since the definition of Spirit is as difficult as is that of Life and Intelligence—as a glimpse into related literatures will readily demonstrate[105]—it seems, in creating a preparatory space for a meaningful discussion of what it could mean that alien life and intelligence is *also* spiritual, a much better approach if we collect the characteristics that typically appear in some selection or in general embrace of the following components in these philosophical and religious discussions.[106] So, in order to count as a spiritual being—Earthly or alien—we should assume:

- *Reason* and *Intellect* (demonstrating systematic knowledge or the ability to systematize knowledge, and the ability to understand matters by content and form)[107]
- *Sentience* and *Compassion* (being able to feel pain and of aiding mutual support)[108]
- *Consciousness* and *Conscience* (exhibiting levels of awareness and self-awareness in environments, local and universal, and acting by ethical impulses)[109]
- A sense of *Aesthetic Realities* (striving for harmonies and depth of feeling) and *Beauty* (demonstrating the ability to transcend mere utility of oneself, others, and scenarios of encountering the world in its variety of phenomena)[110]
- A sense for *Imagination* (exhibiting the power of conceptualizing, independent from given facts, scenarios of alternative situations of living and diverse potentials of altered futures)[111]
- A sense for the *Mystery* of reality (sensing a deeper reality beyond mere rationality or facticity or short-circuits of causal or metaphysical closure, as well as approaching it beyond the ability to systematize and to manipulate)
- A sense of the *Holy* or *Sacred* (experiencing existence as a gift and of uncontrollable, but elevating patterns of meaning and of the presence of a deeper reality)[112]
- A sense of *Healing* (for the activation of changes to situations against perceived predicaments and conditions of natural, social and moral wounding)[113]
- And a sense of *Spirit* (an experiential perception of a metaphysical constitution of reality as meaningful and One)[114]

For the further investigation of the implications and consequences of the conjecture under discussion, namely, that we should expect ETI *also* to exhibit a spiritual dimension, that is, to be an extraterrestrial *spiritual* life-form (ETS),

I have—concordant with the aims of the reflections collected in this book of including spirituality, religion, and philosophy, especially relating to process philosophy and theology and concomitant ideas and thought figures—chosen two of their main exemplars, and widened the field with one other fascinating, but perhaps unexpected, personage, related to a religion neither of the West nor the East (in classical terms), namely the Baha'i Faith.

THREE EXEMPLARS

The figures I will call as witnesses of a cosmology that includes a spiritual force for its understanding, I will, accordingly, draw—in chronological order—on these roughly contemporary figures at the turn of the nineteenth to the twentieth century: 'Abdu'l-Baha (1844–1922); Alfred North Whitehead (1861–1947), and Pierre Teilhard de Chardin (1881–1955). I cannot, in the given space, here, investigate their respective cosmology beyond the aim of this chapter; so, I am neither interested in a comparison of their thought nor a detailed exploration of those cosmologies, but will, instead, directly state their common interest of devising a cosmology that is driven by a spiritual evolution of the cosmos as the basis for my conjecture insofar as their universes of thought counter the above named preconceptions that the Dark Forest hypothesis has made its default state for exploring the existence of and encounters with ETI.

While no further introduction to Whitehead[115] and Teilhard[116] is necessary in our context—and both have to a certain degree been recognized in the cosmological and theological discussion of ETL and ETI[117]—a few remarks may make the choice of 'Abdu'l-Baha understandable.[118] He was the son of the prophet-founder of the Baha'i religion, Baha'u'llah (1819–1892),[119] which, in the middle of the nineteenth century, rose from the orbit of Islam in Persia, much like Christianity appeared from Judaism in ancient times, but developed beyond both Islam and Christianity by way of reclaiming the Persian religious traditions of Zoroaster, Mazdak, and (to a limited extent) Mani (but not Manichaeism), in their own transversal diffusion of Brahmanism and early forms of Hinduism, especially as mediated by Sufism, as a religion of anti-dualism or of unity in multiplicity: of ultimate reality, of religions and of humanity.[120] 'Abdu'l-Baha was, by the Will and Testament of Baha'u'llah, conferred the sole authority of the interpretation of the scriptural verses of Baha'u'llah for the nascent religious community and the function of the living exemplar of its spiritual, intellectual, and social formations of life—a life committed to universal peace and unity on the basis of non-adversarial means of communication and interaction, of the transreligious mutuality of all religions,[121] and the imperative of acting for the good of the whole of humanity, even in local contexts.[122] Like Baha'u'llah, 'Abdu'l-Baha was over the course of his whole life a prisoner of conscience of two empires: the Shi'i dominion of the Persian Shahs and the Ottoman realm of the Calif of Sunni Islam. Yet despite the restrictions pressed upon him, 'Abdu'l-Baha was engaged in wide-reaching conversations with cultural, political and religious, intellectual and scientific personages around the world; and after the constitutional revolution in Turkey of 1908, which facilitated the release from his imprisonment, he journeyed, between 1911 and 1913, to Egypt, Europe, and North America in order to participate in peace conferences in the preheating atmosphere of the conflicts that eventually led to the outbreak of World War I in 1914, in order to propagate a new politics of peace for the secular world-community as well as between religions, fired by a strong anti-racism, and to communicate the new Persian religion to its followers and many and diverse circles, such as NAACP, universities, public intellectual and spiritual settings, and meetings of diverse religious communities.[123] For his preparatory and immediate organization of the survival of the population in the Akko-Haifa area—to which Baha'u'llah was banned in 1868, left to die in prison, where 'Abdu'l-Baha was confined for over forty years, and where the Baha'i Holy Places and World Center are situated until today—and the region which was hard fought over during the world war, the British crown knighted Sir 'Abdu'l-Baha as one of the few non-British individuals in 1921.

What relates the universe of thought of all three figures to one another and makes them exemplars of the idea of a spiritual evolution of the universe,[124] is remarkable, although all three of them had a very different upbringing and carrier—'Abdu'l-Baha, an intellectual and spiritual power of a new religion; Whitehead, a mathematician and philosopher of science turned metaphysician and philosophical cosmologist; and Teilhard, a paleontologist and Catholic priest, a member of the Societas Jesu, the Jesuit Order. All three had strong relationships with science while affirming religion and spirituality: both Whitehead and Teilhard, while being scientists, were extraordinary creative and deep philosophical and theological thinkers; and 'Abdu'l-Baha, being a genuine religious figure himself, promoted nothing less than the harmony of science and religion as a foundational principle of the Baha'i religion.[125] While they represent diverse cultures and experiences—Whitehead, British American; Teilhard, French Chinese; and 'Abdu'l-Baha, Persian Levante—they developed remarkably similar views on the universe in which we live as embodied, but spiritual beings, as well as regarding a universal, evolutionary view, which not only drove their understanding of the emergence of life and intelligence, but of spiritual existence in the wider cosmic environment. Here are a few of their coincident, while inherently differentiated or complementary, views on the matters in which we are interested:

Evolution

All three of these thinkers affirmed a version of the convergent evolution of the universe, but with an infinite diversity of life-forms.[126] In other words, their views are fundamentally non-anthropocentric and –morphic in outlook and feel, intellectual penetration, and spiritual relevance.

Multiverse

For all three of them, the universe is a vast place of movements that mediate between infinite multiplicity and directions of unification. In an ascending order of diversification and of the inherent pluralism that such a cosmology exhibits, we can typologically sense a widening from an infinite multiplicity of worlds in the movement of unification toward a final oneness in Teilhard;[127] to a cyclical universe of infinite cosmic epochs, although, in my understanding, not limited to just one series of universes, in Whitehead;[128] to a breathtaking pluralism of a multiverse of infinite cycles and a diversification into infinite universes or a multiverse in 'Abdu'l-Baha.[129]

Sym-pathy

For all of them, the fundamental force, animating the process of unification and diversification in the universe or multiverse is nothing less than a universal force of love,[130] symbolically expressing itself even already (in an image used by at least two of them) in the coalescence of togetherness, physically exhibited, or symbolized, for instance, by "magnetism";[131] but in a more philosophical rendering, being based on a universal mutuality of all beings,[132] as they are related by sym-pathy[133]—the suffering of being together, even *in* one another, and being drawn by harmonics of composition and coordination.[134]

Universal Spirit

All three of our thinkers affirm the spiritual dimension of virtually all beings as a matter of fact of their very existence, and of a Spirit of life that expresses itself as this force of sym-pathy and coordination, attracting a universal evolutionary process of indefinite unification and diversification throughout all imaginable and unimaginable worlds.[135]

Awakening

Finally, all three of those thinkers are exemplars of an understanding of the existence of the universe, *this* kind of universe, as exhibiting a deeper meaning that allows beings like us to come to insights into its importance and to live a life not exhausting itself in the use of tools for survival, but as being oriented toward deeper values and virtues that arise only with a sense of beauty, harmony, peace, and spiritual awakening—being on a journey in which the universe itself seeks enlightenment or offers it, if one seeks its cosmic Spirit.[136]

In the following sections, I will present evidence for these claims so as to generate the space for the claim, in relation to my conjecture, of the importance of "spiritual evolution" for the meaningful assumption of space-communicating and -faring ETI and in preparation for an encounter of them as ETS.

THE SPIRITUAL DIMENSION

When our three thinkers engage in matters of the spiritual dimension of humanity, what they are after is the recognition of the deep roots of human motives for acting and reflection, which not only define humanity as such but, in some meaningful sense, reveal, corroborate, and evoke a deeper dimension of cosmic becoming as such, beyond and even independently from humanity and life on Earth, pregiven to its appearance and preparatory for a further, as of yet unwritten story even beyond any human horizon.[137]

We can categorize this spiritual dimension with *four forms of self-transcendence*. The term "self-transcendence"[138] was introduced by the Catholic theologian Karl Rahner (1904–1984)[139] in the context of an evolutionary understanding of the emergence of mind/spirit (the German term *Geist* indicates both at once) on a cosmological scale, that is, it was not reduced to the appearance of humanity alone.[140] It indicates that the latency of mind/spirit in a material universe, which he, at one time, characterized as "frozen mind/spirit," leads to the emergence of its own spheres and dimensions beyond the causal nature of matter through its internal drive toward that appearance *and* by the intimate involvement of a divine attraction toward such an appearance.[141] This is, obviously, although expressed in different terms and on different philosophical backgrounds, what our three thinkers also affirmed: that nothing is really without mind/spirit[142] and, hence, involves a self-transcending movement of the development of its *inherent* potentials toward ever-new actualizations *through* an involvement of divine attraction toward actual forms of such a realization, of which humanity is an important, but by no means the only exemplification.[143] Here are four such forms.

Self-transcendence of dimensions

The first form of this spiritual self-transcendence (as the subject, Spirit drives this self-transcendence; as the object, it is the aim of such a process) is that of dimensional interaction. Matter and mind/spirit are not derived from, or reduced to, one another, but mutually conditioning the appearance of evolutionary processes.[144] For Whitehead, everything has a mental dimension (mental prehensions) and is interacting through the dipolarity of every event as well as in the diversified common space of their constitution of the cosmos—initiated by the attractive immanence of God in every event and throughout its nexic interactions, social constitution, and the layers of fields pervading the cosmic community.

> The initial phase of each fresh occasion represents the issue of a struggle within the past for objective existence beyond itself. The determinant of the struggle is the supreme Eros incarnating itself as the first phase of the individual subjective aim in the new process of actuality. Thus in any two occasions of the Universe there are elements in either one which are irrelevant to the constitution of the other.[145]

The same is true for Teilhard, for whom "Creative union is the theory that accepts this proposition" that "everything happens as though the One were formed by successive unifications of the Multiple—and as though the One were more perfect, the more perfectly it centralized under itself a larger Multiple,"[146] *but* really, it attracts the Multiple to itself by complexification and diversification. This attracting One is but "the influence of a spiritual energy immanent in the whole (some soul of the world)."[147] And 'Abdu'l-Baha insists that all "we see around us is the work of mind"; and every being is to its own extent an expression of mind: "mind in the herb and in the mineral."[148] And so is everything, from matter to God, generated and infused with the pervading Spirit, attracting to the actualization of its own diversified potentials.[149] Whitehead even recognizes that all great eventful (social or evolutionary) changes happen through the interaction of these physical and spiritual dimensions.

> The great transitions are due to a coincidence of forces derived from both sides of the world, its physical and its spiritual natures.[150]

Self-transcendence of becoming

All three thinkers affirm the process of the interaction of spirit and matter to be one of evolution, becoming, and temporal development. Whitehead promotes a theory of evolutionary emergence of new layers of being and becoming as the "creative passage of nature,"[151] the "creative emergence of novelty,"[152] in a process of the dipolar interaction of mental and physical feelings in every event and between them in their nexic organization through their contrasting activities, whereby Whitehead subscribes to a "doctrine of emergent evolution."[153] 'Abdu'l-Baha speaks of the "evolution or progress of the spirit" in nature, which knows of no "absolute repose," because "everything moves" and tends toward the emergence of spiritual beings[154] and, as in Teilhard, drives a progression toward God's presence in it.[155] Teilhard says it best:

> Matter and Spirit are not opposed as two separate things, as two natures, but as two directions of evolution within the world.[156]

Self-transcendence toward the greater world

Connectivity, coordination, harmonics of community of being and mutuality are a characteristic of self-transcendence, as can especially be witnessed in all serious spiritual paths throughout human history; and they appear as central philosophical conditions of the evolution of "the universe as One"[157] in Whitehead, too. Teilhard speaks of a "much wider and more spiritual field"[158] that the unity of creative becoming is approaching by integrating all "limitations and diminishments," which hamper the "general progress of the cosmos"—"by integrating them (yet without changing them) in a higher spiritualization of our being" so that the "divine Master of the world . . . moulds us, empties us of our self-love, and penetrates into us"[159] in order to prepare us for a "creative union" in which "everything becomes one by becoming self,"[160] but connects in a universal "synthesis," in a universal "communion" of "spiritualized matter."[161] Unity as the coinherence of a common space of mutuality beyond self-serving interests for a universal peace, motivated by the mutuality of a divine love that is infinite and universalizing the self-giving throughout nature and society, is a central concern of 'Abdu'l-Baha message that reflects the foundational principles of the evolutionary process,[162] namely, that the "universal inclination shall be toward peace, love and unity."[163] And when Whitehead speaks of peace, he has exactly the spiritual motivation in mind that facilitates this widest transformation of self-transcendence toward universal communication and communal existence:

> Thus Peace is the removal of inhibition and not its introduction. It results in a wider sweep of conscious interest. It enlarges the field of attention. Thus Peace is selfcontrol at its widest,—at the width where the 'self' has been lost, and interest has been transferred to coordinations wider than personality. Here

> the real motive interests of the spirit are meant, and not the superficial play of discursive ideas.[164]

Self-transcendence toward mystery

To evoke the mystery of existence is a common practice of all religious traditions in their mystical core. It does not necessitate world-abandonment, but means "world-loyalty," as Whitehead says;[165] and it does not indicate a disappearance into an isolated (spiritual) interiority, but, with Teilhard, if "mystery is put in its correct setting, it becomes as fruitful as the most thoroughly understood truths."[166] It indicates the openness to unprecedented, novel and uncontrollable (and not by the past controlled) events of "direct insight into depth as yet unspoken,"[167] but also of healing and transformation. *This* is, for Teilhard, the experience of the Spirit, motivating the whole evolutionary process—not for "the material vastness of the universe . . . ever . . . to disappear," but for the "Spirit . . . to move into a different zone, for the whole shape of the world immediately to be changed."[168] 'Abdu'l-Baha speaks of the "infinite Mystery" of the "incomprehensible" divine Reality that—in panentheistic manner—"contains all," but "cannot be contained,"[169] except, as Baha'u'llah states, "in the pure soil of the heart" that reveals the "stream of divine mystery."[170] Its access through meditation, contemplation or ecstasy is an undeniable sign of spiritual beings:

> there is a sign (from God) in every phenomenon: the sign of the intellect is contemplation and the sign of contemplation is silence . . . You cannot apply the name "[spiritual]" to any being void of this faculty of meditation; without it [they] would be a mere animal, lower than the beasts. Through the faculty of meditation [a spiritual being] attains to eternal life; through it [it] receives the breath of the Holy Spirit—the bestowal of the Spirit is given in reflection and meditation.[171]

All this is to say that with these forms of self-transcendence we approach a spirit that exhibits itself in the history of humanity within and without specific religions—in a sense, as a kind of universal religiosity in which the whole universe participates, but which has its conscious perception and recognition at least in humanity, which renders it a spirit-faring species,[172] and perhaps, as my conjecture goes, names a necessary condition for its successful star-faring future. This spirit of religiosity has been beautifully circumscribed by Whitehead with these words:

> Religion is the vision of something which stands beyond, behind, and within, the passing flux of immediate things; something which is real, and yet waiting to be realized; something which is a remote possibility, and yet the greatest of

present facts; something that gives meaning to all that passes, and yet eludes apprehension; something whose possession is the final good, and yet is beyond all reach; something which is the ultimate ideal, and the hopeless quest.[173]

SPIRITUAL EVOLUTION

As already implied by the diverse forms of self-transcendence as marks of spiritual beings, which we may also expect of characterizing ETS, self-transcending is an *evolving process of embodiment*: of diversification, of developments of attraction to deep virtues and values, and as a matter of cosmic universality. The universe is a field of spiritual evolution.[174]

Teilhard develops a grand scheme of "creative evolutionism,"[175] of the whole universe in its splendid multiplicity developing by the immanence of the divine attraction to its aim, the Omega Point,[176] the unification of this process in the form of a universal diversified and intensely connected organism,[177] a pleroma of God and the world, that finally will have become the creation of God.[178] From the future, similar to Whitehead and 'Abdu'l-Baha, the Spirit acts as universal energy of a spiritual emergence wherein the potentials of love and peace are increasingly realized by the overcoming of the struggles and strives of the evolutionary process, and in which the Multiple (itself a boundary notion of indifferentiated chaos) derives its spiritual virtues by the capitalization of the immanent divine.[179] In the case of Teilhard, this force is the cosmic Christ, plunging into the Multiple[180] and, as *Christus Evolutor*, attracting it to complexification, diversification, communication and unification.[181] But in a universal, trans-Christian perspective, this driving force reflects a religiosity that believes in, feels, and is attracted to, a universal Spirit that animates this and every other religious drama, however specifics of its historical differences may symbolize it.

> [The] fundamental principle stands out with all the emphasis of a truth of the first order, that 'All consistence comes from Spirit.' In that we have the very definition of creative union. . . . The truth is that nothing holds together except as the result of a synthesis, which means, in short, however lowly the synthesis, by a reflexion of Spirit. The materialist philosopher, therefore, who looks at a lower level than soul for the solid principle of the universe, grasps no more than dust that slips between his fingers.[182]

In this creative process, matter is spiritualized and Spirit embodied, generating the emergence of a pleroma of beauty, sublimity, and abundance.

> Above all, this principle hallows the Kingship of Spirit; but at the same time it thereby preserves and ennobles matter. Indeed, even if it is Spirit that constantly

> carries matter along and supports it in the ascent towards consciousness, it is matter, in return, that enables Spirit to subsist by constantly providing it with a point upon which to act, and supplying it with nourishment. . . . Its sublimity and richness are tied up with an organic multiplicity that It embraces in its 'solid aspect.'[183]

The outcome of this spiritualization is, for Teilhard, a matter of fantastic speculation, but grounded in his religious conviction that it is a process of the healing of all the contradictions that arise in the evolutionary process, driven by the vision of the Omega, a version of the ancient idea of the world-soul.[184]

> The prospects opened up by the application of the law of recurrence which we have called Creative Union, are convincing enough when we are concerned with the representation of the universe's past; they become somewhat fantastic when we turn to the mysteries of the future. To admit that human monads are elements of a higher organic synthesis, to accept that they are destined to form the body of a soul more spiritual than our own, so outstrips our imagination that we feel we must find some positive evidence upon which to base such disturbing extrapolations.
>
> Many non-Christian mystics have not hesitated, trusting to their own desires and natural predilections, to cast themselves into the delectable abyss of belief in a soul of the world. . . . Let us, for brevity's sake, give the name of omega to the upper cosmic term disclosed by creative union.[185]

In the case of 'Abdu'l-Baha, it is the Logos or the Mind of God, the universal immanent Manifestation of the apophatic ultimate Reality,[186] that moves as this energy through the evolving process of universal pacification.

> It is evident, then, that each elemental atom of the universe is possessed of a capacity to express all the virtues of the universe. This is a subtle and abstract realization. Meditate upon it, for within it lies the true explanation of pantheism. From this point of view and perception pantheism is a truth, for every atom in the universe possesses or reflects all the virtues of life, the manifestation of which is effected through change and transformation. Therefore, the origin and outcome of phenomena is, verily, the omnipresent God; for the reality of all phenomenal existence is through Him. There is neither reality nor the manifestation of reality without the instrumentality of God. Existence is realized and possible through the bounty of God, just as the ray or flame emanating from this lamp is realized through the bounty of the lamp, from which it originates. Even so, all phenomena are realized through the divine bounty, and the explanation of true pantheistic statement and principle is that the phenomena of the universe find realization through the one power animating and dominating all things, and all things are but manifestations of its energy and bounty.[187]

And in Whitehead, it is the initial Eros, inherent in all becoming as its very origination,[188] that attracts to the same goal: the development of ever new, refined processes and orderings of life toward beauty and harmony, paired with intensity and the diversity of experience and character.[189] This goal of the universal process of spiritual evolution is "the aesthetic harmony" that "stands before it as a living ideal moulding the general flux in its broken progress towards finer, subtler issues."[190]

> The metaphysical doctrine, here expounded, finds the foundations of the world in the aesthetic experience, rather than as with Kant in the cognitive and conceptive experience. All order is therefore aesthetic order, and the moral order is merely certain aspects of aesthetic order. The actual world is the outcome of the aesthetic order, and the aesthetic order is derived from the immanence of God.[191]

This is the aesthetic immanence of God not only in all processes, in every atom, as 'Abdu'l-Baha and Teilhard agree, but as the universal horizon of a future that must realize itself from and through a divine vision in the sym-pathy of all beings,[192] that is, the virtue of universal compassion in which the world releases and upholds its binding character—and, regarding my conjecture, must define its very future.

> [God] is the binding element in the world. The consciousness which is individual in us, is universal in [God]: the love which is partial in us is all-embracing in [God]. Apart from [God] there could be no world, because there could be no adjustment of individuality. [God's] purpose in the world is quality of attainment. . . . Every act leaves the world with a deeper or a fainter impress of God.[193]

Now, in this creative process of spiritual evolution, the appearance of a multiplicity of spiritual beings is inherently inscribed.[194] This is important: ETS are an expression of the assumption or faith in the truth of the *aesthetic* experience of the deeper layers of human virtuosity, as it expresses itself in a *religiosity* of self-transcendence and connectivity, which is the prime connotation of the Spirit and the energy driving the whole process of the becoming of the universe.

> One thing I think has been made clear: the initial impulse that sets the Multiple in motion towards unity, the energy at the source that animates the whole sequence of cosmic unification and spiritualisation, is the magnetic force of omega. Without that gratuitous appetite for Being, without that prevenient zest for union, the mechanism of the universe would remain immobile, the elements of the world would never emerge from their infinitely tenuous plurality. But once the 'concept' of an omega has kindled in the [beings] the desire to

> meet with Spirit, they develop a restless mobility and feel themselves impelled towards activity. God's first wish, expressed by the surge of life within us, is that [God's] creatures should increase and multiply.'[195]

Concomitant and in clear terms, Baha'u'llah and 'Abdu'l-Baha declare that, from this perspective of the creative proliferation of life and intelligence in the cosmos through the operation of the Spirit-energy that drives its virtual emergence from strife and warring, not only are the worlds of God infinite, constituting a spiritually infused metaverse, but all worlds are populated with beings that can know God. The multiverse and ETS are revealed certainties in the Baha'i religious cosmology:

> Know thou of a truth that the worlds of God are countless in their number, and infinite in their range. None can reckon or comprehend them except God, the All-Knowing, the All-Wise.[196]
>
> . . . in this universe of God's, which appears in the utmost perfection, beauty, and grandeur, the luminous bodies of the material universe are infinite. Pause to infer, then, how infinite and unbounded the spiritual realms of God, which are the very foundation, must be! "Take ye good heed, O people of insight![197]
>
> Know thou that every fixed star hath its own planets, and every planet its own creatures, whose number no man can compute.[198]
>
> Abdu'l-Baha stated that there are other worlds than ours which are inhabited by beings capable of knowing God.[199]

Cautiously, 'Abdu'l-Baha also suggests that ETI will not be like us, but that diversification is inscribed in the creative process such that the environmental conditions of their specific evolution will be a necessary function in establishing their consciousness and intellect. The acceptance of ETS does not imply anthropomorphism or anthropocentrism—just that their spiritual nature is a binding element between them and all other ETS, such as, from their perspective, humanity.

> The earth has its inhabitants . . . then how is it possible to conceive that these stupendous stellar bodies are not inhabited? Verily, they are peopled, but let it be known that the dwellers accord with the elements of their respective spheres. These living beings do not have states of consciousness like unto those who live on the surface of this globe: the power of adaptation and environment moulds their bodies and states of consciousness, just as our bodies and minds are suited to our planet.[200]

And in a breathtaking vision, 'Abdu'l-Baha announces that the very *motive* for the generation of the multiverse is nothing but the emergence and proliferation of ETS.[201]

> Briefly, there were many universal cycles preceding this one in which we are living. They were consummated, completed and their traces obliterated. The divine and creative purpose in them was the evolution of spiritual [beings], just as it is in this cycle. The circle of existence is the same circle; it returns. The tree of life has ever borne the same heavenly fruit.[202]

However, all three of our thinkers' faith in spiritualization must be seen on a much darker background. All have had their immediate experiences and sufferings of World War I: Whitehead lost his son Eric; Teilhard was an active soldier; 'Abdu'l-Baha not only predicted and worked against this war on an international scene, but suffered though it by proactively supporting the population of the Akko-Haifa area while remaining under permanent death threat of the Ottoman troops if they ever reached the sea. All three thinkers are aware and intensely reflect on the constitution of a world that is supposed to be spiritually driven, but allows for the grave injustices of humanity against itself and the planet Earth with such a ferocity that the general intellectual atmosphere after the Great War was becoming utterly pessimistic and nihilistic. All three of them accept the Darwinian mechanisms of evolution, as known at that time, such as chance mutation and natural selection, which prefers the survival of the fittest populations, with the implications of their refection in social forces such as the arbitrary use of power, the ferocity of a gaging competition, and the mercilessness of adversarial processes of survival in the fight for recourses and self-assertion.

Consider: Teilhard reflects on the ineradicability of the "instruments of dominion" and the "absorbed . . . concern" for food and population distribution,[203] and he pains over the overwhelming massiveness of evolutionary failures.[204] Whitehead muses that "Life" is not only "a bid for freedom" and "originality,"[205] but "is robbery."[206] And 'Abdu'l-Baha admits that "Life" is like "war," a "struggle for existence."[207] Yet, nevertheless, they counterfactually always confirm that the spiritualization of Life is a matter of the *same* evolutionary process, the self-transcendence of this struggle toward peace—not as the absence of war alone, but as a spiritual peace that raises the latent virtues it promotes to a level of activation that is meant to actually transcend its warring *condition*. So can 'Abdu'l-Baha assert that, as a matter of divine revelation, but really as a symbol for the direction humanity must take its own evolution, "the law of struggle for existence is abrogated and the canopy of the oneness of the world of humanity is raised on the apex of the globe, casting its shade over all the races";[208] that, therefore, "peace is mutual aid and cooperation among the peoples of the world and the cause of the good pleasure of the True One in the heavenly realm";[209] and that, hence, "peace is the everlasting life of the world of humanity"[210]—if it can achieve it.

Liberation from the mere "captivity of nature"[211] through the combined forces of science and religion[212]—as the basis for a new civilization beyond those struggles, or, even more profound, a civilization that is *defined* by its transformation beyond the very conditions and manifestations of struggle, robbery and war, and the seeming inescapability of the survival of the fittest caught up in competition between life and death[213]—is the philosophical impulse that animates our thinkers profoundly.[214] As this "liberation" cannot be mounted by our ability to reason and a technological revolution *alone*, including the ability to sound and cross the expanse of space and, perhaps, time, but must be *coextensive* with the spiritual manifestation of the virtues that *alone* can match the devastating powers of instrumental reason, vested with self-interests, and the technological tools available for its terrible misuse—may we not assume that we, or any ETI, developing beyond their point of origin in the cosmos, will fail if this insight goes unanswered by the effort of actualizing the virtues of self-transcendence? Isn't survival of the fittest, then, really *more* (*sur*) than mere survival of a life that in its fulness must recognize its dept to the tolerance of a cosmos for its existence in answering its call to life and intelligence with a spirit of life that alone is fitting for its patience? Isn't, as Whitehead's so firmly states, "*the function of Reason,*" and its abilities to submit or to liberate from powers of subjection and manipulation, "*the promotion of the art of life.*"[215] And that it would be the fundamental error of a civilization on the threshold of becoming space-faring to believe in the "fallacy . . . that fitness for survival is identical with the best exemplification of the Art of Life"?[216] That, instead, the Art of Life is a virtuosity in the art of developing its powers as *spiritual* virtues: "(i) to live, (ii) to live well, (iii) to live better"?[217]

THE RELATIONAL MEDIUM

The perhaps most profound presupposition shared by these three thinkers for their analysis of the spiritual dimension, not only of human life and intelligence but of the cosmos itself, as the animating force of its evolution beyond the Darwinian struggles, is their common reference to the experientiality of a relational medium constituting and pervading universal existence as such. Admittedly, their invoking of physical terms of "energy" in a metaphysical context may be seen as a confusion of categories; yet, all three of them, and Whitehead in full knowledge of the scientific restrictions on such a terminology, are not hindered by "the fallacy of misplaced concreteness"[218] and discuss scientific terms as abstractions from a deeper reality of the complex feelings of their reality, "a common reality of physical and spiritual energy"—of "light-energy" and animalic life-energy,[219] of emotional energy[220] and the "progressive energy" of a culture,[221] a "complex energy, emotional and purposeful"[222]—that speaks to us of a deeper reality from which they appear as specifications. It is all-pervasive—and, divorced from its living, experiential aspect, indicative of nothing but *our* failure of recognizing precisely that sense of Spirit that can only emerge with its recognition.

Many scientists, in the wake of the scientific revolution of the beginning twentieth century have sensed this mystery while pondering the philosophical implications of quantum mechanics—one may think of the often dismissed, but very impactful speculations of some of their champions, such as Max Planck (1858–1947), Erwin Schrödinger (1887–1961), Werner Heisenberg (1901–1976), and Wolfgang Pauli (1900–1958); but also of cosmologists, such as Sir Arthur Eddington (1882–1944), John Polkinghorne (1930–2021), Paul Davies (b. 1946), John Barrow (1952–2020) and Roger Penrose (b. 1931); of philosophers, such as William James (1842–1910), C. D. Broad (1887–1971), Gregory Bateson (1904–1980) and Arthur Peacock (1924–2006); of theologians, such as Bernard Loomer (1912–1985), David Griffin (1939–2022), John Hick (1922–2012), and John Haught (b. 1942); and evolutionary biologists, such as Sir James Jeans (1877–1946), Michael Denton (b. 1943), and Simon Conway Morris (b. 1951).[223] In the literary medium of a grand sci-fi cycle of the *Hyperion Cantos*, Dan Simmons (b. 1948) introduces us such universal medium as "The Void Which Binds," which is not merely physical or mathematical, but consists of the lives and voices, experiences and acts, the beauty and tragedy, of all sentient beings, dead or alive.[224]

Whitehead finds its philosophical rendering in the ancient cosmology of Plato: the mysterious reality of the Place, Khora—the Receptacle of all being and becoming, the in-between, in which all realities meet and interreact—and the Void, Kenon, of Epicurious, alike.

> Plato's doctrine of the real Receptacle [hypodoche and khora], and Epicurus' doctrine of the real Void [to kenon], differ in some details. But both doctrines are emphatic assertions of a real communication between ultimate realities. . . . The Receptacle, as discussed in the *Timaeus*, is the way in which Plato conceived the many actualities of the physical world as components in each other's natures. It is the doctrine of the immanence of Law, derived from the mutual immanence of actualities. It is Plato's doctrine of the medium of intercommunication.[225]

This medium of intercommunications, which is nothing for itself, void, but facilitates the existence, communication, and memory of all becomings in an universal reverberation of their mutuality, can be discovered across the lines of philosophical divides we generally erect for strange reasons, but which are mere abstractions of categorizations that break down when we are willing to see: whether idealist or materialist, whether subjectivist or objectivist, whether religious or not, whether dualist or monist, whether scientist or poet, there is a recognition of such a medium that transcends the bifurcated spaces of physics and metaphysics for themselves.

> Thus finally we can understand that the Receptacle, according to Plato, the Void, according to Lucretius, and God, according to Leibniz, play the same part in cosmological theory. Also in his general scholium, Newton definitely connects the Lucretian Void with the Leibnizian God. For he calls Empty space the 'sensorium of God'. . . . The modern cosmologies are all [their] detailed variations.[226]

While Whitehead calls it the "mutual immanence" of all existence,[227] 'Abdu'l-Baha, speaks of the fundamental "relationship between universal beings" in which "all beings are linked together like a chain; and mutual aid, assistance, and interaction are among their intrinsic properties and are the cause of their formation, development, and growth."[228]

> And if such relationships, such spiritual influences and effects, are found among the various members of the body of [a human being], which is only one particular being among many, then there must assuredly exist both spiritual and material relationships among the countless universal beings. And although our present methods and sciences cannot detect these relationships among the universal beings, their existence is nonetheless clear and indisputable.[229]

In his *Tablet of the Universe*, 'Abdu'l-Baha ascribes to this medium the "universal reality" that operates in all beings[230] and that pervades and guides all diverse aspects of becoming, the physical and the spiritual, the particular and the universal, the manifest and the hidden, in a relativity of relationality, that excludes nothing.

> Know thou that the expressions of the creative hand of God throughout His limitless worlds are themselves limitless. . . . This being the case, how can one, without proof or testimony, conceive of creation being bound by limits? . . . Nay, earthly and heavenly, material and spiritual, accidental and essential, particular and universal, structure and foundation, appearance and reality and the essence of all things, both inward and outward—all of these are connected one with another and are interrelated .... For particulars in relation to what is below them are universals, and what are great universals in the sight of those whose eyes are veiled are in fact particulars in relation to the realities and beings which are superior to them. Universal and particular are in reality incidental and relative considerations. . . . Know then that the all-embracing framework that governs existence includes within its compass every existent being—particular or universal—whether outwardly or inwardly, secretly or openly. Just as particulars

> are infinite in number, so also universals, on the material plane, and the great realities of the universe are without number and beyond computation.[231]

It is this universal ecology, far from any anthropomorphic or -centric minimalization of the universe so as to fit into the current understanding of the human mind, that animates the understanding of our three thinkers as they reflect on the future of humanity: regarding the development of a civilization that is worth of that name—a spiritualized civilization that has overcome the misplaced abstraction of itself from a cosmos that it inherits and inheres for a future in the midst of which it must cope with the consciousness of being but a physical speck, but a precursor of a spiritual ascent[232] that receives its relevance from its self-transcendence—and *thereby* transforms into, or yields to, a spirit-faring civilization *becoming* space-faring, becoming ETS.[233]

SPIRITUAL CIVILIZATION

Before I will attempt a summary of the insights gained by these mutually enhancing connotations of spiritual evolution in our three thinkers and its consequences for my conjecture, there is one last element that is worth mentioning: the implication of the latency of spiritualization of evolution for the future of humanity and, presumably, any ETS. As mentioned earlier, all three thinkers are engaged in a vision of a universal process that allows for a multiverse to establish or fail to initiate this spiritualization, and repeatedly so. All three thinkers think of the future direction of this spiritualization as a conscious endeavor of unification that surpasses the biological conditions of the past and, in some sense, proceed through a cultural threshold of acquiring a unified, but diversified community that increasingly becomes an expression of the universal relationality that drives the whole universal process of cosmic evolution.

Whitehead talks of the this whole process as an Apotheosis of the world in a cycle of mutuality with divine reality,[234] as it incarnates itself in it as Harmony of Harmonies[235] that reflects itself increasingly in this process—*if* humanity becomes a civilization that realizes the grand values of Truth, Beauty, Art, Adventure, and Peace as self-transcending virtues[236] by which it reflects the adventure of the universe as One, consciously integrating Khora and Creativity with the divine Eros and Refection on its achievements.[237]

> Adventure belongs to the essence of civilization. The incompleteness of the concept relates to the notion of Transcendence, the feeling essential for Adventure, Zest, and Peace. This feeling requires for its understanding . . . the concept of an Adventure in the Universe as One. . . . [It] includes the Eros which is the living

> urge towards all possibilities, claiming the goodness of their realization. . . . In this Supreme Adventure, . . . thus enjoyed, [lies] the final Beauty with which the Universe achieves its justification. This Beauty has always within it the renewal derived from the Advance of the Temporal World. It is the immanence of the Great Fact including this initial Eros and this final Beauty which constitutes the zest of self-forgetful transcendence belonging to Civilization at its height.[238]

In remaining an Adventure of Self-Transcendence, its physical realization is no automatism, but a decision and a redirection of virtues that inhabit the structure of such a unified society of spiritualization. It remains an indefinite movement—and it can fail.[239] Democracy has as much failed as the technological revolution in overcoming old and modern slavery.[240] So, Whitehead insists on the necessity that, in order to avoid failure, power must be reconsidered *as* virtue, that is, "not [as] an exhibition of over-ruling power," but a power that "lies in its absence of force," for which figures such as Jesus mark a historical departure.[241] Instead of coercion, Whitehead invokes Plato's pervasiveness of Ideas; and he transforms their potentially and historically effective, ruling character that ideologies can misuse into the attractiveness of divine immanence and the mutual immanence of all in all, as the measure for the new socialization of a spiritualized civilization.[242]

Teilhard envisions the divinization of humanity only through a process of its self-evolution beyond the noosphere into a spiritual sphere that foreshadows the Omega in the form of the body of Christ, expanding through its existence and the whole cosmos. The "hominization" toward the "*parousia-point*," which equals its "divinization," happens though a progress of "*ultra-socialization*" that expands the "Christified portion of the world."[243] It is a civilization beyond the evolutionary brackets of our current state and in vision of the latency of the arrival of "super-mankind"—by a "metamorphosis (a combination of the highest psychic tension with the highest technological organization)" as the "critical point" at which "*maturation*" becomes "super-hominization" and "social super-humanization" as the condition "for the extensions of anthropogenesis in the direction of cosmic diffusion."[244] However, while Teilhard embraces the technological insertion into hominization, including eugenics and encephalization, accepting "spiritualized" versions of weapons[245] and the hyper-intensification of the noosphere—all of which may, as some commentators have observed, function as means of social Darwinism—the cyclical understanding of evolution of spirit from matter, namely, *from* its primary involution, of matter from spirit,[246] may also be understood as counter-agency against such implications, because the divine seeks the "'Kenosis in matter, of human 'co-feeling,'"[247] which indicates the presupposed latency of spiritualization as the *initial* nucleus of the evolving universalization of compassion.

The vision of a unified humanity and a world civilization on the basis of the values and virtues of spiritualization is, of course, the fundamental paradigm that the Baha'i Faith promotes[248]—or humanity may fail. In its roots it consists in the integration of reason and spirit, physical and spiritual forces, science and religiosity, technology and art, so as to overcome their mutually contradictory divergence.[249] Much like Stephen Hawking's speculation on the thermodynamic destabilization of information communication through information evolution (genetic engineering, weaponization of AI), 'Abdu'l-Baha highlights that the incongruence between technology and spirituality may result in the loss of a humanity left to its own devices and its self-destruction; but, in reverse logic: the coextensive development of material and spiritual development will essentially contribute to its ability not only of unifying with itself and the Earth, but becoming star-faring as only ETS could become.

> Material civilization concerns the world of matter or bodies, but divine civilization is the realm of ethics and moralities. Until the moral degree of the nations is advanced and human virtues attain a lofty level, happiness for [humanity] is impossible.
>
> Material civilization is like unto the lamp, while spiritual civilization is the light in that lamp. If the material and spiritual civilization become united, then we will have the light and the lamp together, and the outcome will be perfect. For material civilization is like unto a beautiful body, and spiritual civilization is like unto the spirit of life. If that wondrous spirit of life enters this beautiful body, the body will become a channel for the distribution and development of the perfections of humanity.[250]

This vision of a spiritualization of human civilization should, however, not blind us to misinterpret it as an optimistic fantasy. It is rather a mirror and a warning. All three thinkers are aware and forcefully affirm the fact that it is not only material civilization, technology and science, economy and social systems that contribute to the downfall of its perpetuation in the future—as the use of power, manipulation, and ideological propaganda, have demonstrated—but that the same is, if not more, true for religions. Whitehead speaks of "the horrors which can attend religion: human sacrifice, and in particular the slaughter of children, cannibalism, sensual orgies, abject superstition, hatred as between races, the maintenance of degrading customs, hysteria, bigotry,"[251] and asks: "Must 'religion' always remain as a synonym for 'hatred'?" while the "great social ideal for religion is that it should be the common basis for the unity of civilization."[252] 'Abdu'l-Baha condemns religion if it incites violence and insists that, if religion does not "unite all hearts and cause wars and disputes to vanish from the face of the earth, give birth to spirituality, and bring life and light to each heart," if it becomes "a cause of dislike, hatred and division, it were better to be without it, and to withdraw from such a religion would be a truly religious act."[253] And while Teilhard ponders the necessity of a new religion,[254] a mystical, spiritual religion, operating "in the spiritual energies of Matter,"[255] 'Abdu'l-Baha understands the religion of his father, Baha'u'llah, as such a rebirth of religiosity.[256]

We are bound to ask: What happens to humanity if this co-development of the spiritual and material forces is not in its future? What other event will be conjured up from the failure to reach for it; if this "great transition," which is "due to a coincidence of forces derived from both sides of the world, its physical and its spiritual natures,"[257] does not happen? Is our world doomed to crush into a boundary of irreversible decline or annihilation? Stated in a cosmological context: Is the non-integration of material and spiritual aspects of humanity a threshold that functions as a Great Spiritual Filter for any ETI in its ability to evolve beyond its regional cosmic origins?

A GREAT SPIRITUAL FILTER?

Beginning with a summary of elements, we can gather from the cosmologies of the three thinkers the following parameters for the importance of understanding and supporting a spiritual dimension, contributing to our discussion: ETI *as* ETS latently exist in a multiverse; the multiverse has material and spiritual dimensions; the spiritual dimension is pervading all existence; humanity is not the only spiritual species (on Earth or in the multiverse); there is extraterrestrial religiosity; ETS are emergent features of the aesthetic and spiritual nature of existence; ETS can be of nonhuman nature, e.g., "universal beings";[258] spiritual life civilizes the universe collectively; a universal medium of intercommunication has an evolutionary function that, by describing the intersection of all physical and spiritual universes and their beings, names the condition for, and a scale of the potential of, the evolution of ETI beyond their local existence.

If it has merit to consider that the chaotic threshold between biological and cultural evolution, engaged in genetic technology, nanotechnology, synthetic life, AI, combined with the forces of domination and degeneration, such as resource depletion, environmental disaster, ABC war, will self-destroy any technologically able, star-faring civilization[259]—which has been termed the Great Filter hypothesis[260]—may we not also ask, whether this threshold is, in fact, precisely a Great *Spiritual* Filter? In other words, may we not assume—given our evolutionary past and the present global predicaments, humanity is caught up in, and which, in some real sense, are expressions of the Darwinian struggles of life—that only a civilization that will develop beyond this threshold will, in fact, also survive its own self-destruction?[261] That is, how can we project humanity's survival—supposedly the intelligent life we seek in the cosmos—without seeing the warning signs of what will happen if we don't match technology with aesthetic, intelligence with spirituality, science with virtue, survival with compassion? Further, can any civilization of this kind develop by exclusion of the *aesthetic* side of its nature, that is, reach a stage of technology that can survive a transformation into, or by, AI, and *as such* become star-faring, or is the latency of spiritual life the condition for a cosmic evolution beyond local biological life?[262] Is the knowledge and insight of religiosity into developing the sense and the methods for advancing beyond Darwinian evolutionary mechanisms a condition not only for survival but for the proliferation of life in the cosmos?[263] Would, as the Dark Forest hypothesis assumes, a predatory expansion of a Darwinian ETI even be meaningful?

Consider this literary testament to the contrary. It comes from Greg Egan's (b. 1961) novel *Diaspora*, a fantastic, far-reaching exploration of the multiverse by post-human entities that, however, have not been designed (and are not self-designing) without the best virtues of humanity *in* an, and *as* an, "unchartered territory."[264]

> I keep asking myself, though: where do we go from here? History can't guide us. Evolution can't guide us. [Our] charter says understand and respect the universe . . . but in what form? On what scale? With what kind of senses, what kinds of mind? We can become anything at all—and that space of possible futures dwarfs this galaxy. Can we explore it without losing our way? Fleshers used to spin fantasies about aliens arriving to "conquer" Earth, to steal their "precious" physical resources, to wipe them out of fear for "competition" . . . as if a species capable of making the journey wouldn't have had the power, or the wit, or the imagination, to rid itself of obsolete biological imperatives. Conquering the galaxy is what bacteria with spaceships would do—knowing no better, having no choice.[265]

Are we really imagining ourselves, and all ETI, according to the Dark Forest hypothesis, just as mindless bacteria with spaceships, ravaging the interstellar space with Darwinian impulses of exploitation, survival, and disrespectful derogation of Life? *Should* we even try to export our unreformed human condition to space?[266] It is, of course, not impossible that the development of near-future scientific breakthroughs and still fantastic technologies may allow us to leave our planet—as current enterprises are definitively attempting, and in clear propagation of the current crisis into such an expansion: exploitation of minerals and energy sources, colonization of other habitats, inflating the weapons capability for reasons of a local hegemony of rivaling superpowers and the ones aspiring to become the ones of the future to the expense of all others. But isn't that exactly what will also contribute to the magnification of the *crisis* of human existence: war, occupation, exploitation, annihilation? Would such a "revolution," expanded by nanotechnologies, genetic engineering, and a potential AI singularity, not, as the metaphor goes, "eat its children"?[267]

Conversely, if ETS exist in the far reaches of the galaxy or beyond, that is, in reasonable distance for some kind of communication (even if this is, in the current situation, only a hypothetical or even fantastic possibility): Would its knowledge of how to become a spiritual civilization not be the most important wisdom to obtain and distribute? Would this wisdom not become the most important motive for exploration and contact? Would it not mean avoiding colonization for reasons of survival, and instead, initiating universal exploration as a means for civilizing the cosmos? So, if they exist, should this wisdom not already have been distributed? However, what if ETS civilizations may have become content with the insight that such a distribution of their wisdom beyond the threshold of the Great Spiritual Filter must imply the futility of communication by any technological means, which may equal avoiding any materially recognizable communication at all?

So, perhaps, if the Forest is *still* dark, perhaps it is *darker* than we thought: This darkness may indicate that ETS don't exist, that is, no one has overcome the Great Spiritual Filter,[268] *or*, given the assumptions of our three thinkers to the contrary, it may be a veil, indicating *our own inability* of connecting to, with, and through the medium of cosmic relationality: The medium of intercommunication may be *the* filter that is dark as long as it is not invoked, actualized, which means, becoming a spiritualized civilization, but it may also be *the* medium of communication for ETS *when* they have transcended the Great Spiritual Barrier. . . .

So, recollecting the arguments so far: What are the alternatives for the darkness of the forest? Sure, ETS may not exist, otherwise we would already know of them. ETS may exist at *some* time, but have not yet developed. ETS exist, but avoid contact, because self-destructive societies cannot be healed by the knowledge of their wisdom alone: transformation must ensue. ETS may be initiators of, or companions with, *our* spiritual evolution on Earth, but must also follow the rule of *self*-transcendence of spiritual evolution, rather than controlling it. Perhaps, the universe remains dark to the ones that have not reached post-Darwinian spiritual evolution. The forest is dark, because it is up to *us* to develop as an ETS, a civilization in concordance with our spiritual life beyond Earth. It is *our* darkness! In this sense, the Dark Forest is a metaphor that mediates a *spiritual* insight: It indicates a *spiritual crisis*—transformation or failure.

We cannot know what form such a spiritual transition beyond the crisis will take, whether it will tend to embody a universal wisdom or diverse modes of being spiritual and religious. But the principles of pleroma and decentralization may allow us to take seriously that, among its "characteristics," such a spiritual renewal will have to take into account *both* that the

Oneness of Reality *is* a Mystery, that is, apophatically open, *and*, therefore, is in its approximation infinitely renewable.[269] A cosmotheology that reduces projections in either direction—that of oneness as mediocrity without pleroma (without polyphilic proliferation) and that of decentralization without convergence (without approximation to apophatic oneness)—may fail to integrate the insights of our three exemplars of spiritual life in the cosmos, as would any naturalism without the mystical insight into the apophatic nature of Reality.[270] So, we may understand the "dark forest" as an expression of the apophatic insights of such alien civilizations that have overcome any motive of expansion for reasons other than the advance of this wisdom and its pleroma.[271] But this is a bet! Future will show.

LITERARY ARCHETYPES

If we ask where the image of the "dark forest" comes from, we will, besides perhaps psychological states, expressing themselves in experiences of a real dark forest or their similarity with hallucinations evoked in a sensory-depriving tank, and the highly symbolically encoded Grimm's Fairy Tales of lost children preyed on by wicked witches, find it to be a profound spiritual archetype, elucidating the insight into spirituality as a crisis of the spirit for finding orientation, transformation, and meaning. The works of two men of letters may demonstrate this function: Dante Alighieri (1265–1321)[272] and Leon Battista Alberti (1404–1472).[273]

Consider that the whole dramatic journey into the afterlife of Dante Alighieri's *Divine Comedy* (1300) begins literarily with him entering a dark forest and falling into a state of ecstasy that in its terror shakes him out of any security of the meaning of existence, engaging him in a far-reaching journey of the soul through the realms of human history, its failures and advances. This is how Dante describes his entrance into the forest:

> When I had journeyed half of our life's way,
> I found myself within a shadowed forest,
> Ah, it is hard to speak of what it was,
> that savage forest, dense and difficult,
> which even in recall renews my fear[274]

And in the mysterious *Hypnerotomachia Poliphilii* (1499), Leon Battista Alberti, ensuing on a journey through long lost civilizations, mythological specters, and secrets of wisdom, enters the dark forest with these impressions:

> Thus I steered my ignorant course toward a thick wood that lay opposite. Scarcely had I entered it than I realized that I had lost my way, I knew not how. A sudden fear entered my hesitating heart. . . . I realize that no track or side-path was to be seen in the thorny wood—nothing but dense thickets, sharp brambles, the wild ash that vipers shun, rough elms that suit the fruitless vine, thick-barked cork-oaks . . . that did not allow the sun's welcome rays to reach the damp soil, but covered it like a vaulted roof with dense leaves that the nurturing light could not penetrate. Thus I found myself in the . . . sylvan darkness.
>
> This made me suspect . . . that I had arrived at the vast Hercynian Forest, where there are nothing but the lairs of the dangerous beasts and caverns of obnoxious creatures and fierce monsters. I was defenseless and terribly afraid.[275]

Yet both literary figures find the crisis to be most illuminating, as it leads them to aesthetic and spiritual insights beyond the darkness of their disorientation and lostness. In the end, Dante's figure finds a guide toward the threshold of divinity which is veiled by beauty and a multitude of impenetrable light. Alberti's figure finds a clearing at a clear stream with refreshing water where he falls asleep, dreaming of art, architecture, and the treasures of wisdom.

In a philosophical rendering of the image, it is interesting to note that one of the most important notions of the late work of Martin Heidegger (1889–1976) is based on the dark forest: *die Lichtung des Seins*[276]—Being (*Sein*) becoming self-aware in our existential awakening (*Dasein*), imagined symbolically as a "clearing" (*Lichtung*) in the forest . . .

PROPOSITIONS

So, where has this journey through the Dark Forest and ETS brought us? Here are a few final propositions that answer the Dark Forest hypothesis and the Fermi paradox from the perspective of the assumption of ETS and the Great Spiritual Barrier for star-faring civilizations:

> *Proposition 1*: ETS is defined by the civilizational harmonics of a concordance between science and spirituality, individuality and society, organic and algorithmic existence, reason and mystery, wisdom and compassion.

> *Proposition 2*: If ETS do not develop, no star-faring civilization may have survived its own self-destructive use of the evolutionary mechanisms and transcended intellectual abilities to technological self-destruction. ETI may have fallen prey to the Great Spiritual Filter.

> *Proposition 3*: If ETS develop, however, it may be assumed that they solved the problems related to the Great Spiritual Filter and are able to expand beyond

their habitat with the spiritual nature of this endeavor in mind, constraining the runaway effects of evolutionary strives, material needs and technological abilities and potentials.

Proposition 4: Post-biological algorithmic intelligences (AI) will also be subject to the same self-destructive mechanisms if they were not to become aware of the aesthetic and spiritual nature of evolution and participate in the transition implied by this insight.

Proposition 5: Civilizational transition to a post-Darwinian evolution will include both achievements and failures: adventure will remain. We cannot know what form such a spiritual development will take, but we may assume that it includes an apophatic and a polyphilic element, that is, the Mystery of unity and a plurality of expressions.

Proposition 6: If the forest is dark, as we have not encountered any ETI or ETS yet, either they don't exist—and hence no ETI has crossed the Rubicon of space exploration—or they have found their spiritual integration to be unsuitable for expansion with technological means, or their silence is an exclamation mark on the imperative that only a self-transcending civilization will be able to communicate with ETS.

In the end: Are we approaching *Lichtung* or Darkness, an even Darker Forest? In any case, this conjecture is open to experimental testing: Humanity is the test subject—except "Fact Two" can be established. The questions, however, that the Spiritual Dark Forest addresses are these: Will humanity be able to overcome its predicaments, resulting from its persistent inability for transforming itself toward the harmonization of knowledge, technological capacities, and aesthetic and ethical concerns, environmental destruction, social disintegration, and the always latent global warfare? May philosophical and religious resources be a foe or a powerful ally in this proposed transformation? Or will we carry our evolutionary heritage to the stars as predators, exploiters, and enslavers, including economic, political, and religious competition and warfare—and survive such an attempt?

May our future decide . . .

NOTES

1. Cf. Dick, *The Biological Universe*, 73–76; *Many Worlds*, 12, 74–76, 123–26, 179, 192.

2. Philosophical and religious, secular and utopian speculations have always permeated the vision of the future of humanity and planet Earth regarding a transformation that will overcome the predicaments of the human condition and the pains

memorized in historical consciousness. Not all of them are optimistic as regarding the extent of such a transformation, if at all it is assumed, but generally, a cosmological background was always admitted as furthering or hindering such a process. This is the context in which humanity seeks its worth in a universe that is beyond its control: making sense of it all by referring to deeper layers of comic harmonies and rhythms, be it dialectical materialism or the ethical Dharma, salvation history or social utopias. And it will be driving the discussions that follow.

3. A detailed report of Intergovernmental Science-Policy Platform on Biodiversity and Ecosystem Services understands the intricate link between biodiversity and extinction of species as indicating the endangerment of one million species on Earth today, See UN Report, "Nature's Dangerous Decline:" https://www.un.org/sustainabledevelopment/blog/2019/05/nature-decline-unprecedented-report/.

4. It stands to reason that, if evolution includes a tendency to develop complex forms of matter and life, humanity, in the differentiation of the diverse hominids that roamed the Earth, is not the only species that could develop culture to the extent of a self-conscious embrace of coevolution or the transcendence of the mechanism that has produced it, transcending humanity in its ability to not only survive but live differently. Cf. Genta, *Lonely Minds in the Universe*, 154–56; Dick, *The Impact of Discovering Life beyond Earth*, 103; *Space, Time, and Aliens*, ch. 8; Faber, *Ocean of God*, ch. 13.

5. Cf. Ashkenazi, *What We Know About ETI*, 65–68; Ćirković, *The Great Silence*, 6:7; Genta, *Lonely Minds in the Universe*, 101–2. Already in 1961, the astrophysicist Sebastian von Hoerner estimated that the development of technology will be driven by superiority and easiness of life, which will, in a feedback loop, destroy the basis for its own development; cf. Hoerner, "The Search for Signals," 1829–1843. If technology is the potentiation of animal drives such as aggression, subjection of the "other" and domination, culture fulfills evolutionary necessities of survival as ultimate means of its own destruction by invariably binding cultural potentials to the natural mechanisms of self-fulfillment. But this is the question: Are we bound to these mechanisms or can we transform culture into a different character that is driven by insight and choice rather than necessity, or, in another sense, by beauty rather than domination? Cf. Odin, *Tragic Beauty in Whitehead and Japanese Aesthetics*.

6. Cf. Genta, *Lonely Minds in the Universe*, 154–64.

7. Cf. Haught, *The New Cosmic Story*, ch. 1. Whether such forces of composition instead of decomposition, and in the midst of the always changing character of such compositions, evolutionary and culturally, socially and politically, indicate any "progress" of civilization or just a sideways movement that, eventually, will lead to self-destruction, is the open question. But to ignore signs of such human virtues and the way they touch us when we find them practiced, and which fill our histories as extraordinary events or as stories that grant identity to families, societies, and cultures, would be equal to denying the aesthetical impact of experience and the ethical dives in human expressions of life.

8. Indeed, the very stunning wonder of the sky above, its virtual (at least felt) infinity, as present and deeply touching us in our heart, suggest the spiritual dimension of the exploration of the universe; cf. Peters, *Astrotheology*, 10–11.

9. Cf. Faber, *The Cosmic Spirit*; *Divine Appearances*, ch. 4. As an example of such rather novel visions that bind themselves to the Earth and the transformation of humanity in companionship with the Earth, rather than its exploitation, reaches back, before the ecological end ecophilosphical awakening in the middle of the twentieth century, to such thinkers and sages as Sri Aurobindo (1872–1950), indicating a spiritual evolution instead of a mere material evolution; cf. Aurobindo, *The Life Divine*; Satprem, *Sri Aurobindo*; O'Neil, *Toward the Life Divine*. The figures that this chapter will feature later entertain the same vision.

10. Cf. Genta, *Lonely Minds in the Universe*, chs. 3–4.

11. Ibid., 217–22; Ashkenazi, *What We Know About ETI*, ch. 4.

12. Cf. Dick, *The Impact of Discovering Life Beyond Earth*, Part 2; Ashkenazi, *What We Know About ETI*, ch. 9; Genta, *Lonely Minds in the Universe*, ch. 1; Guthke, *The Last Frontier*, ch. 4.

13. Cf. Stenger, *God and the Multiverse*; *The Unconscious Quantum*; Churchland, *Matter and Consciousness*; Haught, *Science and Religion*; Nagel, *Mind and Cosmos*; van Huyssteen, *Alone in the World?*, 17, 55–57, 84–94, 102–5, 112, 159, 218–19, 258–70, 277, 323; Richerson and Boyd, *Not by Genes Alone*, 57, 163, 169; Wills, *Children of Prometheus*, 22, 50–53.

14. Cf. Wilkinson, *Science, Religion and the Search for ETL*, ch. 5; Dick, *The Impact of Discovering Life Beyond Earth*, ch. 2 (Iris Frey); *Space, Time, and Aleins*, ch. 9; Genta, *Lonely Minds in the Universe*, ch. 3; Ashkenazi, *What We Know About ETI*, ch. 4; Haught, *Is Nature Enough?*; Davies, *Are We Alone?*, 73–87. The background for this assumption is, of course, the victory of the paradigmatic assumption that we live in a fundamentally biological universe; cf. Dick, *The Biological Universe*, ch. 11; Vakoch, *Astrobiology, History, and Society*, ch. 9.

15. Cf. Dick, *Many Worlds*, Parts 1–2.

16. Cf. Herrick, *The Making of the New Spirituality*; *Scientific Mythologies*.

17. Cf. Ashkenazi, *What We Know About Extraterrestrial Intelligence*, ch. 8; Haught, *The New Cosmic Story*; Dick, *The Impact of Discovering Life Beyond Earth*, chs. 8 (John Traphagan) and 10 (Mark Lupisella).

18. Cf. Faber, *The Cosmic Spirit*.

19. Cf. Webb, *If the Universe Is Teeming with Aliens*, Solution 24, 31–32; Genta, *Lonely Minds in the Universe*, 209–11, 249–54; Forgan, *Solving Fermi's Paradox*, ch. 20:3; Davies, *Are We Alone?*, ch. 2; Wilson, *The Meaning of Human Existence*, chs. 8–9; Vakoch, *Astrobiology, History, and Society*, ch. 11.

20. Cf. Davies, *Are We Alone?*, ch. 4; Weintraub, *Religions and Extraterrestrial Life*, ch. 1; Bostic, *The Fermi Paradox*, ch. 1.

21. Cf. Ashkenazi, *What We Know About ETI*, 14–16; Wilkinson, *Science, Religion and the Search for ETI*, 39–41; Genta, *Lonely Minds in the Universe*, 26, 165, 179; Dick, *Many Worlds*, 145, 166; *The Impact* of *Discovering Life* 293, 326; *Space, Time, and Aliens*, ch. 7; Davies, *Are We Alone?*, 139–40; Willis, *All These Worlds Are Yours*, 172–75; Forgan, *Solving the Fermi Paradox*, 7–11; Ćirković, *The Great Silence*, ch. 3:9; Traphagan, *ETI and Human Imagination*, ch. 3; Harrison and Vakoch, *Civilizations Beyond Earth*.

22. Cf. Genta, *Lonely Minds in the Universe*, 47–50; Bodin, *Fermi's Paradox*, ch. 10; Ward and Brownlee, *Rare Earth*, 267–68; Vakoch, *Astrobiology, History, and Society*, ch. 13.
23. More conservative estimates using this formula compute 30,000 or 10,000 civilizations eventually being able to communicate; cf. Genta, *Lonely Minds in the Universe*, 166. Carl Sagan estimated one million civilizations in our galaxy; cf. Wilkinson, *Science, Religion and the Search for ETL*, 140; Dick, *Many Worlds*, 166.
24. Cf. Vakoch and Dowd, *The Drake Equation.*
25. Cf. Genta, *Lonely Minds in the Universe*, 167; Ashkenazi, *What We Know about ETI*, 16–17.
26. Cf. Genta, *Lonely Minds in the Universe*, 164–76; Ashkenazi, *What We Know about ETI*, ch. 2:2; Dick, *The Impact of Discovering Life Beyond Earth*, ch. 19 (John and Julian Traphagan); *Life on Other Worlds*, ch. 7; Davies, *Are We Alone?*, 1, 139–40; Forgan, *Solving The Fermi's Paradox*, ch. 2.
27. Cf. Bodin, *Fermi's Paradox*, chs. 6–8.
28. Cf. Willis, *All These Worlds Are Yours*, 172–75; Hanlon, *The Drake Equation.*
29. Cf. Willis, *All These Worlds Are Yours*, 172–73; Genta, *Lonely Minds in the Universe*, 165–67; Wilkinson, *Science, Religion and the Search for ETL*, 40; Ashkenazi, *What We Know about ETI*, 14–16.
30. Cf. Dick, The *Impact of Discovering Life Beyond Earth*, 46–51, 189–204; *Space, Time, and Aliens*, ch. 5:5.
31. Cf. Bowman, *AI & The Fermi Paradox*; Forgan, *Solving Fermi's Paradox*, chs. 12–15.
32. Cf. Webb, *If the Universe Is Teeming with Aliens*, Solution 32.
33. Cf. Bodin, *Fermi's Paradox*, ch. 13; Webb, *If the Universe Is Teeming with Aliens*, Solutions 18–20; Zhu, *On the Origin of ET Industrial Civilizations*, ch. 9.
34. Cf. Vakoch and Dowd, *The Drake Equation*, chs. 11 (Florence Raulin Cerceau), 12 (Seth Shostak).
35. Cf. Dick, *Many Worlds*, ch. 10 (Jill Tarter).
36. Cf. Forgan, *Solving Fermi's Paradox*, part 3; Webb, *If the Universe Is Teeming with Aliens*, Solutions, 60–62; Genta, *Lonely Minds in the Universe*, 86–96. For the Doomsday argument of Brandon Carter and Richard Gott, cf. Wilkinson, *Science, Religion and the Search for ETL*, 107–9; Forgan, *Solving Fermi's Paradox*, ch. 7; Davies, *Are We Alone?*, 48, 61–66, 81; Dick, *Many Worlds*, ch. 8 (John Leslie).
37. Cf. Webb, *If the Universe Is Teeming with Aliens*, Solution 58; Zhu, *On the Origin of ET Industrial Civilizations*, ch. 8.
38. Cf. Vakoch and Dawd, *The Drake Equation*, chs. 13 (David Nunér), 14 (Gary Chick); Dick, *Many Worlds*, 145; Zhu, *On the Origin of ET Industrial Civilizations*, ch. 11:2.
39. Cf. Catling, *Astrobiology*; Vakoch, *Astrobiology, History, and Society*; Dick, *Life on Other Worlds*, ch. 9; *Space, Time, and Aliens*, ch. 2:4; Peters, *Astrotheology*, ch. 1.
40. Cf. Dick, *The Impact of Discovering Life Beyond Earth*, 45, 65, 287; Genta, *Lonely Minds in the Universe*, 229; Forgan, *Solving* Fermi's Paradox, ch. 25:2: Davies, *Are We Alone?*, 132–36; Ćirković, *The Great Silence*, 1.

41. Cf. Ashkenazi, *What We Know about ETI*, 6, 14. Forgan, *Solving Fermi's Paradox*, ch. 2 calls it "Fact A."

42. Cf. Crow, *The ETL Debate, Antiquity*; *The ETL* Debate, 1750; Dick, *Life on Other Worlds*; *Plurality of Worlds*; *Space, Time, and Aliens*; *The Biological Universe*; *The Impact of Discovering Life Beyond Earth*; *Many Worlds*; Weintraub, *Religions and ETL*.

43. Cf. Webb, *If the Universe Is Teeming with Aliens*; Bostic, *The Fermi Paradox*, ch. 4; Ćirković, *The Great Silence*; Bodin, *Fermi's Paradox*, ch. 11; Forgan, *Solving Fermi's Paradox*, 3–7; Dick, *Many Worlds*, 120–23; Weintraub, *Religions and ETL*, 5, 28.

44. For a classification of solutions to Fermi's Paradox, differentiating Rare Earth, Uncommunicating, Catastrophic, Hard and Soft solutions, cf. Forgan, *Solving Fermi's Paradox*, 43–45. For the whole array of seventy-five solutions; cf. Webb, *If the Universe Is Teeming with Aliens*, chs. 3–5.

45. Cf. Webb, *If the Universe Is Teeming with Aliens*, ch. 5; Forgan, *Solving Fermi's Paradox*, chs. 5–6.

46. Cf. Bodin, *Fermi's Paradox*, ch. 14; Vakoch, *Astrobiology, History, and Society*, ch. 8.

47. Cf. Webb, *If the Universe Is Teeming with Aliens*, Solutions 51–57; Dick, *The Biological Universe*, ch. 4; *The Impact of Discovering Life Beyond Earth*, 32, 47–49, 51; Wilkinson, *Science, Religion and ETL*, ch. 5; Willis, *All These Worlds Are Ours*.

48. Cf. Webb, *If the Universe Is Teeming with Aliens*, Solutions 25–31, 68–73; Forgan, *Solving Fermi's Paradox*, chs. 19, 22; Ćirković, *The Great Silence*, ch. 9:4–5; Zhu, *On the Origin of ET Industrial Civilizations*, ch. 8:6.

49. Cf. *Rare Earth*, ch. 8:2; Forgan, *Solving Fermi's Paradox*, chs. 19, 21.

50. Cf. Forman, *Solving Fermi's Paradox*, chs. 15, 23; Bodin, *Fermi's Paradox*, 14:11; Ćirković, *The Great Silence*, ch. 9:2.

51. Cf. Davies, *Are We Alone?*, 144; Forgan, *Solving Fermi's Paradox*, ch. 25:1.

52. Cf. Webb, *If the Universe Is Teeming with Aliens*, Solution 50; Forman, *Solving Fermi's Paradox*, ch. 22.

53. Cf. Forgan, *Solving Fermi's Paradox*, part 2; Ward and Brownlee, *Rare Earth*, ch. 5; Zhu, *On the Origin of ET Industrial Civilizations*, ch. 8.

54. Cf. Dick, *The Impact of Discovering Life Beyond Earth*, 268, 276. For arguments for and against METI, cf. Forgan, *Solving Fermi's Paradox*, ch. 21:4; Dick, *Space, Time, and Aliens*, ch. 15.

55. Cf. Bodin, *Fermi's Paradox*, ch. 5:6; Forman, *Solving Fermi's Paradox*, chs. 20:3, 21.

56. Cf. Forman, *Solving Fermi's Paradox*, chs. 15, 20; Ćirković, *The Great Silence*, ch. 4.

57. Cixin, *The Dark Forest*, 484.

58. Cf. Forgan, *Solving Fermi's Paradox*, chs. 12–14.

59. At least since the vision of *The War of the Worlds* (1879) by H. G. Wells and its actualization by Orson Welles in a 1938 radio piece that allegedly, at least in hindsight and media coverage, which is itself remarkable, created panic, evil aliens have become a staple in science fiction literature and medial presentation, not only

projecting human conflicts from history and prophesies of invasions of earthly powers at that time, such as diverse massive colonizations and territorial wars, into the future, but mirroring this future images as motifs of the critique of current social realities to us—perhaps, as moral allegories, if not as apocalyptic warnings; cf. Guthke, *The Great Frontier*, ch. 5:3. Darwin's delineation of evolutionary competition, natural/cultural selection, and the eternal fight for superiority as perceived from human history had its influence on this kind of vision, as had the anti-imagination of pacifism—we should not forget that H. G. Wells was adamant about the future of humanity as a cosmopolitan community if it wanted to avoid destruction; cf. Wells, *The War of the Worlds*; *The Rights of Man*; *The New World Order*; *The Last Books*; Wilkinson, *Science, Religion, and the Search for ETL*, 176.

60. Cf. Bodin, *Fermi's Paradox*, ch. 10; Forgan, *Solving Fermi's Paradox*, chs. 16–17; Ashkenazi, *What We Know about ETI*, chs. 9–10; Wilkinson, *Science, Religion, and the Search for ETL*, 125–26, 176; Davies, *Are We Alone?*, 66–70; Dick, *Life on Other Worlds*, ch. 4.

61. Cf. Ćirković, *The Great Silence*, ch. 8:2.

62. Cf. Forgan, *Solving Fermi's Paradox*, 255, 270, 282–83; Bodin, *Fermi's Paradox*, 212, 281–83, 290; Genta, *Lonely Minds in the Universe*, 155, 161, 213, 220, 252; Davies, *Are We Alone?*, 68–73: The exploration of self-replicating machines would, given the mathematical estimates of Frank Tipler, not take as long as it would be to estimate the time that it needs for a civilization to develop as ours without, in fact, assuming that they must have already arrived.

63. Cf. the Zoo hypothesis that could be a peaceful reminder that we are not ready to join the intergalactic community or a sign of a hunter waiting until the pray is ready to be slaughtered; cf. Bodin, *Fermi's Paradox*, 290; Ćirković, *The Great Silence*, ch. 4:4; Forgan, *Solving Fermi's Paradox*, ch. 23; Wilkinson, *Science, Religion, and the Search for ETL*, 102–15; Webb, *If the Universe Is Teeming with Aliens*, chs. 7–9.

64. Cf. Forgan, *Solving Fermi's Paradox*, ch. 6.

65. Cf. Genta, *Lonely Minds in the Universe*, 17–18, 247–49; Wilkinson, *Science, Religion, and the Search for ETL*, 98–102, 108–9; Dick, *Space, Time, and Aliens*, chs. 12, 19.

66. Cf. Ashkenazi, *What We Know about ETI*, ch. 4; Wilkinson, *Science, Religion, and the Search for ETL*, ch. 2:3; Dick, *The Impact of Discovering Life Beyond Earth*, chs. 5 (Dirk Schulze-Makuch) and 7 (Carlos Mariscal); Genta, *Lonely Minds in the Universe*, ch. 3; Dick, *Many Words*, part 1; Ward and Brownlee, *Rare Earth*, ch. 7; Forgan, *Solving Fermi's Paradox*, ch. 5; Zhu, *On the Origin of ET Industrial Civilizations*, ch. 4; Bodin, *Fermi's Paradox*, chs. 6–8; Wilson, *The Meaning of Human Existence*, ch. 10; Vakoch, *Astrobiology, History, and Society*, ch. 13. Reductionist in several senses, at least, by reducing the selection process to struggle and survival, not taking into account evolvability as itself developing (D. Dennett), self-organization (S. Kaufman and B. Goodwin), genetic assimilation (C. H. Waddington), individual innovation in environments (Hardy), individual state-related developments of organisms (R. C. Lewontin), top-down causation by environmental information flow (Campbell), group selection (D.S . Wilson and E. Sober), gene hopping (G.A. Dover), contextual adaptive impacts on change and stasis (N. Eldredge), molecular

evolutionary change, immune to natural selection (M. Kimura); and, more generally, by not taking into account the creative and cooperative process of environmental mutuality of and with organisms that is misrepresented by reducing it to "struggle for life" and "survival of the fittest": cf. Arthur Peacock's arguments in Dick, *Many Minds*, 96–97; *The Impact of Discovering Life Beyond Earth*, 155, 165–66, 327–31; For the definition of biological Life, cf. Willis, *All These Worlds Are Yours*, 29–31, 47–48, 51, 136–38, 161, 183–85.

67. Cf. Ward, *God, Chance and Necessity*, 173–78; O'Connell and Ruse, *Social Darwinism*; Wilson, *The Meaning of Human Existence*, ch. 6; Ćirković, *The Great Silence*, 6–7; Ashkenazi, *What We Know about ETI*, ch. 10:6–7; Haught, *The New Cosmic Story*, 165–68; *In Nature Enough?*, 6–16, 19, 108–12, 117, 143–47, 153–80; Wilkinson, *Science, Religion and the Search for ETL*, 95–97; Dick, *The Impact of Discovering Life Beyond Earth*, 24–34, 51, 96, 129–41, 184, 299; *Many Worlds*, 94–99, 108–9; Davies, *Are We Alone?*, 74–80; Genta, *Lonely Minds in the Universe*, 76; Bodin, *Fermi's Paradox*, 106–7; Zhu, *On the Origin of ET Industrial Civilizations*, 2; Wills, *Children of Prometheus*, 252–54; Richerson and Boyd, *Not by Genes Alone*, 16; van Huyssteen, *Alone in the World?*, 38, 55, 77–82, 98–99.

68. The Netflix production of a series on *The Three-Body Problem* was canceled when US lawmakers questioned, in September 2020, its production, because Liu Cixin showed support for China's treatment of its Uighur minority in Xinjiang.

69. Cf. Wilkinson, *Science, Religion and the Search for ETL*, 94–95.

70. Cf. Peters, *Astrotheology*, 85; Ashkenazi, *What We Know about ETI*, ch. 8. Generally, assumptions of Intelligence in related ETI literatures are less interested in aesthetic rather than logic, that is, ETI is more imagined in terms of science and mathematics than art and religion, ethics and spirituality—perhaps a bias regarding possibilities of communication, given the diversity of cultural symbolisms invested with the transport of meaning over the assumed unity of physical reality and its mathematical formulation. Cf. Wilkinson, *Science, Religion and the Search for ETL*, 41–44, 76–79, 83–94, 124, 134, 165, 173; Dick, *The Impact of Discovering Life Beyond Earth*, 127–53; Genta, *Lonely Minds in the Universe*, 179; Ward and Brownlee, *Rare Earth*, 250–51; Forgan, *Solving Fermi's Paradox*, ch. 6:3–6; Zhu, *On the Origin of ET Industrial Civilizations*, 14, 221; ch. 10; Bodin, *Fermi's Paradox*, 290, ch. 6:8; Ćirković, *The Great Silence*, xvii, 124; Peters, *Astrotheology*, 421.

71. Cf. Guthke, *The Great Frontier*, ch. 2, 234; Ćirković, *The Great Silence*, ch. 3:4; Dick, *Many Worlds*, 56; *Life On Other Worlds*, 65–67, 95–98, 101–3, 337–38, 411–14, 424, 524.

72. Cf. Ćirković, *The Great Silence*, ch. 9; Ward and Brownlee, *Rare Earth*, xiv, xxvii, 272, 280, 316–17, 322, 331; Forgan, *Solving Fermi's Paradox*, 128; Zhu, *On the Origin of ET Industrial Civilizations*, 5–14, 24, 310–12, 335–37, 374, 548, 575–83; Bodin, *Fermi's Paradox*, xix–xx, 80–88, 180–81, 196–98, 200–9, 218, 278; 24, 78, 86, 260; Ashkenazi, *What We Know about ETI*, 21; Dick, *The Impact of Discovering Life Beyond Earth*, 55; *Many Worlds*, 56; Davies, *Are We Alone?*, 23, 20–30, 44, 71, 143, 147–48; Genta, *Lonely Minds in the Universe*, xxi, 49.

73. Cf. Faber, *The Divine Manifold*, chs. 5, 13; *The Ocean of God*, 44–48; *The Garden of Reality*, 100, 186, 192–94, 251–52, 376, 380; Ashkenazi, *What We Know*

about ETI, 299; Davies, *Are We Alone?*, 3, 22, 124–28, 147–48; Dick, *Many Worlds*, 191; *The Impact of Discovering Life Beyond Earth*, 65, 78, 113–24, 222; Wilkinson, *Science, Religion, and the Search for ETL*, 37, 73, 182; Ward and Brownlee, *Rare Earth*, i–xxix; Traphagan, *ETI and Human Imagination*, chs. 2., 5; *Science, Culture and the Search for Life on Other Worlds*.

74. Cf. Faber, *The Divine Manifold*, ch. 14; "Ecotheology"; Bodin, *Fermi's Paradox*, ch. 10; Forgan, *Solving Fermi's Paradox*, 262; Ward and Brownlee, *Rare Earth*, xiii; Zhu, *On the Origin of ET Industrial Civilizations*, 10, 14; Ćirković, *The Great Silence*, 24, 138, 255; Wilkinson, *Science, Religion, and the Search for ETL*, 22, 26, 37–38, 74, 107; Dick, *The Impact of Discovering Life Beyond Earth*, 3, 27, 32, 66, 77, 96–97, 155, 208–9, 218, 256, 286, 300, 306; *Many Worlds*, vii, x, 10–12, 73–75, 123–26, 178–81, 192, 2041, 204; *Life*, ch. 8:3; Davies, *Are We Alone?*, 48, 61–66, 115–18, 137, 147; Genta, *Lonely Minds in the Universe*, 47; Haught, *The New Cosmic Story*, 13, 34–36, 61, 164.

75. Cf. Dick, *Many Worlds*, 128; Wilkinson, *Science, Religion, and the Search for ETL*, ch. 2:1; Guthke, *The New Fortier*, 39.

76. Cf. Wilkinson, *Science, Religion, and the Search for ETL*, 17, 73–74, 108, 182; Dick, *Many Worlds*, 163; *The Impact of Discovering Life Beyond Earth*, 34, 60, 124–25; *Life*, 17–20, 35–44, 6, 73, 118, 123, 134, 139, 153, 211, 226, 260, 269, 392, 409, 524; Davies, *Are We Alone?*, 3, 22–23, 145; Dick, *Many Worlds*, 163; Weintraub, *Religions and ETL*, 12, 15–21, 28, 158, 201.

77. Cf. Davies, *Are We Alone?*, 77–80; Ashkenazi, *What We Know about ETI*, ch. 10:1; Dick, *Many Worlds*, 6–8, 91–97, 105; *The Impact of Discovering Life Beyond Earth*, 29, 163–66; Wilkinson, *Science, Religion, and the Search for ETL*, 2, 67–76, 106–18; Genta, *Lonely Minds in the Universe*, 76–77, 133, 162, 109, 198, 242; van Huyssteen, *Alone in the World?*, 38, 48, 62, 70–99, 106–7, 101–95, 229–37, 262–66;

78. Cf. Bodin, *Fermi's Paradox*, 181, 213, ch. 14; Faber, *The Ocean of God*, ch. 13; Genta, *Lonely Minds in the Universe*, 102; Dick, *The Biological Universe*, ch. 2:2.

79. Cf. O'Meara, *Vast Universe*, ch. 1; Bodin, *Fermi's Paradox*, chs. 10, 11:3; Zhu, *On the Origin of ET Industrial Civilizations*, 13, 18, 141, 207–9, 582; ch. 7:1; Forgan, *Solving Fermi's Paradox*, 6, 10, 12, 20, 128, 155; Ward and Brownlee, *Rare Earth*, x–xi, 29, 69, 272; Ćirković, *The Great Silence*, 315n93; Peters, *Astrotheology*, x; Genta, *Lonely Minds in the Universe*, 102–3.

80. Cf. Ćirković, *The Great Silence*, xvii.

81. Cf. Pickover, *Keys to Infinity*, ch. 4; Maor, *To Infinity and Beyond*, part 4; Hanlon, *Equation*, 1–14; Bodin, *Fermi's Paradox*, xxi. However, so could a large number also be used against this probability, e.g., by intervening with the large number of steps that would be necessary and hence are improbable in assuming the emergence of intelligent life; cf. Hanlon, *Equation*, 189; Peters, *Astrotheology*, xi; Davies, *Are We Alone?*, 146.

82. 6:2. Cf. Peters, *Astrotheology*, 276; Davies, *Are We Alone?*, 29–30, 145; Crowe, *The ETL Debate, 1750*, 312–13, 461, 531, 553; Dick, *Life on Other Words*. While the unrestrained calculation with great numbers regarding probabilities of life and intelligence may be unscientific (Davies, "Foreword," in Peters, *Astrotheology*, x), any reasonable restraint will produce finite numbers or degrees of infinities different

from a chaotic equilibrium. E.g., when we assume—as an epistemological standpoint or paradigm, embraced by astrobiology today—that the universe is biophilic (Peters, *Astrotheology*, 5–6), this "unscientific" presupposition is a restraint on large number yeading interesting results, at least, the motivation to go on without evidence to search and find out whether this is the case.

83. Cf. Peters, *Astrotheology*, 77, 153–57, 359–61; Zhu, *On the Origin of ET Industrial Civilizations*, 247; Forgan, *Solving Fermi's Paradox*, 128; Ward and Brownlee, *Rare Earth*, 83, 134–35, 155; Ćirković, *The Great Silence*, 88, 198; Hanlon, *Equation*, 189; Dick, *many Worlds*, 105–6; Ashkenazi, *What We Know about ETI*, 43, 230, 310.

84. Cf. Conway Morris, *Life's Solution* and *Runes of Evolution*; Conway Morris and Gould, "Showdown"; Gould, *The Panda's Thumb* and *The Structure of Evolutionary Theory*; Denton, *Nature's Destiny*, *Evolution: A Theory in Crisis*, and *Evolution: Still in Crisis*.

85. Cf. Bodin, *Firmi's Paradox*, 128; Genta, *Lonely Minds in the Universe*, 105–7; Dick, *The Impact of Discovering Life Beyond Earth*, 96.

86. Cf. Wilson, *The Meaning of Human Existence*, chs. 8–10.

87. Cf. Duncan, *Solving Fermi's Paradox*, chs. 11, 15; Ćirković, *The Great Silence*, chs. 3, 8;

88. Cf. Peters, *Astrotheology*, 260, 317; Duncan, *Solving*, ch. 6:5.

89. Cf. Burton, Hollman and Parker, *Nicolas of Cusa*; Peters, *Astrotheology*, 83, 161; Vakoch and Dowd, *The Drake Equation*, 58, 64, 164.

90. Cf. Weintraub, *Religions and ETL*, 40–41.

91. Cf. Guthke, *Frontier*, ch. 1;4, 2; Crow, *The ETL Debate, 1750*, ch. 1.

92. Cf. Faber, *The Cosmic Spirit*, 197.

93. Cf. Bond, *Nicolas of Cusa*, 19–35; Dick, *The Impact of Discovering Life Beyond Earth*, 235; Wilkinson, *Science, Religion and the Search for ETL*, 19; Genta, *Lonely Minds in the Universe*, 8–9; Faber, *The Garden of Reality*, 466; *The Ocean of God*, 22; *God as Poet*, 113.

94. Cf. Duhem, *Medieval Cosmology*, 505–11; Crow, *The ET Life Debate, 1750*, chs. 2–3; Guthke, *The New Frontier*, chs. 1:4, 2; Dick, *Plurality of Worlds*.

95. Cf. Wallace, *The Emergent Multiverse*; Stenger, *God and the Multiverse*; Kraay, *God and the Multiverse*; Rubinstein, *Worlds Without End*; Leslie, *Universes*.

96. Cf. Faber, *The Cosmic Spirit*, ch. 8.

97. Cf., Weintraub, *Religions and ETL*, ch. 5.

98. Cf. Dick, *Many Worlds*, 204.

99. Cf. Davies, *Are We Alone?*, 137–38.

100. Cf. Dick, *Life on Other Worlds*, ch. 5; *The Biological Universe*, ch. 5; Ćirković, *The Great Silence*, ch. 4:2; Forgan, *Solving Fermi's Paradox*, ch. 25:2; Davies, *Are We Alone?*, 12, 132–36; Wilkinson, *Science, Religion and the Search for ETL*, 105–11, 120–21, 166; Ashkenazi, *What We Know about ETI*, 85–89, 95–97, 187; Dick, *The Impact of Discovering Life Beyond Earth*, 45, 65, 287; Herrick, *The Making of the New Spirituality*, ch. 9.

101. Cf. Vakoch, *Astrobiology, History, and Society*, chs. 15, 16, 19; Ashkenazi, *What We Know about ETI*, ch. 11; Wilkinson, *Science, Religion and the Search for ETL*, 40–41, 76, 83–94, 124, 134, 158, 165, 173; Genta, *Lonely Minds in the*

Universe, 179; Dick, *The Impact of Discovering Life Beyond Earth*, 143–53; Guthke, *The New Frontier*, 11–15.

102. Cf. Weintraub, *Religions and ETL*; Faber, *The Ocean God*, ch. 13. There is, of course, a difference between the resources that religions provide per their ideas and practices as inherited by their oral and scriptural basis and the theological and philosophical explorations, except that in certain cases they are identical, such as in the *Upanishads* and certain writings of Baha'u'llah and 'Abdu'l-Baha for the Baha'i religion; cf. Taylor, *One Reality*. There are also differences between the cosmological magnitudes that are presupposed in diverse religious heritage, proposing a small or large universe, infinite in time and space or limited, constraining their ability to imagine or claim the plurality of worlds as one of planets or universes. For the theological developments especially in the Christian infused west, cf. Crow, *The ETL Debate, Antiquity* and *The ETL Debate, 1750*; Dick, *Life on Other Worlds*, *Plurality of Worlds*, and *The Impact of Discovering Life Beyond Earth*; Duhem, *Medieval Cosmology*; Traphagan, *ETI and Human Imagination*. For current theological discussion: Wilkinson, *Science, Religion and the Search for ETL*; Ward, *God, Chance, Necessity*; Peters, *Astrotheology*; Stenger, *God and the Multiverse*; Davies, *Are We Alone?*; Paul, *Science, Religion, and Mormon Cosmology*; Kisak, *Religious Cosmology*; Lewels, *The God Hypothesis*; Rubinstein, *Worlds Without End*; O'Meara, *Vast Universe*; and literature to the three exemplary figures related in the coming parts of this chapter. For an eastern approach, cf. Dick, *The Impact of Discovering Life Beyond Earth*, ch. 19 (John and Julian Traphagan).

103. Cf. Faber, *The Cosmic Spirit*, Preamble, ch. 1; Haught, *The New Cosmic Story*, ch. 1: especially if one follows the idea that the awakening to humanity implies not just symbolism and language, culture and technology, but also an evolutionary development that is significant to humanity itself, namely, the emergence of religion or a spiritual dimension of human existence; cf. Faber, *The Ocean of God*, ch. 1.

104. Cf. Davies, *Are We Alone?*, 131–36; Wilkinson, *Science, Religion and the Search for ETL*, 145, 150.

105. Cf. Ashkenazi, *What We Know about ETI.*

106. Cf. Faber, *The Cosmic Spirit*; Genta, *Lonely Minds in the Universe*, 100–1.

107. Cf. Wilkinson, Wilkinson, *Science, Religion, and the Search for ETL*, 71, 139; Dick, *The Impact of Discovering Life Beyond Earth*, 96–105; Ashkenazi, *What We Know about ETI*, 76, 83, 83, 244.

108. Cf. Genta, *Lonely Minds in the Universe*, 107–8. It is relevant that "Compassion" is generally missing from such lists of ETI, not to speak of ETS, while it may be essential in the definition of ETS. Teilhard speaks of the spiritual character of the ability to transform pain into spiritual energy; cf. Teilhard, *Writings*, 145–48.

109. Cf. Quincey, *Consciousness from Zombies to Angels*; Chalmers, *The Character of Consciousness*; *The Conscious Mind*; Stapp, *Mindful Universe*; Nagel, *Mind and Cosmos*; Genta, *Lonely Minds in the Universe*, 105; Dick, *Space, Time, and Aliens*, ch. 14; Davies, *Are We Alone?*, ch. 5; Dick, *The Impact of Discovering Life Beyond Earth*, 201–13, 175–204; Genta, *Lonely Minds in the Universe*, 131, 203; Ward, *God, Chance, Necessity*, ch. 8.

110. Cf. LeFavre, *The Sound of Ontology*; Dick, *Many Worlds*, 183; Ashkenazi, *What We Know about ETI*, 224, 355–56.

111. Cf. Wilkinson, *Science, Religion, and the Search for ETL*, 139.

112. Cf. ibid., 140–42.

113. It is interesting that this element, the ability and practice of mending woundings of body and soul, is less present in ET discussions, as related both to medicine and soteriology, the latter of which almost exclusively in theological discussions, except in the context of questions of the meaning of existence, a direction of evolution, and the transformation of society; cf. Lieberman, *The Story of the Human Body*.

114. Cf. Faber, *The Cosmic Spirit*, Preamble. As the comparison of Plotinus with the Stoic's cosmology shows, the search for the Oneness of existence is not a matter of religiously infused philosophy alone; the Stoics' Spirit (*pneuma*) is a refined material substance. And this spirit can, on occasion, be merged with a cosmic or divine mind (nous), as visible in Aristotle, Plotinus, Neo-Platonism, Philo of Alexandria, and early Logos-Christology, such as that of Origen, but also Teilhard, *Writings*, part 3. In later trinitarian speculations within Christianity, Spirit can also appear as a more relational category, less related to ideas (mind and intellect) than love, or, as in monistic systems, as the All in its ultimate expression, such as the Dao, the cosmic body of the Buddha (dharmakaya) and the inherent relationality of Reality (*pratitya-samutpada*), or the identity of Brahman and Atman—even the Khora of Plato. Faber, *Depths As Yet Unspoken*; Odin, *Process Metaphysics and Hua-Yen Buddhism*.

115. Cf. Lowe, *Alfred North Whitehead*, vol 1 and 2; Hosinski, *Stubborn Fact and Creative Advance*; Griffin, *Reenchantment Without Supernaturalism*; Kraus, *The Metaphysics of Experience*; Rapp and Wiehl, *Whitehead's Metaphysics of Creativity*; Rose, *On Whitehead*; Shaviro, *Without Criteria*; Stengers, *Thinking With Whitehead*; Faber, *The Becoming of God*, *God as Poet*, and *The Divine Manifold*.

116. Cf. Fabel and St. John, *Teilhard in the 21st Century*, Part 1; Haught, *The New Cosmic Story*; Barbour, "Teilhard's Process Metaphysics"; Duffy, *Teilhard's Mysticism*; Grummet, *Teilhard de Chardin*; King, *Spirit of Fire*; Overzee, *The Body Divine*; Smith, *Theistic Evolution*; Tucker, "The Ecological Spirituality of Teilhard."

117. Cf. Davies, *Are We Alone?*, 75; Genta, *Lonely Minds in the Universe*, 32, 38, 135; Wilkinson, *Science, Religion and the Search for ETL*, 24, 155–56; Dick, *The Impact of Discovering Life Beyond Earth*, 159; Guthke, *The New Frontier*, 25, 325; O'Meara, *Vast Universe*, 16, 44, 68, 88–89; Vakoch and Dowd, *The Drake Equation*, 198, 203; Ćirković, *The Great Silence*, 197–98; Forgan, *Fermi's Paradox*, 237, 408; Dick, *Astrobiology, Discovery, and Societal Impact*, 244; Peters, *Astrotheology*, 239, 306, 310, 324.

118. Cf. Stockman, *Baha'i Faith*, ch. 8.

119. Cf. Momen, *Baha'u'llah*.

120. Cf. Bausani, *Religion in Iran*; Stockman, *Baha'i Faith*, chs. 2–3; Faber, *Divine Appearances*, ch. 4.

121. For the term "transreligious," cf. Faber, *The Becoming of God*, Sphere 5; *The Ocean of God*, ch. 11; *The Garden of Reality*, chs. 5–6; *Divine Appearances*, Introduction.

122. Cf. Buck, *Baha'i Faith*, ch. 4.

123. Cf. Stockman, *'Abdu'l-Baha*; 'Abdu'l-Baha, *The Promulgation of Universal Peace*.

124. Cf. Thatcher, "Three Theologies of the Future"; Smith, *Theistic Evolution*; Haught, *God After Darwin*; *Deeper*; Duffy, *Teilhard's Mysticism*; Mehanian and Friberg, "Religion and Evolution Reconciled"; Brown, *Evolution and Baha'i Belief*; 'Abdu'l-Baha, *Paris Talks*, ##19–20; and St. John, *Teilhard in the 21st Century*, ch. 13 (John Haught). Not to be confused with simplifications of a necessary and linear progression beyond Darwinism; cf. Herrick, *The Making of the New Spirituality*, ch. 6; Ward, *God, Chance, Necessity*, chs. 7–9.

125. Cf. 'Abdu'l-Baha, *Paris Talks*, ##40, 44. This harmony takes the radical form of measuring the religious availability of truth by its reference of science and reasonability: "Whatever the intelligence of man cannot understand, religion ought not to accept. Religion and science walk hand in hand, and any religion contrary to science is not the truth" (#40). "There is no contradiction between true religion and science. When a religion is opposed to science it becomes mere superstition: that which is contrary to knowledge is ignorance" (#44).

126. Cf. Whitehead, *Adventures of Ideas*, 291; Baha'u'llah, *Gleanings*, #82; Fabel and St. John, *Teilhard in the 21st Century*, part 3; Haught, *The New Cosmic Story*, 28–29, 40, 46–47, 181–82, [AQ1]176–90; Birch and Cobb, *The Liberation of Life*; Mehanian and Friberg, "Religion and Evolution Reconciled." Teilhard speaks of "Ultra-Human" emergence; cf. Teilhard, *Activation of Energy*, 269–81.

127. For Teilhard, cf. Barbour, "Teilhard's Process Metaphysics"; Fabel and St. John, *Teilhard in the 21st Century*, 6–7, 29–30, 143.

128. Cf. For both Teilhard and Whitehead, cf. Thatcher, "Three Theologies of the Future"; Faber, *The Divine Manifold*, chs. 7, 13.

129. Cf. For 'Abdu'l-Baha, cf. Taylor, *One Reality*, ch. 4; Faber, *The Ocean of God*, ch. 13; *The Garden of Reality*, chs. 1:5; 5:2; Epilogue:4; *Divine Appearances*, ch. 5; Brown, "A Baha'i Perspective on the Origin of Matter"; "The Beginning that Hath No Beginning"; 'Abdu'l-Baha, *Tablet of the Universe*.

130. Cf. Whitehead, *Process and Reality*, 350; Teilhard, "My Universe," 73; *The Divine Milieu*; Fabel and St. John, *Teilhard in the 21st Century*, 214; 'Abdu'l-Baha, *Paris Talks*, #52;

131. Cf. Teilhard, "My Universe," 67; 'Abdu'l-Baha, *Selections from the Writings of 'Abdu'l-Bahá*, #166; *Some Answered Questions*, #16.

132. Cf. Whitehead, *Adventures of Ideas*, 134.

133. Cf. Whitehead, *Process and Reality*, 162.

134. Cf. Whitehead, *Process and Reality*, 351; Teilhard, "My Universe," 43–44, 55–56; 'Abdu'l-Baha, *The Promulgation of Universal Peace*, ##23, 96, 100, 105, 110, 125; *Selections from the Writings of 'Abdu'l-Bahá*, 225.

135. Cf. Whitehead, *The Function of Reason*, 25; *Religion in the Making*, 160; *Adventures of Ideas*, 285; 'Abdu'l-Baha, *The Promulgation of Universal Peace*, #93; *Some Answered Questions*, #67; *Selections from the Writings of 'Abdu'l-Bahá*, #29; Teilhard, "My Universe," 56–57. In *The Function of Reason*, Whitehead speaks of an evolutionary "general counter-agency" (25); and Teilhard, in *Activation of*

Energy, speaks of the "Phenomenon of Counter-Evolution"; cf. Teilhard, *Activation of Energy*, 181–97.

136. Cf. Whitehead, *Religion in the Making*, 146, 144, 153, 158–59; 'Abdu'l-Baha, *Paris Talks*, #51; *Some Answered Questions*, #29; *Selections from the Writings of 'Abdu'l-Bahá*, #225; *The Promulgation of Universal Peace*, #52; Teilhard, "My Universe," 57, 61; *Writings*, 166–70.

137. Cf. Whitehead, *Religion in the Making*, 160; *Adventures of Ideas*, 295–96; Teilhard, "My Universe," 57; 'Abdu'l-Baha, *Some Answered Questions*, #30; *Selections from the Writings of 'Abdu'l-Bahá*, #21; *The Promulgation of Universal Peace*, ##29, 35.

138. Cf. Rahner, "Christology within an Evolutionary View," 157–92.

139. Cf. Marmion and Hines, *The Cambridge Guide to Rahner.*

140. Cf. Rahner, *Foundations of Christian Faith*, Part 4, ch. 1. Rahner on ETI: Wilkinson, *Science, Religion, and the Search for ETL*, 149.

141. Cf. Rahner, "The Unity of Spirit and Matter," 153–77.

142. Cf. 'Abdu'l-Baha, *The Promulgation of Universal Peace*, ##38, 85; *Some Answered Questions*, #64; Whitehead, *Process and Reality*; Teilhard, "My Universe," 45.

143. Cf. Teilhard, "My Universe," 48; 'Abdu'l-Baha, *Some Answered Questions*, #52; Whitehead, *Process and Reality*, 32–33, 348.

144. Cf. Faber, *God as Poet*, part 2; *The Becoming of God*, Sphere 1; Hosinski, *Stubborn Fact and Creative Advance*; chs. 2–4; Diessner, *Psyche and Eros*; Brown, "The Beginning that Hath No Beginning"; Hazini, "Neoplatonism"; Medina, *Faith, Physics and Psychology*, ch. 2; Fabel and St. John, *Teilhard in the 21st Century*, ch. 10 (Kathleen Duffy); Grummet, *Teilhard*, ch. 1; Teilhard, *The Phenomenon of Man*, ch. 1.

145. Whitehead, *Adventures of Ideas*, 198.

146. Teilhard, "My Universe," 45; *Writings*, 61–65; Grummet, *Teilhard*, chs. 1, 7.

147. Teilhard, "My Universe," 41; cf. *Activation of Energy*, 319–39, 359–73.

148. 'Abdu'l-Baha, *'Abdu'l-Bahá in London*, 95.

149. Cf. 'Abdu'l-Baha, *The Promulgation of Universal Peace*, ##38, 85.

150. Whitehead, *Adventures of Ideas*, 18.

151. Whitehead, *The Concept of Nature*, 73.

152. Whitehead, *Process and Reality*, 46.

153. Ibid., 229.

154. 'Abdu'l-Baha, *Paris Talks*, #29.

155. Cf. ibid., #2; *Some Answered Questions*, #1.

156. Teilhard, "My Universe," 51.

157. Whitehead, *Adventures of Ideas*, 295.

158. Teilhard, "My Universe," 82.

159. Ibid., 72.

160. Ibid., 74.

161. Ibid., 49, 85; cf. *Activation of Energy*, 373–85.

162. Cf. 'Abdu'l-Baha, *Paris Talks*, #58.

163. Cf. Abdu'l-Baha, *The Promulgation of Universal Peace*, #82.

164. Whitehead, *Adventures of Ideas*, 285.
165. Whitehead, *Modes of Thought*, 174; cf. *Religion in the Making*, 60.
166. Teilhard, "My Universe," 52.
167. Whitehead, *Modes of Thought*, 174.
168. Teilhard, "My Universe," 84.
169. 'Abdu'l-Baha, *Paris Talks*, #5.
170. Baha'u'llah, *The Hidden Words*, Persian #87.
171. Abdu'l-Baha, *Paris Talks*, #45.
172. Cf. Faber, *The Cosmic Spirit*, xi, 8–9, 14, 16, 25, 112, 274.
173. Whitehead, *Science and the Modern World*, 191–92.
174. Cf. Teilhard, "My Universe," 76; Whitehead, *Religion in the Making*, 160; 'Abdu'l-Baha, *Paris Talks*, #20; *The Promulgation of Universal Peace*, #309. Both John Leslie and Keith Ward make an argument from Plato's Ideas of the Good that this not only restricts the world created, but reveals the motive of its creative advance; and so does Whitehead, *Religion in the Making*, 153, 159; cf. *Process and Reality*, 32; and for all of them, cf. Davis, *Mind, Value, and Cosmos*; counter Lewis, "Possible Worlds"; *Counterfactuals*; *Plurality of Worlds*; cf. Lycan, "Possible Worlds and Possibilia"; Faber, *The Divine Manifold*, ch. 7.
175. Teilhard, *Christianity and Evolution*, 172.
176. Cf. Teilhard, *Toward the Future*, 185.
177. Cf. Teilhard, *Christianity and Evolution*, 73–78, 122–41.
178. Cf. Teilhard, "My Universe," 85.
179. Cf. ibid., 60.
180. Cf. ibid., 62.
181. Cf. Teilhard, *Christianity and Evolution*, 147.
182. Teilhard, "My Universe," 49.
183. Ibid., 50.
184. Cf. ibid., 79; *The Divine Milieu*, ch. 1; *Science and Christ*, 181.
185. Teilhard, "My Universe," 53–54.
186. Cf. 'Abdu'l-Baha, *Some Answered Questions*, ##30, 49.
187. Abdu'l-Baha, *The Promulgation of Universal Peace*, #96.
188. Cf. Whitehead, *Process and Reality*, 244.
189. Cf. Whitehead, *Adventures, of Ideas*, 11, 148, 198.
190. Cf. Whitehead, *Science and the Modern World*, 18.
191. Whitehead, *Religion in the Making*, 105–6.
192. Cf. Whitehead, *Process and Reality*, 162.
193. Whitehead, *Religion in the Making*, 158–59.
194. Whitehead only tentatively refers to the ETI dimension of this question, speaking of "all reasonable beings on Earth, in every planet, and in every star-system": Whitehead, *Adventures*, 291; cf. Price, *Dialogues*.
195. Teilhard, "My Universe," 66–67.
196. Baha'u'llah, *Gleanings*, #79.
197. 'Abdu'l-Baha, *Some Answered Questions*, #81.
198. Baha'u'llah, *Gleanings*, #82.

199. Taylor, *One Reality*, 144; Troxel, "Extraterrestrial Life"; "Intelligent Life in the Universe." We should recognize that the relatively new discussion, especially within Christianity and furthered by the ETI dimension, about multiple "incarnations" or the presence of the Logos in multiple personages, Earthly and alien, is something that the Baha'i Faith presupposes as its essential insight and identity, scripturally, philosophically, and theologically; cf. Faber, *The Garden of Reality*, chs. 7–8; *Divine Appearances*; Peters, *Astrotheology*, part 4; Wilkinson, *Science, Religion and the Search for ETL*, ch. 10.

200. 'Abdu'l-Baha, *Divine Philosophy*, 114–15.

201. It is important to recognize, here, that 'Abdu'l-Baha speaks in terms of a multiverse, that is, in terms of spatial distance, which cannot be mapped on spiritual altitude, as in many ancient religious cosmologies.

202. 'Abdu'l-Baha, *The Promulgation of Universal Peace*, #79.

203. Teilhard, "My Universe," 82.

204. Cf. Teilhard, *Science and the Modern World*, 181.

205. Whitehead, *Process and Reality*, 104.

206. Ibid.,105.

207. 'Abdu'l-Baha, *The Promulgation of Universal Peace*, #120.

208. 'Abdu'l-Baha, *Tablets of the Divine Plan*, #4.

209. 'Abdu'l-Baha, *Selections from the Writings of 'Abdu'l-Bahá*, #227.

210. Ibid.

211. Ibid.

212. Cf. 'Abdu'l-Baha, *Paris Talks*, ##40, 44.

213. Cf. 'Abdu'l-Baha, *The Promulgation of Universal Peace*, ##49, 120. There are reasons to believe that the warring reality of humanity may not be a natural necessity, historically even contingent, that is culturally embedded, but in principle not unavoidable; cf. Horgan, *The End of War*. 'Abdu'l-Baha also relates to this spiritual self-transcendence of humanity in natural, cultural, psychological and religious terms with reference to the classic prophetic vision of the day of YHWH in Isaiah 2 and 11, and Micah 4, where wisdom is nonviolence on all levels of existence, war is banned, and life is without fear of mutual destruction. Cf. 'Abdu'l-Baha, *Paris Talks*, #42; *The Promulgation of Universal Peace*, ##47, 60; *Some Answered Questions*, #11.

214. Cf. Whitehead, *Process and Reality*, 15.

215. Whitehead, *Function of Reason*, 4.

216. Ibid.

217. Ibid., 18.

218. Whitehead, *Science and the Modern World*, 51.

219. Whitehead, *Adventures of Ideas*, 247.

220. Ibid., 285.

221. Ibid., 18; cf. Faber, Bell and Petek, *Rethinking Whitehead's Symbolism*

222. Whitehead, *Adventures of Ideas*, 186.

223. Cf. Wilber, *Quantum Questions*; Faber, *God as Poet*, §44; *The Cosmic Spirit*, chs. 8–9; Penrose, *The Emperor's New Mind* and *Large*; Polkinghorne, *The Faith of a Physicist*, *Science and Religion*, and *The Work of Love*; Bateson, *Mind and Nature* and *Steps to an Ecology of Mind*; Barrow and Tipler, *The Anthropic Cosmological*

Principle; Barrow, Davies and Harper, *Science and Ultimate Reality*; Clayton and Davies, *The Re-Emergence of Emergence*; Clayton and Peacock, *In Whom We Live*; Davies, *God and the New Physics*, *The Goldilocks Enigma*, and *The Mind of God*; Davies and Gribbin, *The Matter Myth*; Davis and Clayton, *How I Found God*; Hick, *Death and Eternal Life* and *The New Frontiers of Science and Religion*; Loomer, "A Process-Relational Conception of Creation" and "The Web of Life"; Griffin, *Panentheism and Scientific Naturalism*; *Parapsychology, Philosophy and Spirituality*; *Reenchantment without Supernaturalism*; and *Unsnarling the World-Knot*; Haught, *Is Nature Enough?*, *The New Cosmic Story*, and *Science and Religion*; Denton, *Nature's Destiny*; Conway Morris, *Life's Solution*; James, *Essays in Psychical Research*, *A Pluralistic Universe*, *The Principles of Psychology*, and *Varieties of Religious Experience*; Stapp, *Mindful Universe*; "A Quantum-Mechanical Theory"; Ward, *God, Chance, Necessity*; Epperson, *Quantum Mechanics*; Goswami, *Science and Spirituality*; Jammer; *The Philosophy of Quantum Mechanics*; Malin, *Nature Loves to Hide*; Ricard and Thuan, *The Quantum and the Lotus*; Rosenblum and Kuttner, *Quantum Enigma*.

224. Cf. Simmons, *Hyperion Cantos*. In interesting correlation to recent research on emotional impacts of past places that are sites of events of intense "negative energy" (such as the holocaust) on the emotional states of visitors, even if they are reduced to bio-chemical indicators, as well as the more exotic assumption of a formative resonance of what Whitehead has called the "subjective form" of the transmission of energy in a nexus of events (cf. Whitehead, *Adventures of Ideas*) by Rupert Sheldrake, we may find some similar hypothesis open to scientific experimentation.

225. Whitehead, *Adventures of Ideas*, 134.

226. Ibid., 135.

227. Whitehead, *Adventures of Ideas*, 201; Faber, *Prozeßtheologie*, §21; *God as Poet*, part 4 and § 34; "Immanence and Incompleteness," 91–110; *The Becoming of God*, Exploration 10; *Depths As Yet Unspoken*, 12, ch. 1; *The Cosmic Spirit*, chs. 6, 8–9; Davis, *Mind, Value, and Cosmos.*

228. 'Abdu'l-Baha, *Some Answered Questions*, #46.

229. Ibid., #69.

230. Ibid., #21.

231. 'Abdu'l-Baha, *Tablet of the Universe*, §§5–8; cf. Faber, *The Garden of Reality*, 468–69.

232. Cf. Whitehead, *Religion in the Making*, 160.

233. Cf. Faber, *The Cosmic Spirit*, Prologue; ch. 9.

234. Cf. Whitehead, *Process and Reality*, 350.

235. Cf. Whitehead, *Adventures of Ideas*, 285.

236. Cf. ibid., 294–95.

237. Cf. Whitehead, *Adventures of Ideas*, 295–96.

238. Ibid., 295.

239. Ibid., 82.

240. Cf. ibid., 23–27.

241. Whitehead, *Religion in the Making*, 57; Cf. *Adventures of Ideas*, 167.

242. Cf. Whitehead, *Adventures of Ideas*, 168–69.

243. Teilhard, *Toward the Future*, 157.
244. Ibid., 151–52.
245. Teilhard, "My Universe," 82.
246. Cf. ibid., 80; cf. *Toward the Future*, 151, 191.
247. Teilhard, My Universe," 62.
248. Cf. 'Abdu'l-Baha, *Paris Talks*, #42; *The Promulgation of Universal Peace*, ##5, 104; *Selections from the Writings of 'Abdu'l-Bahá*, ##77, 225.
249. Cf. Peters, *Science, Theology, and Ethics*; Polkinghorne, *The Faith of a Physicist*; Clayton and Peacock, *In Whom We Lived*; Teilhard, *The Phenomenon of Man*, 203.
250. 'Abdu'l-Baha, *The Promulgation of Universal Peace*, #5.
251. Whitehead, *Religion in the Making*, 37.
252. Whitehead, *Adventures of Ideas*, 172.
253. 'Abdu'l-Baha, *Paris Talks*, #40.
254. Cf. Teilhard, *Letters to Léontine Zanta*, 114 (letter of 26 January, 1936).
255. Teilhard, *The Heart of the Matter*, 96.
256. Cf. 'Abdu'l-Baha, *Some Answered Questions*, #43.
257. Whitehead, *Adventures of Ideas*, 18.
258. In the publication of not authorized dialogues with Whitehead, he refers to alien beings and to such universal beings, organisms that are on a different magnitude than humanity; cf. Price, *Dialogues of ANW*. The classical idea of angels has also had the connotation of them being a representation of a cosmocratic domain; cf. Teilhard, "My Universe," 51; 'Abdu'l-Baha, *Selections from the Writings of 'Abdu'l-Bahá*, #39; Teilhard refers to "Superlife" and the "Hyper-personal" development of humanity; cf. *The Phenomenon of Man*, 167, 180.
259. Cf. Ćirković, *The Great Silence*, 25, 87.
260. Cf. Circovic, *The Great Silence*, 237; ch. 8:4; Forgan, *Solving Fermi's Paradox*, 12–14, 43–45, 153, 308, 368–69, 390, 412, ch. 1:3; Peters, *Astrotheology*, 367–68.
261. Cf. Vakoch and Dowd, *The Drake Equation*, 17, 44, 279–83; Ćirković, The Great Silence, 87, ch. 6:4; Bodin, *Fermi's Paradox*, 112; Forgan, *Solving Fermi's Paradox*, chs. 9, 12–15; Peters, *Astrotheology*, 368–73; Fabel and St. John, *Teilhard in the 21st Century*, ch. 9 (Brian Swimme). Baha'u'llah has left a series of warnings regarding the potential of human self-destruction if the material developments, including religiously motivated warfare, are not met with spiritual investment into civilization; cf. Baha'u'llah, *Kitab-i-Aqdas*, *Note* 137; *Tablets*, #3. One ominous prophecy was related to fatal power of the still undiscovered devastating effects of atomic power: "A strange and wonderful instrument exists in the earth; but it is concealed from the minds and souls. It is an instrument which has the power to change the atmosphere of the whole earth, and its infection causes destruction"; in Hornby, *Lights of Guidance*, #1585. 'Abdu'l-Baha in an interview with the Japanese Ambassador in Paris 1913 struck in the same vein: "Scientific discoveries have greatly increased material civilization. There is in existence a stupendous force, as yet, happily, undiscovered by man. Let us supplicate God, the Beloved, that this force be not discovered by science until Spiritual Civilization, i.e., the Kingdom, shall dominate the human mind. In the

hands of men of lower material nature, this power would be able to destroy the whole earth"; in Taherzadeh, *The Revelation of Baha'u'llah*, 225.

262. Cf. Faber, *The Ocean of God*, 182–83; Ward, *In Defense of the Soul*, 145–46; Ćirković, *The Great Silence*, 140, 270, ch. 3:8; Bodin, *Fermi's Paradox*, chs. 6:7, 14:11; Zhu, *On the Origin of ET Industrial Civilizations*, 476; Forgan, *Solving Fermi's Paradox*, ch. 15; Peters, *Astrotheology*, 257. "Superintelligences" would also not count for the integration of spiritualization and civilizational development per se—cf. Dick, *The Impact of Discovering Life Beyond Earth*, ch. 12 (Susan Schneider); Davies, *Are We Alone?*, 137—if they do not cross the divided between mind and spirit; cf. 'Abdu'l-Baha, *Paris Talks*, ##29–31, 54.

263. Cf. O'Meara, *Vast Universe*, ch. 3; Peters, *Astrotheology*, 142; Ward, *God, Chance, Necessity*, ch. 7; Davies, *Are We Alone?*, 76–87.

264. Egan, *Diaspora*, 33.

265. Ibid., 164.

266. The many ethical questions related to current space exploration have also the spiritual implication, noted by the three exemplars highlighted in this chapter, namely, that their very impulse may only be transformative if it allows for the universal aesthetic recognition of self-transcendence in its diverse forms and dimensions, such as universal sym-pathy or compassion of existence, to be its motive-force.

267. Cf. Bowman, *AI & The Fermi Paradox*, chs. 8–9; Forgan, *Solving Fermi's Paradox*, ch. 14.

268. Given the speculation that the contact with such civilizations would presuppose its long-term existence in the state of L of the Drake equation, and, hence, that we would have to assume that if they exist, they would already have been detected either with signals or by probes, this is, of cause, and remains, a deep alternative.

269. Cf. Faber, *The Ocean of God*, chs. 3–4, 13–15.

270. Cf. Faber, *The Cosmic Spirit*, ch. 9. Even if we assume that not religious pluralism, that is, a manifold of religious and spiritual perspectives, is the mirror of the plurality of worlds and their diverse evolutions (cf. Faber, *The Ocean of God*, chs. 2, 14–15), but a "universal religion" (Dick, *Many Worlds*, ch. 10/Jill Tarter; Wilkinson, *Science, Religion, and the Search for ETL*, 125)—counter to the principles of plenitude and decentralization, as explored earlier (!)—we may *not* assume that, even if these alien civilizations would exhibit many advanced characteristics that we could not readily differentiate from our classical theistic understanding of God or the Buddha (Dick, *Many Worlds*, ch. 13/Ted Peters, especially 204–5), or tend to a panentheism in terms of spatial categories (Clayton and Peacock, *In Whom We Live*; Walker and Wickramasinghe, *Big Bang and God*; Griffin, *Panentheism and Scientific Naturalism*), the importance that it is the *apophatic* element, which expresses the most advanced spiritual measure of unity in all mystical traditions of the Earth, can be forgotten or erased (Faber, *God as Poet*, §40; *Depths as Yet Unspoken*, ch. 1; *The Garden of Reality*, ch. 6; *The Ocean of God*, ch. 3).

271. Cf. Faber, *The Ocean of God*, ch. 13; *The Garden of Reality*, chs. 6, 9; *Divine Appearances*, ch. 5.

272. Cf. Barbero, *Dante*.

273. Cf. Lefaivre, *Hypnerotomachia Poliphilii*, Introduction, ch. 4.

274. Dante, *The Inferno*, Canto 1,4.
275. Godwin, *Hypnerotomachia Poliphilii*, 14–15.
276. Cf. Krell, *Martin Heidegger*, ch. 11.

BIBLIOGRAPHY

‘Abdu’l-Baha. *‘Abdu’l-Bahá in London*. Chicago: Baha’i Publishing Trust, 1982.
———. *Divine Philosophy*. Edited by I.F. Chamberlain, Boston: The Tudor Press, 1918.
———. *Paris Talks: Addresses Given by ‘Abdu’l-Baha in 1911*. Wilmette: Baha’i Publishing, 2011.
———. *Selections from the Writings of ‘Abdu’l-Bahá*. Wilmette: Baha’i Publishing, 2014.
———. *Some Answered Questions*. Newly translated. Haifa: Baha’i World Center, 2014.
———. *Tablet of the Universe* (Lawh-i- Aflákiyyih). In *Makátib-i ‘Abdu’l-Baha*, Vol. 1. Translated Anonymous, 13–32. 1997. Access April 6, 2023. https://bahai-library.com/abdul-baha_lawh_aflakiyyih
———. *The Promulgation of Universal Peace*. Wilmette: Baha’i Publishing, 2012.
Alighieri, D. *Paradiso*. Translated by A. Mandelbaum. New York: Bantam Classics, 1986.
———. The *Inferno*. Translated by H.W. Longfellow. New York: Barnes & Noble, 2003.
Ashkenazi, M. *What We Know About Extraterrestrial Intelligence: Foundations of Xenology*. Cham: Springer, 2017.
Aurobindo, S. *The Life Divine*. Pondicherry, India: Sri Aurobindo Ashram Trust, 2006.
Baha’u’llah. *Gleanings from the Writings of Baha’u’llah*. Wilmette: Baha’i Publishing, 1976.
———. *Kitab-i-Aqdas*. India: Bahá’í Publishing, 1993.
———. *Tablets of Baha’u’llah revealed after the Kitab-i Aqdas*. Wilmette: Baha’i Publishing, 1994.
———. *The Hidden Words*. Wilmette: Baha’i Publishing, 2002
Barbero, A. *Dante: A Life*. New York: Pegasus Books, 2022.
Barbour, I. “Teilhard’s Process Metaphysics.” In *Process Theology: Basic Writings*, edited by Ewert Cousins. New York: Newman Press, 1971.
Barrow, J., and F. Tipler. *The Anthropic Cosmological Principle*. Oxford: Oxford University Press, 1986.
Barrow, J. et al., eds. *Science and Ultimate Reality: Quantum Theory, Cosmology, and Complexity*. Cambridge: Cambridge University Press, 2004.
Bateson, G. *Mind and Nature: A Necessary Unity*. Advances in System Theory, Complexity, and the Human Sciences. Creskill: Hampton, 2002.
———. *Steps to an Ecology of Mind: Collected Essays in Anthropology, Psychiatry, Evolution, and Epistemology*. Chicago: University of Chicago Press, 2000.
Bausani, A. *Religion in Iran: From Zoroaster to Bahá’u’lláh*. Winona Lake: Bibliotheca Persica Press, 2000.

Birch, C., and J.B . Cobb Jr. *The Liberation of Life*. Denton: Environmental Ethics Books, 1990.

Bodin, M. *Fermi's Paradox: Cosmology and Life*. Bloomington, IN: Traford Publishing, 2014.

Bond, L, ed. *Nicolas of Cusa: Selected Spiritual Writings*. Translation and introduction by H. Bond. New York: Paulist Press, 1997.

Bostic, A. *The Fermi Paradox: Where Are All the Aliens*. Independently published: D. Alexander Bostic, 2020.

Bowman, J. *AI & The Fermi Paradox: Human Immortality or Civilization's Inevitable Destruction?* Independently published: Jon Bowman, 2016.

Brown, K. "A Baha'i Perspective on the Origin of Matter." In *The Journal of Bahá'í Studies* Vol 2, No. 3 (1990): 15–44. https://doi.org/10.31581/jbs-2.3.423 (1990).

Brown, K. ed. *Evolution and Baha'i Belief: 'Abdu'l-Baha's Response to Nineteenth-Century Darwinism*. Los Angeles: Kalimat Press, 2001.

Brown, V. "The Beginning that Hath No Beginning: Baha'i Cosmogony." In *Lights of Irfan* Vol. 3 (2002): 21–40. Accessed April 6, 2023. https://bahai-library.com/brown_cosmogony.

Buck, C. *Baha'i Faith: The Basics*. New York: Routledge, 2021.

Burton, S. J. G., et al., eds. *Nicolas of Cusa and the Making of the Early Modern World*. Leiden: Brill, 2019.

Catling, D. C. *Astrobiology: A Very Short Introduction*. Oxford: Oxford University Press, 2013.

Chalmers, D. *The Character of Consciousness*. Oxford: Oxford University Press, 2010.

———. *The Conscious Mind: In Search of a Fundamental Theory*. Oxford: Oxford University Press, 1996.

Churchland, P. *Matter and Consciousness: A Contemporary Introduction to the Philosophy of Mind*. Cambridge, MA: MIT Press, 2013.

Ćirković, M. M. *The Great Silence: The Science and Philosophy of Fermi's Paradox*. Oxford: Oxford University Press, 2018.

Cixin, L. *The Dark Forest*. Translated by J. Martinsen. New York: Tor Books, 2015.

Clayton, P., and A. Peacocke, eds. *In Whom We Live and Move and Have Our Being: Panentheistic Reflections Presence in a Scientific World*. Grand Rapids: Eerdmans, 2004.

Clayton, P., and P. Davies, eds. *The Re-emergence of Emergence: The Emergentist Hypothesis from Science to Religion*. Oxford: Oxford University Press, 2006.

Conway Morris, S. *Life's Solution: Inevitable Humans in a Lonely Universe*. Cambridge: Cambridge University Press, 2003.

———. *The Runes of Evolution: How the Universe Became Self-Aware*. West Conshohocken, PA: Templeton Foundation Press, 2015.

Conway Morris, S. and S. J. Gould. "Showdown on the Burgess Shale." *Journal of Natural History* 107 (1998): 48–55.

Crowe, M. J. *The Extraterrestrial Life Debate, 1750–1900: The Idea of a Plurality of Worlds from Kant to Lowell*. Mineola, NY: Dover Publications, 1999.

———. *The Extraterrestrial Life Debate, Antiquity to 1915: A Sourcebook*. Notre Dame: University of Notre Dame Press, 2008.

Davies, P. *Are We Alone? Philosophical Implications of the Discovery of Extraterrestrial Life.* New York: Basic Books, 1995.
———. *God and the New Physics*. New York: Simon & Schuster, 1983.
———. *The Goldilocks Enigma: Why Is the Universe Just Right for Life?* London: Penguin, 2006.
———. *The Mind of God: The Scientific Basis for a Rational World.* New York: Touchstone, 1992.
Davies, P., and J. Gribbin. *The Matter Myth: Dramatic Discoveries That Challenge Our Understanding of Physical Reality*. New York: Simon & Schuster, 1992.
Davis, A. M. *Mind, Value, and Cosmos: On the Relational Nature of Ultimacy*. Lanham, MD: Lexington, 2020.
Davis, A. M. and P. Clayton, eds. *How I Found God in Everyone and Everywhere: An Anthology of Spiritual Memoirs*. Rhinebeck: Monkfish, 2018.
Denton, M. *Nature's Destiny: How the Laws of Biology Reveal Purpose in the Universe*. New York: Free Press, 1998.
———. *Evolution: A Theory in Crisis*. Chevy Chase, MD: Adler & Adler, 1985.
———. *Evolution: Still a Theory in Crisis.* Seattle: Discovery Institute, 2016.
Dick, S. J. *Life on Other Worlds: The 20th Century Extraterrestrial Life Debate*. Cambridge: Cambridge University Press, 1998.
———. *Plurality of Worlds: The Origins of the Extraterrestrial Life Debate from Democritus to Kant*. Cambridge: Cambridge University Press, 1982.
———. *Space, Time, and Aliens: Collected Works on Cosmos and Culture*. Cham: Springer, 2020.
———. *The Biological Universe: The 20th Century Extraterrestrial Life Debate and the Limits of Science*. Cambridge: Cambridge University Press, 1996.
Dick, S. J., ed. *Many Worlds: The New Universe, Extraterrestrial Life, and the Theological Implications.* Philadelphia: Templeton Foundation Press, 2000.
———. *The Impact of Discovering Life beyond Earth*. Cambridge: Cambridge University Press, 2015.
Diessner, R. *Psyche and Eros: Baha'i Studies in a Spiritual Psychology*. Oxford: George Ronald, 2007.
Duffy, K. *Teilhard's Mysticism: Seeking the Inner Face of Evolution*. Maryknoll, NY: Orbis Books, 2014.
Duhem, P. *Medieval Cosmology: Theories of Infinity, Place, Time, Void, and the Plurality of Worlds*. Chicago: University of Chicago Press, 1987.
Egan, G. *Diaspora*. New York: HarperPrism, 1998.
Epperson, M. *Quantum Mechanics and the Philosophy of Alfred North Whitehead.* New York: Fordham University Press, 2004.
Fabel, A., and D. St. John, eds. *Teilhard in the 21st Century: The Emerging Spirit of Earth.* Maryknoll, NY: Orbis Books, 2003.
Faber, R. "Ecotheology, Ecoprocess, and Ecotheosis: A Theopoetical Intervention." *Salzburger Zeitschrift für Theologie* 12 (2008): 75–115.
———. "Immanence and Incompleteness: Whitehead's Late Metaphysics." In *Beyond Metaphysics? Conversations on A. N. Whitehead's Late Thought*. Edited by R. Faber, B. Henning, and C. Combs, 91–110. Amsterdam: Rodopi, 2010.

———. *Depth As Yet Unspoken: Whiteheadian Excursions Mysticism, Multiplicity, and Divinity*, edited by A. M. Davis. Eugene: Pickwick, 2021.

———. *Divine Appearances: Transreligious Aspects of the Baha'i Discourse on the Manifestation of God*. Los Angeles: Kalimat Press. 2022.

———. *God as Poet of the World: Exploring Process Theologies*. Louisville: Westminster John Knox, 2008.

———. *Prozeßtheologie: Zu ihrer Würdigung und kritischen Erneuerung*. Mainz: Matthias Grünewald, 2000.

———. *The Cosmic Spirit: Awakenings at the Heart of All Religions, the Earth and the Multiverse.* Eugene: Cascade Books, 2021.

———. *The Divine Manifold*. Lanham, MD: Lexington, 2014.

———. *The Garden of Reality: Transreligious Relativity in a World of Becoming*. Lanham, MD: Lexington, 2018.

———. *The Ocean of God*. London, Anthem, 2019.

Faber, R., B. Henning, and C. Combs, eds. *Beyond Metaphysics? Conversations on A. N. Whitehead's Late Thought*. Amsterdam: Rodopi, 2010.

Faber, R., J. Bell, and J. Petek, eds. *Rethinking Whitehead's Symbolism: Thought, Language, Culture*. Edinburgh: Edinburgh University Press, 2017.

Forgan, D. *Solving Fermi's Paradox*. Cambridge: Cambridge University Press, 2019.

Genta, G. *Lonely Minds in the Universe*. Springer Praxis Publishing, 2010.

Godwin, J., ed. *Hypnerotomachia Poliphilii: The Strife of Love in a Dream*. New York: Thames & Hudson, 2003.

Goswami, A., and M. Goswami. *Science and Spirituality: A Quantum Interpretation*. New Delhi: Project of History of Indian Science, Philosophy, and Culture, 1997.

Gould, S. J. *The Panda's Thumb: More Reflections in Natural History*. New York: Norton, 1992.

———. *The Structure of Evolutionary Theory*. Cambridge: Belknap Press of Harvard University Press, 2002.

Griffin, D. R. *Panentheism and Scientific Naturalism: Rethinking Evil, Morality, Religious Experience, Religious Pluralism, and the Academic Study of Religion*. Claremont: Process Century Press, 2014.

———. *Parapsychology, Philosophy and Spirituality: A Postmodern Exploration*. Albany: SUNY Press, 1997.

———. *Reenchantment without Supernaturalism: A Process Philosophy of Religion*. Ithaca: Cornell University Press, 2001.

———. *Unsnarling the World-Knot: Consciousness, Freedom, and the Mind-Body Problem*. 1998. Reprint, Eugene: Wipf & Stock, 2007.

Grummet, D. *Teilhard de Chardin: Theology, Humanity and Cosmos*. Leuven: Peeters, 2005.

Guthke, K. *The Last Frontier: Imagining Other Worlds from the Copernican Revolution to Modern Science Fiction. Ithaca:* Cornell University Press, 1990.

Hanlon, J. *The Drake Equation: A Brief Description of Stars, Planets, and the Existence of Life*. 2021.

Harrison, A., and D. Vakoch. *Civilizations Beyond Earth: The Search for Extraterrestrial Intelligence as an Interdisciplinary Effort*. New York: Berghahn Books, 2011.
Haught, J. *Deeper than Darwin: The Prospect for Religion in the Age of Evolution*. Philadelphia: Westview Press, 2004.
———. *God After Darwin: A Theology of Evolution*. Boulder: Westview, 2007.
———. *Is Nature Enough? Meaning and Truth in the Age of Science*. Cambridge: Cambridge University Press, 2006.
———. *Science and Religion: From Conflict to Conversation*. New York: Paulist, 1995.
———. *The New Cosmic Story: Inside Our Awakening Universe*. New Haven: Yale University Press, 2017.
Hazini, N. "Neoplatonism: Framework for a Bahá'í Metaphysics." Presented at the Texas Regional Bahá'í Studies Conference, Austin, Texas, November 1995.
Herrick, J. *Scientific Mythologies: How Science and Science Fiction Forge New Religious Beliefs*. Downers Grove, IL: IVP Academic, 2008.
———. *The Making of the New Spirituality: The Eclipse of the Western Religious Tradition*. Downers Grove, IL: InterVarsity, 2003.
Hick, J. *Death and Eternal Life*. Louisville: Westminster John Knox, 1994.
———. *The New Frontiers of Science and Religion: Religious Experience, Neuroscience, and the Transcendent*. New York: Palgrave Macmillan, 2006.
Horgan, J. *The End of War*. San Francisco: McSweeney's, 2012.
Hornby, H. *Lights of Guidance: A Baha'i Reference File*. New Deli: Baha'i Publishing, 2010.
Hosinski, T. *Stubborn Fact and Creative Advance: An Introduction to the Metaphysics of Alfred North Whitehead*. Lanham, MD: Rowman & Littlefield, 1993.
James, W. *A Pluralistic Universe*. Lincoln: University of Nebraska Press, 1996.
———. *Essays in Psychical Research*. Cambridge: Harvard University Press, 1986.
———. *The Principles of Psychology*. Mineola, NY: Dover, 1950.
James, W. *The Varieties of Religious Experience: A Study in Human Nature*. New York: New American Library, 1958.
Jammer, M. *The Philosophy of Quantum Mechanics*. New York: Wiley, 1974.
King, U. *Spirit of Fire: The Life and Vision of Teilhard de Chardin*. Maryknoll, NY: Orbis, 2003.
———. *Towards a New Mysticism: Teilhard de Chardin and Eastern Religions*. New York: Seabury, 1980.
Kisak, P. S., ed. *Religious Cosmology: Religious Explanations for the Origin of the Universe*. CreateSpace Publishing, 2016.
Kraay, K., ed. *God and the Multiverse: Scientific, Philosophical and Theological Perspectives*. New York: Routledge, 2015.
Kraus, E. *The Metaphysics of Experience: A Companion to Whitehead's Process and Reality*. New York: Fordham University Press, 1998.
Krell, D., ed. *Martin Heidegger: Basic Writings*. New York: HarperCollins, 1993.
LaFave, K. *The Sound of Ontology: Music as Model for Metaphysics*. Lanham, MD: Lexington, 2018.

Lefaivre, L. *Leon Battista Alberti's Hypnerotomachia Poliphilii: Re-Cognizing the Architectural Body in the Early Italian Renaissance.* Cambridge: MIT Press, 1999.
Leslie, J. *Universes*. London: Routledge, 1996.
Lewels, J. *The God Hypothesis: Extraterrestrial Life and its Implications for Science and Religion.* WildFlower Press, 2005.
Lewis, D. "Possible Worlds." In *Contemporary Readings in the Foundations of Metaphysics*, edited by Stephen Laurence and Cynthia Macdonald, 96–102. Malden, MA: Blackwell, 1998.
———. *Counterfactuals*. Malden, MA: Blackwell, 2001.
,,,. *On the Plurality of Worlds*. Malden, MA: Blackwell, 2001.
Lieberman, D. *The Story of the Human Body: Evolution, Health, and Disease*. New York: Vintage, 2014.
Loomer, B. "A Process-Relational Conception of Creation." In *Cry of Environment: Rebuilding the Christian Creation Tradition*. Edited by Philip Joranson and Ken Butigan, 312–28. Santa Fe: Bear Books. 1984
Loomer, B. "The Web of Life." In *The Nature of Life*, edited by William Heidcamp, 93–109. Baltimore: University Park Press. 1978.
Lowe, V. *Alfred North Whitehead: The Man and His Work, vol. 1, 1861–1910*. Baltimore: Johns Hopkins University Press, 1985.
———. *Alfred North Whitehead: The Man and His Work, vol. 2, 1910–1947.* Edited by J. B. Schneewind. Baltimore: Johns Hopkins University Press, 1990.
Lycan, W. "Possible Worlds and Possibilia." In *Contemporary Readings in the Foundations of Metaphysics*, edited by S. Laurence and C. Macdonald, 83–95. Malden, MA: Blackwell, 1998.
Malin, S. *Nature Loves to Hide: Quantum Physics and the Nature of Reality; a Western Perspective.* New York: Oxford University Press, 2001.
Maor, E. *To Infinity and Beyond: A Cultural History of the Infinite*. Princeton: Princeton University Press, 1987.
Marmion, D., and M.E . Hines, eds. *The Cambridge Guide to Karl Rahner*. Cambridge: Cambridge University Press, 2005
Medina, J. F. *Faith, Physics and Psychology: Rethinking Society and the Human Spirit*. Wilmette, IL: Baha'i Publishing Trust, 2006.
Mehanian, C., and S. Friberg. "Religion and Evolution Reconciled: 'Abdu'l-Baha's Comments on Evolution." *Journal of Baha'i Studies* 13, no. 1–4 (2003): 55–93.
Momen, M. *Baha'u'llah: A Short Biography*. Oxford: Oneworld, 2007.
Nagel, T. *Mind and Cosmos: Why the Materialist Neo-Darwinian Conception of Nature Is Almost Certainly False*. Oxford: Oxford University Press, 2012.
O'Connell, J., and M. Ruse. *Social Darwinism*. Cambridge: Cambridge University Press, 2021.
O'Meara, T. F. *Vast Universes: Extraterrestrials and Christian Revelation*. Collegeville, MN: Liturgical Press, 2012.
O'Neil, T. *Toward the Life Divine: Sri Aurobindo's Vision*. New Delhi: Manohar, 1979.
Odin, S. *Process Metaphysics and Hua-Yen Buddhism: A Critical Study of Cumulative Penetration vs. Interpenetration*. Albany: SUNY Press, 1982.

Odin, S. *Tragic Beauty in Whitehead and Japanese Aesthetics*. Lanham, MD: Lexington, 2016.

Overzee, A. H. *The Body Divine: The Symbol of the Body in the Works of Teilhard de Chardin and Ramanuja*. Cambridge: Cambridge University Press, 1992.

Paul, E. R. *Science, Religion, and Mormon Cosmology*. Urbana: University of Illinois Press, 1992.

Penrose, R. *The Emperor's New Mind: Concerning Computers, Minds, and the Law of Physics*. New York: Oxford University, 1991.

Penrose, R., et al. *The Large, the Small and the Human Mind*. Edited by Malcolm Longair. Cambridge: Cambridge University Press, 1999.

Peters, T. *Astrotheology: Science and Theology Meet Extraterrestrial Life*. Eugene: Cascade Books, 2018.

———. *Science, Theology, and Ethics*. Farnham: Ashgate, 2003.

Pickover, C. *Keys to Infinity*. New York: John Wiley & Sons, 1995.

Polkinghorne, J. *Science and Religion: An Introduction*. Minneapolis: Fortress Press, 1998.

———. *The Faith of a Physicist*. Princeton: Princeton University Press, 1994.

Polkinghorne, J., ed. *The Work of Love: Creation as Kenosis*. Grand Rapids, MI: Eerdmans, 2001.

Price, L. *Dialogues Of Alfred North Whitehead*. Boston: Little, Brown and Co, 1954.

Quincey, C., ed. *Consciousness from Zombies to Angels: The Shadow and the Light of Knowing Who You Are*. Rochester, VT: Park Street, 2009.

Rahner, K. "Christology within an Evolutionary View of the World." In *Theological Investigations*, Vol. 6, 157–92. Baltimore: Helicon Press, 1969.

———. "The Unity of Spirit and Matter in the Christian Understanding of Faith." In *Theological Investigations*, Vol. 6, 153–77. Baltimore, MD: Helicon Press, 1969.

———. *Foundations of Christian Faith: An Introduction to the Idea of Christianity*. Chestnut Ridge, NY: Crossroad, 2005.

Rapp, F., and R. Wiehl., eds. *Whitehead's Metaphysics of Creativity*. Albany: SUNY Press, 1990.

Ricard, M., and T. X. Thuan. *The Quantum and the Lotus: A Journey to the Frontiers Where Science and Buddhism Meet*. New York: Crown, 2001.

Richerson, P., and R. Boyd. *Not by Genes Alone: How Culture Transformed Evolution*. Chicago: University of Chicago Press, 2006.

Rose, P. *On Whitehead*. Belmont, CA: Wadsworth, 2002.

Rosenblum, B., and F. Kuttner. *Quantum Enigma: Physics Encounters Consciousness*. New York: Oxford University Press, 2006.

Rubinstein, M. J. *Worlds Without End: The Many Lives of the Multiverse*. New York: Columbia University Press, 2014.

Satprem. *Sri Aurobindo or the Adventure of Consciousness*. Valley Cottage, NY: Discovery Publishers, 2015.

Shaviro, S. *The Universe of Things: On Speculative Realism*. Minneapolis: University of Minnesota Press, 2014.

———. *Without Criteria: Kant, Whitehead, Deleuze, and Aesthetics*. Cambridge: MIT Press, 2009.

Simmons, D. *Hyperion Cantos*: 4 vols: *Hyperion* (1995); *The Fall of Hyperion* (1995); *Endymion* (1996); *The Rise of Endymion* (1998). New York: Bantam Books, 1995–1998.
Smith, W. *Theistic Evolution: The Teilhardian Heresy*. Tacoma, WA: Angelico, 2012.
Stapp, H. "A Quantum-Mechanical Theory of the Mind/Brian Connection." In *Beyond Physicalism: Toward Reconciliation of Science and Spirituality*, edited by E. F. Kelly, A. Crabtree, and P. Marshall, 157–94. Lanham, MD: Rowman & Littlefield, 2015.
Stapp, H. *Mindful Universe: Quantum Mechanics and the Participating Observer*. Berlin: Springer, 2011.
Stenger, V. J. *God and the Multiverse: Humanity's Expanding View of the Cosmos*. Amherst, NY: Prometheus, 2014.
———. *The Unconscious Quantum: Metaphysics in Modern Physics and Cosmology*. Amherst, NY: Prometheus, 1995.
Stengers, I. *Thinking with Whitehead*. Cambridge, MA: Harvard University Press, 2010.
Stockman, R. *'Abdu'l-Baha in America*. Wilmette, IL: Baha'i Publishing Trust, 2012.
———. *The Baha'i Faith: A Guide for the Perplexed*. New York: Bloomsbury Academic, 2013.
Taherzadeh, A. *The Revelation of Baha'u'llah*. Vol. 4. Oxford: George Ronald, 1988.
Taylor, B. *One Reality: The Harmony of Science and Religion*. Wilmette: Baha'i Publishing, 2013.
Teilhard de Chardin, P. *Activation of Energy*. San Diego: Harvest Book, 1978.
———. "My Universe." In *Science and Christ*. New York: Harper & Row, 1965, 37–85.
———. *Christianity and Evolution: Reflections on Sciences and Religion*. San Diego: A Harvest Book, 1969.
———. *Letters to Léontine Zanta,* New York: Harper & Row, 1965.
———. *Science and Christ*. New York: Harper & Row, 1965.
———. *The Divine Milieu*. Brighton: Sussex Academic Press, 2004.
———. *The Heart of the Matter.* San Diego: Harvest Books, 1978.
———. *The Phenomenon of Man.* Brighton: Sussex Academic Press, 2003.
———. *Toward the Future*. San Diego: Harvest Book, 1975.1
———. *Writings*. Introduction by U. King. Maryknoll, NY: Orbis Books, 1999.
Thatcher, A. "Three Theologies of the Future: Moltmann, Teilhard de Chardin and A. N. Whitehead." *Baptist Theological Quarterly* 25 (1974): 242–50.
Traphagan, J. *Extraterrestrial Intelligence and Human Imagination: SETI and the Intersection of Science, Religion, and Culture*. Cham: Springer, 2015.
———. *Science, Culture, and the Search for Life on Other Worlds*. Cham: Springer, 2016.
Troxel, D. "Intelligent Life in the Universe and Exotheology in Christianity and the Bahá'í Writings." 1996. Accessed April 6, 2023. http://bahai-library.com/troxel_extraterrestrials_exotheology.
———. "Extraterrestrial Life." 1996. Accessed April 6, 2023. http://bahai-library.com/compilation_extraterrestrial_life.

Tucker, E. "The Ecological Spirituality of Pierre Teilhard de Chardin." *Spiritus*, Vol. 7, No. 1 (2007): 1–19. https://doi.org/10.1353/scs.2007.0020.

Vakoch, D., and M. Dawd, eds. *The Drake Equation: Estimating the Prevalence of Extraterrestrial Life through the Ages*. Cambridge: Cambridge University Press, 2015.

Vakoch, D., ed. *Astrobiology, History, and Society: Life beyond Earth and the Impact of Discovery*. Cham: Springer, 2013.

van Huyssteen, J. W. *Alone in the World? Human Uniqueness in Science and Theology*. Grand Rapids: Eerdmans, 2006.

Von Hoerner, S. "The Search for Signals from Other Civilizations." *Science* 134, no. 3493 (1961): 1839–1843. https://doi.org/10.1126/science.134.3493.1839.

Walker, T., and C. Wickramasinghe. *The Big Bang and God: An Astro-Theology*. New York: Palgrave Macmillan, 2015.

Wallace, D. *The Emergent Multiverse: Quantum Theory According to the Everett Interpretation*. Oxford: Oxford University Press, 2014.

Ward, K. *God, Chance and Necessity*. Oxford: Oneworld, 2009.

———. *In Defense of the Soul*. Oxford: Oneworld, 1998.

Ward, P., and D. Brownlee. *Rare Earth: Why Life Is Uncommon in the Universe*. New York: Copernicus Books, 2004.

Webb, S. *If the Universe Is Teeming with Aliens . . . Where Is Everybody? Seventy-Five Solutions to the Fermi Paradox and the Problem of Extraterrestrial Life*. New York: Springer, 2015.

Weintraub, D. A. *Religions and Extraterrestrial Life: How Will We Deal with It?* Cham: Springer Praxis Books, 2014.

Wells, H. G. *The Last Books of H. G. Wells: The Happy Turning: A Dream of Life & Mind at the End of Its Tether*. Rhinebeck, NY: Monkfish Books, 2006.

———. *The New World Order*. Orkos Press, 2014.

———. *The Rights of Man*. New York: Vintage Classics, 2015.

———. *The War of the Worlds*. Reader's Library Classics, 2022.

Whitehead, A. N. *Adventures of Ideas*. New York: Free Press, 1967.

———. *Modes of Thought*. New York: Free Press, 1968.

———. *Process and Reality: An Essay in Cosmology.* Corrected edition edited by D. R. Griffin and D. W. Sherburne. New York: Free Press, 1978.

Whitehead, A.N. *Religion in the Making*. New York: Fordham University Press, 1996.

Whitehead, A.N. *Science and the Modern World*. New York: Free Press, 1967.

———. *The Concept of Nature*. Cambridge: Cambridge University Press, 1993.

———. *The Function of Reason*. Boston: Beacon Press, 1929.

Wilber, K., ed. *Quantum Questions: Mystical Writings of the World's Greatest Physicists*. Boulder: Shambhala, 2001.

Wilkinson, D. *Science, Religion, and the Search for Extraterrestrial Intelligence*. Oxford: Oxford University Press, 2013.

Willis, J. *All These Worlds Are Yours: The Scientific Search for Alien Life*. Yale: Yale University Press, 2016.

Wills, C. *Children of Prometheus: The Accelerating Pace of Human Evolution*. Reading, MA: Perseus Books, 1998.

Wilson, E. O. *The Meaning of Human Existence*. New York: Liveright, 2014.
Zhu, S. S. *On the Origin of Extraterrestrial Industrial Civilizations: A Mathematical Resolution to the Fermi Paradox and Implication on the Sustainability Industrial Civilization*. Independently published: Steven Suan Zhu, 2021.

Epilogue

Theological Reflections on Extraterrestrial Life (1968)[1]

Lewis S. Ford

In some respects, a good deal of contemporary theology has been rather parochial. This is particularly true of "existential" theology, which received its major impetus from Rudolph Bultmann and his program of demythologization. Bultmann argues that the Christian proclamation contained in the New Testament was enshrouded in myth, for the worldview of first century man included a three-storied world with heaven above and hell beneath, the miraculous intervention of supernatural powers in the course of evils, demon possession, and the like. He proposes that if the New Testament message is to be heard by contemporary man, it must be freed from this mythical framework. Now for Bultmann this does not simply mean an updating of that worldview, but the realization that our contemporary scientific worldview and the Christian faith really have little to do with one another:

> Faith itself demands to be freed from any world-view produced by man's thought, whether mythological or scientific. For all human world-views objectivize the world and ignore or eliminate the significance of the encounters in our personal existence.[2]

Faith, for Bultmann, must be purely existential; it must affirm the meaningfulness of our own chosen life without any cosmological props.

> The word of God. . . . calls man to his true self. For the self of man, his inner life, his personal existence, is also beyond the visible world and beyond rational thinking. The Word of God addresses man in his personal existence and thereby it gives him freedom from the world and from the sorrow and anxiety which overwhelm him when he forgets the beyond. By means of science men try to take possession of the world, but in fact the world gets possession of men.

> Faith is the abandonment of man's own security and the readiness to find security only in the unseen beyond, in God. This means that faith is security where no security can be seen; it is, as Luther said, the readiness to enter confidently into the darkness of the future.[3]

Scientific statements are objective, factual, uttered without respect for the personal involvement of the speaker. Theological statements, on the other hand, must be purely confessional expressions of the speaker's own self-understanding, and not about the world at all.

> Only such statements about God are legitimate as express the existential relation between God and man. Statements which speak of God's actions as cosmic events are illegitimate. The affirmation that God is creator cannot be a theoretical statement about God as creator mundi in a general sense. The affirmation can only be a personal confession that I understand myself to be a creature which owes its existence to God.[4]

Obviously, there is much truth in this. The question of God's existence is not like that of the Crab Nebulae, but has a far deeper personal import. Faith in God is not primarily a matter of intellectual assent, but of trust, loyalty, and commitment. On the other hand, is it so clear that theoretical and cosmological considerations can be entirely excluded?

Bultmann is quite eloquent in attacking any faith which seeks security in "objectifying knowledge,"[5] but neglects to mention that his own existential theology has been made forever secure against any kind of scientific scrutiny. By radically distinguishing existential self-knowledge and scientific worldviews, Bultmann need never worry about the impact of these particular worldviews upon the Christian faith, for they have been ruled to be irrelevant a priori. The patient effort to reconcile religion and science need not be undertaken, for the two have really nothing to say to one another.

This, I maintain, is a parochial view, for it restricts the relevance of theology solely to man. If the only meaningful assertions I can make about God are confessional statements concerning God's relatedness to my personal existence, then theology can only make sense for human beings. If God is known to me only as he affects me, then it would be impossible to know how God might have any role for any other being, save by analogy. This works roughly for other human persons, but what about intelligent beings in far distant worlds? Unless I can make some kind of cosmological statements about God's relationship to the world, to space and time, to physical processes, can I say anything at all about how God might be related to such beings?

Interestingly enough, it seems to have been such cosmological speculation that first enabled the Israelites to discern other roles for God beyond that of being their national protector. Ancient Israel from the beginning experienced and confessed God as their Redeemer from bondage in Egypt, but for centuries they limited their understanding of God solely to this particular existential relationship. God was the God of Israel, and few questions were raised about God's relations to other peoples; after all, they had their own gods. Little by little, however, it gradually dawned upon them that their God who had delivered them out of Egypt was also the Creator of the world. This faith bursts forth most vigorously and vibrantly in the poetry of Second Isaiah, and it is no accident that he discerns a much enlarged role for Israel beyond that original ethnocentricity:

> It is too light a thing that you should be my servant to raise up the tribes of Jacob and to restore the preserved of Israel; I will give you as a light to the nations, that my salvation may reach to the end of the earth. (Isaiah 49:6)

Once God is acknowledged as creator of heaven and earth, and not simply as standing in the existential relationship of Redeemer to Israel, then it is possible to become aware of God's concern for all peoples. In our contemporary efforts to safeguard the existential relationship between man and God, we take too easily for granted that salvation applies to all men; originally, it was understood to apply only to Israel. Now, however, if we realize that God is creator of the innumerable galaxies and solar systems that populate the heavens, we must recognize that salvation is not just limited to men, but applies to all intelligent beings wherever they may dwell.

Since World War II, there has been a dramatic upsurge of interest in life on other worlds. For many people, particularly those engaged in the natural sciences, the notion of extraterrestrial life is no longer merely an exotic possibility, but virtually a certainty. The issue chiefly turns on our confidence in the regularity of planetary development and of evolutionary growth. If both of these occur regularly, spontaneously, then we should expect the universe to be populated with myriads of planets sustaining life, many of which could be technical civilizations far in advance of ours. On the other hand, if the origin of life, or the formation of planets, is a chance, freak occurrence, then we may well be alone in the universe.

Both views of planetary development have been with us for a long time. The French naturalist Georges de Buffon (1707–1788) proposed that the birth of the planets results from a glancing collision of our sun with a passing comet. In contrast, Immanuel Kant and Pierre de Laplace argued that planetary development was part of a normal process to be expected in the life of almost every star: they assumed the young sun was surrounded by a thin lens-shaped gaseous envelope (solar nebula) which later condensed into planets. During the late nineteenth century, the Kant-Laplace hypothesis was severely criticized by the British physicist Clerk Maxwell, who argued that the forces of differential rotation between parts of the solar nebula would break up any such condensation as soon as it began to form. In the face of this objection, which seemed quite decisive at the time, cosmologists increasingly turned to some version of Buffon's glancing collision. Forest Ray Moulton and Thomas C. Chamberlin in the United States supposed that the sun, under the gravitational pull of some passing star, erupted gigantic globs of matter which in time formed planets, and a comparable theory was proposed by Sir James Jeans and H. Jeffreys. Yet the collisions or near misses dictated by these theories are inherently very improbable. Sir James Jeans calculated that any given star would make a sufficiently close approach to another star only once in 500,000 billion (5×10^{17}) years, or only about ten for the entire life of our galaxy during the past five billion years.[6] With so few planets in existence, we could hardly assume that there would be much life elsewhere, at least not in our galaxy.

Both types of theories developed difficulties, but Maxwell's objections to the Kant-Laplace hypothesis were overcome by C. F. von Weizsäcker toward the end of World War II, who argued that the original objections were based on the assumption that the chemical composition of the sun resembled that of the earth. We now know that the heavier, terrestrial elements only compose about 1 percent of the sun's mass, the rest being essentially a mixture of the two lightest elements, hydrogen and helium. Von Weizsacker argued for a differential treatment between the hydrogen-helium and heavier elements with respect to the angular momentum of original solar mass. With the issue thus resolved in favor of a regular formation of planets, the chances of there being other planets capable of sustaining life is so high as to be practically certain. Even if only one planet out of every 150,000 contained life, there would be one million life-worlds in our galaxy, some of which we can reasonably assume contain intelligent life, for whom, we presume, God would also be concerned.

As Paul Tillich has seen,

> a question arises which has been carefully avoided by many traditional theologians, even though it is consciously or unconsciously alive for most contemporary people. It is the problem of how to understand the meaning of the symbol "Christ" [or any other man-centered religious symbol, for that matter] in the light of the immensity of the universe, the heliocentric system of planets, the infinitely small part of the universe which man and his history constitute, and the possibility of other "worlds" in which divine self-manifestations may appear and be received.[7]

Now what can we say about God's relation to such intelligent beings on other planets? Here our approach cannot be existential, for we cannot participate in the self-understanding such beings might possess, nor can our reflections significantly influence, or be derived from, our own quest for a meaningful selfhood. The issue is quite theoretical, but it is one which helps us to see the boundaries and implications of one particular faith-stance in the wider context of others. Christian thinkers have reflected on these boundaries with respect to their fellow human beings in other cultures, and even with respect to the other animals which share our planet, but rarely with respect to the rest of life populating the universe.

Moreover, the issue entails cosmological assertions bearing on our scientific understanding of the universe. For we must show the possibility of God's involvement in the emergence of other forms of intelligent life before any claim can be entertained concerning their existential standing before God, and this task invites dialogue with scientific accounts of evolutionary processes.

While it may not express itself in these terms, the scientific community has become increasingly confident in the tremendous potential inherent in the universe for evolutionary growth. If the proper conditions are present, e.g., surface temperatures permitting large bodies of liquid such as water or ammonia or methane, atmospheres permitting of energy-exchange, some source of light, etc., we expect that sooner or later life will emerge. The extreme resiliency and buoyancy of the evolutionary thrust makes it unlikely that if all necessary environmental conditions are met that macromolecules of organic compounds will not eventually emerge, to be followed by some form of life.

In 1953, Stanley L. Miller, a collaborator of Harold C. Urey at the University of Chicago, prepared a mixture of methane, ammonia, water, and hydrogen in simulation of the primitive atmosphere postulated for the earth. Stimulated by an electrode discharge passing through the mixture, it yielded a variety of organic molecules within a week: amino acids, acetic acid, simple sugars—all building blocks used in the formation of living cells. If such dramatic growth were possible in such a short space, under the proper conditions we can expect the same sort of process to occur on other worlds. This does not mean to imply that precisely these organic compounds must first be synthesized in order to allow life to emerge, but that these could have been the compounds used here on earth for the formation of life. On other worlds it is conceivable that an original atmosphere rich in hydrogen cyanide would have produced other organic building blocks. We anticipate some sort of growth toward increased complexity: increasingly larger organic macromolecules, then the convergence of many macromolecules to constitute a simple living system, either as a cell with its protective wall and vital nucleus or as some functional analogue, then the convergence of many cells to form larger organisms.

Since Darwin, this process of evolutionary growth, whereby levels of increasing complexity are seen to emerge from simpler ones, has been explained in terms of the double mechanism of natural selection and chance variation. Natural selection affords a measure of stability and durability, for those populations which happen to be best adapted to their environments continue to survive as other populations tend to die out. By itself, however, natural selection provides for no evolutionary advance, for it introduces no novelty, and hence no possibility of anything more than that which already has been. We must recognize that in this context "adaptation" is strictly defined in terms of survival-values and that generally speaking, it is the simpler forms of organization that possess the greatest staying power: living systems, no matter how fantastically intricate and well-organized they might be, have a much shorter span of existence than, say, a rock crystal, or a single stable atom.[8]

The stability of natural selection must be balanced by the novelty of chance variation, which permits the introduction of new forms of existence. In principle this is as far as a scientific explanation can go if it proceeds by strict limitation to efficient causal explanation. Seen most broadly, any efficient causal explanation restricts itself to that which is traditional, for it explains the present in terms of the past. Efficient causality is the way in which the past persists into the present, and the task of scientific analysis is to discover whatever regularity exists in this transfer from past to present. Everything that happens either follows regularly established patterns (whether absolute or merely statistical in nature) or just happens quite accidentally. Ultimately, then, regularity or chance are our only options, and chance signifies little more than the absence of scientific causal explanation. Yet without chance, nothing new could ever occur, i.e., new in the sense of establishing novel causal patterns and forms of organization. If everything happened strictly according to deterministic physical laws, there would be no possibility for the emergence of life, assuming that the organization of life, while dependent upon physical principles, is not reducible to them. Fortunately, physical laws are probabilistic, with an indeterminacy that permits the emergence of novelty. Without chance, there can be no evolutionary advance, yet, strictly speaking, chance explains nothing. It is merely the absence of any efficient causal explanation.

Now the evolutionary process is essentially the emergence of new levels of complexity. Given the character of scientific explanation in terms of efficient causes, it is quite understandable that such evolutionary advance should be explained in terms of natural selection and chance variation as the best possible scientific theory. Chance supplies the novelty, while natural selection permits the consolidation of gains. Nevertheless, as an account of the whole story, it is quite incredible.

Essentially, what is lacking is any account as to why new levels of complexity should ever be achieved. Chance variation will produce novel forms of organization which may be more or less complex than that which preceded it. But should there be any greater tendency for the more complex rather than for the less complex to persist from such variation? Random activity should actually tend to favor the less complex for three reasons. (1) The simpler depends upon fewer specific conditions and has fewer, simple energy needs: human life, for example, depends upon so many more factors than single-celled marine life does. (2) The less complex is more in accord with entropy, the principle that any closed system tends to decrease in order over time. (3) With time, the possibility that random variation should produce anything with greater order should decrease as entropy increases.

God's cosmological function consists in supplying that impetus toward greater complexification we discover operative throughout the natural order. This does not mean that God acts efficiently as one of the causal antecedent conditions out of which the present event emerges. We do not wish to repeat the fundamental error of those who portray God as the maker or mechanic or artisan of the world, for all these images imply that God forms the world through force or coercion. Rather he serves as a lure for actualization, providing novel possibilities of achievement, thereby persuading each creature to create itself. Persuasion entails response, not conformation, and the response is free either to embrace or to reject the novel aim. We do not mean to suggest that there is much free response in the universe: atoms and molecules are extremely traditional in their habits, behaving largely as they have always behaved. Plants, animals, and even humans are not much better, blindly reiterating that which went on before. Yet if there is to be any emergence of greater complexity, then there must be at least a modicum of spontaneous response possible even on the atomic and molecular levels, occasionally permitting the actualization of some evolutionary advance. Divine persuasion is the urge to maximize the possibilities inherent in such indeterminate response.

In a very real sense this theory of divine initiative and creaturely response commits us to some form of neo-Lamarckism, for we are affirming that the inheritance of acquired characteristics is fundamental to evolutionary advance. In appropriating the divine possibility as its own aim, the creature is acquiring some characteristic which is then transmitted by means of efficient causality to subsequent generations. Free response becomes blind habit; novelty becomes tradition; final causality passes over in efficient causality. As we have noted, the realm of activity not wholly governed by efficient causal patterns may be vanishingly small on the simplest levels of existence, but that quantitatively negligible amount is all important in furthering any increased complexity. Random mutation is incapable of explaining the directedness of evolution such response can introduce. Once introduced, however, the new characteristic may be simply transmitted through blind habit.[9]

Generally speaking, neo-Lamarckism has been properly discredited, but for the wrong reasons. It is not the inheritance of acquired characteristics which is erroneous, but the failure to distinguish properly between "levels" of response. How a given person or animal responds to his environmental situation will not affect the genetic makeup of his descendants, for that genetic makeup is determined on the cellular, perhaps even on the molecular, level. Personal response is inherited rather in terms of the habits or dispositions for action we acquire during our lifetime; free decision shapes character. If sufficiently original, human response may shape the common culture inherited by our fellows, for every tradition blindly received originally had its purpose and justification, however feeble that might have been. If we restrict ourselves to biological inheritance, then we must examine cellular and molecular response, ignoring all higher responses on the level of the total organism. Here we can only conjecture, for it is difficult to appreciate what aims DNA molecules may strive to actualize. It may be doubted, however, that such aims would embrace even the relevant aims of the cell to which the DNA molecules belong, let alone the aims of the total animal body. Precisely in this sense, we can say that any gene mutations introduced by novel actualizations by such DNA molecules are random with regard to the future of the total organism in much the same way that the realization of our personal goals is usually random with respect to the future of the human race as a whole.[10]

Creaturely response through the appropriation of a novel aim supplied by God, on whatever level, whether atomic, molecular, cellular, or organismic, becomes the chief means whereby divine purposes become affective in the world. This does not mean, however, that the created order proceeds according to some set plan. Divine persuasion is highly opportunistic, seeking to maximize the possibilities for increasing complexity as are consistent with the actual conditions imposed by the past through efficient causation. Moreover, this persuasion is not coercive, so there is no necessity that every creature must embody the maximum of his potentialities. Whatever happens, happens as the result of the creature's self-activity in using its causal conditions to achieve its ends, but God is everywhere and at all times seeking that which is best, given the circumstances. Such gracious activity will not always be thwarted, so that evolutionary advance, as actualized through free creaturely response, gradually comes into being. Thus, for the universe at large, divine persuasion seeks to evoke life wherever possible, in the form appropriate to particular local conditions.

We assume that there are common physical laws for the entire observable universe, and these laws yield universal laws of chemical bonding. We can be reasonably certain that the way molecules are formed is invariant throughout the universe, and that the most promising chemical elements for organic evolution will be the lighter elements toward the center of the periodic table capable of very supple and complex covalent bonding. From this point on, however, the evolutionary process may branch in many directions, for the macromolecules formed out of these elements will vary considerably depending upon the composition and distribution of such elements in the early stages of that world. One-celled microorganisms, in developing their metabolism, will depend in turn upon whatever macromolecules are available, so we should expect every world to have its own way of organizing simple living systems. As we know from our knowledge of the past history of evolution on this earth, the development of multicelled organisms with or without central coordination (i.e., animals and plants) can take many different routes, but these routes might be even more varied if the basic cellular structure were also radically different. Cell division and bisexual reproduction are the two basic strategies adopted by terrestrial life for its propagation. If, however, bisexual reproduction proved to be unfeasible for other kinds of living cells, other methods of propagation would have to be invented. Thus, there might be worlds in which sex plays no function at all, or where all reproduction requires the cooperation of three sexes acting in concert. The possibilities are naturally enormous. In *Science and the Theory of Value*, Peter Caws suggest that:

> Freedom in the most general sense—that is, the ability not necessarily to suspend the causal order but at least to halt or reverse temporarily and locally the universal trend toward disorder—is to be regarded as an unconditional imperative for any conscious production of value.[11]

We may generalize this contention, and see in the increase of freedom a divine purpose appropriate to all worlds. Freedom is essentially self-creation, requiring both the absence of restraint and the introduction of order by the free agent. On the atomic and molecular level there is a minimum of spontaneity, for each response is overwhelmingly the product of blind habit, endlessly reiterating the same pattern of activity. If we think of a molecule as a society of atoms, then molecules form some of the most tradition-ridden societies in existence, since organic molecules have been discovered to have preserved their structures intact for a good billion years. The decisive difference between living and lifeless matter, I submit, is the difference between novel and habitual response. This may well be a matter of degree, such that what we designate as living may simply be those instances where novelty dominates over habit. Homeostatic adjustment within the living cell requires that it respond to its surroundings in original ways which supersede the customary behavior of molecules. At a higher level, motility frees animals from spatial confinement and renders them open to a great variety of situations to respond to. Yet without sentience, such motility would simply lead to random behavior; sight and smell and touch enable animals to achieve purposeful results meaningful to them. Intelligence is simply the next step in the quest for greater freedom; imagination increases our field of action by including the possible as well as the actual, while reason enables us to order these possibilities in significant ways.

As the capacity for novelty expands, consciousness emerges. We cannot define consciousness in terms of a centralized nervous system, for there is a great deal of neurological activity lying below the threshold of consciousness. Habitual patterns of response, such as getting dressed, riding a bicycle, using a typewriter, so painfully and self-consciously learned at the time, become quite unconscious.[12] A centralized nervous system or its close analogue may be the necessary basis for consciousness, but consciousness itself is the inner concomitant of the presence of some novelty which has not yet faded into the background through incessant repetition.

Once intelligence appears, cultures may develop, varying enormously among themselves, but sharing the common biological inheritance and common general environment for that planet. Thus, we should expect cultural differentiation on other worlds, but despite their diversity all would be characteristically stamped by that biological situation, just as all our cultures by contrast will be found to have certain peculiarly "terrestrial" features. We may leave open the question whether intelligent cultures must further develop into technical civilizations, that is, into civilizations seeking to transform environmental conditions to suit their own purposes and needs. It may be that human technology is essentially an accident owing to the fact that man is biologically so poorly adapted to his natural surroundings. It is quite possible, for example, that the dolphins possess a highly refined culture transmitted orally from generation to generation, but that they have developed no technology because they have no need of it. Man is a tool-using animal, but it may not be necessary that all intelligent species must also become tool-using animals.

With the emergence of conscious alternatives of action, ethics becomes possible, for now some alternatives may be experienced as better and others as worse. What goodness means for other intelligent beings may well be beyond the bounds of our imagination, but it might be just possible to define a general criterion underlying all concrete embodiments. That which fosters the expansion of freedom and the intensity of experience may be regarded as good, although the ways in which freedom and intensity are fostered will depend upon the biological, psychological, social, cultural, political, economic, and possibly religious situations particular intelligent beings find themselves in.

We have been brought up on cultural relativity, but interstellar communication may confront us with biological relativity as well. Sexual morality would be largely incomprehensible to a sexless race, while tri-sexual behavior would introduce incredible complications. The main issue, however, does not center around the means of reproduction but upon the needs of child-rearing. If intelligent beings elsewhere are born largely self-sufficient, then no provisions for close-knit familial contact may be necessary. On the other hand, some species might be biologically self-sufficient from birth and yet be highly dependent emotionally upon one another, all of which would color the ethical character of behavior. Biological heritage also shapes economics and politics. The allocation of scarce resources depends upon what is scarce: up till now, at least, we have no economic policy with respect to the air we breathe, and there are no sanctions against "stealing" air. In an atmosphere poor world, or in a world whose organisms require enormous amounts of oxygen, each individual might jealously guard his own farm of plants for converting carbon dioxide to oxygen, and a whole corpus of law meting out death to oxygen-robbers might well have to be enforced. Man has concentrated his economic concern on goods and services, but in another world it might be time, particularly time with the computer, or knowledge, or intensity of religious experience, that would have to be rationed. Politics could also be radically different on planets containing several intelligent species, where one species might voluntarily and quite spontaneously allow itself to be governed by another species it recognized to be better suited to the task. Depending upon biological circumstance, some societies might have no recognizable form of government at all, while for others a ruthless totalitarianism might mark the difference between life and death. Such a totalitarianism, however, might not be perceived by its subjects as ruthless, nor might the absence of government necessarily seem quite so utopian to its members.

Throughout this multifarious universe the divine creativity is operative, evoking greater and greater levels of complexity, thereby permitting the expansion of freedom and the emergence of intense conscious experience. With consciousness it becomes possible for creatures to be aware of God's directing activity, though on earth this seems to be generally rather sporadic and intermittent. In general, the basic way in which God acts on the human level is through ethical persuasion; it is by the worthiness, the attractiveness, the importance of specific ethical ends envisioned that God lures us on to actualize a world better than what we have known. Such divine persuasion can be effective wherever man is willing to be ethically sensitive, quite apart from whether he consciously affirms or denies the existence of such divine reality.

In this context, we may define God as that dynamic source of values which lures the evolutionary process to an ever-richer complexity productive of increasing freedom and intensity of experience. As such God is necessarily operative in the development of every life and in every culture, whether terrestrial or extraterrestrial. It is another question, however, to what extent extraterrestrial intelligent life would be conscious of God. Here we must be extremely tentative, for religious awareness is inextricably bound up with cultural understanding and ethical sensitivity, and these in turn depend upon psychological, historical, economic, and political factors that must vary immensely with the basic biological structure of each world. It may be, for example, that life is so well attuned for certain intelligent species that they have little ethical awareness at all, having no need for it, and hence no religious awareness of any transcendent source for such values. On the other hand, another species might experience direct, continuous contact with God through some means analogous to mental telepathy.

We must leave the question whether interstellar communication could prove or disprove the existence of God a totally open question. We might encounter an extraterrestrial civilization that had either never known of, or had outgrown any need for, God, one capable of explaining our situation so thoroughly and so well that we would see that terrestrial religious experience was either fundamentally misguided or a very inadequate way of expressing the truth. On the other hand, we should be very wary of any hasty judgments about the religious orientation and attitudes of other intelligent beings. The field of extraterrestrial comparative religion might well take centuries to explore before we could have any degree of confidence in our conclusions; after all, we do not yet know very much about the religions of India and China today after a century and a half of intensive exploration, and much of what passes for knowledge is a Western projection upon an Eastern ignorance. We must also remember that our awareness of Western religion is largely the result of the historical accident that our culture is essentially Graeco-Roman while its religious roots are Hebraic. In other societies we may expect their religious dimension to be so subtly pervasive and so inextricably bound up with their highly exotic culture, as well as being so diverse from our own terrestrial modes of symbolization, that we would be apt to overlook or to misinterpret its importance and function.

Naturally the possibilities for religious expression are endless, but let us mention three of particular interest: (1) On a planet with several intelligent species, one species might be intensely devout, yet channeling all of its energies into the worship of another species it deems more worthy than itself, envisaging nothing higher. (2) On another planet the intelligent race may have "come of age," grateful to God in having matured the evolutionary process making its own existence possible, but now fully capable of generating its own values and aims in full independence from divine guidance. Terrestrial civilization may now be on its way toward such a "coming of age," yet without having achieved full maturity, still full of adolescent rebelliousness toward God intermixed with mature moments of grateful appreciation all terribly confused with infantile dependence. (3) There may be an extraterrestrial society in constant direct communication with God, but, however, with a large subsociety quite antagonistic toward the realization of divine aims, practicing a kind of radical anti-theism. This subsociety might fortify itself with its own symbol-system, its own interpretation of the evolutionary process, its own history, and its own self-justification, perhaps drawn from some theory analogous to Sartre's doctrine of radical freedom, maintaining that obedience to divine demands thwarts all realization of genuine freedom. In such a situation the overriding religious question for sensitive souls would not be: "Does God exist?" but rather, "Ought God be obeyed?"

Now from all these speculations, admittedly quite conjectural, let us come back to earth. We have sketched a liberal theology suitable for the cosmos, but is it also appropriate for our existence here and now? In particular, can such a broad evolutionary theism really do justice to the Christian claim that God acted decisively in Jesus of Nazareth two thousand years ago, or are we being called upon to surrender that heritage in the name of some unproven conjecture?

Consider, if you will, the opening verses of the Fourth Gospel:

> In the beginning was the Word, and the Word was with God, and the Word was God. He was in the beginning with God; all things were made through him, and without him was not anything made that was made. . . .
>
> And the Word became flesh and dwelt among us, full of grace and truth; we have beheld his glory, glory as of the only Son from the Father. . . . And from his fullness have we all received, grace upon grace. For the law was given through Moses; grace and truth came through Jesus Christ. (John 1:1–3, 14, 16–17)

Are we then to conclude that God's only Son became uniquely incarnate once and for all on the third planet of a rather ordinary outlying star of a thoroughly undistinguished galaxy? Paul Tillich argues to the contrary:

> Incarnation is unique for the special group in which it happens, but it is not unique in the sense that other singular incarnations for other unique worlds are excluded. Man cannot claim that the infinite has entered the finite to overcome its existential estrangement in mankind alone. Man cannot claim to occupy the only possible place for Incarnation.[13]

In terms of our text from John, it makes all the difference in the world—or rather, between worlds—if we affirm that Jesus as the Christ is the Word of God or if we say that he is the Incarnation of the Word of God. John, let it be noted, says the latter, not the former.

In his Prologue John uses the Greek word logos for "word;" carrying with it all the philosophical connotations from Heraclitus and the Stoics of a fixed structuring principle by which the cosmos is ordered. Without this Logos, the world decays into chaos. To get at John's intent, however, we should consider his Hebraic antecedents, particularly the more dynamic expression of the Word in the first chapter of Genesis: "And God said, 'Let there be light'; and there was light" (Genesis 1:3). The priestly writers of this creation story block out a scant eight divine directives as being all that is necessary for the entire creation; we extend their number indefinitely, but the principle remains the same. God's speaking symbolizes the divine activity whereby new possibilities for the emergence of greater complexity become lures of feeling for further actualization. What God says depends upon the particular situation confronting that given individual in his own world, and the speech appropriate to macromolecules capable of converging into the nucleus of a cell is characteristically different from the ethical imperative addressing twentieth-century Americans frustrated by a Vietnam War abroad and civil injustice at home, and both differ sharply from that Word addressing intelligent tri-sexual creatures breathing ammonia gas in a supercivilization centuries in advance of our own. God's dynamic Word knows no single form, but assumes that character expressive of God's general aim at intensity of experience appropriate to each circumstance. By this Word, the worlds were created, and by this Word also God has sought the salvation of his people Israel.

To affirm that Christ is the Incarnation of this Word is to confess that in Jesus of Nazareth we behold the embodiment of the divine intent of mankind. The Word becomes incarnate by becoming fully actualized in the words, deeds, and suffering of the Christ. God has spoken before and since to man, with fragmentary success; his Word has come down again and again, but never before has it so taken root and become flesh. For it must be recognized that the divine Word depends upon creaturely response for its actualization. All of God's urging on civil rights will do no good unless we act; but then again, we would not be inclined to act at all unless our ethical sensibilities were aroused.

Christ is the Incarnation of God's Word for confessing Christians, but should we say he is the only Incarnation for all mankind? Tentatively we would rather claim that Christ is the decisive incarnation of divine purpose for Western man outside of Israel. Israel already received her Word in the event of the Exodus which created her communal existence. Christ was the intensification of that divine purpose as culminating in the life of one man, whose life and death made the salvation already inherent in Israel available beyond its borders. The only real competitor for such realization of divine purposes in our Western heritage seems to be Socrates. Socrates, however, labored under the handicap of emerging from a community, Athens, which did not have a history of responding to the divine Word as Israel had. Without the prior history and expectations of Israel, even Jesus could not have become the Christ. We must leave it an open question whether God's purpose became fully incarnate in others outside of Western culture. Here Gautama the Enlightened One, the Buddha, is our most promising candidate, yet his cultural heritage is so radically different from ours that it becomes difficult to judge.

The early church also claimed that Jesus was raised from the dead and lives on the right hand of God. I think we can no longer interpret this claim by identifying Christ with God and allowing Christ to assume the everlastingness belonging to God. For at best Christ is the Incarnation of God's word for man, not for other intelligent species. If Christ lives, he lives as an active channel through which God's purposes may be effectively communicated to man. Here our best analogy is to be found in the relation between the mind and the body, for the cells in my hands may become the effective means of actualizing a divine purpose, not merely on the level appropriate to cells alone, but as communicated to my hands by my mind seeking to accomplish some good end. The eleventh-century Hindu theologian Ramanuja best defined the nature of what it means to be a body: any substance or actuality "which a sentient soul is capable of controlling and supporting for its own purposes, and which stands to that soul in an entirely subordinate relation, is the body of that soul."[14] Such a general definition of the body neither specifies its components as individual cells nor does it insist upon any necessary spatial contiguity of its members. If Christ lives, then, he lives by being effectively active through his Body, a Body composed of all those persons who in faith seek his continuing guidance in the actualization of his aims. Thus, we take Paul's sayings about the Body of Christ with literal seriousness. The resurrection of Christ's Body began on Easter morn when Peter was transformed by his direct encounter with the risen Christ and became the first member of that living body, one that has been growing down through the ages.

The Body of Christ has become problematic in our age, however, for at least in the visible Body, the Church, there seems to be more signs of senility than of vitality. To be sure, the Body of Christ could never be identical with the institutional church, though institutions, like the protective structures of biological organisms may be quite necessary for continued survival. Animals require bones as well as muscle and blood and brains. If it did not sound quite so blasphemous, we should say that the institutional church forms the bones of Christ's Body, and the real question confronting the church is whether those bones encase living flesh, or does only the skeleton remain? Essentially the Christian commitment is a bet that there is still life in that Body to which one is joined, as one seeks to live Christ's life through one's own life. In any event, whether living or dead, the emergence of the Body of Christ in the resurrection was a decisive act, for here, for the first time in earth's history, has a hyperconsciousness emerged capable of embodying mankind, however fragmentarily.

I have used Teilhard de Chardin's term "hyperconsciousness" intentionally, for I find in his vision an understanding of the future consistent with evolutionary growth. At each level we find convergence: elementary particles converge to form atoms, atoms to form molecules, macromolecules to form cell-nuclei, cells to form organisms. At each level, the characteristic properties of the individual components are preserved while being taken up into a richer complexity capable of a much more refined response to its situation. Need we assume that this process of convergence has exhausted itself? Teilhard envisages a final convergence of all mankind in love and freedom forming a single organism with Christ as its head, as the sensitive, directing agency mediating God's purpose to man more intensively than would be possible on the human level alone. This is at one and the same time the final consummation, the kingdom of God, and the full actualization of the Body of Christ.

We might extrapolate one step further, and envisage the convergence of world-civilizations, each having already achieved an organic unity analogous to the Body of Christ, to form a second-level superorganism. At the present time this is highly improbable, to say the least, given the immense distances existing between stars, and hence between planetary civilizations. But if the universe is pulsating, that is, alternating between periods of expansion (as at present when all the galaxies are receding from one another) and periods of contraction (when all the matter of the universe moves toward a single center), then such convergence would be possible during the latter stages of a contraction.

The issue as to whether the universe is indefinitely expanding or pulsating is still an open question for physical cosmology. Carl Sagan writes:

> While we do not yet know whether our universe must experience a contraction phase, we are on the verge of finding out: Does the universe expand forever, or are we trapped in a vast cycle of cosmic deaths and re-births?[15]

He personally finds the pulsating alternative quite unattractive:

> A universe which grinds to ashes 50 to 100 billion years of galactic, stellar, planetary, biological, and cultural evolution is an uncomfortable one to live in, and not only for Marxists. Some religions, such as the Hindu, find no difficulties in the notion of a cyclical cosmos, but the pointlessness of such a universe is awesome.[16]

We must realize, however, that we are trapped by cosmic death in any case. By the second law of thermodynamics, the amount of available energy for the creation and maintenance of complex organization is constantly decreasing. In an ever-expanding universe all available energy will be used up beyond all possible recall, and all life and structure will slowly crumble to nothingness. In a contracting universe, in its very last phase, all existing life will be crushed together into a single highly dense mass, but that convergence of all matter would also make energy available once again for the creation of a new universe.

Expansion and contraction is only pointless if we look to the outcome and not to the process, and it is ultimately only in the process that life and the actualization of value takes place. A final phase of contraction would not only permit the final convergence of this universe into one gigantic superorganism and the renewal of its energy supply, but also the possibility of the realization of values forever excluded from this universe. For as was argued above, the physical and chemical principles we know are constant for this entire cosmic epoch, even though biological organization may vary from one inhabitable planet to another. But in the cataclysmic destruction of such a final contraction, even these habitual patterns of atomic and molecular behavior might be destroyed to make room for a new universe operating on quite different physical and chemical principles, generating radically different biological systems and cultures. The possibilities an everlasting God can actualize in due time through countless cycles of expansion and contraction are simply inexhaustible.

NOTES

1. Ford cites this article as "originally given as the Faculty Research Lecture for the Spring of 1968 at Raymond College of the University of the Pacific, Stockton, California, and published in *The Raymond Review* 2/2 (Fall 1968), 1–14." Ford, *Lure of God*, xi. However, no online record of this article or the *Raymond Review* can be found. As preserved in the Center for Process Studies archives, it is republished here with minor edits.
2. Bultmann, *Jesus Christ and Mythology*, 83.
3. Ibid., 40f.
4. Ibid., 69.
5. Ibid., 84.
6. Shklovskii and Sagan, *Intelligent Life in the Universe*, 166.
7. Tillich, *Systematic Theology*, 95.
8. Individual cells within a complex organism, to be sure, apparently have a shorter life-span than their host. It may be, however, that the kind of organization whereby the individual cells are knit together is itself simpler than the organization of the cell.

Bureaucracies and institutions are less organically structured than individual human beings, yet they can outlast a score of human life-spans. If survival, i.e., that persistence of a given state of organization, is our sole criterion of value, then there is a lot to be said for institutional inertia. It is highly adaptive in its ability to survive most anything.

9. Hardy, *The Living Stream*.

10. The argument of this paragraph is heavily dependent upon Overman, *Evolution and the Christian Doctrine of Creation*, 203–11.

11. Caws, *Science and the Theory of Value*, 103.

12. Schrödinger, *Mind and Matter*, 4f.

13. Tillich, *Systematic Theology*, 96.

14. Quoted by Helfer, in "The Body of Brahman According to Ramanuja," 44.

15. Shklovskii and Sagan, *Intelligent Life in the Universe*, 142.

16. Ibid.,136.

BIBLIOGRAPHY

Bultmann, R. *Jesus Christ and Mythology*. New York: Charles Scribner's Sons, 1958.

Caws, P. *Science and the Theory of Value*. New York: Random House, 1967.

Hardy, A. *The Living Stream: Evolution and Man*. London: William Collins Sons & Co., Ltd., 1965.

Helfer, J. S. "The Body of Brahman According to Ramanuja." *Journal of Bible and Religion* 31, no. 1 (1964): 43–46.

Overman, Richard H. *Evolution and the Christian Doctrine of Creation*. Philadelphia: Westminster Press, 1967.

Schrödinger, E. *Mind and Matter*. Cambridge: Cambridge University Press, 1959.

Shklovskii, I. S., and C. Sagan. *Intelligent Life in the Universe*. San Francisco: Holden-Day, Inc., 1966.

Tillich, P. *Systematic Theology*, vol. 2. Chicago: University of Chicago Press, 1957.

Index

About the Contributors

Andrew M. Davis is a process philosopher, theologian, and scholar of cosmological wonder. He is program director for the Center for Process Studies at Claremont School of theology, where he earned his PhD in religion and process philosophy. As program director, Davis researches, writes, and organizes conferences on various aspects of process-relational thought in dialogue with current issues in science, philosophy, and religion/theology. An advocate of metaphysics and meaning, he approaches philosophy as the endeavor to systematically think through what reality must be like because we are a part of it. His research has led him to a variety of religious contexts and communities around the world, including India, Israel-Palestine, and Europe. He received the 2013 Award for Excellence in Biblical Studies, the 2017 fellowship with FASPE (Fellowships at Auschwitz for the Study of Professional Ethics) and the 2020 Presidential Award for Academic Excellence. He was recently nominated and elected as a fellow for the International Society of Science and Religion (ISSR). He is the author, editor, and coeditor of several books including *How I Found God in Everyone and Everywhere: An Anthology of Spiritual Memoirs* (Vol. 1, 2018; Vol. 2, forthcoming, 2024); *Propositions in the Making: Experiments in a Whiteheadian Laboratory* (2019); *Depths as Yet Unspoken: Whiteheadian Excursions in Mysticism, Multiplicity, and Divinity* (2020); *Mind, Value, and Cosmos: On the Relational Nature of Ultimacy* (2020, nominated for the ISSR 2022 Book Prize); *Nature In Process: Organic Proposals in Philosophy, Society, and Religion* (2022); *Process Cosmology: New Integrations in Science and Philosophy* (2022); and *Metaphysics of Exo-Life: Toward a Constructive Whiteheadian Cosmotheology* (2023). Follow his work at https://www.andrewmdavis.info/.

Constance (Connie) M. Bertka received her PhD in Geology from Arizona State University and an MTS, Master of Theological Studies, from Wesley Theological Seminary in Washington DC. In addition to her research in planetary sciences, Dr. Bertka has had a long-term scholarly and pragmatic interest in the relationships between science and religion and their influence on public understanding of science. Previously she directed the Program of Dialogue on Science, Ethics, and Religion at the American Association for the Advancement of Science. Prior to that she was a senior research associate at the Carnegie Institution of Washington's Geophysical Laboratory where she also served as program director of the Deep Carbon Observatory. She has also taught on contemporary issues in science and religion at Wesley Theological Seminary. Currently she is a consultant with Science and Society Resources LLC. Dr. Bertka cochairs the Broader Social Impacts Committee of the Smithsonian's National Museum of Natural History's Human Origins Program. She is the lead author on the Human Origins Program's Cultural and Religious Sensitivity Teaching Strategies Resource (2015) and part of the team facilitating public programming for the Smithsonian's traveling exhibit *Exploring Human Origins: What Does It Mean to Be Human?* She is editor of *Exploring the Origin, Extent, and Future of Life: Philosophical, Ethical and Theological Perspectives* (2009).

Bruce Damer is director and chief scientist at the BIOTA Institute. He has spent his life pursuing two great questions: how did life on Earth begin, and how can we give that life (and ourselves) a sustainable pathway into the cosmos? A decade of scientific research with his collaborator Prof. David Deamer at the UC Santa Cruz Department of Biomolecular Engineering resulted in the Hot Spring Hypothesis for an Origin of Life published in 2020 in the journal *Astrobiology*. When he is not bending over a hot spring pool with a pipette testing the chemical capacity of these environments to start life, he is refining complex models for the mechanisms of the emergence of life from inanimate matter and pondering their philosophical significance. This curiosity led to a multiyear collaboration with Whiteheadean scholar Matt Segall in which the two thinkers placed this new science into a "cosmological context" with Whitehead's process thought and uncovered several novel linkages and insights. It is their hope that this work will continue to partner philosophy and science through a revolutionary period in the quest for the generative source of biology.

Ilia Delio holds the Josephine C. Connelly Chair in Christian Theology at Villanova University. Her area of research is systematic-constructive theology with a focus on evolution, quantum physics and artificial intelligence and the import of these for Christian doctrine and life. She holds a doctorate in pharmacology from Rutgers University Graduate School of Biomedical Sciences and a doctorate in historical theology from Fordham University. She is the author of twenty-three books including *Making All Things New: Catholicity,*

Cosmology and Consciousness, a finalist for the 2019 Michael Ramsey Prize, *The Unbearable Wholeness of Being: God, Evolution and the Power of Love* for which she won the 2014 Silver Nautilus Book Award and a 2014 Catholic Press Association Book Award in Faith and Science, and most recently *The Not-Yet God: Carl Jung, Teilhard de Chardin, and the Relational Whole* (2023). She is founder of the Center for Christogenesis, an online spiritual and educational resource for the integration of science, religion, and culture, based on the work of Teilhard de Chardin.

Lewis Ford was Louis I. Jaffe Professor of Philosophy at Old Dominion University. He is the author of *The Emergence of Whitehead's Metaphysics, 1925–1929*, *The Lure of God: A Biblical Background for Process Theism*, and coeditor (with George L. Kline) of *Explorations in Whitehead's Philosophy*. From 1971 to 1996, he was the editor of the scholarly journal *Process Studies*.

Steven J. Dick served as the NASA Chief Historian and Director of the NASA History Office from 2003 to 2009. Prior to that he was an astronomer and historian of science at the U.S. Naval Observatory for more than two decades. He was the 2014 Baruch S. Blumberg NASA/Library of Congress Chair in Astrobiology at the Library of Congress's John W. Kluge Center. In 2013 he testified before the United States Congress on the subject of astrobiology. From 2011 to 2012 he held the Charles A. Lindbergh Chair in Aerospace History at the National Air and Space Museum. He is the author or editor of twenty-five books, including most recently *Astrobiology, Discovery, and Societal Impact* (2018), *Classifying the Cosmos: How We Can Make Sense of the Celestial Landscape* (2019), and *Space, Time, and Aliens: Collected Works on Cosmos and Culture* (2020). In 2006, Dick received the LeRoy E. Doggett Prize from the American Astronomical Society for a career that has significantly influenced the field of the history of astronomy. He is the recipient of the NASA Exceptional Service Medal, the Navy Meritorious Civilian Service Medal, and the NASA Group Achievement Award for his work on astrobiology. He has served as president of the History of Astronomy Commission of the International Astronomical Union and as chair of the Historical Astronomy Division of the American Astronomical Society. He has been elected a Fellow of the American Association for the Advancement of Science and a Fellow of the American Astronomical Society. Minor planet 6544 Stevendick was named in his honor.

Brianne Donaldson holds the Shri Parshvanath Presidential Chair in Jain Studies at the University of California, Irvine. Her research explores how implicit metaphysical beliefs inform social inclusion and ethical action toward plants, animals, and marginalized people. She is the author of *Creaturely*

Cosmologies: Why Metaphysics Matters for Animal and Planetary Liberation (2015), and coauthor of *Insistent Life: Principles for Bioethics in the Jain Tradition* (2021). She is the editor or coeditor of *Beyond the Bifurcation of Nature: A Common World for Animals and the Environment* (2014), *The Future of Meat Without Animals*, and *Feeling Animal Death: Being Host to Ghosts.*

Timothy E. Eastman is an independent researcher whose scholarly works have encompassed research and consulting in space physics, plasma applications, space science data systems, space weather, public outreach, and education since the 1970s. By working with NASA spacecraft observations of space plasmas, in 1975 he discovered the low-latitude boundary layer of Earth's magnetosphere and has since published over 100 research papers in space physics and related fields. While serving as a program officer at NASA (1985–1988) and the National Science Foundation, NSF (1991–1994), he was co-coordinator for major international and interagency projects, including the International Solar Terrestrial Physics program, the Interagency Space Weather Program, and NSF's Basic Plasma Science and Engineering program. Beginning in the mid-1990s, he created and has continued to maintain a major website for plasma science and applications (plasmas.org). Having realized the pervasive impact of multilevel processes and emergence in the natural world, he has newly been assessing possibilities for transforming worldviews. Toward this goal, he has carried out research and publications in philosophy: these include over thirty philosophical papers and reviews, editing of four special scholarly journal issues on process thought and natural science, and serving as lead editor of *Physics and Whitehead* (SUNY Press, 2004) and *Physics and Speculative Philosophy* (de Gruyter, 2016). For this and related scholarly publications and workshop leadership, he was awarded the Creative Advance Award in 2009 by the International Process Network. Current research is focused on a synthesis of recent developments in philosophy, physics, logic, biosemiotics, and process thought. His integration in *Untying the Gordian Knot: Process, Reality, and Context* (Lexington, 2020) contributes to a new natural philosophy for the twenty-first century.

Roland Faber is Emeritus Kilsby Family/John B. Cobb, Jr., Professor of Process Studies at Claremont School of Theology, the founder and executive director of the Whitehead Research Project, and codirector of the Center for Process Studies. His research and publications encompass the fields of Whitehead's philosophy, process philosophy and process theology; (de)constructive theology; poststructuralism (Gilles Deleuze); transreligious discourse (epistemology of religious relativity and unity) and interreligious applications (e.g., Christianity, Buddhism, Bahá'í Faith); comparative philosophy and mysticism (Meister Eckhart, Nicolas of Cusa, Ibn 'Arabi), and theopoetics (an approach to poststructuralist and process theology, which addresses the liberating necessity of multiplicity). His publications include *God as Poet of the World* (2008), *The Divine Manifold* (2014), *The Becoming of God* (2017), *The Garden of Reality* (2018), *The Ocean of God* (2019), *Depths As Yet Unspoken* (2020), *The Cosmic Spirit* (2021), *Divine Appearances* (2022), and *The Mind of Whitehead* (2023).

Chelsea Haramia is Associate Professor of Philosophy at Spring Hill College and Senior Research Fellow with the Center for Science and Thought at the University of Bonn, Germany. Her research focuses on the intersections of science, technology, and values, and she is the author of several articles and book chapters on space exploration, astrobiology ethics, and the search for extraterrestrial technology, as well as public philosophy and outreach. She holds a PhD in philosophy and a graduate certificate in gender and women's Studies from the University of Colorado, Boulder. She is coeditor of the journal *1000-Word Philosophy* and an international affiliate of the SETI Post-Detection Hub. Follow her work at http://chelseaharamia.com/.

Noreen Herzfeld is the Nicholas and Bernice Reuter Professor of Science and Religion at St. John's University and the College of St. Benedict. She holds degrees in computer science and mathematics from the Pennsylvania State University and a PhD in theology from the Graduate Theological Union, Berkeley. Herzfeld is the author of *The Artifice of Intelligence: Divine and Human Relationship in a Robotic Age* (2023), *In Our Image: Artificial Intelligence and the Human Spirit* (2002), *Technology and Religion: Remaining Human in a Co-Created World* (2009), and *The Limits of Perfection in Technology, Religion, and Science* (2010) and editor of *Religion and the New Technologies* (2017) and *Religious and Cultural Implications of Technology-Mediated Relationships in a Post-Pandemic World* (2023). Herzfeld is also a senior research associate at the Institute for Philosophical Studies, ZRS Koper, Slovenia.

Wahida Khandker is Senior Lecturer in Philosophy at Manchester Metropolitan University. Her research focuses on intersections between process thought, particularly the work of Henri Bergson, and the philosophy of biology. She is the author of *Philosophy, Animality and the Life Sciences* (2014) and *Process Metaphysics and Mutative Life: Sketches of Lived Time* (2020). She also coedits a book series, Palgrave Perspectives on Process Philosophy.

Jeffery D. Long is the Carl W. Zeigler Professor of Religion and Philosophy at Elizabethtown College, in Pennsylvania, where he has taught since receiving his doctoral degree from the University of Chicago Divinity School in 2000. He has spoken in such prestigious venues as the University of Chicago, Yale University, Princeton University, and the United Nations. He is the author of *A Vision for Hinduism*: *Beyond Hindu Nationalism* (2006), *Jainism: An Introduction* (2009), the *Historical Dictionary of Hinduism* (first and second editions, 2011 and 2020), and *Hinduism in America: A Convergence of Worlds* (2020) as well as being the editor of the volume *Perspectives on Reincarnation: Hindu, Christian, and Scientific* (2019) and a coeditor of the Buddhism and Jainism volumes of the *Springer Encyclopedia of Indian Religions* (2017), and the volume *Beacons of Dharma: Spiritual Exemplars for the Modern Age* (2019). He is also the series editor of *Explorations in Indic Traditions: Ethical, Philosophical, and Theological*, for Lexington Books. In 2021, Elizabethtown College gave Long its Ranck Award for Research Excellence, and in 2022, his book *Hinduism in America: A Convergence of Worlds* was given the Rajinder and Jyoti Gandhi Award for Excellence in Philosophy and Theology by the Dharma Academy of North America. He is an initiated member of the Vedanta Society, established by Swami Vivekananda in 1894, and is active in the Hindu community in North America.

Mark Lupisella received his MA in philosophy of science and PhD in evolutionary biology from the University of Maryland at College Park. He founded the Horizons Project, which aims to improve humanity's ability to address long-term survival challenges. He's worked for NASA for thirty years and presently serves as the Exploration Research and Development Manager at the Goddard Space Flight Center. He has worked on the Hubble Space Telescope, Exploration and Constellation Programs (with an emphasis on human mission architectures and advanced systems), and areas in astrobiology such as planetary protection, artificial life, and societal issues. Dr. Lupisella assisted the Secure World Foundation in its initial stages of development and also helped lead the development of an Ethics Committee for Planet Labs (now called Planet). He also serves on the Advisory Council for METI International (messaging extraterrestrial intelligence). He has authored over thirty-five published works ranging from human Mars missions to the search for extraterrestrial life and space ethics. He is coeditor of the NASA book *Cosmos and Culture: Cultural Evolution in a Cosmic Context* (2011) with previous NASA dhief hHistorian Steven Dick, and recent author of *Cosmological Theories of Value: Science, Philosophy, and Meaning in Cosmic Evolution* (2020).

Derek Malone-France is an Associate Professor of Religion, of Philosophy, and of Writing at George Washington University in Washington, DC. He is also founding codirector of GW's French Embassy Center of Excellence and a member of GW's Academy of Distinguished Teachers (for which he served as the Inaugural Chair from August 2012–August 2014). Derek's research and teaching areas include: philosophy of religion, metaphysics and epistemology, the philosophical and religious implications of astrobiology and human space exploration, political and legal philosophy, political and religious rhetoric, and educational theory and practice. He received a bachelor's degree in philosophy from Wofford College in 1995 and a doctorate in philosophy of religion and theology from the School of Religion at the Claremont Graduate University in 2001. Outside of GW, Derek serves as a Trustee and member of the Board of Directors for METI International, the nonprofit Messaging Extraterrestrial Intelligence Initiative, and as Director of Academic Design for the NASA Astrobiology Debates educational outreach project. He has also previously served as Seminar Director for the Blumberg Dialogues in Astrobiology and Human Society, jointly sponsored by NASA and the Library of Congress. He is an elected research member of the International Whitehead Research Project. He is editor of a two-volume anthology titled *Political Dissent—A Global Reader* (2011); *Deep Empiricism: Kant, Whitehead, and the Necessity of Philosophical Theism* (2006); and *Faith, Fallibility, and the Virtue of Anxiety: An Essay in Religion and Political Liberalism* (2012).

Dr. Ted Peters received his PhD from the University of Chicago and directs traffic at the intersection of science, religion, and ethics. Peters is an emeritus professor at the Graduate Theological Union, where he coedits the journal *Theology and Science*, on behalf of the Center for Theology and the Natural Sciences, in Berkeley, California. Along with Martinez Hewlett, Joshua Moritz, and Robert John Russell, he coedited *Astrotheology: Science and Theology Meet Extraterrestrial Life* (2018). Along with Octavio Chon Torres, Joseph Seckbach, and Russell Gordon, he coedited *Astrobiology: Science, Ethics, and Public Policy* (2021). He is author of *The Cosmic Self* (1993) as well as *UFOs: God's Chariots? Spirituality, Ancient Aliens, and Religious Yearnings in the Age of Extraterrestrials* (2014). Follow his work at TedsTimelyTake.com.

Matthew David Segall is a transdisciplinary researcher, author, and teacher applying process philosophy across the natural and social sciences, including the study of consciousness. He is a faculty member in the Philosophy, Cosmology, and Consciousness graduate program at California Institute of Integral Studies in San Francisco. He is the author of several books including *Physics of the World-Soul: Whitehead's Adventure in Cosmology*

(2021) and *Crossing the Threshold: Etheric Imagination in the Post-Kantian Process Philosophy of Schelling and Whitehead* (2023). Follow his work at Footnotes2Plato.com.

Theodore Walker Jr. received his PhD from the University of Notre Dame, 1983. He is Associate Professor of Ethics and Society at Perkins School of Theology with teaching specialties in theology and science, including astro-theology and cosmology; science and ethics, including theory of biology and ethics (general bioethics); neoclassical natural scientific theology and scientific metaphysics (A. N. Whitehead, Charles Hartshorne, Schubert M. Ogden); and African American liberation theology (Martin Luther King Jr., Malcolm X, James H. Cone) and ethics, including womanist/sisterist thought (Alice Walker, Karen Baker-Fletcher). He is the author of *Mothership Connections: A Black Atlantic Synthesis of Neoclassical Metaphysics and Black Theology* (2004), *Empower the People: Social Ethics for the African-American Church* (1991, 2001, 2023); and coauthor with Chandra Wickramasinghe of *The Big Bang and God: An Astro-Theology* (2015).

www.ingramcontent.com/pod-product-compliance
Ingram Content Group UK Ltd.
Pitfield, Milton Keynes, MK11 3LW, UK
UKHW041448071026
15587UKWH00010B/663